THE ECONOMICS OF JOHN STUART MILL
VOLUME I   THEORY AND METHOD

STUDIES IN CLASSICAL POLITICAL ECONOMY/III

SAMUEL HOLLANDER

The economics of John Stuart Mill

Volume I
Theory and Method

UNIVERSITY OF TORONTO PRESS
Toronto and Buffalo

© Samuel Hollander 1985
First published 1985

Basil Blackwell Ltd.
108 Cowley Road, Oxford OX4 1JF, UK

First published in Canada and the United States
University of Toronto Press
ISBN 0-8020-5671-7

Canadian Cataloguing in Publication Data

Hollander, Samuel, 1937–
  The economics of John Stuart Mill.

(Studies in classical political economy; 3)
Bibliography: v.
Includes index.
Contents: v. 1. Theory and method – v. 2. Political economy.
ISBN 0-8020-5671-7 (Set).

1. Mill, John Stuart, 1806–1873. 2. Mill, John Stuart, 1806–1873. Principles of political economy.
3. Economics. I. Title. II. Series.

HB161.M6H64   1985   330.15′3′0924   C85-099087-4

Phototypeset by Dobbie Typesetting Service, Plymouth, Devon
Printed in Great Britain by T.J.Press Ltd, Padstow

Dedicated to my son Isaac
with love and admiration

' "Very well, my dear," Landowska was overheard saying. "You continue to play Bach your way and I'll continue to play him *his* way" '

(H. C. Schonberg, *The Great Pianists*, New York, 1963, 399).

CONTENTS

PREFACE

This study is the third part of my series on classical political economy. The story of classical economics is carried from 1823, the year of Ricardo's death until the so-called 'marginal revolution' of the early 1870s with special reference to John Stuart Mill. The emphasis, as before, is upon economic theory and policy but with more attention now given to methodology considering Mill's own preoccupations.

Sir John Hicks, in a recent reappraisal, has observed of the classical economists that while 'there is no question, in more modern times, about the standing of Smith and Ricardo . . . Mill, as an economist, seems to have been de-throned' (1983, 60). Hicks himself believes that Mill's economic writings 'deserve more attention than they currently receive'; indeed he regards Mill 'as the most undervalued economist of the nineteenth-century' (1982, 325n). It is in this positive spirit that I approach my task. And throughout I have tried to adhere to Marshall's recommendations made in 1876:

Those . . . who wish rightly to construe any of Mill's economic doctrines must learn the special part which he intended that doctrine to perform, to the end that they may not demand from it the discharge of functions which he has assigned to some other portion of his system; and they must remember that he is not always careful to repeat an indication that he has once given of the special application which he intends to make of a word or a phrase in a particular discussion. They must, therefore, consider each passage in connection with its context; and when its interpretation cannot by this means be conclusively settled, they must with generous caution reject any rendering of it which is inconsistent with the general purport of his writings (1925, 121).

Mill himself, in 1832, had made a general appeal for fair interpretation considering the characteristic ambiguity of 'political terms' (*CW* XVIII, 1–13).[1]

1 Throughout, *CW* refers to *Collected Works of John Stuart Mill*, ed. J. M. Robson, Toronto.

Hicks is surely right, at least concerning Mill's reputation. Thus like many others of his day (cf. Schwartz, 1972, 243-4) Edwin Cannan regarded the *Principles of Political Economy* as a book behind its time (1917, 308). Even Marshall conceded to Jevons in 1875 that 'Mill was not a constructive genius of the first order' (Jevons, 1977, IV, 100) and wrote to Bonar in 1898: 'I incline to regard Petty and Hermann and von Thünen and Jevons as classical, but not Mill' (Marshall, 1925, 374) – in the sense of one who 'has stated or indicated architectonic ideas in thought or sentiment, which are in some degree his own, and which, once created, can never die but are an existing yeast ceaselessly working in the Cosmos'. In our own day Thomas Sowell charges Mill with 'successfully [turning] the clock back on fundamental developments in economics' (1972, 164). Schumpeter, who was sympathetic to Mill the man, complained that 'his attitude of speaking from the vantage ground of definitively established truth . . . besides being ridiculous, made for sterility and – yes – superficiality' (1954, 530). Schumpeter placed Mill, below par, on a Smith–Marshall line, but conceded none the less a claim on Mill's part to novelty on specific points. This kind of concession characterizes the evaluation by Stigler (1965), an evaluation which has done much to rehabilitate Mill by portraying him as 'one of the most original economists in the history of the science (7; cf. also Stigler, 1968). For this high praise applies to 'identifiable theories', a qualification referring to Mill's apparently limited objective – that of making marginal improvements to the 'Ricardian' theoretical scheme rather than undertake a full-fledged revision.

Schwartz, in his important study of Mill, sets out to explain 'why [Mill] thought that the Ricardian analytical fabric could be left to stand intact and that it was worthwhile to employ his great analytical faculties in correcting the defects of Ricardian analysis' (1972, 236). Schwartz's problem is certainly a legitimate one. A sound evaluation of Mill's place in the history of economic thought must indeed turn upon the relation of Mill to Ricardo. But if the underlying conception of Ricardian economics should be wanting, the reading of Mill will inevitably be distorted. The present work constitutes an interpretation of Mill undertaken in the light of the conception of appropriate method – and of substantive theoretical propositions – revealed in our earlier investigation of Ricardo. The extent and nature of Mill's 'genius' is of no concern for us. Mill, like Everest, is a fact of life. In any event innovation for the mere sake of innovation is no virtue; Ricardianism, so it turns out, served his broad social purposes to perfection.

Chapter One sets the stage. Here we argue the case for Ricardo's appreciation of the provisional nature of the axiomatic foundation of political economy and rejection of a body of doctrine of 'universal' relevance, his insistence upon the distinction between 'science' and 'art', his concern for lags and frictions in the application of theory to policy, and his rejection of the notion of economics as an engine for historical prediction. In these respects a sharp distinction must be made between James Mill and Ricardo. An exposition of the growth models of Malthus and Ricardo – shown to be similar but not identical – is followed by a summary view of reactions by some contemporaries, with special attention accorded Nassau Senior, Thomas Chalmers and Mountifort Longfield. The resiliance of 'Ricardianism' understood as a perspective on economic growth divorced from historical predictions of a pessimistic order, is confirmed, reinforcing an earlier demonstration of rather widespread support for the inverse wage–profit relation even when expressed in value terms.

In Chapter Two we proceed to Mill's own views on scientific method in general, his case against purely experimental procedure in social science, his legitimization of the wealth-maximization axiom and – with special reference to the problem of 'disturbing causes' – of a quasi-independent science of political economy, his allowances for model improvement by way of 'verification', his denial that political economy constituted a predictive science and his broad interest in 'ethology' and the 'general science of society'. Mill's economic methodology as outlined in the essay devoted to the subject (written c. 1830) and in the *System of Logic* (1843) – the championship of a nice blend of the 'inductive' and 'deductive' rather than deduction *tout court* – thus turns out to be in line with that of Ricardo as revealed in the latter's practice. At the same time, and notwithstanding appearances, much common ground may be discerned with the Cambridge 'inductivist' critics (such as William Whewell) of 'deductive' economics.

A repeatedly encountered charge against Mill of inconsistency between his methodological pronouncements and his practice in the *Principles* is shown by the investigation of Chapter Two to have doubtful validity. This is further confirmed by the materials of the subsequent chapter where we treat the transition to the *Principles*. The essay with its claims for a 'mixed' method is shown in fact to be essential reading for an appreciation of the major text. The character of the *Principles* is then further analysed with reference to its Smithian form of presentation yet Ricardian substance, the impact of Harriet Taylor and Auguste Comte, and the strategy entailed by adoption of a sequence of books in the order Production, Distribution and Exchange. Mill intended pre-eminently the defence of Ricardian economics against charges that it constituted a 'conservative' *apologia* while at the same time pointing out the implications – by no means 'pessimistic' – of the universal constraint imposed by land scarcity, and also insisting upon (this was to be standard Marxian practice) a variety of technical or analytical characteristics common to differing distributional arrangements. Above all, the postponement of a discussion of exchange must not be understood as signalling a divorce of production and distribution from valuation in the specific context of a capitalist economy; the problem of capitalist distribution, as Ricardo had so well understood, was to be analysed in terms of a general theory of pricing.

With an eye always to the formal methodological pronouncements we proceed in Chapter Four (the first substantive chapter on the *Principles*) to Mill's 'inductive' analysis of the sources of increased efficiency over space and time, specifically the isolation in the historical and contemporary record of the qualitative and quantitative determinants of productivity. Here *inter alia* we take up the empirical basis for the law of diminishing returns, and elaborate further upon the contrast between the 'immutable' laws of production and the 'malleable' laws of distribution. This dichotomy is shown to have little substance – as Mill realized – once allowance is made on the one hand for the impact of changes in productive organization and knowledge on production, and on the other for the 'necessary' consequences of distributive changes. The significance of organizational arrangement for efficiency is illustrated by a section on the impact of land-tenure arrangement upon motivation. We shall also survey Mill's – by-and-large optimistic – evaluation of contemporary and prospective economic progress.

Chapter Five on 'Allocation, Trade and Distribution' constitutes the first of a trio of theoretical chapters. Our analysis of some major issues of domestic and international price theory – including a demonstration of a profound appreciation of alternative cost and derived factor demands – leads us to reject the widely held view that the principles of allocation theory were limited by Mill to the celebrated chapter on foreign trade. Mill's remarkably original treatment of competitive price formation, and his recognition of a 'monopolistic competition' model provide further major themes. Consideration of the fundamental theorem on distribution in terms of allocation analysis brings the chapter to a close.

Our second theoretical chapter takes up problems relating to capital, employment and growth of a largely non-monetary order. This chapter analyses the determinants of aggregate employment capacity implied by the proposition that 'industry is limited by capital' or its corollary 'the demand for commodities is not demand for labour'. In this context we treat Mill's version of the Ricardian problem of 'machinery', as well as aspects of the wages–fund theory including its relationship with the principles of allocation and distribution discussed in the previous chapter. The precise nature of the celebrated recantation from that theory of 1869 constitutes a further major theme.

Mill's famous contrast between 'statics' and 'dynamics' provides an introduction to his theory of economic growth with special reference to the wage- and profit-rate paths. Here we shall demonstrate that Mill followed Malthus in all major respects, adopting the analysis according to which the profit rate declines with the real wage proceeding unchanged although not along the 'subsistence' path. The allowance for a stationary state at a 'high' level of real wages is also seen to originate with Malthus. We shall then consider some specific applications. Mill (like Malthus) did not concern himself with simplification for pedagogical purposes. His model carried a fundamentally important policy message – to encourage the emulation by the unskilled of the 'prudential' behaviour already practised by skilled industrial workers, to forestall the fall in wages which *must* occur upon any decline in the rate of accumulation should population growth continue at its maximum physiological capacity irrespective of the wage. Yet this outcome was not seen to be an imminent prospect; the dynamics of contemporary capitalism were patently clear to Mill giving the lie to the commonly expressed opinion that he was a poor observer in this respect.

Complications for the secular path of the return on capital reflecting cyclical disturbance and various qualifications to the law of markets bring the chapter to the close. Here we conclude that 'Keynesian' overtones should not be exaggerated at least in so far as concerns *secular* analysis.

These latter issues preface the substance of Chapter Seven devoted to Money and Banking, the first chapter of Volume II. The breakdown into volumes is dictated by publishing considerations. But the second volume, as the subtitle indicates, is predominantly devoted to economic policy and its philosophical underpinnings. Chapter Seven has close links with the preceding chapter on growth as far as concerns pure theory, yet considering the attention we accord therein to banking policy, a home can be justified for it in Volume II.

More specifically, Chapter Seven traces the treatment of the law of markets through Mill's various periodical formulations to the *Principles*, with appropriate attention accorded the Quantity Theory and the monetary aspects of interest-rate determination. The qualifications allowed the strictly defined law of markets and quantity theory constitute the major theoretical theme. A second outstanding feature of Mill's analysis is his keen awareness of the problem facing monetary managers in achieving balance – the avoidance of secular inflation yet the need to intervene in order to dampen cyclical instability – and the problem of state control of central banking. Throughout we attempt to define Mill's precise position in the great Currency–Banking School debates.

Chapter Eight attempts to discern in Mill's social philosophy the underpinnings for his approach to the problem of social control. Here we examine the broad issues of 'Utilitarianism' and 'On Liberty' with special attention to Mill's changing attitude towards Jeremy Bentham. A 'qualitative' dimension to Bentham's utilitarianism is demonstrated such that Mill's later incorporation of justice 'within' utility turns out to be a return to Bentham rather than a new discovery.

The role of government is discussed in Chapter Nine. We consider both matters of general principle relating to the theory of economic policy – the legitimate role for the state in economic life based upon the broad rules outlined in the previous chapter – and some specific case studies: education, foreign trade, economic development, population control and poor relief, labour market intervention, colonization, monopoly control and patent protection. The chapter concludes by observations on the probable impact of 'progress' on the extent and nature of legitimate state intervention.

A companion chapter (Ten) takes up Mill's indictment of capitalism and his analysis of alternative institutional arrangements with special reference to his celebrated self-description as 'socialist'. The outstanding feature of the record is a strong championship of competition on grounds of justice, allocative efficiency and motivational force but within an *ideal* co-operative solution to social organization.

Mill took a hard-headed view of possibilities for profound social change in the near future. Chapter Eleven considers *immediate* reform proposals. Here we touch on public finance, in which context a version of diminishing utility is conspicuous. Here too the desirability of economic development in relation to the celebrated pronouncements on the stationary state is considered, as well as Mill's approach to the so-called 'poverty trap'. This topic extends to his evaluation of the prospective contribution of trade unionism towards a solution of the condition-of-the-people issue. Various charges against Mill of 'bourgeois bias' and 'paternalism' are shown in this context to be untenable.

In a concluding chapter we attempt to arrive at a general summary by reference to a number of central issues encountered during the course of our investigation. Here we demonstrate that on matters of doctrine and method Mill at no time – even in his last years – came to share what has been termed 'the disillusion and impatience' with Ricardian tradition which originally owed so much to his own prestige, as some historians have recently maintained. His objections were directed always at second-raters whose apologetic exercises had damaged the reputation of the subject. The orthodox principle of the relativity of axioms and the consistency of Ricardianism with social conscience and with a concern for institutional change plays a large part in the story, the latter characteristics explaining also Mill's appeal for 'historicists' such as Cliffe Leslie.

In this chapter we demonstrate further the inappropriateness of the notion of a demand-oriented 'revolution' in the 1870s, supplementing our argument by a rationalization of Mill's neglect of marginal utility as a matter of principle rather than as a 'failure' as some modern commentators would have it, and a reflection of his anxieties regarding the prospective development of economics as applied mathematics.

We proceed then to reiterate the invalidity of the charge that Mill perpetuated a population doctrine which had been discredited theoretically and empirically, and show that specious charge to be based on a faulty view of his theory, purpose and method. From this same perspective we are able to deal with the complaint of 'Scientism' directed at Mill by Hayek and Knight, namely the charge that he had been infected early on with the belief that scientific and engineering techniques may be applied to the study of society with equally promising results.

A final section summarizes the related matters of the 'reality' of axioms and model improvement.

ACKNOWLEDGEMENTS

It is a pleasure to express my gratitude to the Social Sciences and Humanities Research Council of Canada for continued aid in shifting out my budget constraint, as in the case of the earlier volumes in the present series. My Mill research has also been generously supported by the award of a Lady Davis Fellowship held at the Hebrew University, Jerusalem (1979–80), and a Connaught Senior Fellowship (1984–5).

The Master and Fellows of Trinity College, Cambridge, have kindly permitted citations from the Whewell papers.

Friends and colleagues in Toronto – Spencer Davis, Bob Fenn, Sue Golding, Bob Greenspan, F.E. Sparshott, and Fred Wilson – have provided intellectual and moral support. The staff of the Mill Project under the leadership of Jack Robson has been unfailingly helpful. I am indebted to R.D.C. Black, Walter Eltis, Scott Gordon, David Levy, Neil de Marchi, Shocken Mawatari and P. L. Williams for most helpful criticism at various junctures; also to my doctoral students Evelyn Forget, Rick Kleer, Tom Kompas, Sandra Peart, and Margaret Schabas. Marion Halmos, my research assistant, has proved absolutely indispensable throughout the complex procedure of taking a vast manuscript to final completion. She has my deepest appreciation for accepting this responsibility. Left-handed Kate Franklin efficiently undertook the function of being Miss Halmos' right-hand man. The final touches to this work were made at La Trobe University, Melbourne. I appreciate greatly the opportunities provided under its Distinguished Visiting University Professor Scheme by Vice-Chancellor Scott, Dean Horrigan (Economics, La Trobe), and my colleague Michael Schneider. Gloria Rowe, my typist, has continued to amaze me by never once protesting at an almost illegible and apparently never ending manuscript. For proof-reading, my sincere thanks go to Sarah Anson-Cartwright, Evelyn Forget, Kate Franklin, Laura

Haferkorn, Lisa Haferkorn, Marion Halmos, Rick Kleer, Alexander Klip, Rohinton Medhora, Myer Rosen, Gloria Rowe, Julie Smith, Richard Streiling and Dale Tomlinson.

I like to think that the late Lord Robbins, with whom I discussed the broad outlines of the present work, would have found the outcome a positive contribution to the history of economic thought. My wife has shared the heavy burdens imposed by the preparation of these volumes; for 'of making many books there is no end; and much study is a weariness of the flesh' (*Ecclesiastes*, 12:12).

ONE

The methodological heritage

I INTRODUCTION

A sharp distinction must be made between the methodological orientations of James Mill and David Ricardo. It is James Mill who was guilty of what has been termed by Professor Schumpeter 'The Ricardian Vice'. To tar Ricardo with the same brush is to distort Ricardo's position and the record of the entire post-Ricardian period. It turns out that J. S. Mill's reaction against his father on methodological matters in the early 1830s implies not (as is sometimes suggested) the construction of a new political economy, but rather a return to Ricardo. But we must proceed with caution. We are claiming a commonality of approach, for it is difficult to be precise regarding the matter of actual 'debt'. Dugald Stewart has recently been cited as a probable source for J. S. Mill (de Marchi, 1983b; below pp. 14–15); and while this should not be precluded there are, we shall see, other parallel cases, using similar formulations, all perhaps drawing from common sources. Our case is simply that Mill opposed his father's methodology as formally presented in published writings and that he did not identify James Mill and Ricardo.

As defined by Schumpeter the Ricardian 'vice' refers to 'the habit of piling a heavy load of practical conclusions upon a tenuous groundwork, which was unequal to it yet seemed in its simplicity not only attractive but also convincing.' More specifically:

[Ricardo's] interest was in the clear-cut result of direct, practical significance. In order to get this he cut that general system to pieces, bundled up as large parts of it as possible, and put them in cold storage – so that as many things as possible should be frozen and 'given'. He then piled one simplifying assumption upon another until, having really settled everything by these assumptions, he was left with only a few aggregative variables beween which, given these assumptions, he set up simple one-way relations so that, in the end, the desired results emerged almost as tautologies . . . The habit of applying results of this character to the solution of practical problems we shall call the Ricardian Vice (1954, 472–3; cf. 541, 618, 653n, 668, 1171).

It is as a 'predictive engine' that Schumpeter conceived of Ricardian theory – albeit a defective one. And the 'Ricardian Vice' is best illustrated by the 'pessimistic' conception of economic development attributed to Ricardo (and also to Malthus, West and James Mill) – namely population pressure and diminishing agricultural returns entailing a declining rate of profit and approximate constancy of real wages – upon which (supposedly) so much weight was placed in policy pronouncement (571).

This same view, with special reference to James Mill as the source of 'the Vice', plays a central role in T. W. Hutchison's important work *On Revolutions and Progress in Economic Knowledge* (1978). After a reference to Ricardo's formal description of aggregate distribution as 'the principal problem in Political Economy', Hutchison proceeds to answer in the affirmative the question raised in the title of his second chapter: 'James Mill and Ricardian Economics: A Methodological Revolution?'

But it can be argued that of *much* more fundamental and lasting significance than the shift of interest or priorities regarding the subject of distribution, was the *methodological* claim that problems in political economy are problems of '*determining laws*'. Moreover, of equally fundamental and lasting significance was the *method* of extreme abstraction (or 'strong cases') by which, in his *Principles* Ricardo sought to 'determine' the 'laws' of political economy, which he claimed his new Science was establishing. The transformation in method and epistemology between *The Wealth of Nations* and Ricardo's *Principles* is profoundly significant because it altered the mood in which the 'problems' of political economy were treated and in which 'theories', and policy-recommendations were put forward (26–7).[1]

Similarly, Hutchison complains of 'the tendency to generalize about the English "classical" economists' which 'has fostered the erasing of vital distinctions, notably, for example, of this important methodological contrast between *The Wealth of Nations* and Ricardo's *Principles*. The fruitful combination of history and empirically significant theory in *The Wealth of Nations* was broken. History was largely extruded from the orthodox conception of the subject for decades to come. The legacy of the Mill–Ricardo methodological revolution was one of insufficiently controlled

---

1 Hutchison draws attention to Walter Bagehot's (1895) view that 'the true founder of abstract Political Economy is David Ricardo' (197).

abstraction and over-simplification on the one hand, and of over-confident pretensions on the other hand' (56).[2]

On Hutchison's reading it was James Mill who was responsible for the line followed by Ricardo: 'It has long been known that James Mill played a considerable part in the intellectual development and political career of Ricardo. But the full extent of Mill's influence, and just how vital and substantial it was for Ricardian political economy, in a crucial, perhaps "revolutionary" phase of the subject, and for the subsequent development and methodology of political economy and economics, does not seem to have been fully recognized' (26–7).[3] Precisely the same view has also been expressed by Professor Winch in his account of James Mill on scope and method: 'there is little doubt that Ricardo's eagerness to believe in the applicability of clear-cut principles made him the perfect subject for Mill's teachings . . . Ricardo and Mill were completely at one on questions of method . . . The only difference between Mill and Ricardo was that whereas Ricardo's confidence in "strong cases" shows itself mainly on economic questions, Mill's confidence extended to every subject upon which he wrote' (1966a, 368–9).[4]

The foregoing provides a convenient statement of a number of propositions which to my mind, are suspect:

1. That the 'true founder' of abstract political economy was Ricardo. It was, in fact, Adam Smith who, recognizing that the induction of unique cause–effect relations is all but impossible, provided the archetype of 'abstract' theorizing for the nineteenth century. It was Smith too who

2 Again approving reference is made to Bagehot's (1895) pronouncements: 'It must be remembered that . . . Ricardo . . . had no large notion of what science was . . . To the end of his days, indeed, he never comprehended what he was doing. He dealt with abstractions, without knowledge that they were such; he thoroughly believed that he was dealing with real things. He thought that he was considering actual human nature in its actual circumstances, when he was really considering a fictitious nature in fictitious circumstances. And James Mill, his instructor on general subjects, had on this point as little true knowledge as he had himself. James Mill, above all men, believed that you could work out the concrete world of human polity and wealth from a few first truths' (205).

3 Bagehot (1895) is again cited: 'If Ricardo had never seen James Mill he would probably have written many special pamphlets of great value on passing economic problems, but he would probably not have written *On the Principles of Political Economy and Taxation*, and thus founded an abstract Science' (204).

4 Cf. also: 'All of Mill's writings as a political scientist, educationist, historian and an economist were based on deeply-held convictions concerning the importance of "theory" or abstract principles; and there is good reason to believe that he had considerable influence on the methodological thinking of two of the major contributors to the classical tradition, namely Ricardo and John Stuart Mill' (367). Winch alludes approvingly to Elie Halévy's conclusion that James Mill 'did not so much give [Ricardo] a doctrine as develop in him the doctrinal leaning and make him a doctrinaire' (369). Winch (1983) has recently moderated his position regarding Ricardo, to the extent of conceding that the case against Ricardo is 'not proven' (520).

anticipated the Ricardians in the rejection of predictive accuracy as a means of testing theory on grounds of the almost inevitable intervention of 'disturbing causes'. The quality of Smithian analysis is remarkable. But in one respect Smith was less impressive. He was guilty of careless transfer from pure theory, based on 'strong cases', to application without due allowance for frictions and timelags. It may be unfair to indict Smith of a general predilection of this order, but it is a valid charge as far as concerns his analysis of the contemporary corn laws, and the policy recommendations based thereupon, and he was taken to task by critics for the practice in that context (see Hollander, 1979, ch. 1).[5]

2. That the foregoing charge applied to Ricardo – his supposed formulation of abstract models incorporating 'strong-case' assumptions and the unqualified application of deductions derived therefrom. The fact is that Ricardo's 'strong cases' were intended as analytical simplifications introduced to clarify the argument for better comprehension, but to be appropriately modified in the treatment of real-world problems. The 'principles' of political economy as such could provide but limited guidance in applied economics where a variety of conflicting causal influences is at play and where allowance is required for time-lags and frictions. This balanced perspective is most conspicuous in the contexts of agricultural protection, the return to gold and poor law policy.[6] The charge against him appears better substantiated in the monetary context – the refusal to admit excess aggregate supply in the post-Napoleonic period – but here it reflects less a methodological predisposition than a horror of inflation.

Bagehot's representation of Ricardo as a methodological-neophyte is a parody. Ricardo was keenly conscious of the limitations imposed upon political economy by the very nature of its data; he shared with Smith,

---

5 Schumpeter provides a splendid statement of the matter at issue: 'A. Smith's work looks less "abstract" because it includes so much factual information that the *specialized* later works on economic theory did not include – but left for other specialized works to provide. But where he does move within the orbit of economic theory, his reasoning is not less abstract than is, say, Ricardo's. With the latter, "abstractness" shows more because he confines himself to topics of an "abstract" nature, and does not provide illustrative foliage, but this is all' (538n).

6 The following note by Hutchison (1978), 53n, expresses a dilemma which his interpretation creates for him: 'A contemporary observer, J. L. Mallet, noted this contrast in Ricardo: "It is impossible to be in company with Ricardo and not to admire his placid temper, the candour of his disposition, his patience and attention, and the clearness of his mind; but he is as the French would express it *"hérissé de principes"*, he meets you upon every subject that he has studied with a mind made up, and opinions in the nature of mathematical truths . . . His entire disregard of experience and practice . . . makes me doubtful of his opinions on political economy" . . . Could this contrast be one between Ricardo's own innate nature and the results of Mill's influence? Alfred Marshall's religious or racial explanation does not seem very convincing: "The faults and virtues of Ricardo's mind are traceable to his Semitic origin; no English economist had a mind similar to his".'

for example, the view that 'disturbing causes' disqualified the subject as a predictive science. And he was aware of the historical relativity of the institutional and behavioural axioms adopted in the treatment of the capitalist-exchange system.

3. That James Mill was responsible for diverting political economy from the empirico-historical lines followed by Adam Smith to the pure abstractions and irresponsible applications of Ricardo. The errors in this regard are implied in what has just been said under the first two headings. Mill found much to emulate in the *Wealth of Nations* – the use of extremely abstract models in direct policy pronouncement; while he was unable to exert any effective methodological influence upon Ricardo.

These matters will be taken up in the present chapter. I shall also deal briefly with the 'inductivist' reaction to Ricardian method. Richard Jones drew a picture of Ricardo's procedure as one involving the derivation of generalizations from axioms supposedly of 'universal' relevance but in fact only casually supported by locally obtained observation and even introspection. The picture is distorted – Ricardo had no intention of establishing universally-applicable axioms. It will also become clear that the entire challenge was an abortive one since Jones himself offered no alternative to deductive theorizing. The weakness of his attack on Ricardo was well understood by contemporaries – including Malthus, Longfield, Whately and Torrens.

The chapter proceeds to consider political economy as a 'predictive' engine. It was appreciated by post-Ricardian economists – even those who questioned the trend patterns of declining wages and profit rates – that the classical growth model incorporating the principle of diminishing returns retained its validity as an analytical engine despite allowance for the countervailing force of new technology. What emerges then is the retention of the main features of Ricardo's own position regarding the use of economic theory.

II JAMES MILL

James Mill had nothing but praise for Smith's system-building propensity. His evaluation is brought out clearly in a contrast drawn in 1806 between Smith and Sir James Steuart:

To Sir James's eye the subject presented itself as a rude chaos; and he found himself unable to reduce it to light and order . . . He explained some old errors, and established some new truths. But his opinions have no general bearing. The mind is bewildered in following Sir James's speculations. The general principles of Political Economy seem to become more obscure in his hands than they were before . . . There is no combination of principles in his volumes which can be called a system at all . . . Dr. Smith reared the study to the dignity of a science. He explained the real sources of wealth, which till his time had been so grossly misunderstood; and conferred as great a benefit upon Political Economy, as was conferred on Astronomy by those philosophers who first confuted the perplexed doctrine of the cycles and epicycles, and established the simple principles of the Copernican system (1806, 231-2).

Mill's commendation is all the more important since his reviews are replete with criticisms of those he considered insufficiently speculative. As Professor Winch has put it: 'he regarded the inability to generalise or to move beyond immediate experience as an "infirmity of the mind"; and it was for his own highly-developed powers of "ratiocination" that his works were praised by admirers and denounced by opponents' (1966, 367).

In his early essay of 1804 on the corn bounty, Mill attacked James Anderson, Malthus and others for reliance on 'experience' (1804, 6). He himself subsequently denied, in an unnecessarily sarcastic fashion, any conflict between 'abstract speculation' and 'experience' on the grounds that 'good abstract principles are neither more nor less than the accurate results of experience, presented in an exceedingly condensed and concentrated state.'[7] It may be claimed that experience, provided it is 'comprehensive and profound' cannot conflict with theory; for theory is the abstracted-from-detail rationalization of the condensed and concentrated experience. But it is easy to see how, without due care, such grandiose declarations might lead to the unqualified application of the results of abstract reasoning. The analysis of the corn-export bounty, which runs along Smithian lines in nearly all respects, is particularly revealing.

Appreciation of the case against the bounty turned upon 'a single principle . . . that the nature of the farmer's business is altogether different . . . from all other trades.' Here we meet the typically uncompromising phraseology

---

7 Mill (1813), 411–12, regarding Robert Grant's *The Expediency maintained of continuing the System by which the Trade and Government of India are now regulated* (1813): 'Let us contemplate a few of the author's general maxims before we enter on his argument. – In disposing of political questions, it is either folly or wickedness, he says, to have any thing to do "with abstract or elementary principles". The reader will see nothing very new in this remark, since it is the common declamation of all those who wish for the reputation of wisdom without the trouble of deep thought, and is the miserable artifice by which the superficial hope to disguise and cover their shallowness. When abstract principles are mentioned, good abstract principles, of course, are those which are meant. Now, good abstract principles are neither more nor less than the accurate results of experience, presented in an exceedingly condensed and concentrated state. If Mr. Grant, therefore, wishes political questions to be decided without abstract principles, he wishes them to be decided without the benefit of experience. Accurately speaking, he would have them be regulated by a narrow and empirical instead of a comprehensive and profound experience; by a very imperfect instead of a very perfect guide.'

that characterizes so much of Mill's writings: 'No proposition is better established than this, that the multiplication of the human species is always in proportion to the means of subsistence. No proposition too is more incontrovertible than this, that the tendency of the human species to multiply is much greater than the rapidity with which it seems possible to increase the produce of the earth for their maintenance.' The consequence was that 'the production of corn creates the market for corn'. Conversely, 'to send away any part of the regular produce of the country' – a consequence of the bounty – 'however rapidly that produce may be increasing, is just to cut short a proportional part of the natural population of the country' (23–4).[8]

Mill next considered the argument of proponents turning upon the net stimulus to corn output exerted by the bounty. He conceded that 'agriculture is a little animated for a few years' but only 'till things find their proper level'; which occurs when the temporarily raised return on agricultural capital is reduced by competition for land and higher rents (28). He further conceded that capital would be attracted from other sectors thus reducing the agricultural rate to equality with a manufacturing rate somewhat raised in consequence of reduced 'competition of capitals' (32). By making this latter concession Mill had unwittingly allowed much more than he intended, for clearly a permanent expansion of production is involved.

Finally, we have the full-fledged Smithian analysis of the consequence of the bounty on general prices by way of the effect upon money wages:

> But though the bounty produces no good effects, it is not altogether without effect. We must next advert to the view which Dr. Smith has exhibited of this subject, a view which any one can affect to treat lightly only from not understanding it. No proposition is established more thoroughly to the conviction of those who have studied the scientific principles of political economy than this; that the money price of corn, regulates the money price of every thing else. The wages of the common labourer may in general be reckoned his maintenance. He must earn a sufficient quantity of corn to feed himself, otherwise he cannot exist. If he is paid in money, the sum of money he daily receives must always be equivalent to the quantity of corn he must use. If the price of corn is high he must receive the greater sum of money, as his day's wages, to buy with. This is so obviously necessary, that we need spend no more time in proving it. The money price of labour therefore is entirely regulated by the money price of corn.
>
> Let us next see how the money price of corn affects that of every thing else. It is evident that it must regulate the price of all other products of the earth, as the culture of corn will encroach upon them till they become equally profitable with

---

8 The reference to 'regular' produce reflects the qualification that 'a foreign market can never be necessary, but to take off the surplus of an extraordinary year'.

itself. 'It regulates, for example,' says Smith, 'the money price of grass and hay, of butcher's meat, of horses, and the maintenance of horses, of land carriage consequently, or of the greater part of the inland commerce of the country.'

All the commodities of any country consist either of the rude produce of the land, or of manufactured goods. We have seen that the money price of the rude produce of land is altogether determined by the money price of corn. The price of manufactured goods may be resolved into three parts; 1st, The price of the raw material; 2d, The wages of labour; 3d, The profit of stock. The money price of the first two, we have already seen, is altogether regulated by that of corn.

The quantity of circulating stock in every manufacture is in proportion to the value of the raw material, and the wages of the manufacturer. But we have seen that the price both of the raw material, and the wages of the labourer in all manufactures, are raised in exact proportion to the price of corn. More circulating capital, therefore, is wanted in that proportion to carry on every manufacture, and the reasonable profit upon this additional capital must be added to the price of the manufactured commodity. Every one of the three constituent parts of the price of all manufactured commodities receives then an increase by every increase in the price of corn; and thus the price of all manufactured commodities must rise in a much greater proportion than the price of corn. The price therefore of labour, and of every thing which is the produce of land and labour, every exchangeable commodity which the country produces, is altogether determined by the price of corn.

Nothing then can be more incontrovertible than the proposition of Smith, that 'the real effect of the bounty is not so much to raise the real value of corn, as to degrade the real value of silver; or to make an equal quantity of it exchange for a smaller quantity, not only of corn, but of all other commodities' (36–8).[9]

Professor Hutchison argues that in *The Impolicy of a Bounty on the Exportation of Grain* (1804) and *Commerce Defended* (1808) 'Mill closely follows Adam Smith, at times almost reproducing Smith's arguments word for word. But, as he does so, Mill drastically and decisively sharpens and hardens Smith's theories, putting them in a much starker and more unqualified form, imbuing them with his own particular confident dogmatism, and giving them a much more definite cutting-edge in terms of policy applications, to the extent of making them new and different doctrines' (1978, 29). 'It is clear', he proceeds, 'that several of the key concepts, assumptions, and building blocks of the Ricardian models and "theories" are presented, trimmed, polished and ready for use, in this essay of 1804 by James Mill: notably (1) the simplified concept of "corn" as a short-hand for workers' subsistence, a unique commodity in that it is a raw material for all

---

9 Mill does not refer to Smith's (1937) allowance in the third edition of the *Wealth of Nations* that the main proposition applies to 'home-made commodities' only (476).

The argument of paragraph four is an addendum to the Smithian analysis.

production; (2) an extremely drastic, hard-line version of the Malthusian proposition regarding the relations between population changes and wages; and (3) the idea that adjustments take place so rapidly and completely that lags, or "disequilibria", can be left out of the argument' (31).

It is difficult to see the justification for this sharp distinction between Smithian and Millian method. (1) The simplified concept of 'corn' as a shorthand for workers' subsistence – 'a unique commodity in that it is a raw material for all production' – reflects, without question, Smith's position in this context of the Corn Laws – it is essential to his central argument. (2) The population doctrine is used in the same way by both economists in the analysis of the effects of a change in corn production upon population growth; despite Mill's appeal to Malthus nothing was added to Smith's proposition that one effect of the bounty is to check the growth rate of population. In the analysis involving the linkage of general prices to the price of corn, Smith (unlike Mill), it is true, presumed a constant corn wage at the level – 'liberal, moderate or scanty' – dictated by 'the advancing, stationary, or declining circumstances of society' (1937, 476). But it is the constancy of the corn wage that provides the key to the main argument, not its particular level. (3) That adjustments take place so rapidly and completely that lags, or 'disequilibria', can be left out of the argument is precisely the feature of Smith's analysis which troubled late eighteenth-century critics.

* * *

I proceed to consider Mill's position on other central issues: the agricultural model of growth and the law of markets.

In *Commerce Defended* (1808) James Mill developed a formulation of the law of markets in opposition to William Spence's doctrine relating to the social desirability of landlords' luxury expenditure. Spence's position, a criticism of Adam Smith's argument 'that the practice of parsimony is the most effectual way of accumulating national riches,' was that 'expenditure, not parsimony, is the province of the class of land proprietors; and that it is upon the due performance of this duty by the class in question, that the production of national wealth depends' (cited by Mill, 66). Mill's response was initially formulated in terms of a model involving an annual production period; output in any given year depends upon that portion of the previous year's output now devoted ('destined') to 'reproduction' as distinct from 'unproductive consumption': 'The whole annual produce of every country is distributed into two great parts; that which is destined to be employed for the purpose of reproduction, and that which is destined to be consumed. That part which is destined to serve for reproduction, naturally appears again next year, with its profit. This reproduction, with the profit, is naturally the whole produce of the country for that year. It is evident, therefore, that the greater the quantity of the produce of the preceding year, which is destined to administer to reproduction in the next, the greater will naturally be the produce of the country for that year' (70). (The same conception was adopted in the *Elements of Political Economy*.) Arguing within the framework of this 'wages fund' model, Mill asserted that the process of savings entails (productive) consumption – the Smithian 'savings are consumed' theorem – and that self-interest assures against the presence of unsold stocks, as working capital will be devoted either to consumption or accumulation:

Let [Spence] rest in perfect assurance, that the whole annual produce of the country will be always very completely consumed, whether his landholders choose to spend or to accumulate. No portion of it will be left unappropriated to the one species of consumption, or to the other. No man, if he can help it, will let any part of his property lie useless and run to waste. Nothing is more clear, than that the self-interest of men, ever has impelled and ever will impel them, with some very trifling exceptions, to use every particle of property which accrues to them, either for the purpose of immediate gratification, or of future profit. That part, however, which is destined for future profit, is just as completely consumed, as that which is destined for immediate gratification (71).

As in Smith's formulation the emphasis is not upon the net addition to the stock of capital goods created by savings, but rather upon the process of savings.[10] (It may be added that increased savings was seen by Mill as entailing the absorption of labour from the unproductive sector to the productive sector presumably at an unchanged real wage.)

Mill, however, did introduce a supplementary argument and it is one which, at least at this stage of the proceedings, is formally treated apart from the principle that saving entails consumption.[11] I refer to the theorem that 'the production of commodities creates, and is the one and universal cause which creates a market for the commodities produced.' This principle – one version of the law of markets – was designed to refute the notion that capital might 'increase too fast' or that 'the production of commodities should be too rapid' since there is 'a market for a given quantity of commodities, and if you increase the supply beyond that quantity you will be unable to dispose of the surplus.' The law of markets is phrased in the following terms:

No proposition however in political economy seems to be more certain than this which I am going to announce, how paradoxical soever it may at first sight appear;

10 In Smith's case it seems clear that the result of the savings process in additions to the real stock of capital was fully appreciated (see Hollander, 1973a, 188f). Mill had little to say on fixed capital in *Commerce Defended*.
11 Cf. Winch (1965), 84: 'Say's Law may have developed from Smith's "saving is spending" theorem, but Smith himself did not take this step.'

and if it be true, none undoubtedly can be deemed of more importance. The production of commodities creates, and is the one and universal cause which creates a market for the commodities produced . . . It is obviously . . . the collective means of payment which exist in the whole nation that constitute the entire market of the nation. But wherein consist the collective means of payment of the whole nation? Do they not consist in its annual produce, in the annual revenue of the general mass of its inhabitants? But if a nation's power of purchasing is exactly measured by its annual produce . . . the more you increase the annual produce, the more by that very act you extend the national market, the power of purchasing and the actual purchases of the nation. Whatever be the additional quantity of goods therefore which is at any time created in any country, an additional power of purchasing, exactly equivalent, is at the same instant created; so that a nation can never be naturally overstocked either with capital or with commodities; as the very operation of capital makes a vent for its produce . . . (81).

In the foregoing formulation it is stated that demand cannot exist without production since production generates the income with which commodities are purchased – a Keynesian-type proposition – and, further, that not only is the 'power or purchasing' expanded by production but also the 'actual purchases of the nation'. The 'demand of a nation', Mill concluded, 'is always equal to the produce of a nation'. This latter assertion appears to flow from a conception of the economy in terms of barter, for 'the idea of money frequently tends to perplex':

When money is laid out of the question, is it not in reality the different commodities of the country, that is to say, the different articles of the annual produce, which are annually exchanged against one another? Whether these commodities are in great quantities or in small, that is to say, whether the country is rich or poor, will not one half of them always balance the other? and is it not the barter of one half of them with the other which actually constitutes the annual purchases and sales of the country? Is it not the one half of the goods of a country which universally forms the market for the other half, and vice versa? And is this a market that can ever be overstocked? Or can it produce the least disorder in this market whether the goods are in great or in small quantity? (82)

Mill's position that increased incomes are 'always' spent might at first sight appear to be merely an empirical observation which allows for the (exceptional) possibility in practice of hoarding – an excess demand for money to hold. But his recommendation to envisage the economy in barter terms suggests a much stronger version of the law of markets, that version according to which the excess demand for money is zero under all circumstances. In brief, money *per se*, it is implied, has no utility.

The proposition regarding the impossibility of a general excess supply of commodities is carefully contrasted with the possibility of a partial excess. But, Mill insisted, such distortions are rapidly corrected by a movement of resources between industries (84–5). Mill has sometimes been given credit for introducing the condition that the composition of output must be adapted to the tastes of consumers and investors in which vital respect his statement is said to stand far and away above that of Say (Stigler, 1965, 313). But in fact as early as 1803 Say referred to the need for reallocation of resources to reflect the pattern of demand and the likelihood of rapid correction of any maldistribution in normal conditions (Hollander, 1979, 79f). There is no substantive difference here, although it is true that Mill tended to give the condition a particularly strong emphasis.

There is, however, a related issue wherein a very sharp distinction between the two formulations may be discerned. For Mill 'the production of commodities creates . . . a market for the commodities produced,' while for Say 'a product once created affords . . . a market for other products . . .' The latter formulation allows for excess capacity which will be corrected only if and when other products are brought into existence: 'beaucoup de gens ont moins acheté, parce qu'ils ont moins gagné; et ils ont moins gagné, parce qu'ils ont trouvé des difficultés dans l'emploi de leurs moyens de production . . .' (1814, 149). On balance it appears correct to say, with Chipman, that 'for the sharp and excessively doctrinaire version of the principle, priority must certainly go to Mill – doubtless a dubious distinction' (1965b, 709n).

Secondly, Say sought to rationalize the prompt expenditure of sales receipts in terms of the avoidance of any loss of interest. In later editions there are allusions to the 'perishability' of the value of money. Also of great importance, Say was prepared to specify exceptions to the rule – exceptions which Adam Smith invited by his early formulations. These characteristics also lend support to the view that Say's version was less doctrinaire and rigid than Mill's. In Mill's case we find only a brief reference to the proposition that 'no man, if he can help it, will let any part of his property lie useless and run to waste,' and a formal recommendation to conceive of the system in barter terms which implies that money holdings *per se* yield no utility.

* * *

A word finally regarding the *Elements* itself. In his classified catalogue McCulloch had some sharp words for this textbook, amounting to a charge of excessive abstraction rendering the arguments unhelpful in practical application. What he was prepared to allow Ricardo's *Principles* – with good reason we shall see – he found unacceptable in the *Elements* in view of its objective:

This work, by the distinguished author of the 'History of British India,' is a *résumé* of the doctrines of Smith and of Ricardo with respect to the production and distribution of wealth, and of those of Malthus with respect to population. But it is of too abstract a character to be either popular or of much utility. Those secondary principles and modifying circumstances, which exert so powerful an influence over general principles, are wholly, or almost wholly, overlooked by Mill. But though their consideration might be omitted in an original work like that of Ricardo, it is not so easily excused in an elementary treatise. The object of the latter is not to make discoveries, but to exhibit the connexion, dependence, and real influence of the principles, whether primary or secondary, that are known to be in operation; and this is not to be done by looking only at one set and neglecting the others. The science is very far from having arrived at the perfection Mr. Mill supposed. Any one, indeed, may state generally how certain principles operate; but in the vast majority of instances it is extremely difficult, and sometimes quite impossible, to foresee the mode or the degree in which they may be countervailed by others, or to conjecture what may be the remote results of their combined action (1845, 17–8).

Professor Winch has pointed to Mill's treatment of population and wages as an example of the penchant for 'strong cases', and their use in direct policy application, which characterizes the *Elements*: 'Throughout his discussion of these matters,' Winch observes, 'he dramatised the remedy which he favoured for "the condition of the people," by putting forward an apparently water-tight argument which minimised the value of alternative solutions or outcomes' (1966, 193). Winch's full statement will prove useful for what follows in the next section:

In a wage economy, wages depend on the ratio of population to capital, i.e. the wage-fund doctrine, which Mill expresses in its most unqualified form. He now sets out to show that population has a universal tendency to outpace capital accumulation, and his first argument is an assertion to the effect that the consequences of this tendency are plainly in evidence throughout the world; the living standards of the masses are actually at the physical subsistence minimum. Here Mill typically fails to distinguish between what could happen according to his model, and what has in fact occurred: he moves directly from a *ceteris paribus* situation to the real world. Unlike Smith and Ricardo, Mill makes no allowance for the possibility that real wages may remain above the psychological or physical minimum for what amounts to a lengthy short run.

The next stage of the argument is based on the notion that the biological factors underlying population growth are more persistent in their operation than the social and psychological factors which determine capital accumulation. The exercise in deductive sociology used to support the proposition that capital accumulates slowly constitutes the most interesting part of the whole chapter. . . .

The rest of Mill's chapter on wages is devoted to a demonstration of the futility of all remedies for the population problem other than the one which he favoured (194–5).

As a further example of a willingness to make direct and unqualified practical use of the conclusions drawn from highly abstract models we may refer, following Winch, to Mill's attempt (as Chief Examiner at the India Office) 'to implement the radical conclusions which he drew from Ricardo's interpretation of the rent doctrine' – namely, land nationalization. In the British case – the issue discussed in the *Elements* – rent taxation was the solution proposed, whereby 'Mill was merely taking to its logical political conclusion, Ricardo's argument that there was an inherent conflict between the interests of land-owners and the rest of the community . . . Ricardo's rent doctrine merely provided scientific support for Mill's long-standing antagonism towards the land-owning classes; it accorded completely with his view that the politics of the day were dominated by the clash of selfish interests' (197–9). It seems clear that the reason for Mill's predilection for the use of strong cases was in part their suitability in the advancement of his political programme.[12]

Excellent indications of Mill's intentions are also provided by his advice to Ricardo not to publish the *Notes On Malthus*, and by his decision not to publish the manuscript of Ricardo's *Absolute and Exchange Value* for fear of confusing the public's conception of 'correct principles' (191–2). Similarly, his attitude towards the function of the Political Economy Club as a medium for the propagation of accepted doctrine points to the same end: 'Do you know anything in the legislation or practice of this country, not recently under consideration of this Society, peculiarly at variance with the principles of Political Economy . . .?' The contrast with Ricardo as far as concerns the use of economic theory, and more generally, will become apparent in what follows.

Professor Neil de Marchi accepts by and large the charges directed against James Mill that he 'indulged in severe abstraction without seriously investigating the empirical basis of his assumptions'; and that 'armed . . . with suspect science and bad (because uncorrectably narrow and abstract) economics he displayed the utmost pretentiousness in applying his conclusions directly to policy' (1983b, 158). At the same time de Marchi argues the case for a more balanced James Mill, revealed especially in various (unsigned) contributions to *The Literary Journal* from 1803 to 1806 – 'a rather less dogmatic, more circumspect, Mill than we know from his later published

---

12 Cf. his article 'Education' (1824), for the political nature of his arguments. Robert Fenn has suggested to me that Mill's training in pulpit rhetoric may help account for his manner of argumentation.

writings' (160).[13] These general methodological *obiter dicta* – which include allowance for disturbing causes – are said to be in the spirit of Dugald Stewart; and a liaison is traced from Stewart to J. S. Mill of the essay on method both directly and indirectly via James Mill (169f).

The following considerations are pertinent. First, James Mill's early contributions also show signs of the more characteristic Mill, particularly the analyses of the corn-export bounty and the law of markets and their applications to policy. Secondly, the direct and irresponsible use of narrowly constructed theory in policy application which characterizes the public Mill is all the more reprehensible since he apparently knew better. A 'dogmatic lapse' due to a propagandizing propensity and an idiosyncratic personal character doubtless explains much, but the outcome is none the less a defective methodology with which Ricardo took issue and which J. S. Mill was to repudiate.[14]

III RICARDO'S METHOD: CONTRASTS WITH JAMES MILL

There is, of course, no formal discourse on methodology by Ricardo. But to say, as did Bagehot, that Ricardo 'had no larger notion of what science was', that 'he never comprehended what he was doing' is a gross distortion. Ricardo was sufficiently sophisticated to appreciate the limited scope of political economy in general and of economic theory – 'scientific' economics – in particular. His understanding of the historical relativity of the behavioural and technological axioms of his models is impressive. He rejected over-simplified models on empirical grounds. He was keenly aware of the specifically pedagogic role played by his strong-case assumptions. He was conscious to a high degree of the clash between the short-run and long-run implications of policy proposals; and also of the *ceteris paribus* conditions upon which his models were constructed rendering them inappropriate as predictive engines and unambiguous guides to policy. I turn now to some of these matters.

In a few brief remarks Ricardo made clear the limited scope of the whole subject matter of political economy. Following J. B. Say he took the position that 'it is not the province of the Political Economist to advise: – he is to

13 In a recent paper, Robert Fenn (1982) also refers to James Mill's 'deliberate use of rhetorical structures which obscure his real beliefs'; and to the 'systematic distortion he imposed on his published works for reasons of advocacy'. Thus the case for theory in 'Theory and Practice: A Dialogue' (1836) 'is severely truncated in order to make an attack on the major political sects of his day. A false accuracy of argument is paraded in order to deceive the reader as to what is being left out . . . – the political use of a speculative concept is the true heart of his doctrine of theory'.

14 A sceptical view of the impact of Stewart's lectures on James Mill emerges from Davis (1981).

tell you how you may become rich, but he is not to advise you to prefer riches to indolence, or indolence to riches' (1951, II, 338). Here we have the fundamental contrast between choice of ends and the effects of alternative means – of 'normative' as distinct from 'positive' economics. Moreover, Ricardo brought into question the entire issue of the desirability of economic growth as it applied to contemporary 'underdeveloped' economies such as New Spain – especially in the light of the uncertain distribution of an expanded GNP: 'Happiness is the object to be desired, and we cannot be quite sure that provided he is equally well fed, a man may not be happier in the enjoyment of the luxury of idleness than in the enjoyment of the luxuries of a neat cottage, and good clothes. And after all we do not know if these would fall to his share. His labour might only increase the enjoyments of his employer' (1951, VII, 185). As for already advanced economies we find the same sophisticated appreciation of the broader questions of welfare. Thus 'systems of equality' – such as Owen's – were not rejected on grounds of impracticality but rather because they implied growing numbers at low living standards, a blueprint scarcely conducive to increased 'happiness' (1951, IX, 49–50). Also income stability was even given greater weight than average magnitude – and thus manufacturing in this regard preferred to agriculture – from a welfare perspective: 'We cannot, I think, doubt, that the situation of mankind would be much happier if we could depend with as much certainty on a given quantity of capital and labour producing a certain quantity of food, as we can depend upon the same quantity of capital and labour producing a certain quantity of manufactured goods' (237).

These are not the formulations of a person who 'never comprehended what he was doing'; Ricardo was thoroughly aware of the limited scope of the pronouncements of political economists. That he also appreciated the restricted scope of pure theory is clear from his pronouncements regarding the axiomatic foundation of the subject.

I have in mind here the postulate of 'rational' decision making which was expressed as follows in the *Principles*. 'Whilst every man is free to employ his capital where he pleases, he will naturally seek for it that employment which is most advantageous; he will naturally be dissatisfied with a profit of 10 per cent., if by removing his capital he can obtain a profit of 15 per cent. This restless desire on the part of all the employers of stock, to quit a less profitable for a more advantageous business, has a strong tendency to equalize the rate of profits of all . . .' (1951, I, 88). However, he allowed for 'ignorance' in decision making, giving it a place in applied rather than theoretical economics, as is clear from the correspondence of 1811 with Malthus:

I wish to prove that if nations truly understood their own interest they would never export money from one country to another but on account of comparative redundancy. I assume indeed that nations in their commercial transactions are so alive to their advantage and profit, particularly in the present improved state of the division of employments and abundance of Capital, that in point of fact money never does move but when it is advantageous both to the country which sends and the country that receives that it should do so. The first point to be considered is, what is the interest of countries in the case supposed? The second what is their practise? Now it is obvious that I need not be greatly solicitous about this latter point; it is sufficient for my purpose if I can clearly demonstrate that the interest of the public is as I have stated it. It would be no answer to me to say that men were ignorant of the best and cheapest mode of conducting their business and paying their debts, because that is a question of fact not of science, and might be urged against almost every proposition in Political Economy (1951, VI, 63–4).

The maximization assumption might thus require modification in any application of the theory but served well in scientific investigation. (We shall return presently to the question of the 'reality' of the behavioural axiom.)

The distinction between 'science' and 'fact' will be found in a fundamentally important statement regarding the limited scope of the *Principles of Political Economy*, touching specifically upon the treatment therein of public finance: 'On the subject of taxation a wide field is open for those, who will patiently think, to give instruction to the Public; but the first step must be to make the first principles of Political Economy known, and that remains yet to be done. Without correct notions of rent, no man can be made to understand that a land tax does not ultimately fall on the landlord, and it would be in vain to talk to him, till he did admit the new doctrine on the subject of rent' (1951, VIII, 79). The book was not intended to deal with applied public finance but rather with the pure theory of the subject.[15]

Also implying the distinction between 'science' and 'fact' is Ricardo's own rationalization of his use of 'strong cases' in the *Principles*: 'Our differences,' he wrote to Malthus, 'may in some respects, I think, be ascribed to your considering my book as more practical than I intended it to be. My object was to elucidate principles, and to do this I imagined strong cases that I might shew the operation of those principles' (184).[16]

15 But in fact some of Ricardo's analyses only make sense if the contemporary taxation system is taken into account (see Carr and Ahiakpor, 1982).

16 Ricardo complained to James Mill regarding Malthus: 'Another of his great mistakes is I think this; Political Economy he says is not a strict science like the mathematics' (331). (Ricardo might have had in mind the Preface to Malthus's *Principles*, 1820, in Ricardo 1951, II, 5: 'The science of political economy resembles more the science of morals and politics than the science of mathematics.') Much is made of Ricardo's comment by Hutchison (1978) 56, to Ricardo's detriment. ('Malthus had of course,

Interestingly enough, Ricardo even came to the defence of Malthus on occasion, with the distinction in question at hand: 'I do not think that Mr. Malthus is wrong', he wrote to Trower regarding a criticism made of the latter's work; 'I think he means to say that if you diminish the fertility of the land so much that the whole produce must go to the cultivators there can neither be surplus produce to afford profit or rent. If it should be even enough to afford a trifling profit there could be no rent because no worse land could be taken into cultivation. Now says Mr. Malthus if you diminish the fertility of the land one half you will place us in this condition. This is a question of fact and degree, not of principle, and it is one of my complaints against him that he does not answer your principle but wishes to shew that you have taken your case so wide, that it could under no circumstance exist; but however limited might be your case, the same principle is involved, and it is that which should be answered' (1951, VIII, 234–5).

Ricardo's approach to wage theory can only be understood if we keep to the fore what has just been said. Ricardo frequently utilized the assumption of a constant real wage (although not necessarily a constant wage at subsistence) not because he believed the assumption approximated 'factual' reality, but rather as a deliberate simplification – a working hypothesis – to be relaxed when appropriate, as for example in the examination of the growth process or taxation. His conscious analytical methodology could not be expressed more clearly than in the *Essay on Profits*, at the outset of which he stated his assumptions, namely that agricultural technology is unchanged and that real wage per man is constant, and gave his reasons for thus proceeding: 'We will, however, suppose that no improvements take place in agriculture, and that capital and population advance in the proper proportion, so that the real wages of labour, continue

matured, methodologically, as contrasted with the dogmatic, *a priori* deductivism of his first *Essay*, which was on somewhat Millian–Ricardian lines.') But this is unjustified: Ricardo was without question dealing with matters of 'scientific' – not 'factual' – economics so that the statement as it stands must be understood in the light of what has been said in our text. But in any event too much cannot be made of the complaint in the present context, for we must not neglect the remainder of the sentence – Ricardo was merely appealing for consistency in the use of terms: '. . . he [Malthus] thinks he may use words in a vague way, sometimes attaching one meaning to them, sometimes another and quite different. No proposition can surely be more absurd.'

It should not be taken for granted that Malthus held strictly to his approach to economics as a science of 'morals and politics' rather than of 'mathematics'. For he wrote to William Whewell that 'you have arrived at pretty just conclusions, and . . . that mathematical calculations may in some cases be introduced with advantage into the science of Political Economy . . . I have long thought that there are many of the results in political economy which have some resemblance to the problems *de maximis et minimis* . . .' (de Marchi and Sturges, 1973, 387).

uniformly the same; – that we may know what peculiar effects are to be ascribed to the growth of capital, the increase of population, and the extension of cultivation, to the more remote, and less fertile land' (1951, IV, 12). On this basis he constructed his well-known model, relating to the agricultural sector, which yields a declining profit rate as capital and population expand and are applied to increasingly disadvantageous plots of land, without the slightest suggestion that the model based on a constant real wage might be used without correction in application. Matters of 'fact' required specific attention to actual values of the variables which a statement of principle might neglect.

The illustration from wages entails the adoption of an extreme elasticity value – in this case infinite elasticity of labour supply – in order to focus entirely on the implications for the profit rate of recourse to increasingly inferior land. Examples may also be given of the adoption for theoretical purposes of extreme values because the primary implications of the model are unaffected by more complex (albeit more realistic) assumptions. The problem of 'substitutability' provides an illustration.

Both factor and commodity substitution were recognized even in the first edition of the *Principles* (Hollander, 1979, 223f). Yet the basic model of distribution developed in the chapter 'On Profits' neglects entirely the phenomenon, proceeding on the assumption of fixed factor proportions and uniform factor ratios in all sectors. It seems likely that the model was developed despite awareness of substitutability in production, on the grounds that its neglect had few practical consequences; a theoretical model which illustrated the effect on the rate of profits of rising wage costs may be satisfactory although its assumptions are 'unrealistic'. For Ricardo's basic 'real world prediction' is that, given technology, increased wage costs must bring about a decline in the general profit rate. Now whatever the assumptions regarding substitution, capitalists must be 'worse off' after a rise in wages; recognition of substitutability does not reverse the 'prediction'. This is the significant fact, brought out by the model of distribution. Ricardo was prepared to proceed as if such complications were absent.

'Strong case' assumptions, needless to say, are the more justifiable the closer they reflect 'reality', although – as in the illustration from the labour market – they may none the less be justifiable on other grounds. In his discussion of the corn market, for example, Ricardo rejected the possibility of a precise specification of the price–quantity relationship, but suggested that the extent of such small responsiveness as exists will depend in practice upon the country's 'means of holding over the superfluous quantity to a future season', and also 'on the opinions formed of the probability of the future supply being adequate or otherwise to the future demand' (1951, IV, 220). There can be no doubt at all, therefore, that he was quite aware of the empirical responsiveness of quantity demanded to price. But what evidently struck him was the limited extent of the response, small enough to justify neglect in the construction of theoretical models designed to capture the prime interrelationships – as distinct from applied studies of the corn market – for in such contexts he almost invariably took for granted that the demand for corn is a unique function of population size. Zero demand elasticity constitutes a simplifying assumption comparable to that made in dealing with value theory, namely of adopting a strict labour-input account of relative price movements despite clear recognition of other sources of disturbance, on the grounds that variation in labour input is overwhelmingly the most significant quantitative determinant.

* * *

Thus far we have considered some of the grounds used by Ricardo to justify his use of 'strong case' assumptions. Let us look now a little more closely at the basic behavioural axiom of the entire body of doctrine.

There can be no doubt that the assumption of maximization of money returns reflected, for Ricardo at least, a close first approximation to contemporary business reality. The statement cited above (19–20) from the *Principles* proceeds to describe the institutional arrangements characterizing advanced ('rich') economies whereby advantage may be taken of profit opportunities. Similarly, the early statement of 1811 refers to 'the present improved state of society', while the continuation of the relevant passage – 'It rests with you therefore to prove that a case can exist where it may become the *interest* of a nation to pay a debt by the transmission of money rather than in any other mode, when money is not the cheapest exportable commodity . . .' – implies that the institutional framework would have to be a very different one for an assumption other than maximization of net returns to be relevant.

The *Reply to Bosanquet* points to the same conclusion. Ricardo was troubled by Bosanquet's use of 'the vulgar charge, which has lately been so often countenanced, and in places too high, against theorists' (1951, III, 160). This reaction is sometimes taken to be evidence of an 'extreme' Mill-like position (Hutchison, 1978, 33). It is certainly true that Ricardo here denounced the man 'who is all for fact and nothing for theory'. But there is nothing patently unreasonable about the argument. His complaint was that, without an initial framework of theory, apparent factual evidence might be quite misleading: Those who condemned theory 'can hardly ever sift their facts. They are credulous, and necessarily so, because they have no standard of reference' (1951, III, 181). He appealed accordingly for investigators to follow 'the sober paths of practice and experience' rather than indulge in 'speculations the most wild, and dreams the most chimerical'

(239). Assuming the empirical validity of the income-maximizing assumption, as it applied in the international money market, it was inconceivable, for example, that Bosanquet was correct in his insistence upon differential exchange rates between major international centres lasting for some four years – for what was involved was 'a trade, the slightest fluctuations of which are watched by a class of men proverbial for their shrewdness, and in which competition is carried to the greatest extent' (181).

With a sound empirical basis for the axioms, the argument could not be said to be 'wholly theoretical' (160). There is certainly no question of universally valid axioms. Ricardo was quite aware that the appropriate behavioural assumptions varied geographically and temporally. His consciousness of the relativity of the behavioural assumptions and also the technological assumptions – and accordingly the conditional character of the conclusions of political economy – emerges particularly clearly in the context of alternative solutions to poverty within various sociological contexts. Here differential patterns of behaviour on the part of labourers provide the key, the policies appropriate for raising general standards differing entirely from case to case.

Thus the problem of poverty in 'poor' countries – characterized by abundant fertile land – derives in the final resort from 'the ignorance, indolence, and barbarism of the inhabitants'. Because of the presence of plentiful land resources the potential for substantial increase in labour productivity and in accumulation is excellent and far in excess of any probable population growth rate; excessive population growth is not the issue in 'poor' countries. What is required is a solution to the very low productivity of labour as a means of raising the surplus available for accumulation. By contrast there are those countries where the potential force of diminishing returns is such as to impose serious limits on prospective accumulation relative to population growth, and the solution is control over the latter (1951, I, 99). Similarly, on the basis of his conception of the British labourer, as distinct from those with a low level of aspiration for goods, Ricardo based his position that accumulation does not imply a loss of purchasing power since any increase in labourers' real incomes will be spent if not on corn (assuming an initially unchanged population) then on luxuries (1951, VIII, 268, 275). He also rejected Malthus's case against the proposition that a capitalist with 'food and necessaries' in hand would 'not be long in want of workmen who would put me in possession of some of the objects most useful or most desirable to me', on the grounds that Malthus's objections turn on the sociological characteristics of individuals in South American countries, and are 'little applicable to countries with a dense population abounding in capital, skill, commerce, and manufacturing industry, and with tastes for every enjoyment that nature, art or science will procure' (1951, II, 340–1).

Ricardo's optimistic evaluation of the actual and prospective growth rate of capital and of labour demand in contemporary circumstances is also pertinent here. Although not a 'new' country, Britain retained all the features characteristic of an 'improving' country wherein prospects for labour were satisfactory, at least in the event that artificial or institutional encouragement to excess population growth is eradicated. To some extent Ricardo's optimism regarding Britain reflects his evaluation of the weak force of diminishing returns in that country. Thus at the conclusion to the chapter 'On Profits' Ricardo distinguished between 'the effects of accumulation . . . in different countries': 'However extensive a country may be where the land is of a poor quality, and where the importation of food is prohibited, the most moderate accumulations of capital will be attended with great reductions in the rate of profit, and a rapid rise in rent; and on the contrary a small but fertile country, particularly if it freely permits the importation of food, may accumulate a large stock of capital without any great diminution in the rate of profits, or any great increase in the rent of land' (1951, I, 126). Even with protection in the case of a small but fertile country – an obvious reference to Britain – there is little emphasis on a falling rate of return. It is important to note that the weak force of diminishing returns in an economy such as that of Britain is not even explained here by reference to the adoption of new technology. But a few pages earlier we do read that 'this tendency, this gravitation as it were of profits, is happily checked at repeated intervals by the improvements in machinery, connected with the production of necessaries, as well as by discoveries in the science of agriculture' (120). These evaluations of technological conditions and prospects go far to explain the policy position adopted by Ricardo regarding Corn Law repeal.

That the particular institutional framework assumed governs the appropriateness of the axioms is well clarified in the context of 'systems of equality'. Ricardo did not oppose such systems on grounds of the inability of a restructured economic system to provide additional food for a growing (artificially stimulated) population – even into a far distant future. (Whether growth, which merely implied larger numbers, was desirable was a quite distinct matter.) This perspective is implied in remarks to Francis Place regarding the latter's critical evaluation of Owen's scheme: 'First, in the latter part of the first chapter [of Place's *Illustrations and Proofs of the Principle of Population*] it is I think *inferred that under a system of equality population would press with more force against the means of subsistence than it now does*. This I do not think is true. I believe, that under such a system, mankind would increase faster than it now does, but so would food also. A larger proportion of the whole capital of the country would be employed in the production of food-necessaries, and a less proportion in the production of luxuries, and thus we might go on, even with an increase of capital, without any increased difficulty, till that distant time, which because of its distance, Mr. Malthus says should not damp our ardour' (1951, IX, 49–50).

* * *

Let us briefly digress from political economy to government for a further indication of Ricardo's insistence upon an appropriate, in the sense of empirical, justification for the axiomatic framework. The episode has a keen biographical interest.

Responding to James Mill who was urging his friend to enter Parliament Ricardo observed in 1815: 'Your favourable opinion of my honesty is in striking contrast with your opinion of the honesty of those who at present constitute the house of Commons. On this subject you are, as I have often thought you, unjustly severe. That there are many venal men in Parliament who get there with no other view but to forward their own personal ends, no one can doubt, – but as a body they have more virtue than you are willing to give them credit for. No other assembly is perhaps so much under the influence of public opinion which you will allow is a great security for virtue' (1951, VII, 263). A little later regarding the same issue: 'I have no doubt that your principle is correct, but I think you apply it too rigidly. It appears to me that you allow too much force to the stimulus of money, and the praise of Princes, and too little to the effect of public opinion and the consciousness of deserving approbation' (310–11).

Examination of Mill's *British India* stimulated a similar response. After reading the second volume Ricardo wrote to his friend diplomatically: 'I wish however that I was a more competent judge, as I should then be more fully assured that the decision which I pronounced' – regarding the quality of the earlier part – 'would be ratified by those who must ultimately decide on the real merits of your performance. . . . It is probable indeed that many may not agree with your notions of government – they may think that you give too much weight to some of the motives to human action, and too little to others . . .' For his part, he questioned the weight placed by Mill on the 'love of patronage' by judges, on the grounds that motives deriving from 'love of ease' and 'fear of censure' were neglected. Similarly, he criticized Mill's observation regarding 'the rare occurrence of parliamentary influence with knowledge and talent, in all places where much either of money or power is to be enjoyed'. His point was that 'if money and power were the only things desirable to man your conclusion could not be denied, but while public opinion and public sympathy are so much valued by all ranks of men, sufficient motives exist for the acquirement of knowledge and talent independent of the power and money which they may chance to bring along with them' (1951, VII, 236–9).

Ricardo sensibly reserved judgement before seeing the third volume, with its 'practical application' of Mill's principles. But his insistence upon a requirement of empirical relevance to be satisfied by the axioms is obvious. Ricardo was unhappy on empirical grounds with James Mill's over-simplified behavioural assumptions (the overwhelming quest for money and power) in the analysis of legislation – although he was prepared to accept the 'maximization' axiom in the analysis of the contemporary British economy as a satisfactory first approximation to reality.

* * *

As we have seen, Ricardo justified his 'strong case' assumptions in economics on analytical and sometimes on empirical grounds. But the degree of legitimate simplification in model construction must, of course, always be a matter of judgement. In Ricardo's view James Mill went too far by adopting simplifications which yielded positively misleading conclusions. Ricardo is better considered a critic than a student of Mill in this regard.

Of the first importance is Ricardo's rejection of the corn model of growth. The general context must be explained.

Ricardo turned down the position that 'a demand for labour is the same thing as a supply of necessaries' (1951, VIII, 258). This objection is amplified in the *Notes on Malthus's 'Principles'*: 'If Mr. Malthus had merely said that with the facility of providing food, population will rapidly increase, because food is one of the most important objects of consumption, it would be impossible to differ with him; but he invariably insists that the increase of population does not depend on the means which we possess of providing for it, or rather which the people themselves have of providing for their offspring, but on the previous provision of food, which is laid up for them' (1951, II, 111).

Food supplies expand along with growth of population to satisfy expanded demand; there was no place for a 'wages-fund' theory if by that term is meant a given stock of wage goods (largely food), on hand – no more and no less – at the outset of the production period. The wage bill in fact includes items other than food which are in elastic supply to the labourers. The assumption that increased quantities of such items are in elastic supply only after several seasons of higher wages – so that it is only after several seasons of high wages that workers' real earnings increase – would retain the rigid concept of the wages fund for any given season; but such a view does not allow for the immediate flow of 'luxury' goods, hitherto available for capitalists' consumption, to satisfy increased demand by labourers. In fact, the composition of the wage basket is determined solely by labourers' taste patterns; the increase in money wages is devoted, in the first instance, largely if not entirely to 'luxuries'; and there is no emphasis whatsoever on a technical inability to vary even the food component of the basket (Hollander, 1979, 326f).

No better indication of the rejection of the wages-fund theory exists than Ricardo's denial of the standard 'predictions' which are derived therefrom. According to Malthus, for example, workers' combinations were 'not only illegal, but irrational and ineffectual', for artificially maintained wages in any sector 'must have the effect of throwing so many out of employment, as to make the expense of their support fully equal to the gain acquired by the higher wages, and thus render these higher wages in reference to the whole body perfectly futile'. A clearer application of the notion of a rigid wages fund is difficult to conceive. But Ricardo objected: 'A combination among the workmen would increase the amount of money to be divided amongst the labouring class', a response which clearly implies an intention to focus on possibilities for real-wage increases (presumably at the expense of profits) (1951, VII, 203).

A second standard application of the doctrine was to poor law policy. Ricardo placed no weight at all upon constraints imposed by the food supply in this context and took Malthus to task for his presumption, in some statements in the *Essay on Population*, that allowances merely raised the number of consumers without increasing available output (1951, VI, 3).

The conceptual framework of an annual agricultural cycle governing the demand for labour was characteristic of James Mill as we have seen. Ricardo was evidently unhappy. In a letter to Mill he explained the response that should be given to those 'general glut' theorists who feared excess commodity supplies in consequence of accumulation, a response which Mill failed to give in the *Elements* because of his agricultural frame of reference: 'If every man was intent on saving, more food and necessaries, (the materials which are chiefly employed in procuring labour), would be produced than could be consumed. The supply above the demand would produce such a glut, that with the increased quantity you could command no more labour than before . . . [But in fact] during the period of very high wages, food and necessaries would not be produced in such quantities as to occasion a glut, for it would be the interest of the producer to produce such things as were in demand, and suited to the tastes of those who had high wages to expend' (1951, IX, 131). This passage figured conspicuously in the *Principles* itself (1951, I, 292–3).

Numerous instances may be cited where Ricardo complained of Mill's over-simplifications in the *Elements* on more specific matters of theory. Some of these detailed criticisms are closely related to the primitive nature of the basic growth model. Thus on the demand for labour: 'You say . . . that the demand for labour and the power of employing it will be in proportion to the increase of capital – I believe I have said the same, and it may perhaps be right to say so in an elementary book, altho' it is not strictly correct.

The power of employing labour depends on the increase of a particular part of capital, not on the increase of the whole capital. (See my Chapter on Machinery)' (1951, IX, 127). Mill made no changes in the subsequent editions leaving untouched an introductory reference to the 'universal' validity of his proposition as he put it (Winch ed., 1966a, 230).[17]

As a second example from a related context, reference may be made to Ricardo's criticism of the discussion in the *Elements* of motives to accumulation, turning on the neglect of technical change: 'You do not speak of the two ways by which capital may be increased by saving; of one, the common and usual way, devoting more of the annual production to productive employments, you do speak, but you say nothing of the great increase which sometimes takes place in capital by the discovery of cheaper modes of production' (1951, IX, 127). The issue, by contrast, was of profound import in the *Principles* as, for example, in the context of foreign trade which 'increases the amount and variety of the objects on which revenue may be expended, and affords, by the abundance and cheapness of commodities, incentives to saving, and to the accumulation of capital . . .' (1951, I, 133). Moreover, the 'advertisement' to the third edition of the *Principles* spells out the general principle that 'the ability to pay taxes, depends . . . on the money value of each man's revenue, compared to the money value of the commodities which he usually consumes'.

However, perhaps the best illustration of Ricardo's warnings against excessive simplification is provided by the theory of value. Commenting on Mill's formulation in the first edition of the *Elements* Ricardo noted: 'If a watch and a common Jack altered in relative value without any more or less labour being required for the production of either of them, could we say that the proposition "that quantity of labour determines exchangeable value" was universally true? What I call exceptions and modifications of the general rule you appear to me to say come under the general rule itself' (1951, IX, 127).[18] Similarly, Ricardo found unacceptable Mill's expansion of the discussion to allow for the 'exceptions' recognized by Ricardo, by adoption of the solution suggested by McCulloch which involved treatment of profits as stored-up labour (377). He also rejected J. S. Mill's juvenile effort (in a paper since lost) to retain the labour theory:

My principal objection against John is that he proposes to prove that commodities

---

17 But when Ricardo complained (1951, IX, 128) of a minimization of the empirical significance of fixed capital later in the *Elements* Mill removed the offending clause (Winch ed., 1966a, 286). Elsewhere, Mill allowed for the substitution of fixed capital for labour (221).

18 Mill had stated (Winch ed., 1966a, 264) that a change in relative values upon an altered structure of wages occurred 'without in the least degree affecting the truth of the proposition, that quantity of labour determines exchangeable values'.

are valuable in proportion to the quantity of labour expended on them, and then proves that capital is another name for labour; profit a name for the wage of that labour; and therefore commodities are valuable in proportion to *wages* of all descriptions *paid* for the production of them. Is not this to say that the value of commodities depends upon the value of wages and profits? If he said so he would say what is true, but the proposition is a different one from that which he first advances (387).

Ricardo's reaction to James Mill's observations on population suggest a further example. At one juncture Mill had observed casually in the *Elements*: 'it is needless to exhibit an accurate calculation, to show that population would double itself in some moderate portion of years. It is evident, at once, that it would double itself in a small numbers of years' (Winch ed., 1966a, 233). Ricardo in response alluded to the complexities which Mill neglected: 'The fecundity of the woman would not be admitted as a conclusive argument, by the objectors to the theory, that the population of a rich, luxurious and populous country could under favorable circumstances, increase in the same proportion as a new and poor country, because they contend, that in consequence of the prevalence of luxury, so many women are withdrawn from the office of childbearing, that there are not a sufficient number left to augment the population in the same proportion as at an earlier period. They contend that the demand for nurses, and female servants of all descriptions, lessen the number of childbearing women' (1951, IX, 126).

The law of markets requires special attention. Did Ricardo adhere to James Mill's version – 'Say's Identity' – or that of Say himself – 'Say's Equality'? In other words did he deny the possibility of excess commodity supply even in the short run or did he intend to define thereby only full equilibrium? The evidence is complex, but on balance I am led to conclude that it is the more qualified version, that of Say, which must be attributed to Ricardo. For the fact is that Ricardo explicitly allowed for temporary excess demand for money in consequence of an initial reduction in the money supply: '[A] reduction in the amount of the circulating medium should speedily operate on prices, but the resistance which is offered – the unwillingness that every man feels to sell his goods at a reduced price, induces him to borrow at a high interest and to have recourse to other shifts to postpone the necessity of selling. The effect is however certain at last, but the duration of the resistance depends on the degree of information, or the strength of the prejudices of those who offer it, and therefore, it cannot be the subject of anything like accurate calculation' (1951, VII, 67). The formulation which is elaborated carefully in the *Principles* itself, where Ricardo identifies an increase in the demand for money (the temporary postponement of expenditures during crisis conditions) and an increase in the supply of money as far as concerns the temporary effect upon the rate of interest (1951, I, 297–8), may be said to represent an embryonic statement of J. S. Mill's famous essay 'Of the Influence of Consumption on Production'. Also it cannot be said that it is merely an 'exception' to the rule which does not merit emphasis, for (as we shall see) Ricardo warned strenuously against any monetary policy which might entail sudden contractions of the supply of money because of the consequences for aggregate activity – a very immediate issue was at stake rather than an exceptional case of purely academic interest.

* * *

There is thus good reason to support Piero Sraffa's view that 'Mill's contribution to the making of the *Principles* was less than might have been expected from his promises and encouragement. On the theory there is little doubt that his influence was negligible; he had been out of touch with Political Economy for some time and his letters to Ricardo contain little discussion of theoretical issues. Mill's letters of this period are full of advice relating to "the art of laying down your thoughts, in the way most easy of apprehension". But despite his repeated assurance that he would see to the order and arrangement . . . it seems likely that in the main the sequence of topics was left as Ricardo had originally worked through them' (Sraffa, 1951, I, xx–xxi). This is not surprising; James Mill was interested in economic theory as a weapon in the service of his political program, and not in the difficult analytical problems which preoccupied Ricardo.

His objective also helps account for the advice he gave to Ricardo regarding matters of organization – his recommendation that the projected *Principles* commence with a discussion of the aggregate income shares (1951, VI, 321). But his advice, which reflects his preoccupation with class conflict and rent taxation, was rejected by Ricardo who chose rather to deal with complex matters of value theory first. Moreover, as I have shown elsewhere, Ricardo was not in actuality concerned with the three-way class distribution issue – despite the Preface which carries traces of Mill's hand (Hollander, 1979, 404f).

Mill's ambition for Ricardo was that he should become spokesman in Parliament for 'correct' principles of political economy in the service of 'good government' (1951, VI, 250f., 306f). But Ricardo did not fill the bill. It is difficult to accept the view that 'there was good reason for [Mill] to believe in the malleability of Ricardo's mind when subjected to the hammer of reason' (Winch ed., 1966a, 186). It is doubtless true that Ricardo's primary interest lay in long-run analysis: 'It appears to me,' he wrote to Malthus, 'that one great cause of our difference in opinion, on the subjects which we have so often discussed, is that you have always in your mind the immediate and temporary effects of particular changes – whereas I put these immediate and temporary effects quite aside, and fix my whole attention on the permanent state of things which will result from them. Perhaps you estimate these temporary effects too highly, whilst I am too much disposed to undervalue them. To manage the subject quite right they should be carefully distinguished and mentioned, and the due effects ascribed to each' (1951, VII, 120). It may well be because of this predilection that the tradition has developed according to which Ricardo was somewhat divorced in his intellectual interests from 'day-to-day' considerations, so that his recommendations lost much of their relevance. But this is going much too far. The contrast in question must be understood at most in a comparative sense, for Ricardo was in fact also preoccupied with 'short-run' matters relating both to theory and policy. For us here the important point is that his concern for the 'short-run' implications of policy diluted markedly the political effectiveness of arguments from political economy.

Thus Ricardo rejected corn law reform in time of depression because any liberalization – desirable in the long term – could only worsen immediate employment opportunities (1951, VIII, 103). He clearly wished to avoid any further disturbances that would magnify the need for reallocation of resources, and therefore ruled out corn law repeal as a helpful step. He only called for legislation in 1821 after the return of prosperity in the manufacturing sector. Moreover, repeal – which was not to be total in any event – should be brought about by gradual steps and after due warning with recognition of the possible need for compensation of those (other than landlords) adversely affected (Hollander, 1979, 517).

Ricardo's position is cited by Professor Hutchison as evidence of 'great wisdom . . . in the realisation of ignorance and the scepticism regarding conclusions deduced from long-run, rapidly self-equilibrating models' (1978, 49–50n). But it is read as an exception which proved the rule: 'It amounts to an extreme contrast with the more typical kinds of Millian–Ricardian policy analysis.' It is not clear at all why at this particular point, and nowhere else, Ricardo should have 'entertained doubts' and Hutchison offers no suggestions. But on our reading of the record there is nothing to account for. Ricardo's position on the Corn Laws was no exception; he was typically preoccupied with the appropriate timing of proposed reforms.

Thus Ricardo's cautious attitude to policy emerges clearly in the context of the poor law issue, where the clash between the desirability of reform from a long-run perspective and the problems which may result therefrom in the more immediate future created a dilemma, the solution to which eluded him. 'It is agreed by all who are most friendly to a repeal of these laws,' he wrote in the *Principles*, 'that if it be desirable to prevent the most overwhelming distress to those for whose benefit they were erroneously enacted, their abolition should be effected by the most gradual steps . . .

No scheme for the amendment of the poor laws merits the least attention, which has not their abolition for its ultimate object; and he is the best friend to the poor, and to the cause of humanity, who can point out how this end can be attained with the most security, and at the same time with the least violence' (1951, I, 106–7). In correspondence he insisted that 'no man in his sober senses would wish for any sudden alteration of the present plan'; and while he was pleased by the general attitude of the House of Commons Select Committee on the Poor Laws in July 1817, he complained of its failure to suggest 'an efficient remedy' to the problem (1951, VII, 248, 219). His own 'remedy' amounts to Malthus's solution to the dilemma, namely the refusal of all parish assistance to those born after a certain date, thus allowing for continued help to those of the present generation actually in need; there can be no doubt that Ricardo did not countenance the immediate abandonment of the able-bodied paupers (1951, IX, 49f).

The resumption of cash payments provides us with yet another important instance. It has been suggested that 'at any rate in the early years' Ricardo had 'no apprehensions about the alleged evils of Resumption' (Sayers, 1953, 82). This is going much too far. Ricardo was troubled by the implications for activity, and accordingly insisted in 1810–11 upon a slow reduction of the note issue: 'I am well aware that the total failure of paper credit would be attended with the most disastrous consequences to the trade and commerce of the country, and even its sudden limitation would occasion so much ruin and distress, that it would be highly inexpedient to have recourse to it as the means of restoring our currency to its just and equitable value . . . Before therefore they can safely pay in specie, the excess of notes must be gradually withdrawn from circulation. If gradually done, little inconvenience would be felt; so that [if] the principle were fairly admitted it would be for future consideration whether the object should be accomplished in one year or in five' (1951, III, 94).

However, not only did Ricardo in 1810–11 argue against an immediate or very early return to cash payments at par, because of his awareness of the potentially 'disastrous consequences to the trade and commerce of the country', he formally recommended to the Prime Minister *temporary devaluation* precisely on these grounds. The solution recommended was a return to gold, thereby assuring against further excessive note issues, but at an altered par – £4.15.0 instead of £3.17.10½ per ounce – for an unspecified period, during which interval the note issue would be slowly contracted and the (temporary) mint price correspondingly reduced. Ricardo himself clearly recognized the implication that his proposal amounted to effective devaluation for a time (1951, VI, 43–4).

When we consider the later period we find the same responsible attitude. Ricardo's proposal for resumption, as formulated in 1819, did not require immediate resumption at par; he was satisfied with resumption at the going market rate of bullion followed by a gradual reduction in the Bank's selling price of the metal. During this interval it was expected that the note issue would be slowly cut back by the Bank authorities to assure that the market price of gold did not exceed the ruling selling price and generate a loss of reserves (1951, V, 440; also 381–2; 450–2). Moreover, the ingot plan formulated in 1811 and 1816 would operate to reduce deflationary pressure (405, 413), and it was presumed that the Bank would avoid complicating matters by heavy purchases on world metal markets (13). For these reasons Ricardo made clear in his evidence before the Commons committee that he expected only a small amount of general deflation from note contraction. The actual extent of note contraction, he added in his evidence, would also be lessened by the 'legitimate' demand for accommodation deriving from 'want of confidence'. This latter observation is not a concession forced from Ricardo by the committee. It reflects precisely the position adopted in a letter of 1811, recognizing the necessity for accommodating demands from the public in circumstances of 'embarrassed credit' (and the possibility of doing so without untoward effects on the exchanges), and also in the *Economical and Secure Currency* of 1816 regarding flexibility of credit expansion to meet 'temporary circumstances' (see Hollander, 1979, 490f).

We come now to a fundamental qualification, made during the period of deflation, to the entire question of the merits of a return to gold at par. The qualification amounts to a recognition that permanent devaluation might be a preferable alternative if the excess of the market over the mint price is of massive dimensions. The allowance is made in correspondence with Wheatley in September 1821: 'I never should advise a government to restore a currency, which was depreciated 30 pc^t., to par; I should recommend, as you propose, but not in the same manner, that the currency should be fixed at the depreciated value by lowering the standard, and that no further deviations should take place. It was without any legislation that the currency from 1813 to 1819, became of an increased value, and within 5 pc^t. of the value of gold, – it was in this state of things, and not with a currency depreciated 30 pc^t., that I advised a recurrence to the old standard' (1951, IX, 73–4). The same allowance was made in Parliament: 'It was from seeing the immense power which the Bank, prior to 1819, possessed . . . it was from the view which he took of the extent of that power of the Bank, that he had rejoiced, in 1819, in the prospect of a fixed currency. He had cared little, comparatively, what the standard established was – whether it continued at its then value, or went back to the old standard: his object had been, a fixed standard of some description or other' (1951, V, 310).

Ricardo's unearned reputation as an irresponsible applied economist evidently derives from a failure by readers to appreciate the nature of the *Principles*. The book has been considered from the beginning as 'more practical' than it was intended to be – as Ricardo himself pointed out to Malthus. For his applied economics we must turn to less formal forums – the pamphlets, correspondence, speeches. A conscious need to make appropriate qualifications to a model in practical applications – especially where policy was involved – becomes clear beyond doubt.

That the charge of 'piling a heavy load of practical conclusions upon a tenuous groundwork' – the Ricardian Vice – cannot fairly be applied to Ricardo as a general rule should by now be obvious. But the evidence pointing to this conclusion is yet more extensive. The corn law episode provides a splendid illustration of Ricardo's refusal to use theory without honest qualification. The fact is that Ricardo felt unable to make his main case against agricultural protection in terms of the secular trend in the rate of return on capital 'predicted' by the growth model – and this despite pressure from McCulloch. Ricardo allowed fully for disturbances to the *ceteris paribus* conditions which must be fulfilled to justify any 'prediction' of a gradual vanishing of investment opportunities, and believed that past experience relating to technical change and future prospects could not justify an argument for repeal based upon the predictions flowing from his growth model under conditions of unchanged technology, or relatively sluggish technological improvement. Technical change, from the standpoint of Ricardian analysis, not only in the wage-goods sector (manufacturing as well as agriculture) but also in the luxury-goods sector, was of profound significance. Precisely because of a keen awareness of the potency of new technology, Ricardo adopted the position that corn law repeal, while certainly desirable (on the grounds that the return on capital and the rate of accumulation would be higher at any point of time in an open than a closed economy), was none the less not the *sine qua non* for continuous rapid expansion into a foreseeable future – by which I mean a time span of concern to drafters of policy, even 'long-term' policy (see Hollander, 1979, 605f).

What has been said is relevant for an evaluation of Malthus's charge (1820) that Ricardo, by his concentration on the short-run distributional effects of corn law repeal and agricultural innovation, was guilty of an uncharacteristic interest in the immediate implications of policy for political ends (1951, II, 213–4). Ricardo himself denied the charge: 'Perhaps in no other part of his book has Mr. Malthus so much mistaken me as on this subject – he represents me as supporting the doctrine that the interests of landlords are constantly opposed to those of every other class of the community, and one would suppose from his language that I considered them as enemies to the state' (117). The fact that this is not the only context where the consequences of a disturbance prior to population expansion are carefully analysed – the analysis of accumulation in conditions of labour scarcity is an obvious illustration; and Ricardo's preoccupation in various contexts with the immediate consequences of policy proposals lend considerable support to his protest against Malthus's charge of the use of theory to support a political bias. The same conclusion is suggested by the great weight placed by Ricardo on the social advantages flowing from security of property. Ricardo should not be identified with James Mill as far as the attitude to landowners is concerned.[19]

* * *

The corn law episode nicely illustrates the Ricardian approach to theory confirmation. Ricardo's primary objective in devising his growth model was to provide an analytical framework whereby to investigate the various factors that play upon the profit rate, rather than one designed to yield a specific prediction. Nothing in the actual course of British experience relating to accumulation and population growth necessarily reflects adversely on his model. This position is clearly indicated by the concession to Malthus, in one of the earliest extant letters dealing with protection, that the profit rate had risen and his insistence that this in no way demonstrated the inadequacy of his theory, in the light of 'technological change':

That we have experienced a great increase of wealth and prosperity since the commencement of the war, I am amongst the foremost to believe; but it is not certain that such increase must have been attended by increased profits, or rather an increased rate of profits, for that is the question between us. I have little doubt however that for a long period, during the interval you mention [probably 1793–1813], there has been an increased rate of profits, but it has been accompanied with such decided improvements of agriculture both here and abroad, – for the French revolution was exceedingly favorable to the increased production of food,

19 The evidence points to a responsible use of economic theory. But it is not as totally one-sided as it may appear from the account up till this point. Thus, to return to monetary policy, while Ricardo's position on resumption was far from being as 'irresponsible' as his critics believed – as is clear from the very considerable weight that he placed on the short-run consequences of major and sudden reductions in the money supply – it is none the less true that he approached the post-war depression in terms of a model which incorporated the full employment assumption and explicitly formulated the 'Treasury view' – and this despite his awareness of the existence of severe unemployment of labour and under-utilization of capacity over the years 1815–21. I would conjecture that Ricardo simply overstated his case and this because of the strength of his objections to those economists who maintained that the post-war crisis reflected characteristics of secular stagnation. Secondly, Ricardo's fears of solutions with possible long-term inflationary implications may have played a part in accounting for the narrowness of his perspective.

  This same fear of inflation might also help us appreciate Ricardo's reluctance to emphasize variations in the exchanges and possibly gold movements generated by disturbances to the balance of payments unrelated to initial changes in the money supply – although he clearly recognized them.

that it is perfectly reconcileable to my theory. My conclusion is that there has been a rapid increase of Capital which has been prevented from shewing itself in a low rate of interest by new facilities in the production of food (1951, VI, 94–5).

Precisely the same issue was raised years later when in his *Principles* Malthus charged that Ricardo had 'never laid any stress upon the influence of permanent improvements in agriculture on the profits of stock, although it is one of the most important considerations in the whole compass of Political Economy.' Ricardo strongly objected: 'Once more I must say that I lay the very greatest stress upon the influence of permanent improvements in Agriculture. The passage quoted [1951, I, 126] refers to a state of things when no improvements are taking place, and therefore the argument built upon it which supposes improvements has no foundation' (1951, II, 293).

Ricardo's position on the foreign exchange also illustrates his approach to theory confirmation. Neither he himself nor the Bullion Report contended, Ricardo insisted, that an increased note issue would necessarily lower the exchange; their contention was that 'such an effect will result from a redundant currency', with allowance to be made for changes in the implicit *ceteris paribus* conditions. It would be premature, therefore, to reject the Ricardian theory simply on the grounds that increases in the note supply had sometimes been accompanied by a rising exchange rate. Correction was necessary for any loss of coin; allowance had to be made for changes in the volume of transactions, harvest conditions, and the relative values of gold and silver; and care had to be taken to allow for the time lag involved following an increased note issue (1951, III, 114, 118; cf. 86, 195).

Much the same general approach is evident during the course of the parliamentary evidence on resumption in 1819. Here too Ricardo insisted upon the principle of multi-causality: 'I am fully aware that there are other causes, besides the quantity of bank notes, which operate upon the exchanges . . .' When asked to account for the empirical fact that during the period 1817–19 an increase in the price of gold accompanied a reduced note issue, he replied: 'It does not in the least shake my confidence in the theory, being fully persuaded that such an effect [a reduced gold price] must have followed, if it had not been counteracted by some of those causes to which I have already adverted' (1951, V, 372, 376). He had in mind a variety of disturbing factors which might have been at work such as an independent (worldwide) rise in the value of gold itself, and variations in 'credit' (substitutes for currency) and in country bank circulation.

It is appropriate here to consider Ricardo's general statement of the methodological issues involved in attempts at theory confirmation. I have in mind his well-known comment to Malthus: 'I should be more pleased that we did not so materially differ. If I am too theoretical which I really believe is the case, – you I think are too practical. There are so many combinations, – so many operating causes in Political Economy, that there is great danger in appealing to experience in favor of a particular doctrine, unless we are sure that all the causes of variation are seen and their effects duly estimated' (1951, VI, 295).

There is a good deal of confusion in the secondary literature regarding the implications of the foregoing statement. It is included by Professor Winch amongst evidence purporting to show that Ricardo and James Mill 'were completely at one on questions of method' – an allusion to Ricardo's (supposed) 'eagerness to believe in the applicability of clear-cut principles' (1966, 368). It is difficult to see how the statement can be given this interpretation – it is rather a most responsible formulation of the problem of multi-causality and the impossibility, under such circumstances, of placing much reliance on appeals to 'experience' in theory confirmation. This is surely a characteristic inherent in all deductive theorizing – as Adam Smith had long before recognized. The same characteristic, of course, renders dangerous the use of theory in simple-minded policy applications – and we have seen, in the corn law context, that this was not Ricardo's practice; Adam Smith and James Mill were the guilty parties.

It is sometimes suggested that J. S. Mill introduced the notion of 'disturbing causes', and that he did so in opposition to his father: 'It must be admitted,' writes Professor Schwartz, who takes this position, 'that though this ensured that Mill's economic system was more realistic than his father's and more flexible in its application to practical cases, it is also true that it made it less susceptible to refutation, for any case that did not fit in with theory could be attributed to the intervention of some cause either unknown or of unforeseeable intensity, as in fact was the case with the theory of historically diminishing returns in agriculture . . . The essence of Ricardian doctrine, namely the two laws of diminishing returns and population (apart from the implicit law of the need of capital for the division of labour) remained intact. But he was to use the model itself more as a working hypothesis, universally valid so long as no disturbing forces existed in reality' (1972, 61).

What requires correction, it will be clear from our discussion, is the assertion that J. S. Mill originated a 'new view' on method. On the contrary, it was a view central to both Smith and Ricardo and reflects a belief that economics is not a predictive science (such as astronomy) amenable to test by the measure of accuracy of prediction, and this because of the almost inevitable intervention of disturbing causes. J. S. Mill, we shall see, himself claimed to be representing the position of 'the best writers in political economy'.

I have given reasons elsewhere why Ricardo adopted an analytical framework incorporating the principle of diminishing agricultural returns when he was so conscious of the historical and prospective intervention of technological change (1979, 640–2). To these we may add the notion expressed in the *Essay on Profits* that since technological progress takes the form of random shocks it was to be accorded secondary status compared to diminishing returns – a force in 'constant operation': 'The causes, which render the acquisition of an additional quantity of corn more difficult are, in progressive countries, in constant operation, whilst marked improvements in agriculture, or in the implements of husbandry are of less frequent occurrence. If these opposite causes acted with equal effect, corn would be subject only to accidental variation of price, arising from bad seasons, from greater or less real wages of labour, or from an alteration in the value of the precious metals, proceeding from their abundance or scarcity' (1951, IV, 19n). A similar notion is implied in a reply to a Parliamentary Committee's request that he explain his insistence that note reduction would tend to lower the value of gold although this had not been the case during the period investigated in consequence of intervening causes: 'Because, in commerce, it appears to me that a cause may operate for a certain time without our being warranted to expect that it should continue to operate for a much greater length of time' (1951, V, 377). Again the disturbances to a clear-cut relation between note contraction and the value of gold is envisaged to be of a random nature.[20]

IV INDUCTIVIST CRITICS OF RICARDO AND THEIR RECEPTION

Criticism of Adam Smith's procedure on grounds of hasty applications of deductions drawn from necessarily incomprehensive models were quite common during the period 1776–1816, and similar objections will also be found after 1817. Sir Edward West formulated the charge thus: 'The opinion that the demand for labour is regulated solely by the amount of capital, has originated in a mode of reasoning very much used by Dr. Smith, which frequently furnishes very beautiful illustrations, often gives a clue to important conclusions, and is always a very excellent test of the truth of

---

20 We shall see that J. S. Mill was not quite satisfied with this line of thought: He insisted that 'disturbing causes' themselves had their own laws which it was the duty of the economist to investigate rather than treat as random events. It would thus be desirable to develop, if we seek an example, a theory of technological change.

   Mill's famous analogy with the mechanical sciences, particularly the theory of motion, lends much support to the retention of a model incorporating diminishing returns despite awareness of repeated shifts in the production function. The point is that, even where well-defined causal influences – in our case acting upon the price of corn – work in opposing directions with roughly equal force, it would still be necessary to envisage each as in operation although its effect cannot be directly observed.

   Ricardo also made the mechanical analogy, referring to the 'gravitation' of the rate of profits to a minimum although checked 'at repeated intervals' by technical change (1951, I, 120–1).

our inductions, but which from an incautious use of it, has led perhaps to more false conclusions in the science than any other cause. I mean the reasoning from an assumed state of society and facts, and considering what the effects of the known or supposed principles of human nature would be in a state of society such as never existed' (1826, 80). John Rae pointed to Smith's excessive system-building which conflicted with genuine inductive procedure: 'I believe it will be found, that the . . . author of the *Wealth of Nations* . . . has himself in all his speculations, adopted the explanatory and systematizing form of philosophizing, instead of the scientific and inductive . . . [His] object being every where to build common facts and familiar observations into a system, not to inquire into the causes or real laws from which they spring, he takes those things for fundamental principles which would present themselves to the inductive inquirer as phenomena, the principles of which his manner of philosophizing would call on him to investigate' (1965 [1834] II, 334–5).

Much of post-Ricardian opinion, however, draws a sharp contrast between Ricardian and Smithian method neglecting the foregoing characteristics of the *Wealth of Nations* on the one hand, while exaggerating, on the other, the purely 'speculative' dimension of Ricardo's economics. An early instance of the contrast is by Sismondi, who compared Smith's view of political economy as 'une science d'expérience' with Ricardo's speculations (1951 [1827] I, 69–70). J. B. Say took the same position, complaining of the defects of Ricardian abstractions: 'La meilleure dialectique aussi bien que le calcul le plus exact, s'ils partent d'une donnée incertaine, arrivent à des résultats douteux. Quand on admet pour fondement, au lieu d'un fait bien observé, un principe qui n'est fondé lui-même que sur une argumentation, on risque d'imiter les scolastiques du moyen-age, qui discutaient sur des mots, au lieu de discuter sur des choses, et qui prouvaient tout, hors la verité' (1841, 15–16).

The charges by British writers against Ricardo during his own life-time are well known. In his famous 1824 review for the *Quarterly Review* of McCulloch's *Encyclopaedia Britannica* article on Political Economy, Malthus criticized the Ricardo school – 'the new school' – for altering Smith's theories 'upon mere speculation' (B. Semmel, ed., 1963, 172). He also apparently had the Ricardians in mind by the following charge, in the second edition of his *Principles*, of over-simplified theorizing and avoidance of empirical testing, although he did not fail to denounce the other extreme:

The principal cause of error, and of the differences which prevail at present among the scientific writers on political economy, appears to me to be a precipitate attempt to simplify and generalize. While their more practical opponents draw too hasty inferences from a frequent appeal to partial facts, these writers run into a contrary extreme, and do not sufficiently try their theories by a reference to that enlarged and comprehensive experience which, on so complicated a subject, can alone establish their truth and utility . . . In political economy the desire to simplify has occasioned an unwillingness to acknowledge the operation of more causes than one in the production of particular effects; and if one cause would account for a considerable portion of a certain class of phenomena, the whole has been ascribed to it without sufficient attention to the facts, which would not admit of being so solved . . . The same tendency to simplify and generalize, produces a still greater disinclination to allow of modifications, limitations, and exceptions to any rule or proposition, than to admit the operation of more causes than one (1836, 4–6).

We recall also the complaint by J. L. Mallet, diarist of the Political Economy Club, who complained as early as 1820 that Ricardo 'is as the French would express it *"herissé de principes"* . . . [H]is entire disregard of experience and practice . . . makes me doubtful of his opinions on political economy' (Political Economy Club, 1921, ix). Henry Brougham objected that Ricardo made his case for repeal of the Corn Laws as if he had been dropped from another planet (Ricardo, 1951, V, 56; cf. 85).

After Ricardo's death a concerted attack upon his method – the method attributed to him which was contrasted with that of the *Wealth of Nations* – was launched by the so-called Cambridge inductivists.[21] In what follows we look more closely at the criticisms and seek to evaluate their reception.

---

21 The Cambridge group included the political economist John Cazenove and the scientists John Hershel, William Whewell and Charles Babbage. The key contribution to economics was Richard Jones's *An Essay on the Distribution of Wealth and on the Sources of Taxation: Part I – Rent* (1831), v–vi, xv.

The most proficient mathematical critic of Ricardian theory, including the relationship between wage increases, profits and general prices, was William Whewell. See also T. P. Thompson's objections to Ricardo's rejection of the Smithian proposition that taxes on wages result in a higher general level of prices, based on the nature of infinite series, in *The True Theory of Rent* (1826), 20. Whewell much admired this work.

The publication dates of Whewell's early mathematical papers are usually given as 1830 and 1833. In fact his first paper entitled 'Mathematical Exposition of some Doctrines of Political Economy' was read before the Cambridge Philosophical Society on March 2 and 14, 1829, and immediately sent to press. It was then distributed privately to various persons, including J. W. Lubbock, Malthus, T. P. Thompson, Dionysius Lardner and Richard Jones.

We cannot be sure whether J. S. Mill or other 'Ricardians' were amongst the recipients in 1829 although this cannot be ruled out since Whewell had written to Jones of his intentions to have the paper printed and sent 'to the Economists whom I know, by way of challenge' which perhaps implies economists from whom he expected objections. [March 5 1829, WPC, Add. Mss.c.51[62]] But in any event, the paper appeared the following year in the *Transactions of the Cambridge Philosophical Society*, Vol. III, Part I, 191–230.

Whewell's second paper – 'Mathematical Exposition of Some of the Leading Doctrines

The interesting outcome is that even Malthus believed that the inductivists went too far, while Mountifort Longfield (who praised Smith for the absence of 'system' and 'long and subtle trains of reasoning')[22] came to the defence of Ricardian deductive method in general, and the differential rent doctrine – the specific object of the inductivist attack – in particular.

According to the Cambridge writers Ricardian procedure entailed hasty and illegitimate generalizations from supposedly universal axioms, but in fact based on casual observation and even introspection. The defective and the legitimate ('inductive') procedures are described in the following characteristic passage by Jones:

It wants no great deal of logical acuteness to perceive, that in political economy, maxims which profess to be universal, can only be founded on the most comprehensive views of society. The principles which determine the position and progress, and govern the conduct, of large bodies of the human race, placed under different circumstances, can be learnt only by an appeal to experience. He must, indeed, be a shallow reasoner, who by mere efforts of consciousness, by consulting his own views, feelings and motives, and the narrow sphere of his personal observation, and reasoning *a priori*, from them expects that he shall be able to anticipate the conduct, progress and fortunes of large bodies of men, differing from himself in moral or physical temperament, and influenced by differences, varying in extent and variously combined, in climate, soil, religion, education and government (1831, xv).

It will be noted that the object of the exercise is the derivation of 'universal maxims' and that this object is also attributed to Ricardo – quite illegitimately it will be seen from our earlier discussion. But the construction of a general 'system of economical truths' was to be 'securely founded on the actual experience of mankind' – 'a comprehensive and laborious appeal to experience; – it must be remembered steadily, that the mixt causes which concur in producing the various phenomena with which the subject is conversant, can only be separated, examined, and thoroughly understood

---

in Mr. Ricardo's "Principles . . ." ' – was read April 18 and May 2 1831 and immediately printed. The paper also appeared in the *Transactions* (1833), Vol. IV, Part I, 155–98.
  For details on the matter of this note see de Marchi and Sturges (1973); Hollander (1983), 131. For technical analysis of Whewell's economics see Cochrane (1970), Henderson (1973), Hollander (1983).
22 'Adam Smith appears not to have possessed much taste or capacity for long or subtle trains of reasoning. The "Wealth of Nations" is written with very little attention to system, and this circumstance has probably tended to increase its utility. It prevented any error from infecting the entire work. An erroneous principle could not lead the author far astray' (1971 [1834], 262). The context is a rejection of diminishing agricultural returns as the 'cause' of a secularly falling profit rate.

by repeated observation of events as they occur, or have occurred, in the history of nations; and can never be submitted (except in cases extremely rare) to premeditated experiment' (xix–xx).[23] As de Marchi has phrased it, the Cambridge critics 'held that the true method of arriving at general axioms was successively to adduce cases, being careful to observe whether, with every extension of data, one's original conception of how the facts could be ordered remained valid or required some modification' (1974, 134).

Let us consider in more specific detail Jones's position on rent. Jones objected to the Ricardian theory interpreted as the proposition that 'a difference in the natural fertility of soils is the sole origin of rent' (1831, 206; cf. 212–13).[24] The major conclusion drawn by Jones from his investigation of pre-capitalist and non-capitalist (largely peasant) societies, was that the institutional forms adopted by the Ricardians in their analysis of rent were relevant to scarcely a 'one-hundredth part of the cultivated surface of the habitable globe' (14; cf. vii, 205). Over much of the world very different contractual arrangements applied.

Despite his rejection of the notion of differential land qualities as the 'sole origin of rent', Jones thus apparently accepted a version of the principle of diminishing returns – although as an exception to the general theory of the subject. But he still denied its contemporary relevance in the light of technological progress: 'That there is a certain point, beyond which human labor cannot be employed upon a limited spot of ground, without a diminished return to its exertions, must be admitted at once. But in the progress of those improvements in the art of cultivation, by which its most profitable amount of produce is approached, it may be very possible, that every successive portion of the capital and labor concentrated on the land, may be more economically and efficiently applied than the last' (199–200). It is strongly implied – quite unfairly – that Ricardo and his school neglected technical change: 'there is no foundation whatever for the opinion, that in every stage of such a process, every portion of additional produce successively got from the same lands, must necessarily be obtained by a less advantageous expenditure of labour and capital' (209; cf. 202).

Even granting, for the sake of argument, the orthodox position regarding

---

23 Cf. *ibid.*, vii, xii, xxii–xxiii, xxvii, xxix–xl for the failure of orthodox economists to build up general principles in responsible fashion. Also Jones (1859 [1833]), xxiv–xv for his recommendation to 'look and see' in seeking for principles which are truly comprehensive.

  N. Senior was included amongst those guilty of what Bacon had referred to as 'the jumping or flying to generalities and the principles of things' on the basis of 'the confined obscurity of a few experiments' (see on this de Marchi and Sturges, 1973, 382n).

24 Miller (1971), 202–2 has observed that infrequently Jones used the more general version of the Ricardian position, namely that rents will not rise until 'additional capital is laid out with a diminished return, either upon fresh lands, or upon some portion of the old land.' (see also on Jones, Miller (1977), 346–65).

the 'sole cause' of rent (207), Ricardo was none the less mistaken, Jones maintained, in the further assertion that rents can only rise in consequence of a rise in the magnitude of the differential. Jones's argument is unconvincing. It is based on counter-examples involving expenditure of doubled capitals on three plots of land of initially differential fertility which yields doubled output (or even less than doubled output) on each plot and increased rent despite unchanged differentials (or even diminished) differentials. The argument emphasizes the absolute increase in rent but neglects to assume an unchanged population, and demand for corn – that is, the *ceteris paribus* conditions for a fair treatment of the issue – so that it is implicitly assumed that the three plots are still required (203–4, 207–8). Shortly thereafter Jones alluded to these conditions in discussing Ricardo's famous proposition that technical change may lower rent in the short run, but he objected to the analysis on the grounds that population growth as well as changing technology are on-going processes rendering inappropriate an analysis based upon conceptions involving 'starts and jerks' – an allusion to Ricardian comparative statics (209–12).[25]

The counter-examples devised by Jones involve a quite arbitrary rewriting of the illustrative figures chosen; they have no economic content whatsoever. Indeed it is clear that he never abandoned the deductive method he so harshly condemned, adopting some of its least defencible attributes. He did not have a distinct inductive approach, failing to subject the Ricardian analysis of rent to adequate test, while himself attempting to fit another set of theories to the 'facts'.[26]

Jones's work was devoted to rent but the Preface contains an observation of interest relating to accumulation. Jones maintained that the 'fall of the rate of profits which is so common a phenomenon as to be almost a constant attendent on increasing population and wealth, is . . . so far from indicating greater feebleness in any branch of industry, that it is usually accompanied by an increasing productive power in all, and by an ability to accumulate fresh resources, more abundantly and more rapidly' (1831, xxxii). Any relationship between high profit rates and rapid accumulation, Jones believed, was the exception rather than the rule as was maintained by orthodox doctrine (xvi). The belief to the contrary reflected 'an almost wilful disregard of experience, and of the testimony which the history and statistical

---

25 Jones's criticisms of the comparative statics form of reasoning in the present context are invalid. There is nothing which prevents a Ricardian analysis in terms of a continuous succession of shifts in the marginal product curve and continuous population growth.

26 Cf. Miller (1971), 204: 'Despite the tenor of some of his pronouncements, Jones did not let facts tell their own story, but tried to impose theories on the factors and even made an effort to use deduction to develop the theories. Unfortunately Jones was less efficient than Ricardo.' Also *ibid.*, 206: 'Despite his expressed belief that induction presented an alternative to deduction, Jones really did not have a distinct approach to offer.'

position of every country in the world bear to the laws really determining the varying powers of communities to accumulate capital' (xii). No empirical evidence for the assertions is provided – although this was supposedly to follow.[27]

* * *

I wish to emphasize now a very strong body of opposition to the inductivist position. Jones himself excluded not only Smith but Malthus from his general indictment of the economists (1831, v–vi); he would doubtless have been surprised by the fact that Malthus – 'the insistent realist, countering Ricardo's "strong cases" with references to specific historical experience, and calling attention to what happens during the "intervals" ignored in Ricardo's long-run equilibrium analysis' – refused to join Whewell and Jones in their effort to discredit Ricardo's deductive methodology.[28] This position is reflected in an insistence upon a wider role than allowed by Whewell for definitions as starting hypotheses in political economy; a defence of the principle of diminishing returns as an explanation of rent (in the context of a critique of T. P. Thompson's *True Theory of Rent* so much admired by Whewell); and a charge that Jones in his *Essay* had neglected 'the most important parts of the subject', namely 'the progress of rent in new colonies' and the Ricardian case of 'farmer's rents in the more improved states of Europe', and moreover had 'gone beyond the truth in his unwillingness to admit the tendency of continued accumulation, and of the progress of population and cultivation to lower the rates of profits and corn wages on the land' (386). Malthus's refusal to join the inductivist camp is also clear enough in the *Principles* itself, where reference was made to the 'more practical opponents' of the deductivists who 'draw too hasty inferences from a frequent appeal to partial facts' (above pp. 37–8).

Mountifort Longfield, too, defended Ricardian rent theory against the inductivist critics: 'This theory of rent is brought forward only to explain

27 The kind of argument formally addressed at the Ricardians by the Cambridge School was not restricted to that group. G. Poullet Scrope (1831), for one, in a review of Jones accused the economists of 'setting off from some imaginary *a priori* assumptions, without troubling themselves with observation or history'; and he contrasted their 'imaginary abstractions' with the 'plain and obvious facts' relating to the effect on rent of agricultural improvements (97). And the interest of the inductivists in a broad range of institutional arrangements was also expressed in other contemporary forums. Thus, for example, Mathias Attwood, in Parliament, listed rights of proprietorship and feudal tenure amongst the key considerations determining land use (cf. Ricardo, 1951, V, 162).

For a broad coverage of attitudes hostile to political economy *c.*1830–*c.* 1848, see de Marchi (1974), 123f.

28 De Marchi and Sturges (1973), 379, based upon newly discovered letters of Malthus to Whewell of 1829, 1831 and 1833.

the causes which determine farmers' rents on the system in which they are demanded in these countries, where the land is let to the tenant as the result of a contract which either party may enter into as he thinks proper, and where the amount of rent is a fixed sum, not dependent upon the success of the tenant in raising great or scanty crops. Hence no objections to this theory can be drawn from the examples of those countries where contracts of this nature are not freely made' (1834, 141–2); tests of the theory drawn from the 'experience' of early states of society or alternative contemporary institutions were irrelevant. Longfield also rejected the common objection that there in fact existed no rent-free land. The reply is a complex one but we would emphasize the assertion that in some cases the contractual rent constitutes merely interest on the landowner's investment rather than an 'annual sum paid for the use of the original inherent powers of the soil'; and secondly the recognition that 'the theory does not depend so much upon the varying fertility of different soils, as upon the continually decreasing returns which land will give to successive equal applications of labour and capital' (147–8).[29] This second point is of very great significance. It represents precisely the answer later given by J. S. Mill in his *Principles* to criticism of the Ricardian position (below, p. 208f); and there can be no doubt that this general defence would have met with Ricardo's approval (cf. Hollander, 1979, 666n).

Robert Torrens too, upon reading J. S. Mill's *System of Logic*, came to the defence of Ricardo's proposition regarding the differential rent theory. He took Jones, and indeed himself in his earlier work, to task for characteristically rejecting Ricardo's conclusions based on the theory 'not because they are incorrectly deduced from his premises, but because they do not coincide with other conclusions deduced from other premises' (1844, xiv). Jones, in fact, was criticized for imagining that 'he refutes the doctrines of that profound and original thinker,' whereas in reality he was simply applying 'to the actual circumstances of society, *with the necessary modifications*, the principles which Ricardo deduced from hypothetical circumstances' (xv).

On the matter of the effects of technological change on rent, Torrens insisted on the necessity of distinguishing the immediate and long-term effects. If this distinction were made 'Ricardo was . . . perfectly correct in affirming the abstract proposition that the effect of improvements in agriculture, *other things remaining the same*, is to diminish rent' (xiv–xv). This answer would apply equally to the strictures directed against Ricardo, for example, by Scrope in the *Quarterly Review* (1831, 95).

---

29 One part of Longfield's defence is particularly original: We know that there exists rent-yielding land, and land which it is not profitable to farm, while since 'there are, apparently, all intermediate degrees . . . it is probable that there exists some land of that particular degree of fertility which would just repay the expense of cultivation, exclusive of rent' (146).

It scarcely seems accurate then to say that Jones's *Essay* 'made an immediate, and on the whole, favorable impression on his contemporaries' (Blaug, 1958a, 152). And with the passage of time – partly in consequence of J. S. Mill – any positive influence it might have had was further diminished. Moreover, the Whewell–Jones relation was by no means an easy one.

There can be no mistaking Whewell's initial enthusiasm for Jones's formal stance: 'I am quite satisfied', he wrote to Herschel in 1827, 'of the truth of Jones's general views, and also that they possess the great property of being proved at any step of their induction between the most general and the most particular. You may laugh at this criterion, but it is capable of being exemplified nonetheless' (Todhunter, 1876, II, 86). He was receptive in 1831 to Jones's idea for the establishment of an 'inductivist' review: 'I like much your aspirations after a reform, or at any rate a trial, in the way of reviewing for ourselves . . . I have a very strong conviction that taking such a line of moral philosophy, political economy, and science, as I suppose we should, we might partly find and partly form a school which would be considerable in influence of the best kind' (118). Whately, Senior and McCulloch figure large as the *bêtes noires* in this episode.[30] In 1832 he wrote encouragingly regarding the reception of Jones's book on rent: 'I am bound to say . . . that it is no bad success for a book not deductive like Ricardo's, but inductive, and in its inductions, as you must allow far from complete, to obtain as much notice as yours has done . . .' (151).

However, Whewell had his doubts, even early on, as to the precise nature of the program, as is clear from a letter to Jones in February 1831:

Of induction applied to subjects other than Nat. Phil. I hardly know one fair example. Your book is one. A good deal of Malthus's population is a beginning of such a process, excluding of course his anticipatory thesis, the only thing usually talked of. What else can you produce? or how can you expect to lay down rules and describe an extensive method with no examples to guide and substantiate your speculations? You may say a number of fine things and give rules that look wise and arguments that look pretty, but you will have no security that these devices are at all accurate or applicable . . . [G]eneral propositions can no otherwise be

---

30 For Whewell's comments on Whately and Senior see letters to Jones, Feb. 1831 and 15 July 1831, in Todhunter (1876), II, 115, 123; and his review of Jones for the *British Critic* (July 1831), 56.

    Whewell complained in the *British Critic*, 58, of McCulloch's review of Jones's *Distribution of Wealth* (for the *Edinburgh Review* (1831), 84–99) that McCulloch 'has not been . . . at all induced to retract his assertion of the universality of the principles which belong to farmers' rents alone . . .'; and that he confused *metayer* and *ryot* rents. Apart from this Whewell objected (on religious grounds) to McCulloch's representation of man as a 'machine' (60). See too letter to Jones, 27 Dec. 1832, in Todhunter, II, 152.

understood than by understanding the instances they include, and how the poor boy is to get any good by your declaiming to him about generalization, when he has not learnt any portion of any science, is past my comprehension (115-6).

Whewell's concern that Jones had failed to provide adequate illustrations of the proposed approach grew with the latter's failure to publish the second part of his work on *Distribution* which was to deal with wages (123).[31] And we find him soon complaining that Jones was 'lapsing into definitions' (135).

One of his problems at this time was the difficulty of conceiving of induction from the facts independently of some preliminary theoretical framework: 'conceptions must exist in the mind in order to get by induction a law from the collection of facts' because of 'the impossibility of inducting or even of collecting without this' (141). Indeed, in private correspondence of 1835 he expressed his hope that he would not be misunderstood as taking too 'heterodox' a position regarding the need for theory in statistical studies:

The Statistical Society in London was engaged before the meeting in drawing up a large collection of questions for circulation. They will, I have no doubt, obtain a great deal of information, but my opinion at present is that they would go on better if they had some zealous theorists among them. I am afraid you will think me heterodox, but I believe that without this there will be no zeal in their labour and no connexion in their results. Theories are not very dangeous, even when they are false (except when they are applied to practice), for the facts collected and expressed in the language of a bad theory may be translated into the language of a better when people get it; but unconnected facts are of comparatively small value (To M. Quetelet, 228-9).

The rift which developed in the 1830s between Jones and Whewell on the role of mental constructs, Jones (siding with Herschel) allowing for 'conceptions of the mind' only in the abstract sciences such as geometry, but not in the natural sciences 'may have been quite as important as the better-known fact of Jones's delays in publishing in preventing the development of an effective inductivist alliance against the Ricardian School' (de Marchi and Sturges, 1973, 381n).[32]

---

31 Whewell to Hershel (8 June 1831, WPC.0.15.97[129]): 'I hope he [Jones] will work, for the part he is now employed upon is of more importance both from its novelty and its applicability than the former.'

32 There is a further methodological issue that must be attended to. It is commonly asserted that Richard Whately's (1832) celebrated suggestion to rename political economy 'Catallactics', or the science of exchanges (6), implies a significant indication of a deflection of classical economics from its primary preoccupations to an emphasis upon the 'sphere of circulation' (Dobb, 1973, 109). But this interpretation does not accurately reflect Whately's intentions. His purpose in proposing the change of nomenclature was to assure

We shall return to some of the issues raised above in Chapter Two when we focus upon J. S. Mill's place in the story.

V THE CLASSICAL GROWTH MODEL

In the Ricardian growth model, which utilizes the axiom of diminishing agricultural returns, the real wage rate (w) tends downwards, the profit rate falling notwithstanding this decline. The 'subsistence' wage rules in the stationary state alone and is reached simultaneously with that rate of profit corresponding to zero net capital accumulation. The secular wage movements are governed by deviations between the capital and labour growth rates. It will be helpful to have in mind Ricardo's formulation of the downward path of wages, which presumes a constant population growth rate and a declining capital growth rate:

Although, then, it is probable, that under the most favourable circumstances, the power of production is still greater than that of population, it will not long continue so; for the land being limited in quantity, and differing in quality, with every increased portion of capital employed on it, there will be a decreased rate of production, whilst the power of population continues always the same (*Works*, 1951, I, 98).

In the natural advance of society, the wages of labour will have a tendency to fall, as far as they are regulated by supply and demand; for the supply of labourers will continue to increase at the same rate, whilst the demand for them will increase at a slower rate. If, for instance, wages are regulated by a yearly increase of capital, at the rate of 2 per cent, they would fall when it accumulated only at the rate of 1½ per cent. They would fall still lower when it increased only at the rate of 1, or ½ per cent, and would continue to do so until the capital became stationary, when wages would also become stationary, and be only sufficient to keep up the numbers of the actual population (101).

Ricardo's presumption that 'the supply of labourers will continue to increase at the same rate' notwithstanding reduced wages, implies a precipitous fall in the growth rate of labour demand (g_k) from some

that a study of the wealth of nations should be limited to exchangeable commodities and exclude, for example, the analysis of problems relating to a 'Robinson Crusoe' type economy (Whately, 1832, 8). That the emphasis was not on the phenomenon of exchange as such is clear from his rejection of the term 'Philosophy of Commerce' as a designation of the subject matter (9). Whately's suggested redefinition cannot be interpreted as a sophisticated attempt to redefine the scope of political economy. In the final resort he remained concerned with 'the nature, production, and distribution of wealth' (26), and regarded the study of what wealth 'consist[s] in' and what are 'the fundamental laws that regulate its distribution' as essential for an understanding of topical questions (223).

positive constant to zero when the wage reaches subsistence (w_s). There is, however, reason to believe that allowance was made for some response in the growth rate of population (g_L) to wage reductions from levels above but close to subsistence. For Ricardo (101–2) seems to have been aware of the fact that otherwise 'the labourer . . . would soon be totally deprived of subsistence', and saw some deceleration in the decline as realistic, a deceleration that can only be assured by a reduction in g_L. A generalized positively sloping g_L-w relationship might have been utilized throughout without changing the substance of the argument.

We should note at once that commentators were not all agreed regarding the interpretation of Ricardo. Mountifort Longfield, we shall see, appreciated that the 'subsistence' wage was relevant only in the stationary state (1834, 184–5). Similarly, according to the report by Mallet of the Political Club meetings in 1831 Ricardo 'looks forward . . . to the gradual lowering of wages and profits . . .' (1921, 225). But others attributed to Ricardo a full-fledged subsistence theory.[33]

Malthus, in contrast to Ricardo, developed a 'dynamic-equilibrium' wage path where the falling wage trend emerges notwithstanding equality of the capital and population growth rates. The clearest statement of the dynamic-equilibrium or balanced-growth wage path will be found in Malthus's *Principles* (1820). There are four major propositions in the relevant passage, one of the most brilliant early formulations of a growth process: (1) that in a system entailing population growth the real wage must exceed w_s; (2) that in consequence of land scarcity (diminishing returns) the excess $w > w_s$ must fall to zero; (3) that the profit rate declines steadily as *proportional* wages rise, to the minimum (r^*), and (4) that assuming $g_K = g_L$ at all times, this common growth rate must decline:

---

33 Samuel Read, however, attributed to Ricardo a full-fledged subsistence theory. He insisted that Ricardo had illegitimately included the wages of management within profits whereas they should have been categorized together with those of operating labour. The evidence adduced to support the view that wages of management are thus treated apart in Ricardian economics is based upon Ricardo's definition of the natural wages as 'that price which is necessary to enable the labourers, one with another, to subsist, and to perpetuate their race, without either increase or diminution'. Read asserted that 'this definition could only be intended to apply to the very lowest description of labourers, whom it moreover supposes reduced to the very lowest possible or endurable condition . . . by which life can be supported, – a condition to which I believe no considerable number of people or distinct class of labourers were ever reduced in any civilized society, and yet the whole of Mr. Ricardo's system is built and rested upon this supposition!' (1829, 251). What troubled Read was that the return to 'superior' labour – which must exceed the minimum subsistence wage – would be inconsistent with the definition of long-run equilibrium. (It is not certain that there is much justification for this position; it is easy enough to define cultural attitudes for each group of workers such that a wage exists for that group at which population increase comes to a halt. This is the solution suggested by Lloyd, see p. 56.)

. . . the supposition . . . of a constant uniformity in the real wage of labour is not only contrary to the actual state of things, but involves a contradiction.

The progress of population is almost exclusively regulated by the quantity of the necessaries of life actually awarded to the labourer; and if from the first he had no more than sufficient to keep up the actual population, the labouring classes could not increase, nor would there be any occasion for the progressive cultivation of poorer land. On the other hand, if the real wages of labour were such as to admit of and encourage an increase of population, and yet were always to remain the same, it would involve the contradiction of a continued increase of population after the accumulation of capital, and the means of supporting an increase had entirely ceased.

We cannot then make the supposition of a *natural* and *constant* price of labour, at least if we mean by such a price, an unvarying quantity of the necessaries of life. And if we cannot fix the real price of labour, it must evidently vary with the progress of capital and revenue, and the demand for labour compared with the supply.

We may however, if we please, suppose a uniform progress of capital and population, by which is not meant in the present case the same *rate* of progress permanently, which is impossible; but a uniform progress towards the greatest practicable amount, without temporary accelerations or retardations. . . . [If] poorer lands which required more labour were successively taken into cultivation, it would not be possible for the corn wages of each labourer to be diminished in proportion to the diminished produce; a greater proportion of the whole would necessarily go to labour; and the rate of profits would continue regularly falling till the accumulation of capital had ceased.

Such would be the necessary course of profits and wages in the progressive accumulation of capital, as applied to the progressive cultivation of new and less fertile land, or the further improvement of what had before been cultivated; and on the supposition here made, the rates both of profits and of real wages would be highest at first, and would regularly and gradually diminish together, till they both came to a stand at the same period, and the demand for an increase of produce ceased to be effective (in Ricardo, *Works*, 1951, II, 255–7).

Wage movements due to divergencies of g_L and g_K are distinctly contrasted with those reflecting balanced growth and treated in a separate section.

The wage path is represented as a 'supposition' implying thereby a hypothetical reference path. And indeed this path cannot (assuming regular capital and labour-supply growth functions) be achieved in a competitive world where capitalists and labourers act independently. For since the return on capital or $r = \dfrac{F'(L)}{w} - 1$ (r moves inversely to proportional not absolute wages),[34] it follows that, in an expanding system, the initial

34 For the wage per worker, w, plus profits per worker, or wr (in circulating capital models

impact of diminishing returns at the *going* wage is on capital alone, inducing a fall in g_K relative to g_L; and assuming a regular decline in the marginal product the deficiency $g_K < g_L$ is continually reconstituted notwithstanding the wage decline. This was Ricardo's insight. The dynamic-equilibrium path can only be, therefore, a construct of the mind derived by asking what the wage path would be if it is assumed that $g_L = g_K$ under conditions of diminishing returns. Evidently the return to the 'joint' factor declines; but the share of the incidence of declining productivity can be calculated by reference to the capital-supply and labour-supply growth functions, namely the $g_K - r$ and the $g_L - w$ relationships. There is a unique wage rate for each marginal product which is consistent with balanced factor growth. (For a geometrical treatment see Hollander, 1984b.)

In the *Essay on Population* Malthus investigated the implications for wages of 'prudence'. In the following passage we have a contrast between that check to population growth exercised by falling 'corn' wages and that check exercised by deliberate constraint designed precisely to avoid the deterioration of living standards. That the population growth rate must in one way or the other decline is a necessary implication of land scarcity. Correspondingly, two categories of 'stationary state' are defined – the one entailing a 'low' corn wage and the other a 'high' corn wage:

A diminished power of supporting children is an absolutely unavoidable consequence of the progress of a country towards the utmost limits of its population. If we allow that the power of a given quantity of territory to produce food has some limit, we must allow that as this limit is approached, and the increase of population becomes slower and slower, the power of supporting children will be less and less, till finally, when the increase of produce stops, it becomes only sufficient to maintain, on an average, families of such a size as will not allow of a further addition of numbers. This state of things is generally accompanied by a fall in the *corn* price of labour; but should this effect be prevented by the prevalence of prudential habits among the lower classes of society, still the result just described must take place; and though, from the powerful operation of the preventive check to increase, the wages of labour estimated even in corn might not be low, yet it is obvious that, in this case, the power of supporting children would rather be nominal than real; and the moment this power began to be exercised to its apparent extent, it would cease to exist (Malthus, 1890, 420; first introduced in ed. 5, 1817).

where wages are paid one period in advance of the sale of the product), exhausts the marginal product of labour, $F'(L)$, so that

$$w + wr = F'(L)$$

or

$$r = \frac{F'(L)}{w} - 1.$$

The two Malthusian versions were effectively utilized in response to a proposal by Arthur Young 'so to adjust the wages of day-labour as to make them at all times equivalent to the purchase of a peck of wheat' (1890 [1817], 583). This proposal 'in its general operation, and supposing no change of habits among the labouring classes', Malthus objected, 'would be tantamount to saying that, under all circumstances, whether the affairs of the country were prosperous or adverse; whether its resources in land were still great, or nearly exhausted; the population ought to increase exactly at the same rate – a conclusion which involves an impossiblity.' Here we have an application of the 'standard model' – the wage path *must* fall to assure the appropriate deceleration in the population growth rate. But allow for prudence and the picture is transformed, as Malthus proceeds to show: 'If, however, this adjustment, instead of being enforced by law, were provided by the increasing operation of the prudential check to marriage, the effect would be totally different, and in the highest degree beneficial to society. A gradual change in the habits of the labouring classes would then effect the necessary retardation in the rate of increase, and would proportion the supply of labour to the effective demand, as society continued to advance . . . without the pressure of a diminishing quantity of food.' What Malthus has in mind, it appears, are exogeneous upward shifts in the positively sloped $g_L - w$ function.

Ricardo did not formally trace through the implications of prudential behaviour. But there is good reason to believe that he would have applauded Malthus's elaborations, considering his own appeal to the 'friends of humanity' to encourage prudential behaviour (1951, I, 100).

VI ON PREDICTION: 'THE CONDITION OF THE PEOPLE' ISSUE

In 1829 Nassau Senior appended to his *Two Lectures on Population* a correspondence between himself and Malthus in which he admitted candidly to having misread the *Essay on Population*. As he originally understood the *Essay* it conveyed the message that a temporary improvement to wages will inevitably be cancelled in consequence of population pressure – a view he rejected. The problem of population pressure was 'likely to diminish in the progress of improvement' as tastes become more sophisticated; indeed the experience of higher wages would in all probability generate the improvement in tastes (minimum standards) required to render any improvement permanent (1829, 35; cf. 74–5). But this was in fact Malthus's own position (as he explained to Senior in conversation) and the misreading Senior publicly admitted, although he blamed Malthus's form of expression:

I was misled by your use of the word '*tendency*'. I supposed you to believe, that the desire of marriage, which tends to increase Population, is a stronger principle, or, in other words, a principle more efficacious in its results than the desire of bettering our condition, which tends to increase subsistence; and, consequently, that in an old country, with a people so fully supplied with necessaries as to make it possible for population to increase in a greater ratio than food, such an increase would, in the absence of disturbing causes, be a more probable event than the opposite event; namely, than an increase of subsistence in a greater ratio than that of population. I believe that I was led into this error principally by the conduct of all those writers who, since the appearance of your work, have written on Population. The multitudes who have followed, and the few who have endeavoured to oppose you, have all assumed this to be your opinion. And yet when I recur to your writings, I see how inconsistent it is with your uniform statement, that the pressure upon subsistence is almost always the most severe in the rudest states of society, where the population is the least dense, and the means of procuring subsistence, supposing they were employed, would be the greatest in proportion to that population (56–7).

Malthus explained that his formulation had been designed to impress the message that the solution to low wages or the maintenance of temporarily improved wages lay in population control (71–2). But he admitted that he would have been more careful had he been aware of the danger of misinterpretation (60). All he intended was the notion 'that population was always ready, and inclined, to increase faster than food, if the checks which repressed it were removed' (61) – whereupon Senior again pointed out that 'many, perhaps the majority of your readers, adopt the proposition without the qualification' believing that 'the expansive power of population is a source of evil incapable not only of being subdued, but even of being mitigated . . .' (79). The attribution to Malthus of the notion that reform was to no purpose had given political economy a bad name, and was the regrettable outcome of the misunderstanding which Senior hoped would be cleared away once and for all (81–2). We should add that in his later *Outline of the Science of Political Economy* Senior insisted that 'the great preventive check is the fear of losing decencies, or, what is nearly the same, the hope to acquire, by the accumulation of a longer celibacy, the means of purchasing the decencies which give a higher social rank' (1836, 38); historically food had increased more rapidly than population. But he quite rightly admitted again that Malthus himself had conceded as much. Writing to J. L. Mallet in 1835 he observed: 'Mr. Malthus has not I think very clearly expressed his opinion – but if you will look at his correspondence with me you will find that he nearly abandoned or rather disavowed the doctrine that population has a uniform tendency (in the sense of probability) to exceed subsistence' (1921, 305).

But not everyone appreciated the true status of Malthus's pronouncements; and considerable scepticism regarding what was supposed to be a 'prediction' of secularly falling commodity wages deriving from the 'Malthusian' doctrine is apparent during the immediate post-Ricardian period. A number of illustrations may be given. We shall then examine some of the implications of the record.

The empirical significance of diminishing returns was questioned by Samuel Read (1829, 253n). Considering the impact of 'embodied' technological progress, agricultural expansion did not necessarily require that workers must suffer falling real wages:

> [It] is always to be recollected, that the great instrument of production, *capital*, is constantly increasing, and constantly producing greater and greater effects in proportion to the labour that assists and guides it, and it is by this means chiefly, (and not, as has been absurdly and inconsiderately maintained, solely and exclusively by the degradation of the labourer, or by the diminution of his wages), that any new difficulty, or additional cost in the acquisition of raw produce, is naturally or usually overcome, or that cultivation is extended to the inferior soils, and to the most distant and inconveniently situated lands (255).

But in fact Read was closer to the full Malthusian view than he realized. For any upward trend in the secular real wage would be contingent on labourers practising 'restraint', Read rejecting what he took to be the argument that 'every society or people has an incomparably greater tendency to increase in numbers than in wealth or capital', as a 'theory . . . opposed to the strongest evidence of facts, to the most uniform experience . . . utterly inconsistent' with the Ricardians' own 'admissions in regard to the force and effects of the principle of *saving* and *frugality*' (258).[35] Wages clearly were still accounted for analytically in terms of the relative growth rates of capital and population whatever the precise 'prediction' may have been.

G. Poullet Scrope took issue with the 'economists of Mr. Ricardo's school' – citing in this context both Malthus and McCulloch – for their attribution of the general downward tendency in the profit rate to 'the necessity of resorting to inferior soils for the production of food' (1831, 25–6).[36] Scrope assured his readers that increases in the cost of

---

35 Read regarded the notion of an empirically high savings propensity to be well accepted and cited not only Smith but also McCulloch – 'one of the most noted of Mr. Ricardo's disciples' – regarding the latter's principle that 'in all tolerably well-governed countries the principle of accumulation has uniformly had a marked ascendency over the principle of expense' (261).

36 Malthus is cited for styling diminishing returns 'a cause of the reduction of profits "of such magnitude and power as finally to overwhelm every other"; and McCulloch for conceiving of the phenomenon as "the great law of nature, from whose all-pervading influence the utmost efforts of ingenuity cannot enable man to escape" and which "is sure, in the long run, to overmatch the improvements that occur in machinery and agriculture" ' (27).

agricultural produce in a 'limited district' such as Great Britain, would be inconsiderable, and – in an open economy – would result if at all only from 'the greater *distance* from which it is necessary to fetch the enlarged supply', while any such 'trifling' disadvantages would be counterbalanced by the various advantages of concentration, including improvements in transport facilities, which tend to reduce the cost of processing raw materials. In the event that rising food prices locally are not counterbalanced by the growing advantages of industrial concentration, Scrope concluded that it always remained open for capitalists to export their capital to more productive agricultural areas so that 'in *no case* can profits be generally and permanently lowered by this cause' (26–7).[37]

The relationship between population and food supplies is none the less described as 'infinitely the most interesting problem in the whole range of the science of Political Economy' (1833, 257). Scrope's objections to what he understood to be Malthus's position on population are expressed with fervour. He refers to the so-called 'tendency of population to exceed the procurable means of subsistence' as 'that most pernicious dogma which has long been palmed upon the public as the fundamental axiom of political economy' (xvi). In his *Quarterly Review* article he complained of 'the interminable fallacies set afloat by the political economists on this subject, such as the far-famed theory of the geometrical and arithmetical ratio, and the population panic which it has propagated, with the mischievous result of fixing the attention of the benevolent on the means to be adopted, not for increasing the supply of food, but for diminishing the number of feeders' (1831, 31). Nevertheless, in the same context he adopted the quite orthodox view relating the source of labour demand to capital and explained general wages in terms of the relative growth rates of capital and population:

the only fund for the employment of labour and payment of wages is capital. The amount of capital seeking active occupation, will be, therefore, the measure of the general demand for labour. Hence, as profits fall, wages will naturally rise, (other circumstances remaining the same,) and from the same cause, the competition, namely, of increased capital. If the capital of a country increases faster than the number of labourers, that is, its population, the rate of wages must increase, and, on the other hand, diminish, if the growth of population outruns that of capital.

37 See also his *Political Economy* (1833) where the empirical significance of diminishing returns is denied in the light of evidence of long-term productivity increase (265–6). Even in the absence of technical progress it would always be open for population to diffuse itself geographically and produce at roughly constant costs rather than concentrate near existing centres: 'There was always at hand more than one simple, easy, and effectual resource for keeping the means of subsistence level, *at least*, with the wants of any possible population' (278–9).

A question debated by the Political Economy Club in 1831 relates to the effects of economic progress on distribution. J. L. Mallet's account of the general evaluation of Ricardo's position at that meeting refers to a rejection of the supposed predictions of a downward trend in real wages (in consequence of a less rapid growth rate of capital than that of population), and in the profit rate; a denial that the interests of landowners conflicted with those of society as a whole; and also an implicit criticism of Ricardo's hostility to the Poor Laws:

[Ricardo] is one of the first who has treated the subject of Taxation, and he always reasons out his propositions, whether true or false, with great logical precision and to their utmost consequences; but without sufficient regard to the many modifications which are invariably found to arise in the progress of Society. One of the errors of Ricardo seems to have been to have followed up Malthus' principles of population to unwarrantable conclusions. For, in the first place it is clear from the progress of social improvement and the bettering of the condition of the people in the greater part of the civilised world, that Capital, or the means of Employment – the fund for labour – increases in a greater ratio than population; that men generally reproduce more than they consume, and the interest of the capital besides, which surplus goes to increase the fund for labour. Then he looks forward from the gradual demand for food and the use of land, to the gradual lowering of wages and profits till nothing remains but rent to the Landlords. But long before that, modification would take place in the state of society which would make such conclusions all wrong. First of all, it is contended that the interest of the Landlords does in fact coincide with those of the other classes; and then we see that in Ireland, where rent is absorbing everything, in consequence of the immense competition for land, a system of Poor Laws is likely soon to equalise the division (1921, 225).

At a subsequent meeting in 1835 Senior maintained that while population had the power to increase at a geometric rate, the advance of numerous nations from a barbaric to a civilized state constituted clear-cut evidence that, usually, subsistence increased more rapidly than numbers. This factual evaluation was supported, according to Mallet's report, by Tooke, McCulloch and Torrens; indeed, 'the whole artillery of the Club' was directed against the *Essay on Population* (265).

Senior's position requires no further comment. As for Torrens, he had indeed asserted explicitly in 1829 that 'there is no tendency in Population to increase faster than Capital, and thus to degrade Wages'. Any such tendency 'would fix the labouring class in a state, not only of hopeless, but of perpetually increasing misery' while in fact living conditions had improved 'with the progress of wealth and civilization' (1829, 473; also 1834, 27f). The upward trend in the relative price of agricultural produce in old countries, it is observed, effectively encouraged a decline in the marriage rate:

The high value . . . given to the things which constitute wages, is of no advantage to the married labourer . . . but is of great advantage to the unmarried labourer, who . . . has a surplus quantity, or the price of a surplus quantity, with which to purchase the finer manufactured goods, and the foreign articles, which have fallen in value as compared with food and with the coarser manufactured goods. Hence the unmarried labourer finds himself in much more affluent circumstances than the married labourer . . . It follows, that as a country approaches the limits of her agricultural resources, marriages become less frequent; and the *power* 'to increase and multiply,' instead of being stimulated to its utmost action by the considerations of prudence, and the desire of bettering our condition, is checked and controlled by the prevailing efficacy of these causes to such an extent, that the tendency in every civilized community is not for population to increase faster than capital, but for capital to increase faster than population (1829, 475-6).[38]

In the third edition of his *Principles* (1842) McCulloch wrote strongly against the pessimistic positions attributed to Malthus and Thomas Chalmers:

The recent history of the theory of population affords a striking instance of the abuse of general principles, or rather of the folly of building exclusively upon one set of principles, without attending to the influence of the antagonist principles by which they are partly or wholly countervailed. The principle of increase, as explained by Malthus, and afterwards by Chalmers, appeared to form an insuperable obstacle to all permanent improvement in the condition of society, and to condemn the great majority of the human race to a state approaching to destitution. But farther inquiries have shown that the inferences drawn by these and other authorities from the principle now referred to, are contradicted by the widest experience; that the too rapid increase of population is almost always prevented by the influence of principles which its increase brings into activity; that a vast improvement has taken

---

38 Increasing returns apply in new countries where a high marriage rate is desirable and encouraged: 'In new countries, like North America, where abundance of fertile land remains to be reclaimed, and where the obstacle to production is the want of a population sufficiently dense for the division of employment, and the co-operation of one distinct branch of industry with another, a large family becomes an important source of wealth; the desire of bettering their condition acts in conjunction with the instinct of nature in impelling the labouring classes to early marriages; and population increases with as much rapidity as the human constitution will admit. Yet in new countries, where population doubles in the shortest possible period, the accumulation of wealth is at least as rapid; and capital continues to bear that proportion to labour which makes actual wages permanently high.'

place in the condition of the people of most countries, particularly of those in which population has increased with the greatest rapidity; and that, so far from being inimical to improvement, we are really indebted to the principle of increase for most part of our comforts and enjoyments, and for the continued progress of arts and industry (1864, xv–xvi; cf. 1845, 260).

In a review of Thomas Chalmers' *Political Economy* in 1832 McCulloch insisted upon a general increase in living standards over the preceding half century in both England and Scotland (1832, 56f), and emphasized the desirable effects of the lag between a wage increase and any consequential expansion of population: 'during all this lengthened period, the labourers would enjoy an increased command over the necessaries and conveniencies of life; their notions as to what was required for their comfortable and decent support, would consequently be raised; and they would acquire those improved tastes and habits that are not the hasty product of a day, a month, or a year, but the late result of a long series of continuous impressions . . . [T]ime is afforded for the formation of those improved habits that are of such essential importance' (55).

Thus far we have reviewed examples from the literature suggesting a decline of confidence in Malthus's population doctrine, understanding thereby the 'prediction' of secularly declining real wages (Longfield's position on the matter will be taken up presently). It should, however, be observed that the picture is by no means totally one-sided. And it remains also to consider the implications of the criticisms.

In his *Edinburgh Review* article on Senior's *Political Economy*, Herman Merivale supported the 'Malthusian' doctrine against Senior's criticisms, insisting on a relatively slower growth rate of capital compared with that of population (1837, 94). In the same journal William Empson warmly accepted the doctrine – 'the importance of the truths announced in the essay on Population [of 1803] must last forever', – and he referred to Scrope as a 'cause monger' (1837, 483, 491). W. F. Lloyd adopted the notion of a subsistence wage with fervour, believing that the 'lower orders' lacked that moral restraint which could improve their position at least in contemporary conditions. He repeated Malthus's warning that 'population is at all times pressing against the means of subsistence', and maintained that the 'leading principles' of the *Essay* 'though not . . . all its conclusions have stood their ground' (1835, 8, 5). In his last published work Lloyd developed a model designed to demonstrate that, given the institutions of contemporary England, unskilled labourers could not be otherwise but at the bare subsistence level. Wage differentials in excess of the minimum were recognized, but of this Lloyd observed that 'the necessary expenses of those who are receiving the greater wages, are also greater' (1837, 68).[39]

39 On this matter see Gordon (1966). On Lloyd's adherence to a rather strict version of Malthusianism see Romano (1977).

The most hostile of the critics of Malthusianism and diminishing agricultural returns was (as Empson's criticism indicates) Scrope. Yet odd as it may appear, he himself took a 'pessimistic' view of British prospects in a closed economy with 'superior facilities' in manufacturing: 'Any restriction on the importation of the means of subsistence can only render the condition of a redundant population, that is, of a people already unable to procure a sufficiency of the means of subsistence by its labour, still more deplorable' (1831, 41). He conceded further that with protection 'the rise in the cost of the necessaries of life required by an increasing population in a limited district may lower profits' (26).

It may, however, be accurate to speak on balance, of a general decline in adherence to 'Malthusianism', if we understand thereby pessimistic predictions regarding living standards. But what are the implications of this fact for the fortunes of Ricardian economics? To the extent that 'the body of doctrine which Ricardo bequeathed to his followers rested on a series of definite predictions about the course of economic events which were subject to empirical verification, in the strictest sense of the term' (Blaug, 1956, 41), the empirical fact of rising living standards over several decades is ruinous. But if, as seems to be the case, Ricardo had no intention of using his growth model to make 'definite predictions' the empirical evidence is less consequential. Particularly pertinent here is Richard Whately's insistence on a distinction between the notion of 'a "tendency" towards a certain result' in the sense of 'the existence of a cause which, if *operating unimpeded*, should produce that result', and in the sense of 'the existence of such a state of things that that result *may be expected to take place*'. There may be, he observed, a 'tendency' in population to grow faster than the means of subsistence in the first sense, while, as was evident from the historical experience of several hundred years, there was no such a tendency in the second sense. To 'inaccuracies of this kind', Whately ascribed the 'discrepancies and occasional absurdities from which some persons infer that Political Economy, is throughout a chimaera' (1832, 249–50; cf. de Marchi, 1970, 260f). No adverse mention is made of Ricardo in this context – for very good reason – while elsewhere Whately denied that Malthus himself was responsible: The pessimistic predictions regarding long-term prospects, he wrote, had 'obtained considerable currency; chiefly, I conceive, on the supposed authority of Mr. Malthus; in whose work however I have never myself been able to find this doctrine' (185n). This comment is of great importance in evaluating the claim that the evidence of rising living standards despite population increase (even in a protected economy) gave the lie to the pessimistic predictions supposedly flowing from the principles of diminishing returns and strictly-defined Malthusianism. It was not Whately's view that the empirical evidence constituted a refutation of the orthodox analysis of trend patterns under *ceteris paribus* conditions and the absence of 'disturbing causes'.

Whately's distinction would have been accepted by Ricardo. It was formally adopted by Senior in the same population context (1836, 47) and McCulloch hinted at the same point in the course of a sharp complaint against Malthus: 'The real difficulty does not . . . lie in discussing matters connected with this science, in the statement of general principles, or in reasoning fairly from them; but it lies in the discovery of the secondary or modifying principles, which are always in action, and in making proper allowance for their influence' (1864, xvi). But more important than the commentators is Malthus himself. For, as Whately realized, Malthus did not employ his model of growth in the service of positive prediction, but to indicate alternative possibilities depending upon the adoption or otherwise of a program of prudential control. And this was to be J. S. Mill's position as we shall see. Nothing can be more misleading than the opinion that, with the 'eclipse' of the Malthusian doctrine in the 1830s, 'the Ricardian system lost much of its logical rigor and became incapable of specifying the strength and direction of movement of the major economic variables with which it was concerned' (Blaug, 1958a, 111).

VII THE CASE OF THOMAS CHALMERS

At this point we must attend to Thomas Chalmers's *Political Economy* (1832). Professor Blaug has asserted of Chalmers that he 'brought the [Malthusian] doctrines into further disrepute by carrying the argument to absurd lengths'; his position is described as a thoroughly pessimistic one, the 'principal conclusion' being 'that the growth of population and the necessary recourse to inferior soils had steadily reduced the English standard of living over the last fifty years' (1958, 114). This is an erroneous picture of Chalmers's book which is in fact written in a distinctly positive tone. For Chalmers did not assert that living standards had declined since 1800 – merely that they had not increased. And, while he referred to ultimate 'stationarity' – 'every country has its limit, and every continent its shore . . . [T]he earth we tread upon, can only be made to yield a limited produce, and so to sustain a limited population' (17) – it is also made clear that 'the time may be indefinitely distant, and indeed may never come, when the absolute and impassable barrier shall at length be arrived at'. The experience of the peasantry over the centuries indicated that 'there had been a general march and elevation in the style of their enjoyments', and that 'cultivation may be extended, without deterioration to the comfort of labourers' as a result (in part) of 'the excitement of new desires, which require exertion ere they can be appeased' (9, 15). This change of habit had 'actually taken place in modern Europe' (15). And technical change is also alluded to in the historical account: 'Men have been at a loss to reconcile the descent of labourers among the inferior soils, with the undoubted rise which has taken place in their circumstances, or in the average standard of their comfort' (5); thus 'there may . . . by a mere improvement in the powers of mechanical labour, be a descent among the inferior soils, and so an extension of agriculture, to afford the increasing population as large and liberal subsistence as before' (13). Indeed Chalmers has been described as the 'first writer of eminence who definitely attacked the belief that the returns to agricultural industry have generally diminished, and continue to diminish, in consequence of the increase of population' (Cannan, 1917, 136).

It would appear to have been Chalmers's opinion that the strength of moral restraint as it existed in contemporary Britain, coupled with improved technology, had been insufficient to assure rising living standards since at least 1800 but that it had served adequately in the past. Specific steps were, therefore, required to strengthen the check; the very object of his work was to encourage the strengthening of Malthusian moral restrain, or the postponement of marriage by measures designed to 'raise [the worker's] standard of enjoyment' (23), for British experience since the turn of the century 'afford[ed] the clearest experimental proof of the little which can be done by mere resources for an increasing population, without restraints on the rate of their increase' (38–9; cf. 539). The promising message is clearly proclaimed beyond a shadow of doubt:

Should there . . . be a high standard of enjoyment among labourers, they will not marry so as to overstock the country with population; and so, just because their taste is high, their wages would be high; thus landing us in the important and delightful conclusion, that the people, collectively speaking, have their circumstances in their own hands – it being at the bidding of their collective will, whether the remuneration for their work shall be a scanty or a sufficient one . . . [There] is no headlong necessity in any state of society, for either a wretchedly low wage or a ruinously low profit. Both, in fact, are dependent on moral causes. There is a moral preventive check, which, if put in steady operation throughout the labouring classes, would keep wages high. . . . There is no irreversible fatality in that march of agriculture among the poorer soils, which has been represented as bearing down profit and wages. Instead of this, profit and wages may each, in any point of the progress, make their own resolute stand, and arrest the march of agriculture (514–5).

Indeed, the consequences of the application of foresight by labourers and capitalists is outlined in terms which imply the kind of 'stationary state' envisaged by Malthus and J. S. Mill, whereby wages and profits might be 'high' at the expense of rent:

Instead of the action and re-action being only between these two elements, so that when the one rises, the other must necessarily fall; the action and re-action are shared among three elements, even profit, wages, and rent – so that the two first may draw indefinitely upon the last; and, with the fall of rent, profit and wages may rise, and that contemporaneously. We are aware, that, on this taking place, there might be a contraction of the agriculture; the lowest land having to yield a larger produce for the now increased remuneration of the farmer and labourers than before. But our decided preference is for a happier and more prosperous, even though it should be a somewhat more limited society (517).[40]

In the 'synoptical view' taken of his work Chalmers formulated the conclusion much as J. S. Mill was later to do: 'That high wages are not necessarily confined to the period when the wealth of society is in a state of progressive increase; and neither does it follow, that, when this wealth has attained its maximum, and become stationary, the wages of labour must be low. That it remains in the collective power of labourers to sustain their wages at as high a level in the ultimate, as in the progressive stages of the wealth of a country' (555).

McCulloch's review may have been partially responsible for Chalmers's reputation as a dyed-in-the-wool 'Malthusian'. For he there erroneously implied that Chalmers envisaged an actual decline of living standards over the preceding half century (1832, 57); and give no hint of Chalmers's optimistic statements regarding the historical record on a longer view, his recognition of manifestations of an altered 'style of enjoyments', his notion of stationarity as a far distant prospect even under contemporary population reaction patterns, and his holding out a prospective state of stationarity entailing 'high' living standards.[41]

VIII THE CASE OF MOUNTIFORT LONGFIELD

The discussion by Longfield of the implications for distribution of economic progress merits special attention considering his reputation as the most remarkable of the 'dissenters' from Ricardianism. (For a very fair account of the secondary literature on this matter see Black, 1983.)

Longfield was clear indeed regarding the key feature of Ricardo's growth

---

40 Cf. 554: '[T]here must be less food raised in virtue of this narrower cultivation; and hence, a somewhat narrower society. But . . . this, with general comfort among the families, is vastly preferable to a more numerous society, with all the consequent miseries of an over-peopled land.'

41 McCulloch himself had little confidence in schemes which merely proposed to instil moral restraint. According to Mallet (1921), 'he derided in his cynical manner the notion of inculcating provident feelings among the poor – said that a poor man had no inducement to be provident' (Political Economy Club, 252).

model – that the commodity wage necessarily exceeds 'subsistence' in a growing system (1834, 184–5). But he was troubled by the further property that the burden of the incidence of land scarcity impinges necessarily both on capitalists and labourers; he pointed to the possibility that the incidence might fall entirely on labour failing apparently to appreciate the Ricardian logic – the wage-population growth relation – which precludes that outcome. He also fell into the serious error – an error common to Malthus in fact – of asserting that Ricardian theorists denied a fall in the profit-rate in the case of expanding capital with population unchanged (183). This critical perspective provided the background for his own celebrated and original approach to profit-rate determination turning upon the efficiency of capital (understood largely as 'machinery') in its least-efficient use. Longfield retained the 'prediction' of a general tendency for the profit rate to decline. This trend turns on an increasing shortage of efficient labourers generated by the pressure of increasing capital, and increasing exhaustion of investment opportunities assuming scarce labour reflected partly in a decline in commodity prices (192f).

As mentioned, Longfield was mistaken in asserting that in the case of capital expansion with unchanged population Ricardo denied any profit-rate decline. His valid claim to fame rests rather on the spelling out of the precise manner in which increasing labour scarcity plays upon the return to capital – a supplementary chapter of considerable importance, vitiated, however, from a Ricardian perspective – though not perhaps for Malthus – by its 'Smithian' notion of pressure on final commodity prices.

But there are severe difficulties with the exposition. The reader is left on his own to fathom the precise scope of Longfield's theory of profits when envisaged within a full growth context involving population expansion. We turn to the substance of his approach as outlined in Lecture XI.

Even in the full growth context the prediction of a general 'tendency' for the profit rate to decline – and the real wage to rise – is retained, although a number of countervailing forces are discussed. These include the impact of a growing population which 'gives a greater scope for the employment of capital, and provides a more extended market for its productions' (228). Thus as the capital–labour ratio turns against labour, so the downward pressure on profits slackens – fully in accord with Ricardo – for population increase has a 'tendency to raise the rate of profits, and thus lower the rate of wages' (235).

A second countervailing tendency is the discovery of 'new employments for capital, and new modes of assisting and superseding labour' (228–9).[42]

---

42 The countervailing tendencies of technical progress and diminishing returns are said to assure that 'the price of corn is kept nearly stationary' (1834, 224). But aggregate rents would tend to increase with population growth, despite constancy of agricultural

An increase in profits achieved in this manner will not damage the labourer 'since it is caused by the increased efficiency of capital in assisting his powers of production'. And there is too the positive impact on labour demand: 'The increase will always leave to [the labourer] at least the same share as before, and generally even a greater, and will lead to a further accumulation of capital, to the additional advantage of the labourer'. A third but minor countervailing tendency raising profits is said to be more significant in its direct effect on wages, namely 'increased skill, and good conduct on the part of the labourer'. Yet for all this Longfield still maintained, on empirical grounds, that 'it is on the whole probable, that capital will accumulate so fast as that profits will decline . . .' This is more 'Ricardian' than Ricardo whose 'prediction' in this respect presumed away new technology.

There is a further complexity. Insofar as profits decline in the course of economic development Longfield's view was that commodity wages would rise at least holding productivity constant: 'wages depend upon the productiveness of labour and the rate of profit, and . . . as capital increases in the country, the rate of profits has a tendency to decline' (235). The relevant variable is, of course, *proportionate* wages not the absolute level of wages – for both the profit and the commodity-wage rates might fall assuming a decline in productivity. (Longfield himself had enduced Torrens to withdraw his strictures against the Ricardian inverse wage–profit relation based upon the measure of value; Hollander, 1977b, 236.) Now there is a revealing implicit concession that in the absence of new technology the commodity wage *will* in fact decline: 'We have also seen that the productiveness of agricultural labour has a tendency to diminish. This is a serious evil. Every man has occasion to use the produce of the earth; and as more labour becomes necessary to produce the same quantity of corn, the wages of the same quantity of labour cannot purchase as much corn as before, and consequently the labourer finds increased difficulty in providing subsistence for himself and his family.'[43] Longfield was apparently forced to conclude that (given technology) there will be a coincidental decline in both profit and wage rates – precisely Ricardo's

prices, although allowance is made for the fact that (in the absence of interference by 'artificial institutions') the number of landlords will also rise, so that the social implications of an increasing unearned income, one which is paid 'without any exertions on the part of the proprietor', are played down (224; cf. 225, 227). Longfield is difficult to pin down on the matter, for he had implied earlier in the lectures that despite technical change the trend of corn prices would be upward although perhaps only slightly so (181).

43 Again it is pointed out that the force of diminishing returns may of course be 'neutralized' by agricultural innovation; while, in any event, technical change in manufacturing means that the worker 'will gain more by the diminished prices of all other articles, than he will lose by the dearness of food and all raw materials' (238).

position which he had earlier rejected. Indeed, this 'concession' is already implicit in a remark that diminishing returns acts on profits at most indirectly by 'retarding the increase of population' (183) – which, one presumes, can only be by way of reduced wages. Similarly, we encounter the concession that 'if [the labourer's] wants and necessities exercise, as they do, some influence upon the wages of his labour, it is indirect and secondary, produced by their effect upon the growth of the population . . .' (206). And this despite the thoroughly misleading formal treatment of the population growth rate as an exogenous matter: 'population naturally increases, and . . . there is little room for the wisdom of legislation, except to provide that no artificial stimulus should be given to its increase' (236). In the end Longfield was led unwillingly back to Ricardo. It would have been preferable had he faced squarely and explicitly Ricardo's notion of a 'natural equilibrium between profits and wages' (1951, I, 226).

IX CONCLUSION

There is much pointing to the conclusion that the 'failure' of specific predictions was not ruinous for the Ricardian theoretical framework. Thomas De Quincey provides a nice illustration. For he denied the validity of 'an immutable law of declension' of the profit rate (ascribed to Ricardo), on the grounds that it neglected the 'eternal counter-movement which tends . . . to redress the disturbing balance' (1970 [1844], 249). Yet he adhered to Ricardo's model of growth and distribution – which, he maintained, placed economists 'irredeemably' in his debt – including the fundamental theorem of distribution based upon the measure of value (Hollander, 1977b, 242–4; for a conflicting view see Groenewegen, 1982). Similarly, Mountifort Longfield denied an upward secular trend in agricultural prices (1834, 140, 224) but still placed diminishing returns and differential rent at centre stage:

I attempted to give an explanation of the manner in which, as population increases, inferior soils must be brought into cultivation, and the soils already cultivated must be forced, at a greater proportional expense, to yield a greater produce. I have shewn that rent has therefore a tendency to increase as population increases, and that the natural price of corn, or its cost of production, may be measured independently of rent; that it will be equal to the cost of raising it on that land which pays no rent, or to the cost of production of that portion which is raised under the most naturally disadvantageous circumstances (132).

Longfield, like Ricardo, thus found the conception of diminishing agricultural returns and differential rent to be valuable analytical devices, and this despite awareness not only of the potentialities of new technology but also – in the case of Ireland – of the existence of a body of yet unapplied knowledge available for adoption in support of a rising population (140).

Ricardo did not in fact emphasize in his policy discussions the declining rate of profit as an empirical phenomenon. But to the extent that profits would decline he, like Malthus, feared adverse effects upon the rate of accumulation. It is here, rather than with regard to the specific prospect for wages and profits, that a significant doctrinal revision – or rather a reversion to the Smithian position (and that of Jones) – can be observed. For Longfield by contrast not only 'forecast' but welcomed the prospective fall in the rate of return. 'The first direct and most striking effect of a habitually low rate of profits', he maintained, 'is to render the future and present period of nearly equal importance in all pecuniary speculations' thus encouraging investment in long-lived ventures (229–30). There was, moreover, little to be feared with regard to the negative effects of a falling rate of return on accumulation; Longfield emphasized the countervailing effects of rising total profits, and prospective changes in the attitude towards saving, namely 'the habits of saving, and the prudent regard to the future' which are likely to rule in a mature economy (234).[44] Even capital exportation is said to be impeded by the nature of the investments undertaken in a mature system (including that in a skilled work force) (234–5). Longfield's optimism with regard to the consequences of declining profits is most striking in the light of his reputation as an 'abstinence' writer.

What of the functional relationship between the wage rate and population growth as an analytical principle, that is to say when changes in the conventional 'minimum of subsistence' and in the institutional framework are ruled out? This issue does not involve the acceptance or rejection of a perfectly elastic supply of labour. Our question relates rather to the acceptance or rejection of a functional dependency of population growth upon the magnitude of the excess of actual (long-run) wages above the 'minimum' existing at any time.

Senior by-and-large divorced his discussion of population from that of the wage rate (Bowley, 1937, 173f). Nevertheless, while there is certainly no concerted discussion of the wage-population relationship it seems to have been taken for granted since a parallel is drawn with the profit–capital supply relationship:

44 A suggestion to the same effect appears in the work of Scrope (1831) despite his emphasis upon interest as a reward for abstinence: 'If the love of accumulation lead capitalists to be contented with a low profit, they have in this their reward – over-balancing in their own estimation the sacrifice of immediate enjoyment; and there remains a greater quantity of produce to be divided amongst the labourers in the shape of wages' (22).

The minimum at which wages can be permanently fixed is of course the sum necessary to enable the existing labouring population to subsist. On the other hand, as the rate of wages depends in a great measure on the number of labourers, and the rate of profit on the amount of capital, both high wages and high profits have a tendency to produce their own diminution. High wages, by stimulating an increase of population, and therefore an increase of the number of labourers, and high profits, by occasioning an increase of capital (1836, 140).[45]

Longfield's position is similarly ambiguous. We have encountered his formal treatment of the growth rate as exogenous but his actual allowance for an impact of the real wage.

The orthodox labour supply relationship is certainly complicated by the kind of consideration noted by McCulloch, namely the lag between a wage increase and consequential population expansion which assured the firm establishment 'of new and improved tastes and habits' (1864, 334). Here it is suggested that the alteration in working-class habits is itself caused by variations in the wage rate. But it is still possible to distinguish between a *long-run* labour supply curve valid under given conditions and shifts over time in the curve. And indeed McCulloch himself retained the conception of the 'natural or necessary rate of wages' and considered the effects of an increase in the price of wage goods in orthodox terms, that is to say in terms of the effect upon labour supply (324f). Moreover, in his review of Chalmers a positive relationship between population growth and real wages operating by way of variation in the general death rate is alluded to (1832, 61).

We conclude that the wage–population growth relationship can scarcely be said to have been abandoned after 1830. But there is a further consideration. Even the rejection of the linkage could not be regarded as ruinous to Ricardianism, to the extent at least that it is legitimate to envisage the core of that system as the inverse profit–wage relationship – and the theory of value upon which it was constructed. For the distribution theorem does not imply any particular theory of wage-rate determination. And Ricardianism in that sense remained firmly entrenched even in the so-called 'dissenting' literature (Hollander, 1977b). This matter will prove to be highly pertinent to our investigation of J. S. Mill.

---

45 Senior's (1836) famous 'elementary propositions' (26f) include a version of the population principle which is consistent with the wage–population relationship: 'That the Population of the world, or, in other words, the number of persons inhabiting it, is limited only by moral or physical evil, or by fear of a deficiency of those articles of wealth which the habits of the individuals of each class of its inhabitants lead them to require.' It is here implied that wages exceeding the cultural minima will generate population increase. But a regular functional relationship is not necessarily intended.

TWO

On scope and method

I INTRODUCTION

A sharp difference is often said to be discernible between J. S. Mill's methodological pronouncements in his essay on method penned about 1830,[1] and his practice in the *Principles of Political Economy*. In some accounts, this 'difference' is interpreted as a matter of inconsistency, in others, as a change of opinion. A brief review of these commentaries on the essay is in order before turning to our own analysis of Mill's formal discussions of method.

According to an early paper by Jacob Viner, Mill conceived of political economy as a psychological science: 'the science relating to the moral and psychological laws of the production and distribution of wealth'. Furthermore, citing the following passage, Viner pointed out that it is not even *all* of human psychology that is involved, or for that matter *real* human psychology, but rather an *assumed* psychology:

Geometry presupposes an arbitrary definition of a line, 'that which has length but not breadth.' Just in the same manner does Political Economy presuppose an arbitrary definition of man as a being who invariably does that by which he may obtain the greatest amount of necessaries, conveniencies and luxuries, with the smallest quantity of labour and physical self-denial with which they can be obtained

---

1 'On the Definition of Political Economy and on the method of Philosophical Investigation in that Science', *London and Westminster Review* (October 1836); reprinted (with some revision) as Essay V in *Essays on Some Unsettled Questions of Political Economy* (1844) (*CW*, IV, 309f). The Preface to the *Unsettled Questions* gives the date of writing as 1829 and 1830; but Mill's bibliography gives it as Autumn 1831. Mill also refers to a rewriting of the essay in the summer of 1833, and we find him in early 1834 asking for suggestions of 'all manner of further developments, clearer explanations, and apter illustrations'. For details see editorial note, *ibid.*, 309.

in the existing state of knowledge . . . Political Economy, therefore, reasons from *assumed* premises – from premises which might be totally without foundation, in fact, and which are not pretended to be universally in accordance with it (1962 [1971], 108–9).[2]

The picture painted by Viner suggests that the postulates of Mill's political economy involve the fiction that persons are ruled by egoism (combined with the further fiction that the means of satisfying such egoism are seen and applied) – the famous fiction of 'economic man'. The rationale for the adoption of such a fiction is that although men are in reality motivated by conflicting desires, the possibilities for direct observation of such desires are limited – human psychology is not readily amenable to inductive treatment.[3] We therefore abstract from all but the economic motive and arrive at 'generalisations of universal application' by deduction 'from our knowledge of the action of this desire obtained through our introspective examination of its operation within ourselves, and our inference – induction! – that it operates likewise within other individuals'. Similarly, by assuming a uniform environment – and again relying on introspection to discover how the 'economic' motive works in such an environment – we save our generalizations, which apply to *all* men, from limitation by the differences which exist in the situations men actually find themselves. 'To modify or reduce the extension of our generalisations, as differences appear in the psychology and environment of different individuals or groups, is to step beyond the bounds of economic science into the field of application of science' (111–12).

From the limited number of hypothetical postulates relating to human psychology and the social environment was to be derived the body of economic doctrine. To compare the doctrines with reality was to step beyond the bounds of economic science – or indeed of any science. As Mill put it: 'To verify the hypothesis itself *à posteriori*, that is, to examine whether the facts of any actual case are in accordance with it, is no part of the business of science at all, but of the *application* of science'. Conceding a place for testing and verifying the hypotheses and bringing them into agreement with reality did not imply conceding a place for induction in economic research (110).[4]

---

2 Cf. Viner (1958 [1949]), 327: 'By the "science" of political economy Mill meant a body of deductive analysis, resting on psychological premises derived from introspection and observation of one's neighbours, and even with resect to these premises abstracting from all aspects of human behaviour except those most intimately and most generally associated with the business of buying and selling'.

3 Cf. also Viner (1962), 114–15, 116 regarding the motive of egoism.

4 Cf. Viner (1962), 109: 'A hypothetical psychology, and a hypothetical environment cannot, of course, be subjected to anything but hypothetical examination or experimentation. So long as the science is built upon such a basis, Mill is justified in

Viner was highly critical of the procedure, for even the mathematical sciences would not get far unless, as in geometry, they introduced new material in the form of postulates, axioms, definitions and hypotheses. What Mill envisaged, by contrast, was 'the possibility of drawing out new knowledge in an endless stream from a given set of initial propositions, as a magician draws endless ribbons of paper from an empty hat'.

This perspective will be found in the literature both before and after Viner's statement. A very early statement is by Richard Jones: 'Young Mill has been publishing a paper to prove that a priori reasoning is not only good in Pol Econ but the *only* good reasoning applicable to it. God help him & those this belief leads to trust in him, his Paper & his school' (cited Rashid, 1979, 168). Nassau Senior took much the same view of the essay (1848, 304), as did Ingram (1888, 150), J. N. Keynes (1955 [1891], 19), and Marshall (1920, 823-4). Recent examples include Schumpeter (1954, 537), O'Brien (1975, 73), Koot (1975, 322), and Hutchison (1978, 55, 63-4).[5]

This view of the essay does not ring true. There is a wealth of evidence in the essay itself – and also Mill's *System of Logic* (1843) – of a profound appreciation of the key role of empirical evidence at various stages of scientific investigation, and the limited scope of purely hypothetical procedure. In taking this view we follow Schwartz (1972) who chose the title of his work to reflect 'the repercussions of Mill's reactions against the Utilitarians on his methodology of economics' – reactions discernable in the essay (59).[6] There are ambiguities of formulation to be sure, but on the whole the formal pronouncements on method constitute an admirably balanced position. I shall show that Mill insisted upon an 'inductive' basis for the behavioural and other axioms of political economy, a basis which involves introspection

claiming deduction as its only possible logical method, and that not a positive but an "a priori" or "hypothetical deduction".'

Viner objected further that universally applicable generalizations based upon unobservable motivations can be of little practical use – the 'excessive generalizations of the pure deductive school . . . are not always capable of being brought into closer touch with actual phenomena, and . . . even if capable of comparison with reality, remain either extremely devoid of content or extremely removed from direct relevance to the problems of this world' (112).

5 For a more satisfactory account which recognizes the need to avoid over-simplifying Mill's position in the essay, see Collini, in Collini, Winch and Burrow (1983), 138f.

6 A good account is provided by Winch (1970). See also Ryan (1970) for an emphasis upon an 'empiricist' dimension to Mill's methodology: 'he rejects also the excessively abstract and hypothetical approach to Economics, which was typified by what he named the "geometrical" method of James Mill's *Essay on Government*' (xviii); cf. 140: 'Mill takes care to point out that a priori does not mean altogether independent of experience; the principles of human nature from which we reason are obtained by inductive inquiry. In the light of Macaulay's jibes at James Mill, this is an essential point to make'.

to some extent. But this stage of investigation was undertaken at a high degree of generality. The generalizations had to be fleshed out and this process was to be undertaken by way of 'verification' of the outcome of the theoretical model against real-world evidence. It was expected, in brief, that the axiomatic basis for 'deductive' reasoning would be subject to continual improvement and modification in the light of empirical evidence; great care must be exercised in the interpretation of Mill's statement that verification was a matter of 'applied' rather than 'scientific' economics. Moreover even 'introspection' was subject to time and space; there is much true empirical content in the introspective foundation of the axioms utilized by the science of 'political economy'. In the most general terms it may thus be said that Mill's appeal was to avoid founding a science upon universally relevant axioms – axioms based upon psychological properties independent of circumstances. The premises had to be built up with an eye to historical and empirical circumstances to avoid Bacon's charge against deductive procedures based upon premises 'hastily snatched up' or 'arbitrarily assumed'. We may add that the thrust of Mill's argument points towards the economic end rather than egoistic motivation as the starting point of deductive analysis – Viner's alleged 'hypothetical psychology' seems totally inappropriate.

It is apparently all too easy to forget that political economy was not envisaged by Mill as the science of wealth. In fact it was one branch only of such an investigation, the decision to proceed by way of quasi-independent disciplines based upon a strategic evaluation of a higher likelihood of insightful results relating to causal processes by an analytical, as distinct from a synthetical, treatment. But the synthesis of causal influences in their entirety was certainly the ideal, while in the interval the very greatest care had to be taken to avoid casual application to policy of the conclusions drawn from necessarily incomprehensive models.

Mill's position may fruitfully be compared with that of Nassau Senior who formally addressed the methodological issue before Mill. Here a contrast recently made should be noted: 'Mill does not say that we should take the whole man as he is, staking our claim on correctly predicting how he will actually behave in economic affairs. This is a theory of the 'real man' that Senior held to throughout his life despite Mill's essay . . .' (Blaug, 1980, 61; cf. also Machlup, 1972) – a standpoint attributed to Marshall and 'all modern economists'. Mill was the odd man out; economic science for him was 'a body of deductive analysis, resting on assumed psychological premises, and abstracting, even in respect of those premises, from all non-economic aspects of human behaviour' (Blaug, 64). But this contrast is misleading if intended to imply a small concern on Mill's part with the empirical accuracy of the behavioural axiom, a championship of a priori procedure based upon unobservable motivation. On the contrary, his case for a specialized political economy is made in part on empirical grounds; and far from being odd man out by adopting a fictional-man position, he was, as he himself insisted, representing the orthodox view – that of Adam Smith and David Ricardo – by his concern with an empirical foundation for the axioms.

Mill's appeal for caution is striking. He proceeded by emphasizing the empirical justification for the maximizing axiom within a specific institutional environment and range of activities, while warning at the same time of the need for vigilance in application of theory even within that constrained environment – for it is average or normal behaviour only that is covered by the axiom, so that in any particular application there would inevitably be at play 'disturbing causes' not allowed for by the model. It is this very attempt at precision that disguised the fact that the axiom itself was justified in empirical terms. Mill's economic man is not far removed from Adam Smith's *entire* man placed within an environment of a kind precluding much opportunity for other-regarding and non-maximizing behaviour in the market place (cf. Hollander 1977a, 139, 143; Viner, 1972, 82).

Mill maintained that the sixth book of his *Logic* – 'On the Logic of the Moral Sciences' – constituted the main objective of the entire work.[7] To the Italian historian Pasquale Villari, he wrote in 1854: 'Vous avez vu, avec raison, dans ce sixième livre, le but principal de l'ouvrage tout entier, qui a été surtout destiné à répandre sur la méthode des sciences morales, les lumières qu'on peut trouver dans les procédés des sciences physiques' (*CW*, XIV, 239).[8] We shall follow Mill therefore, by attempting to place in proper perspective the specific question of appropriate method in the social sciences. The next section is devoted accordingly to a brief summary of some issues of scientific methodology in general. The key to an appreciation of Mill's position, it turns out, lies in the central role allowed for 'induction' within 'deductive' procedure. Section III provides the intellectual background to Mill's new perspective on the social sciences in 1829–30; and is followed by an account of his case against purely experimental method in the social sciences. The behavioural laws adopted by political economy, specifically the legitimization of the wealth-maximization axiom, is taken up in Section V; and in Section VI we discuss the qualified case for an 'independent' science of political economy in the light of the problem of 'disturbing causes'. Section VII considers the role accorded 'verification' by Mill in model improvement. This discussion is

---

7 For accounts of the development of Mill's ideas on logic see Kubitz (1932), McCrae (1973) and Whitaker (1975), esp. 1034–7.

8 Cf. Mill to Gomperz: '[You] have rightly judged that, to give to the cultivators of physical science the theory of their own operations, was but a small part of the object of the book . . .' (238).

followed by a demonstration that political economy was not envisaged by Mill as a predictive science. Section IX is devoted to the implications from this perspective of Mill's interest in a science of human character formation ('ethology') and a general 'science of society'.

A comparison between Mill and Senior is made in Section X in order to underscore the main lines of Mill's position. The extensive common ground between Mill and the Cambridge inductivists despite a formal appearance of *Methodenstreit* is discussed in the following section.

We conclude by distinguishing between Mill's rejection of universally valid axioms of political economy and his enthusiastic references to the universality of the method.

II ON SCIENTIFIC METHOD IN GENERAL

Following Mill any conception of 'induction' and 'deduction' as constituting opposing methodologies is to be avoided. Thus in the essay on method in political economy the a posteriori method is represented as involving induction alone and the a priori or deductive method as 'a mixed method of induction and ratiocination' (*CW*, IV, 325). Similarly, in the *System of Logic* (ed. 1, 1843) it is made clear that 'although all sciences tend to become more and more Deductive, they are not, therefore, the less Inductive', for there must be 'a direct induction as the basis of the whole' (VII, 291, 454).

The distinction between the two categories must none the less be kept in mind. By 'induction' Mill intended the inference of a proposition from propositions '*less general* than itself'; thus, '[w]hen, from the observation of a number of individual instances, we ascend to a more general proposition, or when, by combining a number of general propositions, we conclude from them another proposition still more general, the process . . . is called Induction' (162).[9] The principle or the law inferred from particular instances or individual experiments is, however, 'more than a mere summing up of what has been specifically observed in the individual cases which have been examined; it is a generalization grounded on those cases, and expressive of our belief, that what we there found true is true in an indefinite number of cases which we have not examined, and are never likely to examine' (163). These are cases which resemble the individual instances in which a phenomenon has been observed to occur '[i]n what are regarded as the material circumstances', the generalization in question based upon the assumption that there are parallel cases in nature so that what has happened once will happen again under similar circumstances (306)[10] – 'the belief we

9 Cf. 284: 'For the purposes of the present inquiry, Induction may be defined [as] the operation of discovering and proving general propositions.'
10 Cf. 288: 'Induction . . . is that operation of the mind, by which we infer that what we know to be true in a particular case or cases, will be true in all cases which resemble

entertain in the universality, throughout nature, of the law of cause and effect' (567).[11]

Deduction – based upon induction in a manner to be shortly described – is a logical procedure which entails 'inferring a proposition from propositions *equally* or *more* general . . . When from a general proposition, not alone (for from a single proposition nothing can be concluded which is not involved in its terms), but by combining it with other propositions, we infer a proposition of the same degree of generality with itself, or a less general proposition, or a proposition merely individual, the process is Ratiocination' (162–3).[12] The distinction between the two approaches, using alternative terminology, is conveniently summarized in the essay on political economy: 'Inductive ('empirical' or a posteriori) method entails argument 'upward' from specific experience of particular facts to a general conclusion while deductive (or a priori) method entails argument 'downwards' from general principle to specific propositions (IV, 325; see further below, p. 149).

We turn immediately to the key issue: the scientific problems amenable to treatment by the respective methodologies. The outstanding characteristic of natural – and, we shall see, also of social – phenomena was, in Mill's account, that of 'Composition of Causes' (also referred to as 'Intermixture of Effects'), where 'the effects of different causes are . . . not dissimilar, but homogeneous, and marked out by no assignable boundaries from one another: A and B may produce not *a* and *b*, but different portions of an effect *a* . . .' (VII, 434). This characteristic was said to be typical both of mechanics and chemistry but in strikingly different forms: 'A concurrence of causes, not separately producing each its own effect, but interfering with or modifying the effects of one another, takes place . . . in two different ways. In the one, which is exemplified by the joint operation of different forces in mechanics, the separate effects of all the causes continue to be produced, but are compounded with one another, and disappear in one total. In the other, illustrated by the case of chemical action, the separate effects cease entirely, and are succeeded by phenomena altogether different, and governed by different laws' (440).[13]

the former in certain assignable respects. In other words, Induction is the process by which we conclude that what is true of certain individuals of a class is true of the whole class, or that what is true at certain times, will be true in similar circumstances at all times.'

11 This principle itself is said to illustrate an inductive operation.
12 See especially Book II, chs. 2, 3 dealing with Syllogism.
13 'Composition of causes' should be distinguished from what is called 'plurality of causes', where an effect is 'producible by two or more causes' rather than where every effect is connected exclusively with a single cause. This kind of complication could be accommodated more easily by experimental method (434, 439). Similarly, the problem of 'composition of causes' as manifested in chemistry could be so accommodated. The differences between mechanics and chemistry in this regard are elaborated, 442f.

The composition of causes of the mechanical variety was responsible for an optical illusion. While each (separate) cause–effect relationship continued to operate, it may not, at first sight, appear to be in operation at all. To underscore the continuous operation of the individual 'causes', Mill supported the use of the term *tendency* – a practice which we shall later see opens up a variety of important issues regarding political economy (such as the status of the 'law of diminishing returns') and therefore requires particular attention for our special purposes:

The general idea of the Composition of Causes has been seen to be, that though two or more laws interfere with one another, and apparently frustrate or modify one another's operation, yet in reality all are fulfilled, the collective effect being the exact sum of the effects of the causes taken separately . . . All laws of causation are liable to be in this manner counteracted, and seemingly frustrated, by coming into conflict with other laws, the separate result of which is opposite to theirs, or more or less inconsistent with it. And hence, with almost every law, many instances in which it really is entirely fulfilled, do not, at first sight, appear to be cases of its operation at all . . . To accommodate the expression of the law to the real phenomena, we must say, not that the object moves, but that it *tends* to move, in the direction and with the velocity specified (443–4).

The standard illustration alluded to here is that of motion: 'If, for instance, it were stated as a law of nature that a body to which a force is applied moves in the direction of the force, with a velocity proportioned to the force directly, and to its own mass inversely; when in point of fact some bodies to which a force is applied do not move at all, and those which do move (at least in the region of our earth) are, from the very first, retarded by the action of gravity and other resisting forces, and at last stopped altogether; it is clear that the general proposition, though it would be true under a certain hypothesis, would not express the facts as they actually occur.' The phenomenon might be described by saying 'that the body moves in that manner unless prevented, or except in so far as prevented, by some counteracting cause'. But the term tendency was preferable because 'the body does not only move in the manner unless counteracted; it *tends* to move in that manner even when counteracted; it still exerts, in the original direction, the same energy of movement as if its first impulse had been undisturbed, and produces, by that energy, an exactly equivalent quantity of effect' (444–5).[14]

14 Cf. 372: 'Each agent produced the same amount of effect as if it had acted separately, though the contrary effect which was taking place during the same time obliterated it as fast as it was produced.'

Now one of the primary tasks of science was to determine the effect that will follow a certain combination of causes, and conversely the combination of causes that would produce a given effect (458, 460). More precisely, the problem was 'to trace each effect to the concurrence of causes in which it originated, and ascertain the conditions of its recurrence – the circumstances in which it may be expected again to recur' (446), a formulation consistent with a general conception of causation as 'invariable, certain, and unconditional sequence' (VIII, 837).[15] Inductive procedures could not deal with this kind of problem in the face of 'composition of causes' of the mechanical, as distinct from the chemical, variety even when based on well-established experimental methods of empirical inquiry, namely the methods of agreement, difference, residues and concomitant variations (and various combinations thereof) (VII, 388–406);[16] these cases, which were 'by far the most frequent' eluded treatment by the experimental methods (440).[17] The problem of 'mutual interference of causes, where each cause continues to produce its own proper effect according to the same laws to which it conforms

15 Cf. 838 reference to 'uniformity of order, and capability of being predicted'.
16 For a convenient summary, see Jacob Viner (1962 [1917]), 106: 'The purpose of an induction being supposed to be the proof that A is a cause of B, the problem of induction is, What are the possible methods of obtaining this proof? They all resolve themselves into a demonstration that B is present wherever A is present, and that B is absent wherever A is absent. If, in a number of otherwise diverse situations, B is always found to be present when A is present, we may conclude that A and B are causally connected – the method of agreement. If the addition of A to a situation from which it was formerly absent causes B to appear in the situation, and if the subtraction of A from the situation causes B to disappear, we may conclude that A and B are causally connected – the method of difference. Each of these methods may be improved if in each case we examine all the negative instances of A, and find that B is always absent when A is absent – the methods of double agreement and of double difference. These are the primary methods, from which are derived two other methods, the methods of residues and of concomitant variations.'
The method of concomitant variations presumes the possibility of measurement over time. It is described thus by Mill (VII, 400): 'Though we cannot exclude an antecedent altogether, we may be able to produce, or nature may produce for us, some modification in it. By a modification is here meant, a change in it, not amounting to its total removal. If some modification in the antecedent A is always followed by a change in the consequent a, the other consequents b and c remaining the same, or vice versa, if every change in a is found to have been preceded by some modification in A, none being observable in any of the other antecedents; we may safely conclude that a is, wholly or in part, an effect traceable to A, or at least in some way connected with it through causation.'
The method of residues is described as follows: '*Subtract from any phenomenon such part as is known by previous inductions to be the effect of certain antecedents, and the residue of the phenomenon is the effect of the remaining antecedents*' (398). The presumption that it is by induction that the effects of 'certain antecedents' are known turns out to be of the first importance as we shall see (below, p. 102).
17 Cf. 373–6 for the argument that composition of causes of the mechanical variety was the rule and that of the chemical variety the exception.

in its separate state . . . present[ed] . . . so far as direct induction apart from deduction is concerned, infinitely greater difficulties' than the problem posed by chemical reactions; between the result (the effect) and the causes producing it, 'there is often an insurmountable difficulty in tracing by observation any fixed relation whatever' (443).

Repeatedly and forcefully Mill expressed his scepticism regarding inductive attempts – by simple observation or by more sophisticated experimental methods – to solve the problem of causation as defined above: 'no such attempt can have the smallest chance of success, it has been abundantly shown in the tenth chapter of the Third Book. We there examined whether effects which depend on a complication of causes can be made the subject of a true induction by observation and experiment; and concluded, on the most convincing grounds, that they cannot' (VIII, 880). Scientific use of the methods of experiment, in such cases, was 'out of the question', and 'from the very nature of the case, inefficient and illusory . . .' (VII, 451, 453).[18] After an account of the problem of plurality of causes – the relatively simple problem involving an effect which is producible by more than one cause – Mill summarized his position regarding the main problem: 'We here close our remarks on the Plurality of Causes, and proceed to the still more peculiar and more complex case of the Intermixture of Effects and the interference of causes with one another: a case constituting the principal part of the complication and difficulty of the study of nature; and with which the four only possible methods of directly inductive investigation by observation and experiment, are for the most part, as will appear presently, quite unequal to cope. The instrument of Deduction alone is adequate to unravel the complexities proceeding from this source; and the four methods have little more in their power than to supply premises for, and a verification of, our deductions' (439).

Mill adopted a strict position regarding the isolation of true causal relationships. Let it be supposed that the researcher was fortunate enough to observe 'a real uniformity of nature' revealed spontaneously without formal experiment.[19] Even then, the most that would thus be yielded

18 To proceed inductively by a posteriori method would entail consideration of 'the whole assemblage of concurrent causes which produced the phenomenon, as one single cause' (446) while science desires to distinguish the individual causes and discover the consequences of their union in various combinations.

19 Where feasible, artificial experimentation had the great advantage of permitting examination of 'innumerable combinations of circumstances not found in nature' (382). Formal experimentation was likely to prove most successful where it is possible to 'take a cause, and try what it will produce: but we cannot take an effect, and try what it will be produced by. We can only watch till we see it produced, or are able to produce it by accident' (385). Thus artificial experiments were sometimes impossible to devise (as in astronomy) or at least subject to disadvantages 'equivalent to impracticability' as in mental philosophy, physiology and social science (384).

would be '*invariable* antecedence' (the most common relationship in zoology) rather than true causation or '*unconditional* antecedence'; a particular antecedent can be regarded as a cause, and the induction complete, only when the process has been reversed and the effect produced (artificially) by means of that antecedent: 'Observation . . . without experiment (supposing no aid from deduction) can ascertain sequence and coexistences, but cannot prove causation' (386).

Inductions of relationships of causality from direct experience had, Mill readily conceded, been fruitful in the infancy of the various branches of natural philosophy. But the broad historical trend involved progress away from purely experimental procedure towards the deductive in order to base the science upon 'the fewest and the simplest possible inductions, and to make these, by any combinations however complicated, suffice for proving even such truths, relating to complex cases, as could be proved, if we chose, by inductions from specific experience (218). For the situation arose with scientific progress 'whereby multitudes of truths, already known by induction from as many different sets of experiments, have come to be exhibited as deductions or corollaries from inductive propositions of a simpler and more universal character'. This transition away from purely experimental procedure is represented as 'the greatest triumph of the investigation of nature', for there were 'weighty scientific reasons for giving every science as much of the character of a Deductive Science as possible'.[20] A revolution in science, Mill proclaimed, was in progress reversing the Baconian rejection of deductive method (481–3).

* * *

We must now look more closely at the inductive components of deductive procedure. Formally, the procedure entails three stages, although (as we shall see later), the order thereof was not rigidly specified.

The mode of investigation which, from the proved inapplicability of direct methods of observation and experiment, remains to us as the main source of knowledge we possess or can acquire respecting the conditions, and laws of recurrence, of the more complex phenomena, is called in its most general expression, the Deductive Method; and consists of three operations: the first, one of direct induction; the second, of ratiocination; the third, of verification (454; cf. epigraph from Herschel, 484).

Bacon's objections to deductive procedure had been directed at deductions drawn 'from premises hastily snatched up, or arbitrarily assumed'. They

20 Reference is made to mechanics, optics, acoustics, meteorology and astronomy.

were only valid in the case of principles which were not derived 'by legitimate canons of experimental inquiry', and of results which were not 'tested by that indisputable element of a rational Deductive Method, verification by specific experience' (482).

The first stage – in effect the establishment of the premises or initial principles of the model by use of inductive procedures (observation or more formal experiment) – itself required the satisfaction of certain preconditions. A preliminary mental exercise sets the stage whereby we 'learn to see in the chaotic antecedent a multitude of distinct antecedents, in the chaotic consequence a multitude of distinct consequences' (379). The next step is to discover 'on which of the antecedents each consequent is invariably attendant'; to which end 'we must endeavour to effect a separation of the facts from one another, not in our minds only, but in nature'. Now, '[i]f those antecedents could not be severed from one another except in thought, or if those consequents never were found apart, it would be impossible for us to distinguish (a posteriori at least) the real laws, or to assign to any cause its effect, or to any effect its cause. To do so, we must be able to meet with some of the antecedents apart from the rest, and observe what follows from them; or some of the consequents, and observe by what they are preceded. We must, in short, follow the Baconian rule of *varying the circumstances*' (381).

In some cases, especially physiological phenomena, 'the causes do not suffer themselves to be separated and observed apart . . .' (456).[21] But this was the exception: 'In general, the laws of the [individual] causes on which the effect depends may be obtained by an induction from comparatively simple [empirically observed] instances, or, at the worst, by deduction from the laws of simpler causes, so obtained. By simple instances are meant, of coures, those in which the action of each cause was not intermixed or interfered with, or not to any great extent, by other causes, whose laws were unknown' (458). In brief, the first, inductive, stage involves the isolation of the law 'of each of the concurrent causes: and this supposes

---

21 The problem arose from the peculiar difficulties of separating the different agencies composing an organized body 'without destroying the very phenomena which it is our object to investigate'. For this reason, Mill added: 'I am inclined to the opinion that physiology (greatly and rapidly progressive as it now is) is embarrassed by greater natural difficulties, and is probably susceptible of a less degree of ultimate perfection, than even the social science; inasmuch as it is possible to study the laws and operation of one human mind apart from other minds, much less imperfectly than we can study the laws of one organ or tissue of the human body apart from other organs or tissues.'

Yet at the end of the preceding chapter (451–2), Mill seemed to reach the reverse evaluation: 'If so little can be done by the experimental method to determine the conditions of an effect of many combined causes, in the case of medical science; still less is this method applicable to a class of phenomena more complicated than even those of physiology, the phenomena of politics and history.'

a previous process of observation or experiment upon each cause separately; or else a previous deduction, which also must depend for its ultimate premises on observation or experiment' (454). Only when inductions furnishing the laws of the individual causes have been possible 'has the application of . . . [the deductive] method to the ascertainment of the law of a complex effect, been attended with brilliant results' (458). These causal laws may, but need not, amount to exact numerical laws, as in some branches of mathematics, or to geometrical laws, as in mechanics, optics, acoustics and astronomy (458–9).

The second major stage, that of ratiocination is designed 'to find the law of an effect, from the laws of the different tendencies of which it is the joint result' (454),[22] and (as we have just seen) requires knowledge of the independently operating causes:

Suppose . . . that two different agents, operating jointly, are followed, under a certain set of collateral conditions, by a given effect. If either of these agents, instead of being joined with the other, had operated alone, under the same set of conditions in all other respects, some effect would probably have followed; which would have been different from the joint effect of the two, and more or less dissimilar to it. Now, if we happen to know what would be the effect of each cause when acting separately from the other, we are often able to arrive deductively, or a priori, at a correct prediction of what will arise from their conjunct agency. To render this possible, it is only necessary that the same law which expresses the effect of each cause acting by itself, shall also correctly express the part due to that cause, of the effect which follows from the two together (370).

The third stage, that of verification, is designed to test the ratiocination and its foundation and thereby avoid the kind of objections raised against the purely experimental method:

When in every single instance, a multitude, often an unknown multitude, of agencies, are clashing and combining, what security have we that in our computation a priori we have taken all these into our reckoning? How many must we not generally be ignorant of? Among those which we know, how probable that some have been overlooked; and even were all included, how vain the pretence of summing up the effects of many causes, unless we know the numerical law of each, – a condition in most cases not to be fulfilled; and even when it is fulfilled, to make the calculation transcends, in any but very simple cases, the utmost power of mathematical science with all its most modern improvements (460).

22 Cf. 458: to determine 'from the laws of the [individual] causes, what effect any given combination of those causes will produce'.

To assure that the general conclusions derived deductively are more than mere conjectures they must be shown to 'accord with the results of direct observation wherever it can be had'. If there are known cases where the expected effect has not followed it must be possible 'to show (or at least to make a probable surmise) what frustrated it' (461) – the counterpart to what in the study of the social sciences are referred to as 'disturbing causes', namely causal influences which may be unknown (at least at the outset of the exercise) or, if already known to operate, are none the less not incorporated within the formal model. Here we come face to face with the position that 'the method of all Deductive Sciences is hypothetical. They proceed by tracing the consequences of certain assumptions; leaving for separate consideration, whether the assumptions are true or not, and if not exactly true, whether they are sufficiently near approximation to the truth' (259).[23]

For purposes of verification it was desirable to have readily at hand 'empirical laws' of the effect (possibly only approximate) furnished by observation, and the systematic collection of instances entailing application of the methods of experimental enquiry, particularly 'the method of agreement'. Specifically,

An Empirical Law . . . is an uniformity, whether of succession or of coexistence, which holds true in all instances within our limits of observation, but is not of a nature to afford any assurance that it would hold beyond those limits; either because the consequent is not really the effect of the antecedent, but forms part along with it of a chain of effects, flowing from prior causes not yet ascertained; or because there is ground to believe that the sequence (though a case of causation) is resolvable into simpler sequences, and, depending therefore on a concurrence of several natural agencies, is exposed to an unknown multitude of possibilities of counteraction. In other words, an empirical law is a generalization, of which, not content with finding it true, we are obliged to ask, why is it true? knowing that its truth is not absolute, but dependent on some more general conditions, and that it can only be relied on in so far as there is ground of assurance that those conditions are realized (VIII, 861).[24]

---

23 Cf. VIII, 900: 'All the general propositions which can be framed by the deductive science, are . . . hypothetical'.

24 For a nice example from medicine, see VII, 451:

'Anything like a scientific use of the method of experiment, in these complicated cases, is . . . out of the question. We can generally, even in the most favourable cases, only discover by a succession of trials, that a certain cause is *very often* followed by a certain effect. For, in one of these conjunct effects, the portion which is determined by any one of the influencing agents, is usually, as we before remarked, but small; and it must be a more potent cause than most, if even the tendency which it really exerts is not thwarted by other tendencies in nearly as many cases as it is fulfilled. Some causes indeed there

The best conceivable verification of a theory would amount to a demonstration 'that it led deductively to those empirical laws; that the uniformities, whether complete or incomplete, which were observed to exist among the phenomena, were accounted for by the laws of the causes – were such as could not but exist if those be really the causes by which the phenomena are produced.' For while empirical laws were useless in the discovery of complex causal laws, yet they might serve to confirm those laws. And only in the event of a positive outcome could one presume to apply the causal relation thus confirmed to other cases 'of which our specific experience is to come'. In the reverse case, assuming that we cannot show or convincingly suggest the reason for the disaccord, the theory remains 'imperfect and not yet to be relied upon' (VII, 461–2).

* * *

We have encountered Mill's insistence regarding deductive procedure that 'there must be a direct induction as the basis of the whole'. But he did not intend to be taken literally; the three stages of induction, ratiocination and verification did not have to be undertaken in that precise order.[25] What is referred to as the hypothetical method of deduction – a misleading term since all deductive procedures were said to be 'hypothetical' – 'suppresses the first of the three steps, the induction to ascertain the law; and contents itself with the other two operations, ratiocination and verification; the law which is reasoned from, being assumed, instead of proved' (492). In principle, there would be no reason to rule out any hypothesis imaginable and account for an effect by 'some cause of a kind utterly unknown, and acting according to a law altogether fictitious' (490). But such hypotheses would be implausible and would 'not supply the want which arbitrary hypotheses are generally invented to satisfy, by enabling the imagination to represent to itself an obscure phenomenon in a familiar light'.[26] Accordingly, there probably never existed an hypothesis involving

are which are more potent than any counteracting causes to which they are commonly exposed; and accordingly there are some truths in medicine which are sufficiently proved by direct experiment. Of these the most familiar are those that relate to the efficacy of the substances known as Specifics for particular diseases; 'quinine, colchicum, lime juice, cod liver oil', and a few others. Even these are not invariably followed by success; but they succeed in so large a proportion of cases, and against such powerful obstacles, that their *tendency* to restore health in the disorders for which they are prescribed may be regarded as an experimental truth.'

25 But 'although the process need not necessarily be carried on in this form, it is always susceptible of the form, and must be thrown into it when assurance of scientific accuracy is needed and desired' (203).

26 Adam Smith had also insisted upon the requirement for 'familiar' axioms; cf. Hollander (1977).

both a fictitious 'agent' and a fictitious law of operation; '[e]ither the phenomenon assigned as the cause is real, but the law according to which it acts, merely supposed; or the cause is fictitious, but is supposed to produce its effects according to laws similar to those of some known class of phenomena'. If this is indeed so, it would appear that the hypothetical method does not totally 'suppress' the first stage; and this is indeed confirmed by the detailed argument.

The hypothetical method of deduction – in so far as we are concerned to prove 'the truth' of the hypothesis – could be justified only if the final step of verification 'shall amount to, and fulfil the condition of, a complete induction' (492). For this to be so, it is no longer sufficient that the process generates results (by way of ratiocination) which are 'verified' by the evidence; even 'false' hypotheses might account for known phenomena (500).

Supposing that the hypothesis itself relates to the law of operation or the mode of dependency of a known cause, a positive outcome at the stage of verification proves the hypothetical law to be the 'true' law only where it can be shown that no other law can possibly yield the same conclusions. In this case alone 'what was an hypothesis at the beginning of the inquiry, becomes a proved law of nature before its close' (492).[27] The condition required for this to hold good is that the assigned cause 'should not only be a real phenomenon, something actually existing in nature, but should be already known to exercise, or be capable of exercising, an influence of some sort over the effect'. If this condition is not satisfied, 'it is no sufficient evidence of the truth of the hypothesis that we are able to deduce the real phenomena from it' – the hypothesis remains an hypothesis, neither proved nor disproved by the verification (495).

It was also legitimate to assume a fictitious agent (and not merely ascribe a fictitious law to a known agent). Indeed this step was frequently an essential part of the process of discovery: 'The process of tracing regularity in any complicated, and at first sight, confused set of appearances, is necessarily

27 Mill explains that the 'method of difference' is the relevant category of experimental procedure involved in such demonstrations. The precise conditions required to satisfy the demonstration is that the assigned cause constitutes 'something actually existing in nature', but also it 'should be already known to exercise, or be capable of exercising, an influence of some sort over the effect' (495). In the case of Newton: 'If it had not been previously known that the planets were hindered from moving in straight lines by some force tending towards the interior of their orbit, though the exact direction was doubtful; or if it had not been known that the force increased in some proportion or other as the distance diminished, and diminished as it increased; Newton's argument would not have proved his conclusion. These facts, however, being already certain, the range of admissible suppositions was limited to the various possible directions of a line, and the various possible numerical relations between the variations of the distance, and the variations of the attractive force: now among these it was easily shown that different suppositions could not lead to identical consequences' (493–4).

tentative: we begin by making any supposition, even a false one, to see what consequences will follow from it; and by observing how these differ from the real phenomena, we learn what corrections to make in our assumption. The simplest supposition which accords with the more obvious facts is the best to begin with; because its consequences are the most easily traced. This rude hypothesis is then rudely corrected, and the operation repeated; and the comparison of the consequences deducible from the corrected hypothesis, with the observed facts, suggests still further correction, until the deductive results are at last made to tally with the phenomena' (496–7).[28] But yet an hypothesis might be a false one – as, for example, the vortices of Descartes – even if it is possible to deduce the real phenomena from it. There must be independent evidence that the assumed cause constituted 'a fact in nature' – a 'known cause' – for the hypothesis to be received as true (a 'proved' fact) merely because it explained the phenomena.[29]

The significance of all this lies in Mill's view that by the celebrated declaration 'Hypotheses non fingo', Newton did not intend to rule out assumed causes. Rather the hypothesis, if it was to represent a 'vera causa', must not entail mere supposition, its existence must be detected by 'independent evidence' (495–6), and the ability of the model, based on hypothesized causes to anticipate and predict, did not by itself confirm the causes as vera causa.[30] Thus Whewell was taken to task for his belief to the contrary, that the verification of predictions flowing from a hypothetically based theory constituted conclusive proof of the 'truth' of that theory; in

28 Mill cites Comte: 'Neither induction nor deduction would enable us to understand either the simplest phenomena, "if we did not often commence by anticipating on the results; by making a provisional supposition, at first essentially conjectural, as to some of the very notions which constitute the final object of the inquiry".'

29 'The vortices of Descartes would have been a perfectly legitimate hypothesis, if it had been possible, by any mode of exploration which we could entertain the hope of ever possessing, to bring the reality of the vortices, as a fact in nature, conclusively to the test of observation. The vice of the hypothesis was that it could not lead to any course of investigation capable of converting it from an hypothesis into a proved fact. It might chance to be disproved, either by some want of correspondence with the phenomena it purported to explain, or (as actually happened) by some extraneous fact. "The free passage of comets through the spaces in which these vortices should have been, convinced men that these vortices did not exist" [Whewell]. But the hypothesis would have been false, though no such direct evidence of its falsity had been procurable. Direct evidence of its truth there could not be' (499).

30 This seems to be a stricter condition than that accepted by Adam Smith; cf. Hollander (1977a).

    However, this was no reason to avoid tracing out the consequences of hypothetical assumptions which do not meet such standards, provided it was recognized 'that the utmost they can prove is, not the hypothesis is, but that it may be true' (504). On this matter Mill expressed his disagreement with Comte.

Mill's view this was insufficient since the 'condition of accounting for all the known phenomena is often fulfilled equally well by two conflicting hypotheses' (500).[31] Mill's conditions for the justification of 'assumed causes' were in fact stronger than those of Whewell: 'at the risk of being charged with want of modesty, I cannot help expressing astonishment that a philosopher of Dr. Whewell's abilities and attainments should have written an elaborate treatise on the philosophy of induction, in which he recognizes absolutely no mode of induction except that of trying hypothesis after hypothesis until one is found which fits the phenomena; which one, when found, is to be assumed as true, with no other reservation than that if on re-examination it should appear to assume more than is needful for explaining the phenomena, the superfluous part of the assumption should be cut off. And this without the slightest distinction between the cases in which it may be known beforehand that two different hypotheses can lead to the same result, and those in which, for aught we ever know, the range of suppositions, all equally consistent with the phenomena, may be infinite' (503). We return to this matter of the proof of hypotheses, below (p. 154).

III THE NEW PERSPECTIVE ON POLITICAL ECONOMY

1830 marks a clear watershed as far as concerns Mill's pronouncements on appropriate method in political economy and also his practice. Reacting in 1825 to Malthus's charge that 'the new school' of political economy (the Ricardians) 'altered the theories of Adam Smith upon pure speculation' Mill had written that 'it would, indeed, have been somewhat surprising if they had altered them on any other ground' (*CW*, IV, 25). It is highly unlikely that he would have made this kind of statement much after 1830.

---

31 Thus the Cartesian hypothesis 'in its finally improved state' served as well as Newton's in explaining at least a certain range of phenomena. See also 432–3 for the argument, against Whewell, that it is insufficient merely to have hypotheses which generate results that are not '*dis*proved' by experience – a plea for inductive evidence:

'If discoveries are ever made by observation and experiment without Deduction, the four methods are methods of discovery: but even if they were not methods of discovery, it would not be the less true that they are the sole methods of Proof; and in that character, even the results of deduction are amenable to them. The great generalizations which begin as Hypotheses, must end by being proved, and are in reality (as will be shown hereafter) proved, by the Four Methods. Now it is with Proof, as such, that Logic is principally concerned. This distinction has indeed no chance of finding favour with Dr. Whewell; for it is the peculiarity of his system, not to recognise, in cases of Induction, any necessity for proof. If, after assuming an hypothesis and carefully collating it with facts, nothing is brought to light inconsistent with it, that is, if experience does not *dis*prove it, he is content: at least until a simpler hypothesis, equally consistent with experience, presents itself. If this be Induction, doubtless there is no necessity for the four methods. But to suppose that it is so, appears to me a radical misconception of the nature of the evidence of physical truths' (432).

We also have his description of himself as 'a mere reasoning machine' in consequence of his early training, which he applied only to 'two or three years' of his life (I, III).[32]

Mill's early practice is also revealing of his original perspective. Monetary theory provides a nice illustration. In the defence of the 'new school' Mill precluded the factual possibility of excess commodity supply on purely logical grounds, engaging in a grotesque formal demonstration to prove that 'there cannot be an excess on the whole', adding insult to injury by prefacing his syllogistic account by describing as 'truly amusing' Malthus's proposal to rebut demonstration by testimony' (IV, 41–2).

Attention was, it is true, paid at this period to the importance of factual material. Thus rapid legislative enactments (unless absolutely essential) were, Mill warned, in general dangerous unless based on principles 'universally admitted by all well-informed persons who have directed their attention to the subject; provided, also, that the applicability or inapplicability of those principles to existing circumstances, depends upon facts, all of which are so obvious and familiar, as to be in no danger of being either mistaken or misinterpreted' (78). The second proviso constitutes a limitation on the unqualified applicability of deductions drawn from a body of theory. But it is lip-service only.

Blame for Parliament's error in placing responsibility for the commercial crisis of 1825–6 on the small-note issue is laid at the door of faulty method – a failure to argue from 'general principles' and an appeal to 'experience' (83f., 110f). In this early context Ricardo is praised for his procedures. Mill himself doubted the purported 'facts' that it is possible to experience substantial periods of absolute increase in currency before corrective gold flows occur; or that increases in the small-note issue facilitated the speculation and over-trading of 1824–5 (80); and he complained that Parliament had not attempted to examine the facts of the case in a serious manner, adding that had they done so 'the discordance of the facts with the theory would probably have rendered any other mode of combating it superfluous'. Yet he himself did not then proceed to combat the supposedly fallacious theory from an empirical angle. Rather he approaches the problem 'by reasoning from the nature and properties of currency and trade': 'We are persuaded that even

---

32 In a letter of October 1831 Mill said of himself that 'the only thing that I believe I am really fit for, is the investigation of abstract truth, & the more abstract the better' (*CW*, XII, 78). But this statement was immediately preceded by the observation that the 'principles' (with which he was then preoccupied in a variety of subjects), although 'of use for all times', were to be applied 'cautiously & circumspectly to any'. Shortly after he spoke of this period as one of 'reaction from logical utilitarian narrowness of the very narrowest kind' (12 Jan. 1834; 204). Subsequently he described himself as an example of a logician and political economist interested in more than logic and political economy.

by this means, we shall be able to disprove the theory in every one of its essential parts, and by arguments so cogent, that the non-production of so many facts will only injure the simplicity, without detracting from the conclusiveness, of our reasoning' (80–1). And he reflected on the paucity of data relating to the period 1824–5 with the remark that 'though, for want of evidence, we cannot trace accurately the connexion of the events, we can prove at least, that what is known of them does not afford any ground for adhering to an opinion, the fallacy of which in theory has been so clearly shewn' (86). As we shall see the entire attitude came to be transformed when Mill insisted on the Herschel rule to 'examine your residuals' – an appeal for careful empirical testing against the evidence before any future reliance could be placed in a body of theory.

The primary source of our knowledge regarding Mill's emotional crisis in his twentieth year and recovery upon his encounter with metaphysics and the romantic poets is, of course, the *Autobiography* (*CW*, I, 149–53).[33] Here Mill described his subsequent receptiveness to 'the influences of European, that is to say, Continental thought, and especially those of the reaction of the nineteenth century against the eighteenth' which, as he put it, 'were now streaming in' upon him. From Coleridge, from the Coleridgians, from his reading of Goethe, from Carlyle and from the French literature, he learned – although, he tells us here, he did not adopt the position 'in the exaggerated and violent manner' of the thinkers with whom he compared notes – 'That all questions of political institutions are relative, not absolute, and that different stages of human progress not only *will* have, but *ought* to have, different institutions: That government is always either in the hands, or passing into the hands, of whatever is the strongest power in society, and that what this power is, does not depend on institutions, but institutions on it: That any general theory or philosophy of politics supposes a previous theory of human progress, and that this is the same thing with a philosophy of history' (169).

Some of Mill's complaints against the 'old school' of political economists first appear in print in the 1834 review of Harriet Martineau's *Summary of Political Economy*. The English political economists attempted 'to construct a permanent fabric out of transitory materials' and enunciated propositions pertinent to contemporary Britain 'as if they were universal and absolute truths' (IV, 225); they 'revolve in their eternal circle of landlords, capitalists, and labourers, until it seems to think of the distinction of society into these three classes, as if they were one of God's ordinances, not man's, and as little under human control as the division of day and night. Scarcely any one of them seems to have proposed to himself as a subject of inquiry, what changes the relations of those classes to one another are likely to undergo

---

33 Mill first read Wordsworth in Autumn 1828.

in the progress of society; to what extent the distinction itself admits of being beneficially modified, and, if it does not even, in a certain sense, tend gradually to disappear' (226–7). In the *Autobiography*, Mill alluded to his own originally narrow conception:

In those days I had seen little further than the old school of political economists into the possibilities of fundamental improvements in social arrangements. Private property as now understood, and inheritance, appeared to me, as to them, as the *dernier mot* of legislation: and I looked no further than to mitigating the inequalities consequent on these institutions, by getting rid of primogeniture and entails. The notion that it is possible to go further than this in removing the injustice – for injustice it was whether admitting a complete remedy or not – involved in the fact that some are born to riches and the vast majority to poverty, I then reckoned chimerical; and only hoped that by universal education, leading to voluntary restraint on population, the portion of the poor might be made more tolerable (I, 239).

Writing of the 'common run of political economists' Mill complained of their neglect of the fundamental distinction between the laws of production, which are 'real laws of nature, dependent on the properties of objects', and the 'modes of its distribution', which 'subject to certain conditions, depend on human will' and which are therefore to be treated as 'provisional' and thus 'liable to be much altered by the progress of social improvement'. The common run

confuse these together, under the designation of economic laws, which they deem incapable of being defeated or modified by human effort; ascribing the same necessity to things dependent on the unchangeable conditions of our earthly existence, and to those which, being but the necessary consequences of particular social arrangements, are merely coextensive with these. Given certain institutions and customs, wages, profits, and rent will be determined by certain causes; but this class of political economists drop the indispensable presupposition, and argue that these causes must be an inherent necessity, against which no human means can avail, determine the shares which fall, in the division of the produce, to labourers, capitalists, and landlords (255, 257).

The Saint-Simonians, first encountered in 1829 and 1830, were said to have been the most potent influence at the time: 'They were then only in the earlier stages of their speculations. They had not yet dressed out their philosophy as a religion, nor had they organized their scheme of Socialism. They were just beginning to question the principle of hereditary property. I was by no means prepared to go with them even this length; but I was greatly struck with the connected view which they for the first time presented to me, of the natural order of human progress; and especially with their division of all history into organic periods and critical periods' (171).[34] 'Their criticisms on the common doctrines of Liberalism seemed to me full of important truth; and it was partly by their writings that my eyes were opened to the very limited and temporary value of the old political economy, which assumes private property and inheritance as indefeasible facts, and freedom of production and exchange as the *dernier mot* of social improvement' (173, 175). Also, among the Saint-Simonians, Auguste Comte was spelled out as the most important with particular reference to 'the doctrine, which he afterwards so copiously illustrated, of the natural succession of three stages in every department of human knowledge – first, the theological, next the metaphysical, and lastly, the positive stage', and the view 'that social science must be subject to the same law; that the feudal and Catholic system was the concluding phasis of the theological state of the social science, Protestantism the commencement, and the doctrines of the French Revolution the consummation, of the metaphysical, and that its positive state was yet to come. This doctrine harmonized well with my existing notions, to which it seemed to give a scientific shape. I already regarded the methods of physical science as the proper models for political' (173).[35] But above all, Mill attributed to the Saint-Simonians and Comte a 'clearer conception than ever before of the peculiarities of an era of transition in opinion', whereby he ceased to mistake the moral and intellectual characteristics of such an era, for the normal attributes of humanity.[36]

---

34 Specifically: 'During the organic periods (they said) mankind accept with firm conviction some positive creed, claiming jurisdiction over all their actions, and containing more or less of truth and adaptation to the needs of humanity. Under its influence they make all the progress compatible with the creed, and finally outgrow it; when a period follows of criticism and negation, in which mankind lose their old convictions without acquiring any new ones, of a general or authoritative character, except the conviction that the old are false. The period of Greek and Roman polytheism, so long as really believed in by instructed Greeks and Romans, was an organic period, succeeded by the critical or sceptical period of the Greek philosophers. Another organic period came in with Christianity. The corresponding critical period began with the Reformation, has lasted ever since, still lasts, and cannot altogether cease until a new organic period has been inaugurated by the triumph of a yet more advanced creed.'

35 The references to Comte are to the *Système de Politique Positive* (1823) which Mill apparently first read in 1829 (see *CW*, XII, 34).

36 The Continental influences were first encountered on an extended stay in France, 1820-1: 'The chief fruit which I carried away from the society I saw, was a strong and permanent interest in Continental Liberalism, of which I ever afterwards kept myself *au courant*, as much as of English politics: a thing not at all usual in those days for an Englishman, and which had a very salutary influence on my development, keeping me free from the error always prevalent in England, and from which even my father with all his superiority to prejudice was not exempt, of judging universal questions by a merely English standard' (63).

John Austin certainly deserves mention in an account of the early 1830s. As Mill put it, Austin had 'professed great disrespect for what he called "the universal principles of human nature of the political economists"', and insisted on the evidence which history and daily experience afford of "the extraordinary pliability of human nature" (a phrase which I have somewhere borrowed from him); nor did he think it possible to set any positive bounds to the moral capabilities which might unfold themselves in mankind, under an enlightened direction of social and education influences' (185, 187).[37] These various trends played upon Mill after he had himself isolated the source of what (in the confident mood following his depression) he discerned to be the basic error of his father's *Essay on Government*.[38] At an early stage, Mill had recognized weaknesses in the Benthamite theory of government as a theory of government in general, particularly its neglect of relevant considerations in its axiomatic foundations. But these lacunae, he originally believed, did not reflect matters of principle and could be corrected for in application. In particular, he had believed 'that politics could not be a science of specific experience; and that the accusations against the Benthamite theory of *being* a theory, of proceeding a priori, by way of general reasoning, instead of Baconian experiment, shewed complete ignorance of Bacon's principles, and of the necessary conditions of experimental investigation' (165). At this juncture appeared the celebrated criticisms of James Mill's *Essay on Government* by T. B. Macaulay in the *Edinburgh Review*.[39] He rejected these

---

37 Mill also adopted from Austin the term 'provisional' as a description of existing social arrangements (241).

Mill himself rejected Austin's position on 'the pliability of human nature'. In later correspondence with Harriet Taylor, he emphasized the *difficulty* in the way of 'making people unselfish' (below, p. 774) and there was no change in his position on this issue between 1830 and 1848. On the contrary, what Mill had to say more generally regarding prospects for profound attitudinal change suggests a hard-headed position from an early stage. He himself insisted in his *Autobiography* that his enthusiasm for Owenite, Saint-Simonian and other anti-property doctrines in the 1830s did *not* reflect a belief in their truth (*CW*, I, 179); and he wrote to d'Eichthal: 'It appears to me utterly hopeless and chimerical to suppose that the regeneration of mankind can ever be wrought by means of working on their opinions' – especially in England (9 Feb. 1830; XII, 47).

In the *Principles* Mill also refers critically to the 'old school': He could not, he wrote, 'regard the stationary state of capital and wealth with the unaffected aversion so generally manifested toward it by political economists of the old school. I am inclined to believe that it would be, on the whole, a very considerable improvement on our present condition' (*CW*, III, 753–4). Mention is made specifically of Adam Smith and McCulloch although the complaint is generalized to economists of the 'last two generations'. This issue also reflects Mill's concern to take a broader view of possible forms of economic organization.

38 Supplement to the *Encyclopaedia Britannica*, 5th ed., first published in September 1820. Reproduced in Lively and Rees (1978).

39 *Edinburgh Review* for March (no. xcvii), June (no. xcviii), October (no. xcix); reproduced with James Mill's *Essay on Government* in Lively and Rees (1978).

criticisms, in so far as they constituted a claim for 'the empirical mode of treating political phenomena, against the philosophical', as reflecting an erroneous notion of the logic of politics. Yet he could not deny that 'there was truth in several of [Macaulay's] strictures on [James Mill's] treatment of the subject; that [the] premises were really too narrow, and included but a small number of the general truths, on which, in politics, the important consequences depend. Identity of interest between the governing body and the community at large, is not, in any practical sense which can be attached to it, the only thing on which good government depends; neither can this identity of interest be secured by the mere conditions of election.'

Mill found his father's response to Macaulay unsatisfactory: 'He did not, as I thought he ought to have, justify himself by saying, "I was not writing a scientific treatise on politics. I was writing an argument for parliamentary reform." He treated Macaulay's argument as simply irrational.' The elder Mill's inadequate response suggested that 'there was really something more fundamentally erroneous in [his] conception of philosophical Method, as applicable to politics', than J. S. Mill originally supposed (165). And Mill's researches on logic, under way since the famous 'morning conversations' at Grote's house, suggested to him where the error lay: James Mill rightly insisted upon deductive methodology – for Macaulay's experimental methodology was certainly inappropriate – but his model was that of pure geometry, where the problem of 'Composition of Forces' does not arise, rather than that of the natural sciences:

On examining . . . what the mind does when it applies the principles of the Composition of Forces, I found that it performs a simple act of addition. It adds the separate effect of the one force to the separate effect of the other, and puts down the sum of these separate effects as the joint effect. But is this a legitimate process? In dynamics, and in all the mathematical branches of physics, it is; but in some other cases, as in chemistry, it is not; and I then recollected that something not unlike this was pointed out as one of the distinctions between chemical and mechanical phenomena, in the introduction to that favourite of my boyhood, Thomson's *System of Chemistry*. This distinction at once made my mind clear as to what was perplexing me in respect to the philosophy of politics. I now saw, that a science is either deductive or experimental, according as, in the province it deals with, the effects of causes when conjoined, are or are not the sums of the effects which the same causes produce when separate. It followed that politics must be a deductive science. It thus appeared, that both Macaulay and my father were wrong; the one in assimilating the method of philosophizing in politics to the purely experimental method of chemistry; while the other, though right in adopting a deductive method, had made a wrong selection of one, having taken as the type of deduction, not the appropriate process, that of the deductive branches of natural philosophy, but the inappropriate one of pure geometry, which, not being a science of causation at all, does not require or admit of any summing-up of effects. A foundation was thus laid in my thoughts for the principal chapters of what I afterwards published on the Logic of the Moral Sciences; and my new position in respect to my old political creed, now became perfectly definite (169).

It was with these considerations in mind, Mill tells us, that he composed the essay on method in political economy, the substance of which was subsequently incorporated into the *System of Logic*.

Mill's insistence upon the qualified nature of his attraction to the Saint-Simonians has been noted. But it is unfortunate that he did not emphasize in the *Autobiography* what leaps to the eye in the correspondence for October 1829, namely the striking similarity between the criticisms of James Mill's *Essay on Government* and of the theory of government in Comte's *Système de Politique Positive* (1823):

[It] is a great mistake, a very common one too, which this sect seem to be in great danger of falling into to suppose that a few striking and original observations, are sufficient to form the foundation of a *science positive*. M. Comte is an exceedingly clear and methodical writer, most agreeable in stile, and concatenates so well, that one is apt to mistake the perfect coherence and logical consistency of his system, for truth. This power of systematising, of tracing a principle to its remotest consequences, and that power of clear and consecutive exposition which generally accompanies it, seem to me to be the characteristic excellencies of all the good French writers; and are nearly connected with their characteristic defect, which seems to me to be this: They are so well satisfied with the clearness with which their conclusions flow from their premisses, that they do not stop to compare the conclusions themselves with the fact though it is only when they will stand this comparison that we can be assured the premisses have included all which is essential to the question. They deduce politics like mathematics from a set of axioms & definitions, forgetting that in mathematics there is no danger of partial views: a proposition is either true or it is not, & if it is true, we may safely apply it to every case which the proposition comprehends in its terms: but in politics & the social science, this is so far from being the case, that error seldom arises from our assuming premisses which are not true, but generally from our overlooking other truths which limit, & modify the effect of the former (letter to Gustave d'Eichthal; XII, 35–6).

Mill found particularly unsatisfactory the 'very first and fundamental principle of the whole system, that government and the social union exist for the purpose of concentrating and directing all the forces of society to some one end'. This and similar errors vitiated Comte's attempt to establish a *positive* science of politics although Mill admired the '*partie critique*' of Comte's endeavours.

Mill's objections to Comte extended much further, for he was to charge him with attempting to construct 'universal principles of sociology' on a rigid conception of human behaviour – a pattern dictated by ignorance of 'the laws of human character formation' (see below, p. 174). These reservations were expressed at the time of publication of the *Principles of Political Economy* but they summarized the outcome of years of study of Comte's work and of direct correspondence. In this regard too, there existed close parallels with the charges directed at English social scientists.

The autobiographical account of his father's unsatisfactory reply to Macaulay in 1829 relates to the theory of government, and we cannot presume that propositions made in this context apply automatically to political economy. Yet J. S. Mill's realization – the specific outcome of the Macaulay episode – that politics required not merely deductive method, but specifically that method utilized by natural philosophy, which (unlike geometry) recognizes the principle of Composition of Forces, without doubt applied equally to political economy as is made clear in the essay on method itself. A good case can therefore be made out whereby Mill's essay was formulated as a reaction against his father's 'geometrical' constructions in economics – the failure to recognize the conditional or hypothetical nature of the assumptions which rendered the conclusions of the economist true in the abstract and requiring therefore allowance in application for 'disturbing causes' and an appropriate 'summing-up of effects'.

That this is so is further suggested by the distinction made in the essay between the 'science' and the 'art' of political economy and the allowance for an extensive empirical or inductive dimension at various stages of the deductive process – positions with which it is doubtful whether James Mill would have sympathized. For his approach – at least his 'public' approach – was 'to construct the social science *more geometrico*; starting from self-evident premises, and by a chain of syllogisms working down to universally-valid conclusions' (Schwartz, 1972, 59f; cf. also Winch, 1970, 20).[40] Indeed, his last two articles on political economy together provide the model which J. S. Mill was opposing in the essay on method. (Interestingly, he waited till his father was dead before publishing it.) Thus 'Whether Political Economy is Useful' for the *London Review* asserts that political economy

40 I am grateful to Robert Fenn for putting at my disposal letters from James Mill, dated 1833, regarding 'the usual excellence of Miss M's production's, which afforded him 'the highest satisfaction'. This approbation is revealing, at least with regard to James Mill's conception of the methodology of economics. On Martineau's *Illustrations of Political Economy*, see Scott Gordon (1971), 198: 'She could not distinguish between the "laws of Political Economy" as a simple abstract model whose focus is the heuristic one of assisting one to analyse the complex processes of the real economy; "laws" which are the findings of such an analysis; and "laws" which are ethical precepts . . . the same air of certainty, indeed necessity, is given alike to statements that are logical deductions, empirical predictions, and moral exhortations.'

is the science of both means and ends; that its method entails exhaustive definition; that its truths are certain and known to all experts, only incompetents and the 'interested' disagreeing. The tendency to take premises for granted and to assume that the propositions, theorems and laws in the field had been established is all too apparent. No criteria are offered for the establishment of 'truth', apart from the internal criteria of comprehensiveness and logical elegance. How new evidence might alter a model is certainly not touched upon. There is also a presumption regarding institutional arrangements: 'There is no dispute . . . that the whole of the annual produce falls into three shares – one to the labourers, one to the capitalists, and one to the owners of land' (1836a, 567). In the second paper, 'Theory and Practice: A Dialogue', also for the *London Review* – whose readers are 'directed to a shallow casting about in the history of human experience for regularities of conduct or of observed phenomena' (Fenn, 1982) – comprehensiveness and universality are taken as the signs of 'good theory': 'The whole business of philosophy consists in the endeavour to render each theory as comprehensive as possible . . . the proper business of philosophy is to trace every sequence as far as possible, and ascertain its greatest extent . . . And it is not at all impossible, it is on the other hand very probable, that all the changes which we observe in this world, innumerable as they seem to be, may be the result of a small number of sequences, traceable through all terrestrial beings . . . If philosophy shall ever discover these sequences, all knowledge competent to human nature will be correctly summed up in a few propositions; and mistaken practice will be no longer possible' (1836b, 231–2). This is precisely the 'French' approach that J. S. Mill set out to oppose.

Various criticisms of his father during the early 1830s provide further background to the essay. In an (unsigned) Appendix to Bulwer's *England and the English* (1833) he complained of James Mill's emphasis upon a 'common universal nature': 'We seldom learn from Mr. Mill [*Analysis of the Phenomenon of the Human Mind*] to understand any of the varieties of human nature; and, in truth, they enter very little into his own calculations, except where he takes cognizance of them as aberrations from the standard to which, in his opinion, all should conform . . . I believe the natural and necessary differences among mankind to be so great, that any practical view of human life which does not take them into account, must, unless it stop short in generalities, contain at least as much error as truth; and that any system of mental culture, recommended by such imperfect theory in proportion as it is fitted to natures of one class, will be entirely unfitted for all others' (*CW*, I, 591). It is narrowness of perspective that was particularly reprehensible: 'The faculty of drawing correct conclusions from evidence, together with the qualities of moral rectitude and earnestness, seem to constitute almost the whole of his idea of the perfection of human nature . . . We see no provision in his system, so far as it is disclosed to us, for the cultivation of any other qualities; and therefore, (as I hold to be a necessary consequence,) no *sufficient* provision for the cultivation even of these' (592). Mill conceded that his father's 'powers of metaphysical analysis' had the potential 'to divest our philosophic theories of everything like narrowness'; but, this required 'that those powers of analysis should be applied to the details, not solely to the outlines, of human nature; and one of the most strongly marked of the mental peculiarities of Mr. Mill, is, as it seems to us, impatience of details.' This is fully in line with the observation in the *Autobiography* that James Mill was prone to trust 'too much in the intelligibleness of the abstract, when not embodied in the concrete' (I, 16). (Interestingly, various contemporary observers made the same point about J. S. Mill himself; Thomas, 1979, 183.)

To Carlyle in 1834 he had described his 'intense philosophic intolerance' as a youth, reflecting 'the onesidedness of [his] understanding', as 'a school-boy fresh from the logic-school, [who] had never conversed with a reality; never seen one; knew not what manner of thing it was; had only spun, first other people's & then my own deductions from assumed premises' (XII, 204–5). The *Autobiography* describes James Mill's indignation at his thirteen-year-old son's suggestion that something 'true in theory' 'might be at variance with practice' (*CW*, I, 35). In 1844 he wrote of Molesworth, whose 'intelligence est plus déductive qu'inductive; sa nature est géomètre; il est par nature ce que j'étais il y a quinze ans par mon éducation' (XIII, 638). And we have a comparison in 1833 of Grote with James Mill: 'He is a man of good, but not first-rate intellect: hard and mechanical; not at all quick; with less *subtlety* than any able and instructed man I ever knew: with much logical and but little aesthetic culture; *narrow* therefore; even narrower than most other Utilitarians of reading and education: more a disciple of my father than of any one else . . .' (XII, 170). In 1836 he wrote to Bulwer that his father's death 'has made it far easier to do that, in the hope of which alone I allowed myself to become connected with the [*London and Westminster*] review – namely to soften the harder & sterner features of its radicalism and utilitarianism, both of which in the form in which they originally appeared in the Westminster, were part of the inheritance of the 18th century' (312).

Robson has observed that Mill 'never praised his father's method, being too aware of its limitations . . .' (1964, 267). But allowance should be made for de Marchi's recent argument (above, pp. 14–15) that there was a 'private' James Mill who belied the public image (see also Ryan, 1970, 144–5). For J. S. Mill does remark that his father 'did injustice to his own opinions by the unconscious exaggerations of an intellect emphatically polemical', and that 'when thinking without an adversary in view, he was willing to make room for a great portion of the truths he seemed to deny. I have frequently observed that he made large allowance in practice for considerations which seemed to have no place in his theory' (*CW*, I, 211). In his Introduction to the *Analysis of the Phenomena of the Human Mind* Mill similarly notes that in conversation his father 'removed difficulties which in his writings for the public he often did not think worth while to notice' (1869, I, xvi), although he also refers critically to his peculiar 'mental quality' (xix-xx). These allowances are more than a looking backwards through rose-coloured glasses. For in the 1834 letter to Carlyle itself he had noted that 'even in the narrowest of my then associates, they being older men, their ratiocinative and nicely concatenated dreams were at some point or other, & in some degree or other, corrected and limited by their experience of actual realities . . . (*CW*, XII, 205).

The positive comment in the 1833 Appendix to Bulwer on James Mill's 'powers of metaphysical analysis' is important here. (In the letter to Bulwer of 1836 he had also written of his father as the world's 'greatest philosophical genius'; XII, 312). On another occasion he referred to his father's 'power of recondite analysis' writing of him as 'a man who united the great qualities of the metaphysicians of the eighteenth century, with others of a different complexion, admirably qualifying him to complete and correct their work' ('Bentham', 1838, X, 80). In the *Autobiography* Mill warmly says of his contribution to analytical psychology that it will make him 'known to posterity as one of the greatest names in that most important branch of speculation, on which all the moral and political sciences ultimately rest . . .' (*CW*, I, 213). To Comte he included his father together with the Scottish writers Kames and Ferguson, as belonging to a better School (closer to the French) than the English philosophers (28 Jan. 1843, XIII, 566; cf. also 5 Oct. 1844, 637–8). But all this is consistent with the view of 1833 that Mill's character, particularly his 'impatience of details' vitiated the potential. In any event notwithstanding the allowances, the fact remains that he was dissatisfied with his father's inadequate response to Macaulay and that he did not hesitate to criticize the public image. There is every indication that the essay formulated in 1829–30 was designed to challenge that image in the domain of political economy.

IV THE CASE FOR 'DEDUCTIVE' SOCIAL SCIENCE

The social sciences deal with subject matter characterized more than any other (with the possible exception of physiology) by plurality and composition of causes: 'There, Plurality of Causes exists in almost boundless excess'; worse still, 'effects are, for the most part, inextricably interwoven with one another. To add to the embarrassment, most of the inquiries in political science relate to the production of effects of the most comprehensive description, such as the public wealth, public security, public morality, and the like: results liable to be affected directly or indirectly either in *plus* or in *minus* by nearly every fact which exists, or event which occurs, in human society' (*A System of Logic, CW*, VII, 452).[41] Social phenomena are, in brief, pre-eminently of such a nature which requires 'deductive' treatment, based upon an axiomatic foundation. The closest model for emulation was not geometry (as James Mill implied, above, p. 89) – 'a science of coexistent facts, altogether independent of the laws of the succession of phenomena' – but rather those 'physical Sciences of Causation which have been rendered deductive', precisely because they deal with phenomena subject to causes which counteract and modify each other rendering the net outcome complex in the extreme (VIII, 887).[42] Yet '[h]owever complex the phenomena, all their sequences and coexistences result from the laws of their separate elements. The effect produced, in social phenomena, by any complex set of circumstances, amounts precisely to the sum of the effects of the circumstances taken singly . . .' (895).

We shall show presently that the basic behavioural axiom of political economy was represented by Mill as a realistic description of actual behaviour patterns within a specifically designated area of activity. This position can be generalized to other premises of the model, on grounds which are said to characterize all deductive process in science as a whole, and not only in social science, namely that the individual causal laws governing a particular effect must be obtained 'by an induction from comparatively simple instances', those in which the operation of the individual cause is not disturbed by others whose law is unknown (VII, 458, see above, pp. 77–8). Inductions of this order provide the premises of political economy:

Thus, if the subject be social or historical phenomena, the premises of the Deductive Method must be the laws of the causes which determine that class of phenomena; and those causes are human actions, together with the general outward circumstances under the influence of which mankind are placed, and which constitute man's position on the earth. The Deductive Method, applied to social phenomena, must begin, therefore, by investigating, or must suppose to have been

---

41 Cf. VIII, 833: 'Social phenomena are those in which the plurality [of causes] prevails to the utmost possible extent.' (Mill should have written 'composition' of causes.)

42 Mill had in mind astronomy or natural philosophy by the reference to deductive physical sciences of causation.

    Cf. 895 for the contrast between social science and geometry: 'It infers the law of each effect from the laws of causation on which that effect depends; not, however, from the law merely of one cause, as in the geometrical method; but by considering all the causes which conjunctly influence the effect, and compounding their laws with one another.' The comparison is here also made with astronomy (advanced in the use of deductive procedure) and physiology (at a more elementary stage). The parallel with astronomy is qualified (878) because the data of social science are 'innumerable' and 'perpetually changing' in contrast to astronomy where they are few and change little and that little according to known laws.

already investigated, the laws of human action, and those properties of outward things by which the actions of human beings in society are determined. Some of these general truths will naturally be obtained by observation and experiment, others by deduction: the more complex laws of human action, for example, may be deduced from the simpler ones; but the simple or elementary laws will always, and necessarily have been obtained by a directly inductive process (454–5).

Unfortunately, Mill himself, as he was the first to admit,[43] did not always adequately illustrate his formulations, and thus he failed to specify – apart from the behavioural assumption – the precise axiomatic foundation envisaged for political economy during the course of his formal methodological pronouncements. But presumably he had in mind an inductive basis for the behavioural axiom and for 'technological' axioms such as the law of diminishing returns which must be derivable by 'observation and experiment' in circumstances free from the disturbances exerted by other counteracting influences, or at least from those whose laws of operation are unknown (458). As we explained earlier, the real difficulty was not in defining the individual laws by way of inductive or empirical methods, but 'to find the law of an effect, from the laws of the different tendencies of which it is the joint result' (454; above, p. 78). For this complex task the ratiocination stage of the deductive process was required.

Mill did not spare those who rejected 'abstract' theory in the social sciences in favour of a purely experimental approach – the 'chemical' method – purporting to derive causal relationships by induction from specific experience. At the very best 'such generalizations, even if well established, could not be ultimate truths, but must be the results of laws much more elementary; and therefore, until deduced from such, could at most be admitted as empirical laws, holding good within the limits of space and

43 Discussing the essay on method in correspondence, Mill pointed out that it would note serve for a popular readership unless rewritten: 'This might be done, though it scarcely suits my vocation, which is not for illustration or exemplification; I am always much too dry and abstract' (To J. P. Nichol, 14 Oct. 1834, XII, 235).
   The following observation by J. K. Whitaker (1975), 1037–8, is also pertinent here: 'Understanding of Mill's detailed views on social science method is hindered by the vagueness of his conception of a hierarchy of laws – treated by him as essentially synonymous with a hierarchy of cause and effects. In general, he envisages certain fundamental laws which stand as axioms and from which may be deduced further intermediate laws. Lower-level laws then follow from various combinations of these higher- and intermediate-level laws, and so on, down to explanations of variegated day-to-day occurrences in terms of unique combinations of superior law or causes.' In the behavioural and social sciences, psychological 'laws of the mind' are placed at the top of the hierarchy as 'empirical truths, to be ascertained by direct induction and introspection'. These laws are to be taken as they are found with no further need to rationalize them.

time by which the particular observation that suggested the generalization were bounded' (VIII, 789).

However, Mill was not only concerned by the error, however reprehensible, of 'placing mere empirical laws, and laws in which there is no direct evidence of causation, on the same footing of certainty as laws of cause and effect . . .'. More serious was the error committed by those who drew causal inferences from evidence falling far short even of 'a well-ascertained' empirical law.[44] He had in mind simple-minded attempts 'to draw conclusions of general application from a single unanalysed instance' or 'to arbitrarily refer an effect to one of several antecedents, without any process of elimination or comparison of instances' (880). Such so-called inductions reflected the error of *post hoc, ergo propter hoc* reasoning – 'causation inferred from casual conjunction, without either due elimination, or any presumption arising from known properties of the supposed agent' (792–3).[45]

It should also be added how little respect Mill had for the quality of the empirical evidence characteristically used as the basis for the misbegotten inductions: 'Whoever has carefully examined any of the attempts continually made to prove economic doctrines by . . . recital of instances, knows well how futile they are. It always turns out that the circumstances of scarcely any of the cases have been fully stated; and that cases, in equal or greater numbers, have been omitted, which could have tended to an opposite conclusion (VII, 453n).[46]

The subjects of his displeasure, Mill complained, demonstrated furthermore a total failure to appreciate the nature of the trends in science generally:

So lengthened a discussion would not [be] necessary, if the claim to decide authoritatively on political doctrines were confined to persons who had competently studied any one of the higher departments of physical science. But since the generality of those who reason on political subjects, satisfactorily to themselves and to a more or less numerous body of admirers, know nothing whatever of the methods of physical investigation beyond a few precepts which they continue to parrot after Bacon, being entirely unaware that Bacon's conception of scientific inquiry has

44 Mill sometimes took the position that there are no genuine empirical laws in social science; but this is not always the case as his references to Buckle (below, n 57) suggest.

45 Cf. 789: 'Their fallacy consists in this, that they are inductions without elimination: there has been no real comparison of instances, nor even ascertainment of the material facts in any given instance.' Examples include statements to the effect that England owed its industrial pre-eminence to commercial restriction (792–3; 880).

46 Cf. VIII, 789 and the complaint that there is typically no 'ascertainment of the material facts in any given instance'. See also the reference to the 'worthlessness of much testimony' in the essay on method (IV, 334).

done its work, and that science has now advanced into a higher stage; there are probably many to whom such remarks . . . may still be useful. In an age in which chemistry itself, when attempting to deal with the more complex chemical sequences, those of the animal and even the vegetable organism, has found it necessary to become, and has succeeded in becoming, a Deductive Science – it is not to be apprehended that any person of scientific habits, who has kept pace with the general progress of the knowledge of nature, can be in any danger of applying the methods of elementary chemistry to explore the sequences of the most complex order of phenomena in existence (VIII, 886).[47]

Even if undertaken by investigators familiar with the empirical methods of scientific enquiry, and assuming a good grasp of historical data on their part, the purely empirical method could be of no use. The obstacles in the way of such an approach in the social sciences were outlined in some detail. It must be presumed that the following observations regarding the impossibility or inefficacy of experimental procedures refer to experiments which involve attempts to discover the law of an effect where 'composition of causes' – the 'mutual interference of causes' – is at play; for Mill never ceased insisting that it has to be possible to undertake a process of observation or experiment on each cause separately for deductive work to proceed (above, pp. 97–8).

In the first place, the design of artificial experiments to ascertain the laws of social phenomena was ruled out on practical grounds: 'We cannot adapt our logical means to our wants, by varying the circumstances as the exigencies of elimination may require' (881). In any event, were it possible to contrive an experiment the probability was high that 'before sufficient time had elapsed to ascertain the result of the experiment, some material circumstance would always have ceased to be the same' so that the 'conclusiveness' of the result would be open to question. At best the researcher might fortuitously be provided with an 'experiment' produced, naturally as it were, by current and historical events: 'If the spontaneous instances, formed by contemporary events, and by the successions of phenomena recorded in history, afford a sufficient variation of circumstances, an induction from specific experience is attainable; otherwise not.' But the prerequisites for a satisfactory induction from specific experience provided naturally in this manner, Mill insisted, could not be satisfied. The point

---

47 Mill further observed that such misconceptions were 'chiefly committed by persons not much accustomed to scientific investigation: practitioners in politics, who rather employ the commonplaces of philosophy to justify their practice, than seek to guide their practice by philosophic principles: or imperfectly educated persons, who, in ignorance of the careful selection and elaborate comparison of instances required for the formation of a sound theory, attempt to found one upon a few coincidences which they have casually noticed' (887).

was made strongly in the essay on method: In the social sciences – unlike chemistry (presumably elementary chemistry) and natural philosophy – we are 'confined to the limited number of experiments which take place (if we may so speak) of their own accord, without any preparation or management of ours; in circumstances, moreover, of great complexity, and never perfectly known to us; and with the far greater part of the processes concealed from our observation' (IV, 327–8).

The essential conclusion is that, in such circumstances, inductions regarding causation from 'a comparison of the effects' are ruled out (329).[48] But much more is involved than the difficulties in the way of artificially devising experiments as we shall now show.

To illustrate his case against 'Baconian' induction from specific experience, Mill took as an illustration the (cross-section) analysis of the consequences for national wealth of restrictive commercial legislation – pre-eminently a problem involving 'composition of causes' (and thus confirming that his major objection to experimental procedure did not relate to the observation of individual causes). He proceeded to consider the applicability of the various methods of experimental enquiry (above, p. 74). The *method of difference* – the 'most conclusive' form of investigation based on specific experience – could be ruled out, for what would be required was observation of unequal levels of prosperity in the case of two nations identical in every respect except for the legislation in question, a *desideratum* impossible to satisfy (VIII, 881–2).[49] The next best method – the *indirect method of difference* – is also precluded, on the grounds that even the most satisfactory outcome (and this is also highly unlikely to be encountered) would be inconclusive (883).[50] For it would have to be presumed that national prosperity can be accounted for by a single cause, whereas it is 'the collective result of a multitude of favourable circumstances; and of these, the restrictive nation may unite a greater number than either of the others, though it may have all of these circumstances in common with either one or the other of them'.

48 Mill allows for the possibility of actually 'calculating . . . backward to a cause', but only where 'the nature of the phenomena admits of our obtaining numerical data of sufficient accuracy' (VII, 620; cf. XXI, 236). Von Thünen, however, conceived of the possibility of experiments of a logical order to isolate a cause, even when impossible empirically; cf. Blaug (1979), 27–8.

49 And if such identical cases were found, their legislative policies would be identical too since '[d]ifferences in legislation are not inherent and ultimate differences', but 'the effects of pre-existing causes'.

50 In the indirect method of differences nation A with restrictive policy would be compared with B and C, nations which allow free trade but have nothing else in common. Let it be assumed that B is identical with A in certain respects (x, y, z) only (not in all as in the direct method of difference) and C identical with A with regard to the rest (p, q, r). If it turns out that B and C are poorer than A, it cannot be in consequence of either x, y, z or p, q, r but must be because of the differing trade policy.

The *method of agreement*, which requires the unlikely case of two nations having nothing in common, except that both had protective regulations, also yielded inconclusive results. Assuming that both were observed to be prosperous nations, '[w]hat degree of presumption does this raise, that the restrictive system caused the prosperity?' The basis for such presumption would be that 'some one antecedent is the cause of a given effect, because all other antecedents have been found capable of being eliminated . . .'. But this would be satisfactory 'only if the effect can have but one cause', which was patently not the case in the social sciences where causes were 'infinitely numerous'. 'From the mere fact, therefore, of our having been able to eliminate some circumstance, – in the sense that the effect sometimes exists without it – we can by no means infer that this circumstance was not instrumental to the effect in some of the very instances from which we have eliminated it. We can conclude that the effect is sometimes produced without it; but not that, when present, it does not produce its share' (883–4).

In summary: The methods for discerning cause–effect relationships by reference to 'specific experience' involving comparison of effects are ruled out, both because they would require the artificial formulation of an experiment – the fortuitous existence of an appropriate confluence of circumstance being difficult to conceive – and because the outcome, assuming such unlikely confluences, would in any event be inconclusive in consequence of complex causality.

The foregoing analysis refers to a comparison between national states. To the argument of those who claimed that even in 'politics' there is scope for specific experiments relating to the effects of legislative enactments within given states, Mill objected that the time required for the consequences of the legislation to manifest themselves rendered it 'certain' that other determinants would alter to vitiate the experiment:

The nearest approach to an experiment in the philosophical sense, which takes place in politics, is the introduction of a new operative element into national affairs by some special and assignable measure of government, such as the enactment and repeal of a particular law. But here there are so many influences at work, it requires some time for the influence of any new cause upon national phenomena to become apparent; and as the causes operating in as extensive a sphere are not only infinitely numerous, but in a state of perpetual alteration, it is always certain that before the effect of the new cause becomes conspicuous enough to be a subject of induction, so many of the other influencing circumstances will have changed to vitiate the experiment (VII, 452–3).

To the argument that cases might yet be envisaged where the reactions were instantaneous – for example (citing Bain's *Logic*) 'when an agent suddenly introduced is almost instantaneously followed by some other changes, as when the announcement of a diplomatic rupture between two nations is followed the same day by a derangement of the money-market' – Mill countered: 'this experiment would be quite inconclusive merely as an experiment. It can merely serve, as any experiment may, to verify the conclusion of a deduction. Unless we already knew by our knowledge of the motives which act on businessmen, that the prospect of war *tends* to derange the money-market, we should never have been able to prove a connexion between the two facts . . .' (453n). The deductive science itself must 'stand sentinel' even where conclusions from specific observation existed; for '[i]n making use . . . of even the best statistical generalizations for the purpose of inferring (though it be only conjecturally) that the same empirical laws will hold in any new case, it is necessary that we be well acquainted with the remoter causes, in order that we may avoid applying the empirical law to cases which differ in any of the circumstances on which the truth of the law ultimately depends' (VIII, 908).

Mill's requirements were thus stringent. Even were experimentation feasible, he was unprepared to accept the outcome as a genuine causal law assuming a complexity of conflicting causes, as in the case of the effects of legislation on wealth. Any apparent causal relationship yielded by experiment would lack an 'explanatory' basis which must be provided by the more fundamental behavioural and technological axioms.

* * *

A word next regarding the quantitative experimental methods. First, that of '*Concommitant Variations*'. This method would require that each characteristic in society, such as national wealth, could be attributed to a distinct cause. In this case only might we attribute to each such cause 'that property of the effect which waxed when it waxed, and waned when it waned.' But this too is precluded by the nature of social phenomena: 'every attribute of the social body is influenced by innumerable causes; and such is the mutual action of the coexisting elements of society, that whatever effects any of the more important of them, will by that alone, if it does not affect the others directly, affect them indirectly. The effects, therefore, of different agents not being different in quality, while the quantity of each is the mixed result of all the agents, the variations of the aggregates cannot bear an uniform proportion to those of any one of its component parts' (884). What appears in the twentieth century to be the most important quantitative method of induction was thus ruled out of court by Mill with little ado.

What Mill had to say regarding the *method of residues* is particularly significant because it has been asserted on high authority that on this matter he failed to reconcile apparently conflicting positions. The method requires in principle that we allow for the effects of all the causes whose 'tendencies' are known so that the residue which is not accounted for by those causes must be due to the remaining circumstances – in our present case the prohibitive enactments (884–5).[51] This method, Mill insisted in this context, was *not* in fact a method of observation and experiment at all; 'it concludes, not from a comparison of instances, but from the comparison of an instance with the results of a previous deduction.' For 'it presupposes that the causes from which part of the effect proceeded are already known; and as we have shown, these cannot have been known by specific experience, they must have been learnt by deduction from principles of human nature; experience being called in only as a supplementary resource, to determine the causes which produced an unexplained residue'.[52]

As first formulated, this position is difficult to appreciate for, as Mill repeatedly insisted, the individual causes are, and must be, known from specific experience or, induction; and in fact the general method is described earlier in a fashion explicitly recognizing this allowance: '*Substract from any phenomenon such part as is known by previous inductions to be the effect of certain antecedents, and the residue of the phenomenon is the effect of the remaining antecedents*' (VII, 398; see above, p. 74, n16).[53] Viner suggested that Mill 'retracts his admission that some of the causes can be measured inductively' (1962, 119). But this is most unlikely; to have done so would have been to remove a key principle attributed to the entire deductive procedure – its dependency upon a firm foundation in individual causal tendencies derived by observation and experiment. A more likely resolution is that, by his denial that 'the causes from which part of the effect proceeded' can be known by specific experience, Mill intended the summation, or net outcome, of these individual causes – rather than their respective tendencies – rendering the formulation quite consistent with his general position, according to which, once we are dealing with a problem involving a combination of causes whose tendencies may be in conflict, we must have recourse to ratiocination (deduction) based upon premises each derived from the principles of human nature and (one should add) technological relationships.[54]

---

51 Mill refers to an application of the method by Coleridge to historical matters.

52 Mill asserts that if the principles of human nature sufficed to establish some causes, they would surely suffice for all.

53 Cf. also 397: 'Substracting from any given phenomenon all the portions which, by virtue of preceding inductions, can be assigned to known causes, the remainder will be the effect of the antecedents which have been overlooked, or of which the effect was as yet of an unknown quantity.'

54 A similar problem arises in a different context. The chapter in the *System of Logic* entitled 'Of Observation and Experiment' deals with the possibilities open to investigators seeking to satisfy 'the Baconian rule of varying the circumstances' in order to define individual cause–effect relationships. On the basis of his discussion of this issue, Mill concluded that 'in the sciences which deal with phenomena in which artificial experiments are

To return to the evaluation of the method in the context of legislative enactment, it was Mill's position that ' [i]f it be admissable to say, England must have prospered by reason of [a] prohibitory system, because after allowing for all the other tendencies which have been operating there is a portion of prosperity still to be accounted for; it must be admissable to go to the same source for the effect of the prohibitory system, and examine what account the laws of human motives and actions will enable us to give of its tendencies' (CW, VIII, 885). Thus, although legal enactment is often referred to by Mill as a 'cause', it is clearly not to be classified on a par with the basic principles; rather it alters the conditions within which these premises operate. By this formulation we bring to the fore the fundamental characteristic that whenever the problem is at one remove from the actual establishment of the individual axioms, and therefore by its nature involves complex combinations thereof, the ratiocinative process *must* be brought into play.[55] In any event, the experimental method of residues for isolating the effects due to the legislation alone must fail for 'we shall never succeed in substracting the effect of all causes but one' (886).

* * *

impossible (as in the case of astronomy), or in which they have a very limited range (as in mental philosophy, social science, and even physiology) induction from direct experience is practiced at a disadvantage in most cases equivalent to impracticality: from which it follows that the methods of those sciences, in order to accomplish anything worthy of attainment, must be to a great extent, if not principally, deductive. This is already known to be the case with the first of the sciences we have mentioned, astronomy; that it is not generally recognized as true of the others, is probably one of the reasons why they are not in a more advanced state' (VII, 384).

It is patently untrue, however, that this is the reason generally offered by Mill elsewhere for the unproductiveness of inductive methods. On the contrary, such methods are said to be possible and indeed necessary for the estalishment of the basic premises, or the laws of individual causes, upon which deductive procedure depends.

55 We may, if we wish, formally continue to represent legislation as a cause, remembering, however, Mill's qualifications regarding the derivation of individual causal laws by pure observation and experiment: 'Some of these general truths will naturally be obtained by observation and experiment, others by deduction: the most complex laws of human action, for example, may be deduced from the simpler ones . . .' (VII, 454–5). Similarly: the laws of the individual causes may have to be obtained 'by deductions from the laws of simpler causes' (the latter derived inductively) (458). More specifically for our particular case, it is 'perfectly certain, from theory, what *kind* of effects corn laws more produce, and in what general direction their influence must tell upon industrial prosperity' (VIII, 908–9; cf. Marshall (1925), 168). It may therefore formally be claimed that the effects of legislative enactments (consistently with Mill's assertions regarding the isolation of individual causes) may, after all, be obtained by a form of observation and experiment – albeit indirectly. But this is merely a matter of formal exposition; the substance is not altered – the tendencies engendered by legislative enactments can be discovered only by use of ratiocination based on empirically justified axioms.

There remains a further and (from an historical perspective) profoundly significant procedure which is implied by what Mill had to say in a discussion of 'fallacies of observation' in political economy. Here he alluded to a distinction between surface manifestations of apparent causal linkages and a deeper undercurrent of causation moving in a very different direction (a distinction made much of by Marx): 'The economical workings of society afford numerous cases in which the effects of a cause consist of two sets of phenomena: the one immediate, concentrated, obvious to all eyes, and passing, in common apprehension, for the whole effect; the other widely diffused, or lying deeper under the surface, and which is exactly contrary to the former' (VIII, 781). Examples include 'the common notion so plausible at the first glance, of the encouragement given to industry by lavish expenditure' and 'the common argument against free trade' (782).

Mill reserved his harshest criticisms for 'experimental' method and placed the blame for what he called the 'infant state' of the social sciences upon the failure fully to recognize that 'to accomplish anything worthy of attainment, [scientific methods] must be to a great extent if not principally, deductive' (VII, 384). But some of the weaknesses of experimental procedure had their counterpart in deductive analysis and these will be taken up later.

V 'ECONOMIC MAN': THE CASE FOR SPECIALIZATION

The champions of what Mill labelled the 'chemical' or 'experimental' method in the social sciences recommended resort to 'specific experience' and objected to reasoning based upon 'the principles of human nature' – referring to it as 'abstract theory' (*System of Logic*; *CW*, VIII, 879). In sharp contrast, governing Mill's approach to the (necessarily) 'deductive' investigation of social phenomena, was his belief that such phenomena reflect the 'laws' of human nature, and precisely for that reason were themselves subject to regular and predictable cause–effect relationships: 'All phenomena of society are phenomena of human nature, generated by the action of outward circumstances upon masses of human beings: and, if, therefore, the phenomena of human thought, feeling, and action, are subject to fixed laws, the phenomena of society cannot but conform to fixed laws . . .' (877). The belief, Mill feared, remained only too strongly entrenched that 'no universality and no certainty are attainable in such matters' (876) – that 'human volitions . . . can neither be foreseen, nor reduced to any canon of regularity even after they have occurred' (932).

Mill went to great lengths indeed to establish the precise limits of his position which, he feared, was prone (on a surface view) to misinterpretation as a denial of free-will: '[T]he doctrine of the Causation of human actions, improperly called the doctrine of Necessity, affirms no mysterious *nexus*, or overruling fatality', Mill explained in a summary statement; 'it asserts only that men's actions are the joint result of the general laws and circumstances of human nature, and of their own peculiar characters; those characters again being the consequence of the natural and artificial circumstances that constituted their education, among which circumstances must be reckoned their own conscious efforts' (932). An entire chapter ('Of Liberty and Necessity') was devoted to establish the position – of vital importance from Mill's own personal perspective – that 'moral freedom' did not imply unpredictability, but rather the feeling 'of our being able to modify our own character *if we wish*' (841).[56]

That the actions of human beings were subject to invariable laws – in the limited sense that 'whoever knew the motives, and our habitual

---

56 It is in this chapter that we encounter the definition of 'causation', referred to earlier (above, p. 74) as 'invariable, certain, and unconditional sequence' (837). A distinction, Mill here insisted, had to be made between this conception – which involved only 'uniformity of sequence' – and 'irresistibleness'. Thus 'the given cause will be followed by the effect, subject to all possibilities of counteraction by other causes'; '[w]hen we say that all human actions take place of necessity, we only mean that they will certainly happen if nothing prevents'; 'human actions . . . are never (except in some cases of mania) ruled by any one motive with such absolute sway, that there is no room for the influence of any other. The causes, therefore, on which action depends, are never uncontrollable; and any given effect is only necessary provided that the causes tending to produce it are not controlled' (839). By contrast, '[a] fatalist believes . . . not only that whatever is about to happen, will be the infallible result of the causes which produce it, (which is the true necessitarian doctrine), but moreover that there is no use in struggling against it, that it will happen however we may strive to prevent it'. The individual had it in his power to alter his own character, Mill maintained, in contrast to the view of the Owenites who believed that 'character is formed *for* him, and not *by* him' (840).

In a letter of May 21, 1841 (XIII, 477) Mill wrote of 'that awful shadow' – a reference to the conception of Necessity – that he had himself once been under; and in a draft of a letter to Florence Nightingale (but crossed out by Mill), dated 10 Sept. 1860 (XV, 706n) he wrote: 'It is very agreeable to me that you should have found my Logic of so much use to you, & particularly the chapter on Free Will & Necessity, to which I have always attached much value as being the writing down of a train of thought which had been very important to myself many years before, & even (if I may use the expression) critical in my own development.'

See also the important allusion in Mill's *Autobiography*, 175, 177, '. . . during the later returns of my dejection, the doctrine of what is called Philosophical Necessity weighed on my existence like an incubus. I felt as if I was scientifically proved to be the helpless slave of antecedent circumstances . . . I pondered painfully on the subject, till gradually I saw the light through it. I perceived, that the word Necessity, as a name for the doctrine of Cause and Effect applied to human action, carried with it a misleading association . . . I saw that though our character is formed by circumstances, our own desires can do much to shape these circumstances, and that what is really inspiriting and ennobling in the doctrine of freewill, is the conviction that we have real power over the formation of our own character; that our will, by influencing some of our circumstances, can modify our future habits or capabilities of willing . . . From that time I drew in my own mind, a clear distinction between the doctrine of circumstances, and Fatalism; disregarding altogether the misleading word Necessity'.

susceptibilities to them, could predict how we should will to act' – Mill took for granted as 'a mere interpretation of universal experience, a statement in words of what every one is internally convinced of' (837–8). In the establishment of this position it is difficult to avoid the impression that the notion of 'evidence' (so much favoured by the Physiocrats) intrudes a little too often: The position would be clear, it is asserted, to 'any one who is willing to take the trouble of thinking himself into the doctrine as thus stated'; it would be found to represent 'a faithful interpretation of the universal experience of human conduct, and a correct representation of the mode in which he himself, in every particular case, spontaneously interprets his own experience of that conduct' (932). But apart from such 'introspective' evidence Mill also believed there was a posteriori (statistical) 'verification' of the law of causation as far as concerned human behaviour.[57]

It was an individualist behavioural basis for the process of ratiocination upon which Mill insisted. Thus while 'the laws of the phenomena of society are, and can be, nothing but the laws of the actions and passions of human beings united together in the social state', it was none the less the case that '[m]en . . . in the state of society, are still men; their actions and passions are obedient to the laws of individual human nature' (879). More specifically, 'average' behaviour was relevant rather than the behaviour of specific individuals: 'it is generally enough to know that *most* persons act or are acted upon in a particular way; since [the statesman's] speculations and his practical arrangements refer almost exclusively to cases in which the whole community, or some large portion of it, is acted upon at once . . .' (VII, 603; cf. also VIII, 847, 890).

We come now to the nature and role of the 'wealth-maximization' hypothesis. The subject matter of political economy as outlined in the essay on definition and method encompasses, broadly speaking, the accumulation of wealth and its use in production; the determinants of labour productivity; the legal protection of property; and distribution, or 'settling the division of the produce by agreement, under the influence of competition . . . and employing certain expedients (as money, credit, &c.) to facilitate the distribution' (IV, 322). A formal, though provisional, definition of the subject is 'the science which treats the production and distribution of wealth,

---

57 He had in mind statistical evidence, much of it due to Buckle, that events (such as murder, the ratio of legitimate to illegitimate births, accidents and so forth) which appear uncertain and capricious in individual cases occur in the mass 'with a degree of regularity approaching to mathematical' (932–3). With large enough numbers the variety of individual characteristics is neutralized, and if the time span covered by the data is sufficiently brief the 'state of civilization' or the general circumstances of the country and its inhabitants, will be unchanged; in which conditions 'if human actions are governed by invariable laws, the aggregate result will be something like a constant quantity', an outcome confirmed by the evidence (923).

so far as they depend upon the laws of human nature' (318).[58] Political economy was thus a social science precluding interest in man, the individual, as such – the subject matter of 'pure mental philosophy' as far as concerns his 'purely self-regarding desires'; also precluded are those laws of human nature relating to 'the feelings called forth . . . by other individual human or intelligent beings' (affections, conscience, feeling of duty, love of approbation), and to 'the conduct of man, so far as it depends upon, or has relation to, these parts of his nature' – the subject matter of that range of pure mental philosophy upon which was based the art of morals or ethics (319). Contrasting with the science of '*social economy*' (also referred to as '*speculative politics*' and 'the natural history of mankind') which 'embraces every part of man's nature, insofar as influencing the conduct or condition of man in society', political economy was thus more narrowly conceived: 'What is now commonly understood by the term "Political Economy" is not the science of speculative politics, but a branch of that science' (320). It does not treat of the whole of man's nature as modified by the social state, nor of the whole conduct of man in society. It is concerned with him solely as a being who desires wealth, and who is capable of judging of the comparative efficacy of means for obtaining that end. It predicts only such of the phenomena of the social state as take place in consequence of the pursuit of wealth' (321). The specifications are finally formulated as wealth-maximization subject to 'two perpetually antagonizing principles':

[Political economy] makes entire abstraction of every other human passion or motive; except those which may be regarded as perpetually antagonizing principles to the desire of wealth, namely, aversion to labour, and desire of the present enjoyment of costly indulgences. These it takes, to a certain extent, into its calculations, because these do not merely, like other desires occasionally conflict with the pursuit of wealth, but accompany it always as a drag, or impediment, and are therefore inseparably mixed up in the consideration of it. Political Economy considers mankind as occupied solely in acquiring and consuming wealth; and aims at showing what is the course of action into which mankind, living in a state of society, would be impelled, if that motive, except in the degree in which it is checked by the two perpetual counter-motives above adverted to, were absolute ruler of all their actions (321–2).[59]

Mill's formulation in the closing sentence of the foregoing passage is somewhat loosely phrased, on his own terms, since not only is the science

58 Alternatively phrased: 'The science relating to the moral or psychological laws of the production and, distribution of wealth'.
59 Cf. 323: 'The science which traces the laws of such of the phenomena of society as arise from the combined operations of mankind for the production of wealth, insofar as those phenomena are not modified by the pursuit of any other object.'

based upon a specific behavioural pattern – and as such provisional – but the social phenomena under investigation are themselves limited in range specifically to the production and distribution of wealth. It is not concerned with the entire 'course of action' into which mankind thus motivated would be impelled as a literal reading suggests.[60] Indeed, an intimate connection was drawn between the wealth-maximization hypothesis and the limited range of subject matter treated by political economy. To this central issue, which touches on the matter of specialization within the social sciences, I now turn.

There existed, Mill emphasized, practical limits to the complexity of deductive models: 'the hypothetical combinations of circumstances on which we construct the general theorems of the science, cannot be made very complex, without so rapidly-accumulating a liability to error as must soon deprive our conclusions of all value'. The solution was to limit the range of applicability of the models to classes of social phenomena 'which, though influenced . . . by all sociological agents, are under the *immediate* influence, principally at least, of a few only' (*System of Logic*, VIII, 900). This constrained exercise was by itself a 'sufficiently intricate and difficult business to make it expedient to perform it once and for all, and then allow for the effect of the modifying circumstances'.

It was this practical *desideratum* that dictated the location of disciplinary boundaries in the social sciences. Different classes of social fact, Mill asserted, were dependent 'immediately and in the first resort . . . on different kinds of causes', allowing therefore for 'distinct and separate, though not independent, branches of sociological speculation'. In so far as political economy is concerned, the class of relevant social phenomena was said to be that class 'in which the immediately determining causes are *principally* those which act through the desire of wealth; and in which the psychological law *mainly* concerned is the familiar one, that a greater gain is preferred to a smaller. I mean, of course, that portion of the phenomena of society which emanate from the industrial, or productive operations of mankind; and from those of their acts through which the distribution of the products of those industrial operations takes place insofar as not effected by force, or modified by voluntary gift' (emphasis added). By ratiocination from that law of human nature combined with 'the principal outward circumstances (whether universal or confined to particular states of society) which operate on the human mind through that law, we may be enabled to explain and predict this portion of the phenomena of society, so far as they depend on

---

60 Similarly, the statement that political economy 'predicts only such of the phenomena of the social state as take place in consequence of the pursuit of wealth' is stated too broadly on Mill's own terms (321).

that class of circumstances only; overlooking the influence of any other of the circumstances of society . . .' (900–1).[61]

The same general rationalization for a specialist treatment of wealth based upon a major behavioural trait is given in the essay on method itself. Specialization is again represented as the most practical procedure in the light of causal complexity – with special reference, in the present instance, to plurality of motives:

All these operations [encompassed within 'the production and distribution of wealth'], though many of them are really the result of a plurality of motives, are considered by Political Economy as flowing solely from the desire of wealth. The science then proceeds to investigate the laws which govern these several operations, under the supposition that man is a being who is determined, by the necessity of his nature, to prefer a greater portion of wealth to a smaller in all cases, without any other exception than that constituted by the two counter-motives already specified. Not that any political economist was ever so absurd as to suppose that mankind are really thus constituted, but because this is the mode in which the science must necessarily proceed. When an effect depends upon a concurrence of causes, those causes must be studied one at a time, and their laws separately investigated, if we wish, through the causes, to obtain the power of either predicting or controlling the effect; since the law of the effect is compounded of the laws of all the causes which determine it (IV, 322).[62]

But as in the *Logic*, wealth-maximization was not regarded as one among roughly equal motives. It was the predominating influence governing a specific range of social phenomena:

The manner in which [political economy] proceeds is that of treating *the main and acknowledged end* as if it were the sole end; which, of all hypotheses equally simple, is the nearest to the truth. The political economist inquires, what are the actions

---

61 We shall consider below passages which seem to conflict with the view expressed here, by insisting that 'deductions should be from the whole and not from a part only of the laws of nature that are concerned' (pp. 117–19 below). But we shall argue that the stricter position relates to the general analysis of wealth – pertinent to a wide variety of geographical areas and eras and thus involving differing patterns of average behaviour, whereas the more lenient conditions required in political economy, the narrower behavioural base, may be accounted for by the narrower range of problems that the subject is designed to treat.

62 Thus 'the law of the centripetal and that of the tangential force must have been known before the motions of the earth and planets could be explained, or many of them predicted. The same is the case with the conduct of man in society. In order to judge how he will act under the variety of desires and aversions which are concurrently operating upon him, we must know how he would act under the exclusive influence of each one in particular.'

which would be produced by this desire, if, within the departments in question, it were unimpeded by any other. In this way a nearer approximation is obtained than would otherwise be practicable, to the real order of human affairs in those departments (323; emphasis added).

In the next section we return to the case made out for a specialist branch of political economy and the presumption that since 'the law of the effect is compounded by the laws of all the causes which determine it' there must ideally be an ultimate synthesis of the results of all the specialist branches to form a general theory of wealth.

On one occasion in his essay, Mill suggested what is, from his own perspective, an unfortunate analogy between geometry – which 'presupposes an arbitrary definition of a line, ''that which has length but not breadth'' ' – and political economy which presupposes 'an arbitrary definition of man, as a being who invariably does that by which he may obtain the greatest amount of necessaries, conveniences, and luxuries, with the smallest quantity of labour and physical self-denial with which they can be obtained in the existing state of knowledge'. Both reason 'from assumptions, not from facts'. Similarly, 'Political Economy reasons from *assumed* premises – from premises which might be totally without foundation in fact, and which are not pretended to be universally in accordance with it' (325). The point to be emphasized, however, is that 'arbitrary' must not be read to imply 'purely hypothetical', but only a less than perfect description of actual behaviour; and while the premises in question might be totally without foundation in some circumstances, this was not in fact the case as far as concerned the contemporary science designed as it was, we shall see, to deal with conditions as they existed in contemporary Great Britain.

To confirm this interpretation we might refer back to what Mill had to say in the *Logic* regarding the disadvantages of adopting implausible hypotheses involving both a wholly fictitious agent or cause, and a wholly fictitious law of its operation (VII, 490; above, pp. 80–1). But everything that has been said thus far of the predominating influence attributed to the behavioural axiom in the production and distribution of wealth rules out the notion of an 'arbitrary definition of man' in the literal sense of that term. This Mill confirms himself, quite specifically, in his comments in the essay on method: 'We are aware that the . . . expression is sometimes used to characterize a supposed mode of philosophizing, which does not profess to be founded upon experience at all. But we are not acquainted with any mode of philosophizing, on political subjects at least, to which such a description is fairly applicable' (IV, 325).[63] The wealth-

---

63 Here Mill adds that a posteriori method focusses upon specific experience (rather than mere experience), implying that the evidence appropriate as the basis for deductive

maximization hypothesis indeed entailed an abstraction by excluding all other motivating forces, acting upon wealth; but the results would be trifling 'if the assumed circumstances bear no sort of resemblance to any real ones'. The assumption was correct 'as far as it goes' – differing from the truth 'as a part differs from the whole'. Precisely for this reason the results derived from ratiocination on the basis of the behavioural assumption constituted '*abstract* truth' and when completed by adding or subtracting the effect of the non-calculated circumstances' would be rendered 'true in the concrete' (329). Alternatively expressed: 'In proportion as the actual facts recede from the hypothesis, [the investigator] must allow a corresponding deviation from the strict letter of his conclusion . . . That what is true in the abstract, is always true in the concrete with proper *allowances*. When a certain cause really exists, and if left to itself would infallibly produce a certain effect, that same effect, *modified* by all the other concurrent causes, will correctly correspond to the result really produced' (326–7).

Mill touches here upon the treatment of 'disturbing causes' – in the present case behavioural patterns left out of the formal account. The distinction between this category of causes and those causes incorporated into the model formally as axioms is governed by empirical circumstances. For what distinguishes the latter from 'modifying' or 'disturbing' causes is their predominating influence, in the sense that the class of phenomena under investigation (the production and distribution of wealth, in our case) depends largely upon them, combined with their ubiquity – that they are 'causes common to the *whole class* of cases under consideration' (326).[64] In the *Logic*, Mill observed further that allowance for modifying circumstances at a subsequent stage of investigation was particularly desirable 'as certain fixed combinations of the former' – influences common to all cases – were 'apt to recur often, in conjunction with ever-varying circumstances of the latter class: namely the less important or less ubiquitous influences (VIII, 901).[65] Consistently with this categorization the 'desire for wealth', the

reasoning can be of a higher degree of generality – evidence supporting the general behavioural assumptions for example derived from observation of business activity.

64 On occasion Mill expressed himself much more strongly, implying universal applicability, but this expression is not typical: 'To render [political economy] perfect as an abstract science, the combination of circumstances which it assumes, in order to trace their effects, should embody all the circumstances that are common to all cases whatever, and likewise all the circumstances that are common to any important class of cases' (329).

65 In his review of August Comte, *Cours de Philosophie Positive* in 1865, Mill took a somewhat different though not contradictory position: 'When an effect depends on several variable conditions, some of which change less, or more slowly, than others, we are often able to determine, either by reasoning or by experiment, what would be the law of variation of the effect, if its changes depended only on some of the conditions, the remainder being supposed constant' (X, 309).

basic axiom of political economy, was accorded a predominant role in so far as concerns industrial or productive operations.

* * *

We must now consider the source of our knowledge regarding the relative significance of the various motive forces, and attempt to specify more precisely the intended scope of the subject matter of 'political economy'.

'Introspection' is one form of general experience – the basis for the (individual) causal laws or premises – to which Mill alluded. Deductive theorists based themselves on 'an observation of the tendencies which human nature has manifested in the variety of situations in which human beings have been placed, and especially observation of what passes in our own mind' (IV, 325). Similarly: 'The desires of man, and the nature of the conduct to which they prompt him, are within the reach of our observation. We can also observe what are the objects which excite those desires. The materials of this knowledge every one can principally collect within himself; with reasonable consideration of the differences, of which experience discloses to him the existence, between himself and other people' (329). In the *Logic* Mill wrote more generally, that 'axioms are but a class, the most universal class, of inductions from experience; the simplest and easiest cases of generalization from the facts furnished to us by our senses, or by our internal consciousness' (VII, 252).

However, great care must be exercised in our interpretation of the term 'introspection', for there is more genuine empirical content intended thereby than meets the eye. For introspection must be sharply distinguished from 'intuition'; Mill divorced himself from any such source of knowledge as 'direct intuition': 'I have endeavoured to keep clear so far as possible of the controversy respecting the perception of the highest Realities by direct intuition, confining Logic to the laws of the investigation of truth by means of extrinsic evidence whether ratiocinative or inductive. Still, I could not avoid conflict with some of the subordinate parts of the supersensual philosophy . . .' (letter of Sept. 1839; XIII, 406).[66] It is consistent with

66 Mill was writing of the *Logic* which he expected to complete the following year.
See also letter to Theodor Gomperz, (19 Aug. 1854, XIV, 239–40) regarding the *Logic*: '[T]o give to the cultivators of physical science the theory of their own operations, was but a small part of the object of the book and that any success in that attempt was chiefly valued to me as a necessary means towards placing metaphysical & moral science on a basis of analysed experience, in opposition to the theory of innate principles, so unfortunately patronised by the philosophers of your country, & which through their influence has become the prevailing philosophy throughout Europe. I consider that school of philosophy as the greatest hindrance to the regeneration so urgently required, of man and society; which can never be effected under the influence of a philosophy which makes

this perspective that evidence drawn from introspection varies with time and place and does not reflect some universal psychological propensity acting in a vacuum, as it were, divorced from the peculiar circumstances of time and location. That this was Mill's position is strongly suggested by what he had to say regarding the diverse behaviour patterns to be observed geographically, and the potential malleability of human behaviour.

In the following passage from the *Logic*, relating to the appropriate axiomatic basis in alternative cultural frameworks, the source of knowledge in direct observation could not have been expressed more graphically. The observation is particularly important for us since it relates to 'money-getting' activities:

In political economy . . . empirical laws of human nature are tacitly assumed by English thinkers, which are calculated only for Great Britain and the United States. Among other things, an intensity of competition is constantly supposed, which, as a general mercantile fact, exists in no country in the world except those two. An English political economist, like his countrymen in general, has seldom learned that it is possible that men, in conducting the business of selling their goods over a counter, should care more about their ease or their vanity than about their pecuniary gain. Yet those who know the habits of the Continent of Europe are aware how apparently small a motive often outweighs the desire of money-getting, even in the operations which have money-getting as their direct object (VIII, 906).

Mill also heaped scorn on historians who reasoned as if human behaviour was unchanged and unchangeable over time – 'persons whose acquaintance with moral and social facts is confined to their own age, [but] take the men and the things of that age for the type of men and things in general' (791); and reiterated his opinion regarding 'those who generalize empirically from the people of their own country to the people of other countries, as if human beings felt, judged, and acted everywhere in the same manner'.[67] It is unlikely, under these circumstances, whether the

opinions their own proof, and feelings their own justification. It is, besides, painful to see such a mass of cultivated intellect, and so great an educational apparatus, as exist in your country, wasted in manufacturing a false appearance of science out of purely subjective impressions.'

67 Cf. Mill's allusion to 'the wonderful pliability and amenability to artificial discipline, of the human mind', in his review of Grote's *History of Greece* (1846, *CW*, XI, 302). There is a pertinent observation in the *Principles of Political Economy*, regarding this issue: 'There are so many new elements at work in society, even in those deeper strata which are inaccessible to the mere movements on the surface, that it is hazardous to affirm anything positive on the mental state or practical impulses of classes and bodies of men, when the same assertion may be true to-day, and may require great modifications in a few years time' (II, 346; cf. 351 regarding the growing intelligence of factory workers). There is a striking appeal for open-mindedness in this respect in Mill's Inaugural

'observation of what passes in our own mind' would turn out to be the same if, for example, the observer were an Englishman or a Frenchman (below, p. 135, regarding 864).[68]

Moreover, statistical data revealing regularity of behaviour patterns in the mass, to which we referred earlier (above, p. 106), related to limited geographical areas and time periods. They did not reflect an invariant 'general moral average of mankind' over time and from place to place (935). Nor did they imply that 'great men' or governments could do nothing to influence social progress, including the intellectual development of the race:

> However universal the laws of social development may be, they cannot be more universal or more rigorous than those of the physical agencies of nature; yet human will can convert these into instruments of its designs, and the extent to which it does so makes the chief difference between savages and the most highly civilized people. Human and social facts, from their more complicated nature, are not less, but more, modifiable, than mechanical and chemical facts; human agency, therefore, has still greater power over them. And accordingly, those who maintain that the evolution of society depends exclusively, or almost exclusively, on general causes, always include among these the collective knowledge and intellectual development of the race (936-7).[69]

> Address as honorary president of St Andrews University (1867): 'We are not likely to correct any of our opinions or mend any of our ways, unless we begin by conceiving that they are capable of amendment: but merely to know that foreigners think differently from ourselves, without understanding why they do so, or what they really do think, does but confirm us in our self-conceit, and connect our national vanity with the preservation of our own peculiarities. Improvement consists in bringing our opinions into nearer agreement with facts; and we shall not be likely to do this while we look at facts only through glasses coloured by those very opinions. But since we cannot divest ourselves of preconceived notions, there is no known means of eliminating their influence but by frequently using the differently coloured glasses of other people: and those of other nations, as the most different, are the best' (XXI, 226).

68 There is considerable evidence that introspection was by no means for Mill a casual exercise. Thus his remarks regarding Bentham: 'Bentham's knowledge of human nature is bounded. It is wholly empirical; and the empiricion of one who has had little experience. He has had neither internal experience nor external; the quiet, even tenor of his life and his healthiness of mind, conspired to exclude him from both . . . How much of human nature slumbered in him he knew not, neither can we know. He had never been made alive to the unseen influences which were acting on himself, nor consequently on his fellow creatures' (1838, *CW*, X, 92).

69 Mill's protestation against simple-minded '*post hoc ergo propter hoc*' inductive generalizations referred to earlier (below, p. 117), includes a case against the notion regarding human behaviour that 'whatsoever has never been, will never be' (788). Conversely, he condemned generalizations made from the present to the past, the practice of those who 'take the men and the things of that age for the type of men and things in general, and apply without scruple to the interpretation of the events of history, the empirical laws which represent sufficiently for daily guidance the common phenomena of human nature at that time and in that particular state of society' (791).

This position, too, points away from any conception of Mill's position regarding the behavioural axioms which emphasizes some absolute psychological quality independent of empirical circumstances.

These same considerations also are relevant for the resolution of a perennial difficulty perceived in Mill's studies, namely whether by the central behavioural axiom Mill intended 'self-interested', in the sense of egoistic, motivation or whether the axiom was neutral in this regard, alluding simply to that kind of behaviour dictated by wealth maximization whatever the actual motive might be.

Even the formulation of the axiom as 'a desire for wealth' or as 'a preference for a greater portion of wealth to a smaller' – expressions which seem to have been used interchangeably – in fact bear no necessary egoistic connotation since the use to which wealth is put might imply 'other-regarding concerns'.[70] But equally important is the assertion encountered above that 'the desires of men, and the nature of the conduct to which they prompt him are within the reach of our observation' – as indeed they have to be for the entire deductive operation to proceed; it is direct observation that indicated a sharp contrast between the actual behavioural pattern common in business practice on the Continent, an observed failure to maximize pecuniary gain despite opportunity to do so, and that common in Great Britain – a comparison which, precisely because it is limited to observed activity carries no implications whatsoever regarding motive, whether self- or other-regarding.[71] And as we have seen, this conclusion is unaffected by the allowance for 'introspection' as a source of empirical knowledge.

70 Mill took for granted that 'the strongest propensities of uncultivated or half-cultivated human nature' were 'the purely selfish ones, and those of a sympathetic character which partake most of the nature of selfishness' (926) – a position pointing away from egoism as a central trait in contemporary British conditions.

71 It may be remarked here that Mill repeatedly protested in his writings against the common identification of utilitarianism and selfishness. Thus for example his reactions to James Martineau's review of Bentham's *Deontology*: 'I dissent . . . from his adoption of what is called the selfish system, and which he has put under the same mantle as the utilitarian doctrine' (to Nichol, 14 Oct. 1834, *CW*, XII, 236); and to Sedgwick's 'Discourse on the Studies of the University of Cambridge' (1835): 'What has "calculating the chances of personal advantage" to do with the principle of utility?" ' (X, 71). To Hickson of the *Westminster Review* Mill wrote: 'I wonder that you as the representative of the old sterling Westminster Review opinions, should have allowed to be printed in it vulgar misrepresentation of Bentham, its founder; vilifying a man who has done more for the world than any man in modern times, by talking about "the Gospel according to Jeremy Bentham" as synonymous with the most grovelling selfishness. (There is no selfishness in Bentham's doctrines)' (15 Oct. 1851, XIV, 78).

See also his review of Whewell on Moral Philosophy (1852, X, 183) and *Utilitarianism* (1861, X, 216, 218).

VI SPECIALIZATION AND THE PROBLEM OF DISTURBING CAUSES

The major outcome of our argument is that, far from being represented as of 'universal' relevance, the axiomatic foundation of political economy was said to be pertinent only to well-defined environmental conditions. This same conclusion can be arrived at from another direction; unless this interpretation is accepted, we are forced to conclude that Mill's methodological position suffers from a ruinous internal contradiction. This issue must be carefully laid out.

As we have seen, the outcome of the specialist ratiocinative model constituting political economy – a model based upon the wealth-maximization hypothesis as the 'sole end' of activity – is represented by Mill as an approximation. This approximation has then to be qualified in application to particular cases because of 'disturbing causes', including conflicting behavioural patterns:

So far as it is known, or may be presumed, that the conduct of mankind in the pursuit of wealth is under the collateral influence of any other of the properties of our nature than the desire of obtaining the greatest quantity of wealth with the least labour and self-denial, the conclusions of Political Economy will so far fail of being applicable to the explanation or prediction of real events until they are modified by a correct allowance for a degree of influence exercised by the other causes (*CW*, IV, 323).

When the principles of Political Economy are to be applied to a particular case, then it is necessary to take into account all the individual circumstances of that case; not only examining to which of the sets of circumstances contemplated by the abstract science the circumstances of the case in question correspond, but likewise what other circumstances may exist in that case, which not being common to it with any large and strongly-marked class of cases, have not fallen under the cognizance of the science. These circumstances have been called *disturbing causes* (330).

Indeed, Mill went so far as '[t]o verify the [wealth maximization] hypothesis itself *a posteriori*, that is to examine whether the facts of any particular case are in accordance with it, is no part of the business of science at all, but of the *application* of science' (325). Those behavioural assumptions involving motives other than wealth maximization, which constitute disturbing causes from the perspective of political economy, belonged, however, 'to some other science' (331).

The foregoing position at first sight appears difficult to appreciate. Mill had many complaints against purely experimental procedure; these we have reviewed in detail. But he also took issue with some practitioners of deductive procedure. His major objection was to explanations of complicated social phenomena by too simple a theory, the counterpart in a priori reasoning to the faulty generalization – *post hoc, ergo propter hoc* – of the a posteriori method (VIII, 792–3). The source of the error was seen to be the representation of geometry as the appropriate model for emulation by the social sciences rather than astronomy or natural philosophy (above, p. 89); precisely because the social sciences dealt with phenomena subject to conflicting causes which counteracted and modified each other, any deductive procedure which drew its conclusions from 'the laws of human nature' – in itself of course desirable – but from a too restricted conception thereof would be inappropriate: 'the deductions should be from the whole and not from a part only of the laws of nature that are concerned . . .' (893–4). This was so 'even if those omitted were so insignificant in comparison with the others, that they might, for most purposes and on most occasions, be left out of the account'. But in actuality, 'the phenomena of society do not depend, in essentials, on some one agency or law of human nature, with only inconsiderable modifications from others. The whole of the qualities of human nature influence those phenomena, and there is not one which influences them in a small degree. There is not one, the removal or any great alteration of which would not materially affect the whole aspect of society, and change more or less the sequences of social phenomena generally.' Desirable (deductive) procedure was, therefore, one which – 'conformably to the practice of the more complex physical sciences' – draws its deductions 'from many, and not from one or a few, original premises; considering each effect as (what it really is) an aggregate result of many causes, operating sometimes through the same, and sometimes through different mental agencies, or laws of human nature'.

It was the Bentham school and its 'interest philosophy' – defined by Mill as the one comprehensive premise that '[a]ny succession of persons, or the majority of any body of persons, will be governed in the bulk of their conduct by their personal interest' – that he had in mind (890):[72] 'Although . . . the private interest of the rulers or of the ruling class is a very powerful force, constantly in action, and exercising the most important influence upon their conduct; there is also, in what they do, a large portion which that private interest by no means affords a sufficient explanation of: and even the particulars which constitute the goodness or badness of their

72 Mill claimed that by so phrasing the doctrine he was giving it the benefit of the doubt at various points. In its crudest form it maintained 'that men's actions are always determined by their interests'. Bentham was unclear whether this referred to their 'private, or worldly interest' or 'anything a person likes', but could be interpreted to mean the former; secondly, the doctrine could be read to refer to average behaviour since it was self-evident that specific individuals are not governed in all their actions by their worldly interests.

government, are in some, and no small degree, influenced by those among the circumstances acting upon them, which cannot, with any propriety, be included in the term self-interest' (VII, 891–2). To base a theory of government and make proposals for reform on the assumption of self interest was 'unscientific' (892).[73] The qualified nature of the objection does little to reduce its force:

It is not to be imagined possible, nor is it true in point of fact, that these philosophers regarded the few premises of their theory as including all that is required for explaining social phenomena, or for determining the choice of forms of government and measures of legislation and administration. They were too highly instructed, of too comprehensive intellect, and some of them of too sober and practical a character, for such an error. They would have applied, and did apply, their principles with innumerable allowances. But it is not allowances that are wanted. There is little chance of making due amends in the superstructure of a theory for the want of sufficient breadth in its foundation. It is unphilosophical to construct a science out of a few of the agencies by which the phenomena are determined, and leave the rest to the routine of practice or the sagacity of conjecture. We either ought not to pretend to scientific forms, or we ought to study all the determining agencies equally, and endeavour, so far as it can be done, to include all of them within the pale of science; else we shall infallibly bestow a disproportionate attention upon those which our theory takes into account, while we mis-estimate the rest, and probably underrate their importance (893).

Mill did not object to the Benthamite recommendations regarding Parliamentary reform in the particular circumstances ruling in England and elsewhere in modern Europe. It was the portrayal of the recommendations as deductions from 'a complete theory' that he rejected. And it was a realization of this characteristic in James Mill's work on government and in the French literature that stimulated Mill to compose the essay (above, p. 90).[74]

We are apparently faced by a serious paradox. The entire case for 'distinct and separate, though not independent, branches of sociological speculation', we have seen earlier, was based upon the presumption that different classes of social phenomena were dependent 'immediately and in the first resort, on different kinds of causes. And this, it will be recalled, despite 'the universal

---

73 The theory of government involved two general principles: first, that the actions of the average ruler is determined solely by self-interest; second, that the sense of identity of interest with the governed is producible only by accountability to the governed. Mill objected to both premises.

74 It is not certain that Mill was fair to Bentham considering the latter's *Influence of Time and Place in Matters of Legislation*. It cannot be excluded that Mill was using Bentham as proxy for criticism of his own father.

consensus of the social phenomena, whereby nothing which takes place in any part of the operations of society is without its share of influence on every other part', and despite 'the paramount ascendency which the general state of civilization and social progress in any given society must . . . exercise over all the partial and subordinate phenomena' (900–1; also VII, 452 cited above, p. 100). Yet it is precisely these latter characteristics of social science which govern Mill's criticisms of deductive procedures based upon 'a few [only] of the agencies by which the phenomena are determined'. His appeal was either to abandon a pretext of science or 'study all the determining agencies equally'.[75]

We would be closer to a solution were Mill's strictures against the Benthamites concerned solely with irresponsible applications – particularly advice regarding policy – based upon conclusions drawn by the deductive exercise from partial assumptions without the necessary qualifications. In his own account of political economy, Mill insisted that the policy advisor must have in mind the kinds of complexities created by 'disturbing' causes, warning that the specialist political economist must fail if he attempted application on the basis of his necessarily restricted model: 'No one who attempts to lay down propositions for the guidance of mankind, however perfect his scientific acquirements, can dispense with a practical knowledge of the actual modes in which the affairs of the world are carried on, and an extensive personal experience of the actual ideas, feelings, and intellectual and moral tendencies of his own country and of his own age. The true practical statesman is he who combines this experience with a profound knowledge of abstract political philosophy. Either acquirement, without the other, leaves him lame and impotent if he is sensible of the deficiency; renders him obstinate and presumptious if, as is more probable, he is entirely unconscious of it' (IV, 333). But it is unlikely whether we will find the explanation of our problem in this direction, since Mill specifically conceded that the Benthamites 'would have applied, and did apply, their principles with innumerable allowances'. His argument against them was precisely that 'it is not allowances that are wanted', that '[t]here is little chance of making due amends in the superstructure of a theory for the want of sufficient breadth in its foundations', and that one could not leave the causal agencies neglected by the formal analysis 'to the routine of practice' (VIII, 893).[76]

---

75 For an excellent summary of Benthamite 'ratiocination' in politics, and Mill's objections in the *Logic* see Finer (1982), 32–4.

76 Mill also absolved the Benthamites of confusing 'science' with 'art' – of failing to recognize that rules of conduct must follow, not precede, the ascertainment of laws of nature (889–90). He did complain, however, of attempts by 'a large proportion of those who had laid claim to the character of philosophic politicians . . . not to ascertain universal sequences, but to frame universal precepts. They have imagined some one form of government, or system of laws, to fit all cases; a pretension well meriting the ridicule

We are no closer to a resolution, for this seems to be precisely what he suggested for political economy.[77]

Let us try to get to grips with this apparent contradiction. The main complaint against the Benthamite 'interest-philosophy' was its supposed predominance throughout time and space – its universality. Theoretical work of this order undertaken on such a narrow behavioural basis is represented as involving 'a kind of error to which those are peculiarly liable whose views are the largest and most philosophical: for exactly in that ratio are their minds more accustomed to dwell upon those laws, qualities, and tendencies, which are common to large classes of cases, and which belong to all place and all time; while it often happens that circumstances almost peculiar to the particular case or era have a far greater share in governing that one case' (IV, 333). Qualifications and allowances in application for 'disturbing causes' would not suffice. But this criticism would not apply to political economy, if we envisage it as a science of relatively narrow scope not designed to be of universal relevance – even as a first approximation. The procedure of distinguishing between the wealth-maximization and other behavioural patterns, it will be recalled, reflected a quest for practicality in the face of the problem of 'composition of causes', and was valid only in the light of the empirical 'relevance' of the wealth-maximization hypothesis in that temporal and geographic environment with which the model was designed to deal.[78] The statement that the verification of the hypothesis is 'no part of the business of science, but of the application of science', taken literally, is too loosely formulated, since the model of political economy was indeed constructed to reflect, in its assumptions, an empirical reality. The statement is, however, an accurate reflection of Mill's position if understood as a warning that even within the general environment of an advanced competitive capitalist system for which the model was designed, other motives may be at play in special cases so that the applied economist must be ever on the alert.

Mill himself formulated the precise conditions under which the specialist procedure of political economy is legitimate. Essentially it must be

with which it is treated by practitioners . . .' (876) – an allusion probably in part to James Mill.

77 Cf. also Viner's (1958) reference to Mill's 'dilemma': on the one hand he subscribed to the view that the scientific basis for economics turned on the axiom of 'competition'; on the other hand he conceded that it was not 'philosophic' to base a science on a few agencies alone (328–9).

78 For an excellent statement of the distinction in question see Schumpeter (1954), 429. Here he contrasts the 'unrelieved nonsense' of James Mill's purely speculative rational maximizer of the *Essay on Government* with the maximization axiom of economic theory: '[A]ny theory involves abstractions and therefore will never fit reality exactly, hence economic theory is inevitably unrealistic in this sense; but its premises are induced from realistic observation of the profit-seeking and calculating businessman'.

empirically the case that the range of specialist study encompasses a sufficiently homogeneous pattern of behaviour:

the process of dividing off the social science into compartments, in order that each may be studied separately, and its conclusions afterwards corrected for practice by the modifications supplied by the others, must be subject to at least one important limitation. Those portions alone of the social phenomena can with advantage be made the subjects, even provisionally, of distinct branches of science, into which the diversities of character between different nations or different times enter as influencing causes only in a secondary degree. Those phenomena, on the contrary, with which the influences of the ethological state of the people are mixed up at every step (so that the connexion of effects and causes cannot be even rudely marked out without taking those influences into consideration) could not with any advantage, nor without great disadvantage, be treated independently of political ethology, nor, therefore, of all the circumstances by which the qualities of a people are influenced (VIII, 906).

Similarly, the specialist exercises are 'liable to fail in all cases in which the progressive movement of society is one of the influencing elements', for this movement is implicitly frozen within *ceteris paribus* conditions (916).

To appreciate fully the foregoing condition for specialist social science we must attend to the general context in which it appears, namely that of a plea for the development of a science of human character formation ('ethology'), and of a general science of society – the primary objective of social investigation. To these matters we shall turn presently. For our purposes here, it suffices to emphasize that the separate science of political economy, with its basis in wealth-maximization, was conceived as part of the *first stage* of construction of a general theory of wealth – a first stage allowing for a variety of specialist treatments based on alternative axiomatic foundations – which would incorporate a wide range of behavioural patterns. All this is very clearly expressed in the essay:

The method of the practical philosopher consists, . . . of two processes; the one analytical, the other synthetical. He must *analyze* the existing state of society into its elements, not dropping and losing any of them by the way. After referring to the experience of individual man to learn the *law* of each of these elements, that is, to learn what are its natural effects, and how much of the effect follows from so much of the cause when not counteracted by any other cause, there remains an operation of *synthesis*; to put all these effects together, and, from what they are separately, to collect what would be the effect of all the causes acting at once (IV, 336).[79]

79 Cf. 322 (cited above, p. 109): 'When an effect depends upon a concurrence of causes, those causes must be studied one at a time, and their laws separately investigated, if

Those contemporary political economists who suggested by their practice and preaching that their subject, though founded on axioms reflecting contemporary conditions, was of universal relevance were sharply criticized. This kind of false perspective Mill hoped would weaken with a better appreciation of ethology: 'The more highly the science of ethology is cultivated, and the better the diversities of individual and national character are understood, the smaller, probably, will the number of propositions become, which it will be considered safe to build on as universal principles of human nature' (VIII, 906). But Mill's own conception of political economy eschewed all universalist connotation. At the same time Mill insisted upon the utility of the subject constructed on a relatively narrow axiomatic base, and this for two major reasons: first, because of 'the value of the propositions, considered with reference to the state of society from which they were drawn'; and second because while 'many of its conclusions are only locally true, its method of investigation is applicable universally . . .' (904; see Section XII). Mill was unprepared to countenance criticism of deductive theory simply on grounds of the restricted institutional range of its basic assumptions, while at the same time he himself warned against neglect of this characteristic restriction.

* * *

The problem posed at the outset of this section is thus less serious than it would be had Mill taken the Benthamites to task for constructing a narrowly-based theoretical structure of supposedly universal relevance, at the same time that he himself proposed a general theory of wealth – of universal applicability – based on the wealth-maximization hypothesis.[80] But there is also a second consideration, to which we devote the remainder of this section, which further diminishes the apparent contradiction formulated above.

How seriously must we take the formal demarcation line between those 'causes' to be incorporated within 'political economy' and those which are to be treated as 'disturbances' and the subject matter of other sciences? On close consideration it turns out that the distinction between the categories is far from clear-cut; indeed, Mill himself was uneasy about the very designation 'disturbing' attached to the excluded causal influences. I revert here again to the matter of 'composition of causes', and to the insistence

we wish, through the causes, to obtain the power of either predicting or controlling the effect; since the law of the effect is compounded of the laws of all the causes which determine it.'

80 On this general issue see also the comments by Collini, in Collini, Winch and Burrow (1983), 143.

upon the term 'tendency' to indicate the continuous operation of possibly conflicting causal influences (above, p. 73). 'The habit of neglecting this necessary element in the precise expression of the laws of nature', Mill complained, 'has given birth to the popular prejudice that all general truths have exceptions; and much unmerited distrust has thence accrued to the conclusions of science, when they have been submitted to the judgment of minds insufficiently disciplined and cultivated. The rough generalizations suggested by common observation usually have exceptions; but principles of science, or in other words, laws of causation, have not' (VII, 445). The matter was developed at length in the essay:

With regard to *exceptions*; in any tolerably advanced society there is properly no such thing as an exception. What is thought to be an exception to a principle is always some other and distinct principle cutting into the former: some other force which impinges against the first force, and deflects it from its direction. There are not a *law* and an *exception* to that law – the law acting in ninety-nine cases, and the exception in one. There are two laws, each possibly acting in the whole hundred cases, and bringing about a common effect by their conjunct operation. If the force which, being the less conspicuous of the two, is called the disturbing force, prevails sufficiently over the other force in some one case, to constitute that case which is commonly called an exception, the same disturbing force probably acts as a modifying cause in many other cases which no one will call exceptions (IV, 337–8).

By adopting this position, however, Mill seriously diluted the distinction between the two categories of causal influence. And this is strongly suggested in a passage which to all intents and purposes treats the two on a par:

When the disturbing causes are known, the allowance necessary to be made for them detracts in no way from scientific precision, nor constitutes any deviation from the *a priori* method. The disturbing causes are not handed over to be dealt with by mere conjecture . . . The disturbing causes have their laws, as the causes which are thereby disturbed have theirs; and from the laws of the disturbing causes, the nature and amount of the disturbance may be predicted *a priori*, like the operation of the more general laws which they are said to modify or disturb, but with which they might more properly be said to be concurrent. The effect of the special causes is then to be added to, or subtracted from, the effect of the general ones (330).

Thus disturbing causes might be compared to friction in mechanics: 'they may at first have been considered merely as a non-assignable deduction to be made by guess from the result given by the general principles or the science; but in time many of them are brought within the pale of the abstract science itself, and their effect is found to admit of as accurate an estimation as those more striking effects which they modify' (331). Indeed, those disturbing causes which operate through the desire for wealth might even be absorbed within the abstract model: 'As for the . . . kind of disturbing causes . . . which operate through the same law of human nature out of which the general principles of the science arise, these might always be brought within the pale of the abstract science, if it were worthwhile; and when we make the necessary allowances for them in practice, if we are doing anything but guess, we are following out the method of the abstract science into minuter details; inserting among its hypotheses a fresh and still more complex combination of circumstances, and so adding *pro hâc vice* a supplementary chapter or appendix, or at least a supplementary theorem, to the abstract science'.

The foregoing qualification is, however, apparently of limited applicability. For, Mill proceeds, should the disturbing cause reflect 'some other law of human nature, it . . . belongs to some other science'. Political economy was thus to retain its independent status on the basis of wealth-maximizing behaviour (albeit qualified). Yet it is difficult to take this restriction literally, for Mill himself, earlier in the essay, had recognized explicitly that the axiomatic foundation of contemporary political economy included some behavioural assumptions in conflict with wealth maximization. Thus the approximations based upon the wealth-maximization hypothesis are, as we have seen, 'to be corrected by making proper allowance for the effects of any impulses of a different description, which can be shown to interfere with the result in any particular case'; but Mill then allowed that 'in a few of the most striking cases (such as the important one of the principle of population)' these corrections 'are . . . interpolated into the expositions of Political Economy itself; the strictness of purely scientific arrangement being thereby somewhat departed from, for the sake of practical utility' (323). And, of course, he allowed for the 'two perpetually antagonizing principles to the desire for wealth' – 'aversion to labour' and 'the desire of the present enjoyment of costly indulgences'. It is simply not the case that all motives apart from the desire for wealth – even when they entail conflicting 'tendencies' – are relegated 'to some other science' to be treated as 'disturbing causes' from the point of view of political economy.

It is also pertinent to our theme that Mill seems hesitant to specify the particular specializations which would in principle complement political economy.[81] 'One suspects', one recent commentator has written, 'that he

81 Cf. 'I would not here undertake to decide what other hypothetical or abstract sciences similar to Political Economy, may admit of being carved out of the general body of the social science; what other portions of the social phenomena are in a sufficiently close and complete dependence, in the first resort, on a peculiar class of causes, to make it convenient to create a preliminary science of those causes; postponing the consideration of the causes which act through them, or in concurrence with them, to a later stage of

had no clear idea' (Whitaker, 1975, 1041). As Whitaker suggests (1043–4) 'it cannot be ruled out that when in the *Principles* (II, 239) Mill coined the celebrated distinction between competition and custom and argued that 'only through the principle of competition has political economy any pretension to the character of a science', he had in mind the character of an independent science, the case of custom being the subject of a complementary science based upon an alternative axiomatic basis.[82] An investigation of wealth based on altruistic behaviour might provide a second example. In any event even with these exclusions political economy was sufficiently complex to recognize conflicting 'tendencies' reflecting the problem of 'composition of causes'.

VII 'VERIFICATION' AND MODEL IMPROVEMENT

We continue in this section with the demonstration that the axiomatic foundation envisaged by Mill for political economy was to reflect closely empirical reality. Our special concern here is the role of verification in model improvement.

'Disturbing causes' known at the outset of an investigation of a particular problem in political economy have been discussed above. Those not known to the investigator when approaching a particular case, to which we now turn, were said by Mill to be the source of the 'only uncertainty' to which political economy (and other moral sciences) are subject: 'here only it is that an element of uncertainty enters into the process – an uncertainty inherent in the nature of these complex phenomena, and arising from the impossibility of being quite sure that all the circumstances of the particular case are known to us sufficiently in detail, and that our attention is not unduly diverted from any of them' (*CW*, IV, 330). Such disturbing causes were in principle discoverable by observation, in the sense of verification: 'we cannot . . . too carefully endeavour to verify our theory, by comparing, in the particular case to which we have access, the results which it would have led us to predict, with the most trustworthy accounts we can obtain of those which have been actually realized. The discrepancy between our anticipations and the actual fact is often the only circumstance which would have drawn our attention to some particular disturbing cause which we had overlooked' (332).[83]

---

the inquiry' (VIII, 904–5). Political ethology is an exception. On Mill's conception of a place for a science of government, see Collini in Collini, Winch and Burrow (1983), Chapter IV.

82 This issue is taken up below, pp. 160–1.

83 Mill here explains that if all the facts of the case were known there would be no need for specific experience since 'the causes are human feelings, and the outward circumstances which are fitted to excite them; and these 'might be' familiar to us, in which case 'we

Some have understood Mill to imply that the task of detailed empirical work is limited to the identification of unknown disturbing causes in particular cases leaving intact the basic model itself (cf. Blaug, 1980, 77, 81). Indeed, several of Mill's formulations seem to suggest that the model is impervious to criticism on grounds of inaccuracy of prediction:

Having now shown that the method a priori in Political Economy, and in all the other branches of moral science, is the only certain or scientific mode of investigation, and that the a posteriori method, or that of specific experience, as a means of arriving at truth, is inapplicable to these subjects, we shall be able to show that the latter method is notwithstanding of great value in the moral sciences; namely, not as a means of discovering truth, but of verifying it, and reducing to the lowest point that uncertainty before alluded to as arising from the complexity of every particular case, and from the difficulty (not to say impossibility) of being assured a priori that we have taken into account all the material circumstances (CW, IV, 331).

The point in question had been stated even more sharply a little earlier when Mill wrote, in a statement already encountered above, that 'to verify the hypothesis itself a posteriori, that is, to examine whether the facts of any actual case are in accordance with it, is no part of the business of science at all, but of the application of science' (325; above, p. 67).

It is apparent then that the basic model might indeed be impervious to verification. But it is my contention that this was not all that verification amounted to. It is conceivable that the testing procedure yields new information of general relevance – rather than of particular relevance in a specific case – and if that is so it must have an impact on the model. Mill alluded to this possibility when he questioned the accuracy of the term 'disturbing' to describe causes at work which cannot be regarded as 'exceptions'. The discovery of a hitherto unknown disturbing cause of this order – provided it operates through the same law of human nature as that reflected in the general behavioural axioms – 'might always be brought within the pale of the abstract science' by 'inserting among its hypotheses a fresh and still more complex combination of circumstances . . .' (331; above, p. 124). Indeed, verification – which might draw attention 'to some important disturbing cause which we had overlooked' – often 'discloses to us errors in thought, still more serious than the omission of what can with any propriety be termed a disturbing cause. It often reveals to us that the basis itself of our whole argument is insufficient; that the data, from which we

should become prophets'; testing against specific experience would not then be called for, since we would be sure that we are accurate. Mill goes too far since there remains the question of combining by ratiocination the individual causal tendencies.

had reasoned, comprise only a part, and not always the most important part, of the circumstances by which the result is really determined' (332).[84] In short, verification demonstrates the inadequacy of the axiomatic framework, even as a first approximation, and obliges a reformulation.

Mill was further explicit that verification might reveal the logical ratiocinative process to be defective. This allowance is made in the course of discussion of the conditions to be satisfied before a model can legitimately be used as the basis of policy recommendation; verification by reference to existing data (past and present) provided the sole means of assuring the soundness of a model for 'the guidance of practice':

[The political economist's] knowledge must at least enable him to explain and account for what *is*, or he is an insufficient judge of what ought to be. If a political economist, for instance, finds himself puzzled by any recent or present commercial phenomena; if there is any mystery to him in the late or present state of the productive industry of the country, which his knowledge of principle does not enable him to unriddle; he may be sure that something is wanting to render his system of opinions a safe guide in existing circumstances. Either some of the facts which influence the situation of the country and the course of events are not known to him; or, knowing them, he knows not what ought to be their effects. In the latter case his system is imperfect even as an abstract system; it does not enable him to trace correctly all the consequences even of the assumed premises (335).

Evidently, a failure to confirm the implications derived from a model of this order reveals the need for its reformulation – in this case to enable the theorist 'to trace correctly all the consequences even of the assumed premises'. Indeed, the theorist ('the speculative politician') is obliged to seek the explanation for any failure 'conscientiously, not with the desire of finding his system complete, but of making it so . . .'; and he is duty bound to carry out a verification 'upon every new combination of facts as it arises'. He must allow for 'the disturbing influence of unforeseen causes', but he also 'must carefully watch the result of every experiment, in order that any residuum of facts which his principles do not lead him to expect, and do not enable him to explain, may become the subject of a fresh analysis, and furnish the occasion for a consequent enlargement or correction of his general views' (335–6). It was the ideal that 'the anticipations of the philosopher guide the observation of the practical man, and the specific

---

84 Mill here alludes to the neglect of 'circumstances, almost peculiar to the particular case or era' under investigation, which may 'have a far greater share in governing that one case' than have 'those laws, qualities, and tendencies, which are common to large classes of cases, and which belong to all place and all time'. The observation thus relates to his dissatisfaction with general axioms of supposedly universal coverage.

experience of the practical man warn the philosopher where something is to be added to his theory' (334–5). Theorists must fight against the natural 'reluctance . . . to admit the reality or relevancy of any facts which they have not previously either taken into, or left a place open for in, their system' (336).

Mill's utterances in the *Logic* regarding verification are also relevant to the present issue. In science generally the process of verification entailed 'collating the conclusions of the ratiocination either with the concrete phenomena themselves, or, where such are obtainable, with their empirical laws', to the end of evaluating the accord between the results of the a priori reasoning and the results of a posteriori observation (VIII, 896).[85] In the specialist branches of social sciences, however, there were no genuine empirical laws derivable from specific experience. Thus, for example, regarding the effects of corn laws: 'Though it may be perfectly certain, from theory, what kind of effects corn laws must produce, and in what general direction their influence must tell upon industrial prosperity; their effect is yet of necessity so much disguised by the similar or contrary effects of other influencing agents, that specific experience can at most only show that on the average of some great number of instances, the cases where there were corn laws exhibited the effect in a greater degree than those where they were not. Now the number of instances necessary to exhaust the whole round of combinations of the various influential circumstances, and thus afford a fair average, never can be obtained.' Empirical generalizations with which 'to collate the conclusions of the theory' were largely precluded by the uncertainty of statistical data and the paucity of available cases for observation 'within the limits of the given state of society and civilization which such inquiries always presuppose' (908–9).

Furthermore, direct verification by comparing the theoretical conclusions with the outcome of particular instances was equally problematic '[f]or in order to verify a theory by an experiment, the circumstances of the experiment must be exactly the same with those contemplated in the theory. But in social phenomena the circumstances of no two cases are exactly alike. A trial of corn laws in another country or in a former generation, would go a very little way towards verifying a conclusion drawn respecting their effect in this generation or in this country.'

Now, as in the essay, it is clear that the entire object of verification, as Mill viewed the matter, was to assure the validity of the theoretical model

---

85 Cf. 907: '[I]n most deductive sciences . . . a preliminary work of preparation is performed on the observed facts, to fit them for being rapidly and accurately collated (sometimes even for being collated at all) with the conclusions of theory. This preparatory treatment consists in finding general propositions which express concisely what is common to large classes of observed facts: and these are called empirical laws of the phenomena'.

for purposes of policy proposal; it was certainly not confirmation of the theory for its own sake. But since the most appropriate case study for verification of the predictions of the theory is in all likelihood precisely that for which the proposals, based on the model's predictions, are being made 'the verification comes too late to be of any avail for practical guidance'. Again, as in the essay, only indirect verification of theory was in any way practical or meaningful: 'The test of the degree in which the science affords safe ground for predicting (and consequently for practically dealing with) what has not yet happened, is the degree in which it would have enabled us to predict what has actually occurred.' We must be able to account for the existing state of social phenomena in terms of our model – after allowance for disturbing causes. Otherwise, 'either the facts which ought to be taken into account are not yet completely known to us, or . . . although we know the facts, we are not masters of a sufficiently perfect theory to enable us to assign their consequences. In either case we are not, in the present state of our knowledge, fully competent to draw conclusions, speculative or practical, for that country . . . [We] must turn back, and seek the explanation by an extension and improvement of the theory itself' (909–10).

We cite finally a passage, already alluded to briefly in the account of scientific procedure in general which summarizes beautifully the role of testing in model improvement, and the warning against reliance upon a model which has been found wanting by reference to existing data:

But . . . are not the same arguments by which the methods of direct observation and experiment were set aside as illusory when applied to the laws of complex phenomena, applicable with equal force against the Method of Deduction? When in every single instance a multitude, often an unknown multitude, of agencies, are clashing and combining, what security have we that in our computation *a priori* we have taken all these into our reckoning? How many must we not generally be ignorant of? Among those which we know, how probable that some have been overlooked; and, even were all included, how vain the pretence of summing up the effects of many causes, unless we know accurately the numerical law of each, – a condition in most cases not to be fulfilled; and even when it is fulfilled, to make the calculation transcends, in any but very simple cases, the utmost power of mathematical science with all its most modern improvements.

These objections have real weight, and would be altogether unanswerable, if there were no test by which, when we employ the Deductive Method, we might judge whether an error of any of the above descriptions had been committed or not. Such a test however there is: and its application forms, under the name of Verification, the third essential component part of the Deductive Method; without which all the results it can give have little other value than that of conjecture. To warrant reliance on the general conclusions arrived at by deduction, these conclusions must be found, on careful comparison, to accord with the results of direct observation wherever it can be had. If, when we have experience to compare with them, this experience confirms them, we may safely trust to them in other cases of which our specific experience is yet to come. But if our deductions have led to the conclusion that from a particular combination of causes a given effect would result, then in all known cases where that combination can be shown to have existed, and where the effect has not followed, we must be able to show (or at least to make a probable surmise) what frustrated it: if we cannot, the theory is imperfect, and not yet to be relied upon. Nor is the verification complete, unless some of the cases in which the theory is borne out by the observed result, are of at least equal complexity with any other cases in which its application could be called for (VII, 460–1).

It may be noteworthy for some methodological issues that Mill never refers to the actual abandonment of a theory in consequence of a failure to account adequately for past and present phenomena. But for us here the important point is simply that Mill allowed for alterations in the basic model in consequence of the procedure of verification – testing against specific experience: its 'improvement', 'correction', 'completion', 'extension'.[86]

In the light of all this it might seem difficult to appreciate Mill's formal insistence that 'no general truths can be attained in the affairs of nations by the *a posteriori* road' although the theorist is duty-bound to 'sift and scrutinize the details of every specific experiment'; or his allusion to the desirability of undertaking the scientific task of theory construction 'once and for all', leaving the allowances for modifying causes to a later stage of investigation (IV, 333; cf. VIII, 900, cited above, p. 107). The fact is that theory construction was portrayed as a continuous task turning crucially on verification against evidence. And this, after all, constitutes the substances of the protest against contemporary critics who failed to appreciate that theoretical demonstration amounted only to conditional demonstration – 'a proof at all times liable to be set aside by the addition of a single new fact to the hypothesis' (IV, 334).

But those statements which seem at first glance to point away from this view of the matter are not difficult to appreciate. We must consider that the so-called third stage of verification in the deductive process, although

---

86 Professor Whitaker (1975), 1038–9, 1042–3, has written of Mill's discussion of 'the science of man in society' (VIII, 875f.) that he 'possibly concedes more than he recognizes. For the test of a fruitfully-simplified generalization is now predictive ability, not unique entailment by a full set of preconditions, and certain lines of enquiry prosper only gradually as they lead to successful predictions.' But this is not an accurate evaluation, for it has emerged from our discussion that the need for continuous observational verification pervades Mill's analysis of scientific method in general as well as the social sciences. The evaluation neglects the key role accorded verification in model improvement. For specific examples of model improvement see below, p. 962f.

it may indeed contribute towards establishment of the axioms or correction of the logical process of deduction, is not itself a device for the derivation of complex causal relations in the face of the problem of 'composition of causes' which characterizes the social sciences; that remains the function of ratiocination. Verification contributes only indirectly by indicating the need for improvement in the axiomatic foundation or in the logical process itself. Thus, for example, to be satisfied with the practical relevance of a model (when, in particular, it is proposed as the basis for a policy recommendation) it must be assumed that the axiomatic foundation has been proven adequate for the purpose – that the 'third stage' has been undertaken with reference, of course, to the already existing body of empirical evidence; but it would not be misleading to say that our present axiomatic framework constitutes a first stage for the subsequent ratiocination although it may incorporate refinements shown to have been required by the earlier verification. At this juncture the scientific work at hand might be described as involving ratiocination on the basis of the axiomatic framework – again without reference to the possibility that the framework owes something to a preceding verification, or that further modifications might be proven necessary by a new verification as additional evidence accumulates.

VIII   POLITICAL ECONOMY AND PREDICTION

Mill took pains to distinguish 'science' from 'art'. In his essay on definition and method, he observed that his objective in attempting to coin a definition was dictated by a desire to counteract the 'indefinite, and often erroneous, conceptions of the mode in which the science should be studied' (*CW*, IV, 311);[87] and the celebrated definition ultimately arrived at recognized the distinction between 'science' and 'art' – between the indicative and imperative moods: 'Science takes cognizance of a *phenomenon*, and endeavours to discover its *law*; art proposes to itself an *end*, and looks out for *means* to effect it' (312; cf. VIII, 943f). The definition was designed, in large part, to counter the common view which considered political economy as the science professing to teach how a nation may be enriched: 'If . . . Political Economy be a science, it cannot be a collection of practical rules; though, unless it be altogether a useless science, practical rules must be capable of being founded upon it . . . Rules, therefore, for making a nation increase in wealth, are not a science, but they are the results of science'.[88]

---

87 Cf. 324: '[W]ith the consideration of the definition of a science, is inseparably connected that of the *philosophic method* of the science; the nature of the process by which its investigations are to be carried on, its truths to be arrived at.'

88 Mill suggested that the erroneous definition was 'to some degree' implied by the title and arrangement of the *Wealth of Nations*.

We have encountered in the previous section the strict conditions to be satisfied before a model could legitimately be used for purposes of policy recommendation. But even after satisfying these conditions, the 'guidance of practice' was a hazardous exercise.

The dangers were twofold. First, the intervention of new or unforeseen disturbing causes was always a possibility: 'Effects are commonly determined by a *concurrence* of causes. If we have overlooked any one cause, we may justly reason from all the others, and only be further wrong. Our premises will be true, and our reasoning correct, and yet the result of no value in the particular case' (337). Mental discipline might secure against false premises and faulty logic, but not against the possibility that some causal feature had been neglected. Theoretical demonstration was conditional at best – 'a proof at all times liable to be set aside by the addition of a single new fact to the hypothesis' (see above, p. 130).

This problem was compounded by the fact that by its nature the 'art' of political economy involved much more than the application of the specialist science of political economy:

One of the strongest reasons for drawing the line of separation clearly and broadly between science and art is the following: – That the principle of classification in science most conveniently follows the classification of *causes*, while arts must necessarily be classified according to the classification of the *effects*, the production of which is their appropriate end. Now an effect, whether in physics or morals, commonly depends upon a concurrence of causes, and it frequently happens that several of these causes belong to different sciences. Thus in the construction of engines upon the principles of the science of *mechanics*, it is necessary to bear in mind the *chemical* properties of the material, such as its liability to oxydize; its electrical and magnetic properties, and so forth. From this it follows that although the necessary foundation of all art is science, that is, the knowledge of the properties or laws of the objects upon which, and with which, the art does its work; it is not equally true that every art corresponds to one particular science. Each art presupposes, not one science, but science in general; or, at least, many distinct sciences (331n).

Because of the 'inter-disciplinary' nature of 'art' it followed that the 'mere political economist, he who has studied no science but Political Economy, if he attempts to apply his science to practice, will fail' (331).

However, there were major problems of applied social science deriving from another source. As Mill forcefully clarified in the *Logic*, there could be no practical maxims of general application even if social phenomena conformed to known causal relationships, precluding the problem of 'disturbing' causes: Social phenomena 'might not only be completely dependent on known causes, but the mode of action of all these causes might be reducible to laws of considerable simplicity, and yet no two cases might admit of being treated in the same manner. So great might be the variety of circumstances on which the results in different cases depend, that the art might not have a single general precept to give' (VIII, 877). The problem was that of weighing in the balance the relative force of the numerous (supposedly known) causal influences playing upon the condition and progress of society – influences 'innumerable, and perpetually changing; and though they all change in obedience to causes, and therefore to laws, the multitude of the causes is so great as to defy our limited powers of calculation. Not to say that the impossibility of applying precise numbers to facts of such a description, would set an impassable limit to the possibility of calculating them beforehand, even if the powers of the human intellect were otherwise adequate to the task' (878).

The primary conclusion drawn from these characteristics is that 'Sociology, considered as a system of deductions *a priori*, cannot be a science of positive predictions, but only of tendencies. We may be able to conclude, from the laws of human nature applied to the circumstances of a given state of society, that a particular cause will operate in a certain manner unless counteracted; but we can never be assured to what extent or amount it will so operate, or affirm with certainty that it will not be counteracted; because we can seldom know, even approximately, all the agencies which may coexist with it, and still less calculate the collective result of so many combined elements' (898). Social science, to the extent that it was 'insufficient for prediction' had to be distinguished from astronomy, the data of which were relatively few and stable.

The same position is maintained in the essay. If the processes of analysis and synthesis involved in deductive procedure 'could be correctly performed, the result would be prophecy; but as they can be performed only with a certain approximation to correctness, mankind can never predict with absolute certainty, but only with a less or greater degree of probability; according as they are better or worse apprised what the causes are, – have learnt with more or less accuracy from experience the law to which each of those causes, when acting separately, conforms, – and have summed up the aggregate effect more or less carefully'. Predictions of 'an actual result', were therefore to be avoided; at most 'a *tendency*' to that result – a power acting with a certain intensity in that direction can be 'predicted' (IV, 336–7).

Despite these limitations political economy was not to be disparaged. It 'would have attained a very high degree of perfection, if it enabled us, in any given condition of social affairs, in the condition for instance of Europe or any European country at the present time, to understand by what causes it had, in any and every particular, been made what it was; whether it was tending to any, and to what, changes; what effects each feature of its existing state was likely to produce in the future; and by what means any of the effects might be prevented, modified or accelerated, or a different class of effects superinduced.' There was nothing chimerical in the hope that general laws, sufficient to enable us to answer these various questions' – at least 'for any country or time with individual circumstances of which we were well acquainted' – 'do really admit of being ascertained . . .' (VIII, 878).[89] For social science to justify itself it was thus unnecessary 'to foresee infallibly the results of what we do' whether in social or private affairs: 'We must seek our objects by means which may perhaps be defeated, and take precautions against dangers which possibly may never be realized. The aim of practical politics is to surround any given society with the greatest possible number of circumstances of which the tendencies are beneficial and to remove or counteract, as far as practicable, those of which the tendencies are injurious. A knowledge of the tendencies only, though without the power of accurately predicting their conjunct result, gives us to a considerable extent this power' (898). Similarly, '[e]xtensive and important practical guidance may be derived, in any given state of society, from general propositions such as those before indicated; even though the modifying influence of the miscellaneous causes which the theory does not take into account, as well as the effect of the general social changes in progress, be provisionally overlooked' (903).

Mill's repeated references in the foregoing passages limiting the practical guidance to be derived from political economy to 'the circumstances of a given state of society', to 'any given condition of social affairs', to 'any country or time the individual circumstances of which we are well acquainted' and so forth, are crucially significant. They confirm our interpretation offered earlier regarding the limited scope conceived by Mill for the propositions of the contemporary science. As far as we are concerned here the moral to be drawn is that even the isolation of tendencies – as distinct from the making of specific predictions – required the initial postulation of a constrained enviornment; universally applicable propositions relating to tendencies were not in order. And this because of 'the eminently modifiable nature of the social phenomena, and the multitude and variety of the circumstances by which they are modified; circumstances never the same, or even nearly the same, in two different societies or in two different periods of the same society' (898–9).

IX ON 'ETHOLOGY' AND 'PROGRESS'

In this section we follow Mill by placing the methodological discussions regarding the social sciences in the broader perspective provided by the

---

89 But even in the best conceivable case – where the causes are known – the role of theory would still be limited to 'that of watching the circumstances of the particular case, and adapting our measures to the effects which, according to the principles of the science, result from those circumstances' (877).

so-called 'science of human character formation', or ethology, and by the 'General Science of Society'. The limited range ascribed to the propositions derived from political economy will thereby be confirmed.

The behavioural axioms of political economy or other moral sciences, Mill emphasized, derived from common experience and amounted to 'approximate generalizations'. They served their purpose: '[W]henever it is sufficient to know how the majority of the human race, or of some nation or class of persons, will think, feel, and act, these propositions are equivalent to universal ones. For the purposes of political and social science this *is* sufficient . . . [A]n approximate generalization is, in social inquiries, for most practical purposes equivalent to an exact one: that which is only probable when asserted of individual human beings indiscriminately selected, being certain when affirmed of the character and collective conduct of masses' (*CW*, VIII, 847). But only if a sound causal basis were provided for such empirical generalizations could a science of human behaviour be said to exist:

in order to give a genuinely scientific character to the study, it is indispensable that these approximate generalizations, which in themselves would amount only to the lowest kind of empirical laws, should be connected deductively with the laws of nature from which they result; should be resolved into the properties of the causes on which the phenomena depend. In other words, the science of Human Nature may be said to exist, in proportion as the approximate truths, which compose a practical knowledge of mankind, can be exhibited as corollaries form the universal laws of human nature on which they rest (848).

With such a foundation, it would be possible to define 'the proper limits of those approximate truths' and 'to deduce others for any new state of circumstances, in anticipation of specific experience'. For such purposes we cannot be satisfied with knowledge merely of the consequences of the causal laws: '[I]f we have not yet accounted for the empirical law – if it rests only on observation – there is no safety in applying it far beyond the limits of time, place, and circumstance, in which the observations were made' (862).

A science of ethology, although a possibility, was still a matter for the future. There existed therefore a danger of claiming too much for empirical generalizations ('the common wisdom of common life'); for example,

when maxims of this sort, collected from Englishmen, come to be applied to Frenchmen, or when those collected from the present day are applied to past or future generations, they are apt to be very much at fault. Unless we have resolved the empirical law into the laws of the causes on which it depends, and ascertained that those causes extend to the case which we have in view, there can be no reliance placed in our inferences. For every individual is surrounded by circumstances different from those of every other individual; every nation or generation of mankind from every other nation or generation: and none of these differences are without their influence in forming a different type of character. There is, indeed, also a certain general resemblance; but peculiarities of circumstances are continually constituting exceptions even to the propositions which are true in the great majority of cases (864).

The absence of an ethological foundation for the specialist branches of social science – more specifically the absence of a science of 'political ethology' or of national character – 'vitiates them in their practical applications as branches of a comprehensive social science' (906).

The severe restriction upon the applicability of empirical laws has profound implications. For the basic assumptions regarding human nature accepted as axiomatic for purposes of deductive theorizing in political economy are based upon observation including, as is clear in the essay on definition and method, 'observation of what passes in our own mind'. But even such introspective evidence was itself regarded as relative to time and place thus restricting narrowly the entire range of the subject matter of 'political economy'. It is not merely that the wealth-maximizing hypothesis is only justified if we concentrate upon a particular range of activity where such an assumption constitutes at least an excellent first approximation – activity involving the production and distribution of wealth – but that even this assumption only holds good (again as an approximation) in the context of a specified nation and period.

* * *

Much of what has just been said regarding the limited scope of theorems derived ultimately from axioms which have the validity only of narrowly constrained empirical laws is of the highest relevance to the ultimate objective of social investigation. I have in mind the general science of society.

The narrower disciplines dealt (as in the case of political economy) with issues such as the effect of repealing or imposing agricultural protection within a specified institutional framework – 'in the present condition of society and civilization in any European country, or under any other given supposition with regard to the circumstances of society in general'; the general science, by contrast, would deal with 'the causes which produce, and the phenomena which characterize, States of Society . . .' (911). The nature of the constraints imposed upon the social sciences and the changes therein over time, are matters falling in principle within the domain of a specialist discipline, containing (following Comte) two branches: That of social statics which treats 'the mutual actions and reactions of contemporaneous social phenomena' – and abstracts from changes over time in the whole (918); and that of social dynamics which seeks to discover the laws 'according to which any state of society produces the state which succeeds it and takes it place' (912).[90]

The existence of a 'natural correlation' among the elements constituting each particular state of society – 'Uniformities of Coexistence' binding the social phenomena relating to each state – is taken for granted (912); such a correlation is described as 'a necessary consequence of the influence exercised by every one of those phenomena over every other'. But the precise grounds for this presumption are not made adequately clear. Thus the mutual correlation between the different elements is said to be a 'derivative law, resulting from the laws which regulate the succession between one state of society and another'. But the statement is difficult to comprehend since these so-called 'laws' constitute the subject matter of a science – social dynamics – by no means yet off the ground. It is more reasonable to interpret Mill as relying heavily on empirical evidence: 'the information which we possess respecting past ages, and respecting the various states of society now existing in different regions of the earth, does, when duly analyzed, exhibit uniformities' (912). The 'uniformities' thus yielded amounted only to empirical laws but 'some of them, once suggested, are found to follow with so much probability from general laws of human nature, that the consilience of the two processes raises the evidence to proof, and the generalizations to the rank of scientific truth' (920).[91]

It is a presumption of relevance to the study of social dynamics that the properties of the subject matter of social science in the most basic sense – namely behavioural patterns – are changeable: 'not only the qualities of individuals vary, but those of the majority are not the same in one age as in another'. Indeed by progress is meant – that there is a progressive change both in the character of the human race, and in their outward circumstances as far as moulded by themselves: that in each successive age the principal phenomena of society are different from what they were in the age preceding, and still more different from any previous age . . .' (913–14).[92]

90 Alternatively: the 'science of successions of social phenomena' designed 'to observe and explain the sequences of social conditions' (924). See also VII, 323, for the contrast: 'The phenomena of nature exist in two distinct relations to one another; that of simultaneity, and that of succession. Every phenomenon is related, in an uniform manner, to some phenomena that coexist with it, and to some that have preceded and will follow it'.

91 Mill commended Comte in this context for the proposition – termed a 'natural law' – that the form of government existing at any time is intimately bound up with the contemporary state of civilization. He also cites at great length his own paper 'Coleridge' (1840) regarding government forms, and states of society.

92 In this context a single generation is said to constitute a characteristic period marking a change in the 'principal phenomena of society'; subsequently the change is described as 'gradual' (918–19).

The only rationale given for the presumed alteration in character is that circumstances form human character and humans shape circumstances. Mill conceded that the mutual linkages between character and circumstances might generate cycles rather than progress; but he believed there was evidence for the latter.[93]

To discover the law of progress, Mill noted, was a well-established preoccupation on the Continent.[94] But scientific knowledge had not even there advanced beyond the isolation of empirical laws and even these were not firmly established; moreover, the study laboured under a misconception regarding the scope of such laws:

The misconception consists in supposing that the order of succession which we may be able to trace among the different states of society and civilization which history presents to us, even if that order were more rigidly uniform than it has yet been proved to be, could ever amount to a law of nature. It can only be an empirical law. The succession of states of the human mind and of human society cannot have an independent law of its own, it must depend on the psychological and ethological laws which govern the action of circumstances on men and of men on circumstances (914).

The same error characterized those British thinkers who recognized that human nature and society were 'in a state of necessary progression', but failed to realize that their evidence was merely in the nature of empirical laws:

Accordingly, while almost all generalizations relating to Man and Society, antecedent to the last fifty or sixty years, have erred in the gross way which we have attempted to characterize, namely, by implicitly assuming that human nature and society will for ever revolve in the same orbit, and exhibit essentially the same phenomena; which is also the vulgar error of the ostentatiously practical, the votaries of so-called common sense, in our day, especially in Great Britain; the more thinking minds of the present age, having applied a more minute analysis to the past records of our race, have for the most part adopted a contrary opinion, that the human species is in a state of necessary progression, and that from the terms of the series which are past we may infer positively those which are yet to come . . . [But] we must remember that even this other and better generalization, the progressive

93 Mill raised the question whether 'progress' means 'improvement', and asserted that it is a theorem of the science itself 'that the general tendency is, and will continue to be, saving occasional and temporary exceptions, one of improvement; a tendency towards a better and happier state'.

94 The notion of a science of history – that the course of history or 'the collective series of social phenomena' is subject to general laws – was said to be common on the Continent but almost a novelty in Britain when the *Logic* first appeared. Since then Buckle's work had promoted a change in attitude (931).

change in the condition of the human species, is, after all, but an empirical law: to which, too, it is not difficult to point out exceedingly large exceptions; and even if these could be got rid of, either by disputing the facts or by explaining and limiting the theory, the general objection remains valid against the supposed law, as applicable to any other than what, in our third book, were termed Adjacent Cases. For not only is it no ultimate, but not even a causal law. Changes do indeed take place in human affairs, but every one of those changes depends on determinate causes; the 'progressiveness of the species' is not a cause, but a summary expression for the general result of all the causes (790-1).

The fundamental task thus remained to be undertaken: 'Until that [empirical] law could be connected with the psychological and ethological laws on which it must depend, and, by the consilience of deduction *a priori* with historical evidence, could be converted from an empirical law into a scientific one, it would not be relied on for the prediction of future events, beyond at most, strictly adjacent cases' (915). And the difficulties in the way are not minimized; they explained why little advance had occurred beyond the perception of empirical laws (and these only of the nature of general tendencies): 'these and all such results are still at too great a distance from the elementary laws of human nature on which they depend, – too many links intervene, and the concurrence of causes at each link is far too complicated, – to enable these propositions to be presented as direct corollaries from these elementary principles. They have, therefore, in the minds of most inquirers, remained in the state of empirical laws, applicable only within the bounds of actual observation; without any means of determining their real limits, and of judging whether the changes which have hitherto been in progress are destined to continue indefinitely, or to terminate, or even to be reversed' (925).[95]

It was, however, accepted that the appropriate procedure in the investigation of progress (and of contemporaneous states) necessarily required as the first step the establishment of provisional conclusions from specific experience – the establishment of empirical laws – and the linkage thereof with the principles of human nature as a second step only, the reasoning entailed in the latter process constituting a form of 'verification' of the empirical law. The problem was 'to ascertain [the empirical laws], and connect them with the laws of human nature, by deductions showing that such were the derivative laws naturally to be expected as the consequence

95 The kind of empirical laws Mill had in mind did not in fact amount to the transition of one social state into another; rather the general tendencies perceived were of the order of 'a progressive increase of some social elements, and dimunition of others, or a gradual change in the general character of certain elements' – an increasing prevalence of mental over bodily qualities, a reduction in the proportion of military to productive activity and so forth (924).

of those ultimate ones' (916). This method – termed the inverse deductive method (895f)[96] – involved the opposite sequence from that appropriate in the specialist branches of the social sciences where causal relationships drawn by deductive procedure are then verified by direct observation. The difference turned in part upon the absence of empirical laws in the specialist branches which ruled out the inverse deductive method. But conversely, what was appropriate in the specialist branches – the concrete (physical) deductive method – was ruled out in the investigation of the laws of progress: 'I do not think any one will contend that it would have been possible, setting out from the principles of human nature and from the general circumstances of the position of our species, to determine *a priori* the order in which human development must take place, and to predict, consequently, the general facts of history up to the present time.' The problem here alluded to derived from the very length of the time series involved, namely that what 'we now are and do, is in a very small degree the result of the universal circumstances of the human race, or even of our own circumstances acting through the original qualities of our species, but mainly of the qualities produced in us by the whole previous history of humanity'. It would be impossible to compute laws of progress 'from the elementary laws which produce it' (915–16).

Mill did not, however, expect the inverse deductive method to yield anything like definite results. 'Verification' of the order of the probable or even only the possible was all that could be achieved. It would not be possible to demonstrate a priori that the empirical law regarding progress derived from historical evidence was the only order of succession of events consistent with the laws of human nature, but at best that there were a priori reasons to expect that particular sequence of events given knowledge of 'the nature of man and the general circumstances of his position' (917).[97]

Mill had some positive proposals to make regarding the isolation of empirical laws. He recommended that the contemporaneous set of circumstances characterizing given states of society be considered as a unit and similarly its overall change over time; a fragmentary view involving isolated aspects of the social state and its change over time should be avoided. This recommendation amounted to an appeal to combine the static and dynamic conceptions. But the problem would be vastly eased in the event that there could be discerned a single element in the social state of man 'pre-eminent over all others as the prime agent of the social movement'. If such a prime mover existed 'we could then take the progress of that one element as the central chain, to each successive link of which, the

---

96 Not to be confused with the hypothetical deductive method, discussed above (Section II).

97 A similar position was adopted regarding the coexistence of circumstances at any time.

corresponding links of all the other progressions being appended, the succession of the facts would by this alone be presented in a kind of spontaneous order, far more nearly approaching to the real order of their filiation than could be obtained by any other merely empirical process' (925). Mill confidently suggested, following Comte, that there indeed was discernible a 'predominant, and almost paramount' agent of social progression – namely 'the state of the speculative faculties of mankind; including the nature of the beliefs which by any means they have arrived at, concerning themselves and the world by which they are surrounded' (926). That this was so was not merely shown by 'the evidence of history'; 'the evidence of human nature' pointed towards the same conclusion that 'the order of human progression . . . will mainly depend on the order of progression in the intellectual convictions of mankind, that is, on the law of the successive transformations of human opinions'. Mill did not (apparently) believe that such a relationship yet had the status of an established empirical law, and in any event an empirical law had still to be 'converted into a scientific theorem by deducting it *a priori* from the principles of human nature' (927).[98] But the lines suggested promised the most rewarding research programme.

Mill's interest in such investigations was not academic. The ultimate objective of the exercise was application. In the absence of causal laws of progress, 'prediction' of secular trends was ruled out including relative growth rates of capital and population, and distributive patterns:

Changes do indeed take place in human affairs, but everyone of those changes depends on determinate causes; the "progressiveness of the species" is not a cause, but a summary expression for the general result of all the causes. So soon as . . . it shall be ascertained what causes have produced these successive changes, from the beginning of history, in so far as they have really taken place, and by what causes of a contrary tendency they have been occasionally checked or entirely counteracted, we may then be prepared to predict the future with reasonable foresight; we may be in possession of the real law of the future; and may be able to declare on what circumstances the continuance of the same onward movement will eventually depend. But this it is the error of many of the more advanced thinkers, in the present age, to overlook; and to imagine that the empirical law collected from a mere comparison of the condition of our species at different past

---

98 Mill observed that while he agreed with Buckle 'that the intellectual element in mankind, including in that expression the nature of their beliefs, the amount of their knowledge and the development of their intelligence, is the predominant circumstance in determining their progress', this was not because 'their moral or economical condition' are less powerful or less variable agencies, but because these too are governed by the intellectual condition (935). See the harsh evaluation by Whitaker (1975), 1040, on the role accorded Mill to the speculative faculties of mankind.

times, is a real law, is *the* law of its changes, not only past but also to come. The truth is, that the causes on which the phenomena of the moral world depend, are in every age, and almost in every country, combined in some different proportion; so that it is scarcely to be expected that the general result of them all should conform very closely, in its details at least, to any uniformly progressive series. And all generalizations which affirm that mankind have a tendency to grow better or worse, richer or poorer, more cultivated or more barbarous, that population increases faster than subsistence, or subsistence than population, that inequality of fortune has a tendency to increase or to break down, and the like, propositions of considerable value as empirical laws within certain (but generally rather narrow) limits, are in reality true or false according to times and circumstances (790–1).

In the light of all that was said regarding the impossibility of making specific forecasts even in the narrowly defined science of political economy, it is unlikely that Mill intended positive predictions by his references to the opportunities hopefully to be created 'for the prediction of future events' by the discovery of causal laws of societal development. As in the specialist social sciences, Mill probably had in mind tendencies – in the present case secular tendencies extending beyond the limits of individual countries – knowledge of which would permit policy makers to exercise intelligent judgement. Once confirmed, 'the derivative laws of social order and of social progress' would provide the means of 'looking far forward into the future history of the human race', and of 'determining what artificial means may be used, and to what extent, to accelerate the natural progress in so far as it is beneficial; to compensate for whatever may be its inherent inconveniencies and disadvantages; and to guard against the dangers or accidents to which our species is exposed from the necessary incidents of its progression. Such practical instructions, founded on the highest branch of speculative sociology, will form the noblest and most beneficial portion of the Political Art' (929–30).

X   A COMPARISON OF MILL AND SENIOR

Some of the major themes of Mill's position will be brought sharply into focus by comparison with those of Nassau Senior who had earlier addressed the methodological issue.

In Senior's *Introductory Lecture* of December 1826, political economy was represented as 'the science which teaches in what wealth consists, – by what agents it is produced, – and according to what laws it is distributed, – and what are the institutions and customs by which production may be facilitated and distribution regulated, so as to give the largest possible amount of wealth to each individual'. The science, therefore, incorporates both 'theoretical' and 'practical' branches.

The first branch, which deals with the nature, production and distribution of wealth, is said to rest 'on a very few general propositions, which are the result of observation, or consciousness, and which almost every man, as soon as he hears them, admits, as familiar to his thoughts, or at least, as included in his previous kowledge' (Senior, 1827, 7).[99] The conclusions, in so far as concerns production, are 'universally true'; and while those concerning distribution vary depending upon particular institutional arrangement, 'the natural state of things can be laid down as by general rule, and the anomalies produced by particular disturbing causes can be afterwards accounted for'.[100] The axioms of theoretical economics are thus represented by Senior to be of a self-evident nature, resting partly upon 'consciousness' and partly upon observations of which everyone has experience; and to be of universal relevance, independent of particular institutional arrangement.[101]

99 Richard Whately appended to his *System of Logic* (1826; 2nd ed. London 1827) a list of ambiguous terms in political economy prepared by Senior; the list is introduced by an almost identical statement: 'The foundation of Political Economy being a few general propositions deduced from observation or from consciousness, and generally admitted as soon as stated, it might have been expected that there would be as little difference of opinion among Political-Economists as among Mathematicians; – that, being agreed on their premises, they could not differ in their conclusions, but through some error in reasoning, so palpable as to be readily detected' (309).

100 This matter is amplified in the course of a criticism of those who would appeal not to science but to 'the opinions of practical men, or their own common-sense'. The 'experience' relevant to political economy is familiar to everyone and, as earlier intimated, is precisely that which constitutes the axiomatic foundation of the science as far as concerns behaviour: 'By practical men are meant, I suppose, those who have had experience in the matters which Political Economy considers. But who has not had that experience? The revenue of all men must consist of rent, profit, or wages. They must all exchange it for commodities or services. They all know, or have equally the means of knowing, for it can be discovered only by reflection, why they set a high value upon some things, a low one upon others, and disregard a third class' (24). (To insist upon the universality of the behavioural axioms and yet to specify an institutional organization involving landowners, capitalists and labourers as Senior does in the cited passage seems contradictory; but it is not a matter that seems to have been taken further.)

101 On this characteristic see J. N. Keynes, (1955, [1891]), 15, 224n. Five general propositions served as the axioms of economic theory for Senior (1827), although the first represents the definition of wealth and is thus somewhat out of line with the others: 'Firstly. That wealth consists of all those things, and of those things only, which are transferable; which are limited in quantity; and which, directly or indirectly, produce pleasure or prevent pain: or, to use an equivalent expression, which are susceptible of exchange; . . . or, to use a third equivalent expression, which have value' (35).

The definition of wealth seems to be sufficiently broad to cover services as well as material goods. Earlier in his lecture, however, Senior had alluded to 'the pursuit of wealth' in terms of 'the endeavour to accumulate the means of future subsistence and enjoyment' which of necessity excludes services (12). Moreover, the remaining axioms are formulated in such terms as to imply a preoccupation with material goods.

The second or practical branch of the science – concerned with ascertaining the institutions most favourable to the production of wealth – depends to some degree upon the conclusions of the theoretical branch which it accepts as its premises and to this extent yield conclusions which possess 'equal certainty and universality' (11). But in addition it 'has many [premises] which depend on induction from phenomena, numerous, difficult of enumeration, and of which the real sequence often differs widely from the apparent one. The machinery of civilized society is worked by so many antagonistic springs; the dislike of labour, the desire for immediate enjoyment, and the love of accumulation are so perpetually counteracting each other, and they produce such opposite conduct, not only in different individuals, but in whole masses of people, that we are liable to the greatest mistakes when we endeavour to assign motives to past conduct, or to predict the conduct which a new motive will produce' (8–9).

'Disturbing causes', including 'antagonistic springs' of behaviour, are therefore a matter for applied economics. It is this branch only that 'must sometimes draw its premises from particular facts, respecting particular climates, soils, and seasons; and must sometimes take into account the influence of every human passion and appetite, under every modification of government and knowledge'. Applied economics is represented as, by its nature, subject to a degree of uncertainty, since from 'probable premises', only 'probable conclusions' can be deduced (10).

J. S. Mill is likely to have been fully familiar with Senior's position when he composed his own essay. There are common features: the representation of 'disturbing causes' as a problem in applied economics, and the 'uncertainty' thus created in that realm of discourse. Some of Senior's terminology, specifically 'antagonistic' springs which 'perpetually counteract' each other, may even have been borrowed by Mill, who refers to 'perpetually antagonistic principles' and 'perpetual counter-motives'.[102] But there is a profoundly important difference: Senior conceived the behavioural axiom to be of universal relevance and this, we have seen, Mill did not do either in his essay or in the later *System of Logic*. Mill's 'economic man' was intended

---

Of the four axioms strictly so called, only the behavioural is said to be 'a matter of consciousness'; the others, capital productivity, diminishing returns in agriculture and the population principle, are matters of 'observation' (35–6). Marian Bowley (1937), 49, represents the population principle as having its basis in consciousness, but this was not Senior's position in 1826–7.

102 It should be noted also that various synonyms for disturbing causes used by Mill in 1831 appeared in Malthus's *Principles of Political Economy* (1820). Thus: 'The . . . tendency to simplify and generalize, produces a still greater disinclination to allow for modifications, limitations, and exceptions to any rule or proposition, than to admit the operation of more causes than one' (6–7). McCulloch (1824, 13) warned against the hasty *modification* of 'constant and universally applicable' principles on the basis of special cases.

to describe representative behaviour in the context of a specific temporal and geographic environment – the British capitalist-exchange system, and the notion of 'universality' was avoided. Consistently with this practice, Mill allowed into the behavioural axiom a more complex pattern than did Senior; and he emphasized the need for continuous correction and improvement of the axioms in consequence of the outcome of on-going verification against the facts (the ever changing facts) of the conclusions drawn deductively from the model.

This distinction comes very clearly to light in the course of Senior's remaining lectures of the 1826-7 academic year (unpublished at the time),[103] and in lectures of 1847-8 which were incorporated into a review of Mill's essay on method. We turn now to the elaboration of the behavioural axiom in these materials.

Senior represented the axiom in his lectures of 1827 as 'the cornerstone of the doctrine of wages and profits, and, generally speaking, of exchange', filling the same role in political economy as gravitation in physics – 'the ultimate fact beyond which reasoning cannot go, and of which almost every other proposition is merely an illustration' (*Outline* ed. 6 [1872], 1938, 28). He took pains to circumscribe his position in the following way:

We must not be supposed to mean that every body, or indeed any body, wishes for an indefinite quantity of every thing; still less as stating that wealth, though the universal, either is, or ought to be, the principal object of human desire. What we mean to state is, that no person feels his whole wants to be adequately supplied; that every person has some unsatisfied desires which he believes that additional wealth would gratify. The nature and the urgency of each individual's wants are as various as the differences in each individual character. Some may wish for power, others for distinction, and others for leisure; some require bodily, and others mental amusement; some are anxious to produce important advantage to the public; and there are few, perhaps there are none, who, if it could be done for a wish, would not benefit their acquaintances and friends. Money seems to be the only object for which the desire is universal; and it is so, because money is abstract wealth. Its possessor may satisfy at will his ambition, or vanity, or indolence, his public spirit or his private benevolence; may multiply the means of obtaining bodily pleasure, or of avoiding bodily evil, or the still more expensive amusements of the mind. Any one of these pursuits would exhaust the largest fortune within the limits of individual acquisition; and, as all men would engage in some of them, and many in all, the desire for wealth must be insatiable, though the modes in which different individuals would employ it are infinitely diversified (27).

103 The bulk of which, as far as concerns the discussion of the Fundamental Propositions, was later absorbed into the *Outline of the Science of Political Economy*. Cf. Bowley (1937), 48n, 49.

Senior thus represented the desire for wealth (for its own sake or as an indirect means to other ends) as 'universal' and 'insatiable'. It is characteristic of his later formulations to represent this position as a 'positivistic' doctrine sharply opposed to that of Ricardo and J. S. Mill who, he believed, considered the wealth-maximization axiom in purely 'hypothetical' terms.[104] Accordingly, in his review of Mill's *Essays on Some Unsettled Questions* the famous passage in the essay on method which argues that political economy is 'built upon hypotheses', is taken as representative of a faulty procedure:

But neither the reasoning of Mr. Mill, nor the example of Mr. Ricardo induce us to treat Political Economy as a hypothetical science. We do not think it necessary, and, if unnecessary, we do not think it desirable. It appears to us that if we substitute for Mr. Mill's hypothesis that wealth and costly enjoyment are the *only* objects of human desire, the statement that they are universal and constant objects of desire, that they are desired by all men and at all times, – we shall have laid an equally firm foundation for our subsequent reasonings, and have put a truth in the place of an arbitrary assumption. We shall not, it is true, from the fact that by acting in a particular manner a labourer may obviously obtain higher wages, a capitalist larger profits, or a landlord higher rent, be able to infer the further fact that they will certainly act in that manner, but we shall be able to infer that they will do so in the absence of disturbing causes. And if we are able, as will frequently be the case, to state the cases in which those causes may be expected to exist, and the force with which they are likely to operate, we shall have removed all objections to the positive as opposed to the hypothetical treatment of that science (1848, 302; cf. 1852, 62–3).

That 'wealth and costly enjoyment' are universal and constant objects of desire represents the position adopted by Senior in the 1820s as we have seen.[105] He was now merely spelling out explicitly what he envisaged to be the implications of his earlier formulation and particularly his divergence from Mill.

A number of reasons are offered in justification of the 'positive' method

104 'Among the writers who appear to have taken this view of Political Economy, the most remarkable is Mr. Ricardo. His treatment of it, indeed, is more abstract than that proposed by Mr. Mill. He adds to Mr. Mill's hypothesis other assumptions equally arbitrary. And he draws all his illustrations not from real life, but from hypothetical cases. Out of these materials he has framed a theory as to the distribution of wealth possessing almost mathematical precision' (1848, 301–2). In his lectures of 1847–8 Senior added a reference to Herman Merivale in this context; cf. Senior (1852), 61.

105 Allowing, however, for the added reference to 'costly enjoyment' in the maximand, the significance of which is difficult to evaluate since it appears in a restatement of Mill's formulation.

proposed. In the first place 'no one listens to an exposition of what might be the state of things under given but unreal conditions, with the same interest with which he hears a statement of what is actually taking place'. Secondly, 'a writer who starts from arbitrarily assumed premises is in danger of forgetting from time to time their unsubstantial foundation and of arguing as if they were true' – a charge directed in particular against Ricardo. Thirdly, 'the strangeness' of the conclusions which flow from hypothetical axioms 'gives no warning' since we expect them to differ from what we observe; by contrast, 'when a writer takes his premises from observation or from consciousness, and infers from them what he supposes to be real facts, if he has committed any grave error it generally leads him to some startling conclusions. He is thus warned of the probable existence of an unfounded premise or of an illogical inference, and, if he be wise, tries back until he has detected his mistake' (1848, 302–3; cf. 1852, 62f).[106]

Senior's representation of the Ricardo-Mill method appears to me to be without foundation, for it attributes to Mill the recommendation that political economy adopt unreal, arbitrary and unsubstantial premises. Robert Torrens, who in 1821 had complained of the weak inductive basis to Ricardo's theorizing as distinct from a superb 'ratiocinative' ability (1821, iv) came in the end to defend Ricardo against a variety of charges along the lines suggested by Mill in the *System of Logic*: 'His conclusions are necessary truths, irresistibly commanding assent under the premises assumed, and enabling us, if we will make the necessary corrections for the difference between the hypothetical circumstances and the circumstances which actually exist, to arrive at conclusions practically true under all the varying conditions of society' (1834, xiii). J. E. Cairnes was later to insist upon this very point in his defence of Mill: Mill's premises were factual not hypothetical; the model, however, was hypothetical because the premises while 'true' were 'yet incomplete – true so far as the facts which they assert go, and yet not including all the conditions which affect the actual course of events' (1888, 68). Even the 'historicist' Cliffe Leslie appreciated Mill's position (below, p.927).

That Senior had misunderstood Mill seems clear. But Cairnes did not go far enough in the defence. It seems fair to say that Senior's charges against Mill can be applied against himself at least as far as concerns his formal position. For unlike Mill, who limited the scope of the behavioural axiom to a particular environment and allowed (in principle) for on-going

---

106 Senior illustrated from Ricardian theory, rejecting the following 'premises' attributed to Ricardo: that rent arises from differential fertility; fixed commodity wages; diminishing agricultural returns – 'The corn now raised with the greatest labour in England is raised with less labour than that which was raised with the least labour 300 years ago, or than that which is now raised with the least labour in Poland'; that the share of profits in national product declines secularly.

modifications to the axioms in the light of new empirical evidence, Senior formally insisted upon the universal relevance of the axioms thereby setting the model up in splendid isolation for all time and place. It was Mill not Senior who required a genuine inductive treatment of the premises of political economy, a characteristic of his mentor's position which Cairnes unfortunately failed sometimes to appreciate.[107]

It is pertinent that Cliffe Leslie took Senior, (not Mill) to task for effectively maintaining that the deductions of political economy 'follow from premises obtained without labour of investigation, lying on the surface of the mind or of things . . .' (1870, 357; cf. 379). McCulloch, also, criticized Senior on precisely the same grounds as those directed by Senior against Ricardo,[108] his observation suggesting how Senior was understood by a contemporary to be championing an 'hypothetical' rather than a 'positive' approach (in Senior's sense of the terms). My argument has been precisely that Senior's proposals imply a more 'hypothetical' method than do those of Mill and Ricardo – one more in line with that of James Mill at least in his formal writings.[109]

107 And Cairnes appears continuously to revert in his exposition to the 'self-evident' nature of the basic axioms thus aligning himself rather with Senior's 'positivistic' position. On this matter see Checkland (1951b), 161f; Bordo (1975), 351–2.

108 'Mr. Senior appears to take an erroneous view of the evidence on which its principles and conclusions are founded. He affirms, for example, that the facts on which its general principles rest may be stated in a very few sentences, or rather in a very few words; and that the difficulty is merely in reasoning from them. But while we admit the difficulty of drawing correct inferences we greatly doubt whether the general principles can be so easily established as Mr. Senior supposes. He lays it down, for example, as a general principle, or rather axiom, that supposing agricultural skill to remain the same, additional labour employed on the land will, speaking generally, yield a less return. But though this proposition be undoubtedly true, it is at the same time quite as true that agricultural skill never remains the same for the smallest portion of time; and that its improvement may countervail, for any given period, the decreasing fertility of the soils to which recourse is necessarily had in the progress of civilization. *It would indeed, be easy to show, that the worst lands now under tillage in England, yield more produce per acre, and more as compared with the outlay, than the best lands did in the reigns of the Edwards and the Henrys.* It is, therefore to no purpose to say, that the science rests on principles of this description. They, no doubt, form a part of its foundation; but as they are modified in different degrees by others, the only general principles of any practical value are those deduced from observations made on their combined action; or, in other words, on the phenomena really manifested in the progress of society' (1965 [1842], xiii–xiv).

The italicized passage (added emphasis) should be compared with Senior's statement regarding Ricardo, cited above (n106). On McCulloch's methodological position see O'Brien (1970), 96f; (1975), 68; Hirsch (1980), 103, 113.

109 Senior derived much of his methodology from Archbishop Whately (cf. Schumpeter, (1954), 483–4), and Whately reads like James Mill. Thus he defended deductive economics against those who appealed to 'the experience of practical men' or 'common sense' (1832, 61–3, 75), on the grounds that no progress is possible without 'fixed principles by which to regulate their judgement on each point' (76). He took issue with

In practice the differences may not be so great. Senior had to make allowance for 'disturbing causes' conceding that the wealth-maximizing axiom even from his perspective did not permit the inference that the various agents would certainly act to maximize wealth and that the principle of diminishing returns was not in fact universally valid (1836, 85–6). (In his early lectures he had spoken of the uncertainty of applied economics for this very reason.) The three arguments adduced by Senior to support his own supposedly more 'positivistic' approach are quite irrelevant from this point of view.

XI THE CAMBRIDGE (INDUCTIVIST) CRITICS

Something of a 'mystery' surrounds the origins of Mill's essay on definition and method. The problem concerns the relationship between the essay and John Frederick William Herschel's *Preliminary Discourse on the Study of Natural Philosophy* of 1830, William Whewell's papers on mathematical economics and economic method, first published in 1829 and 1831, and his review for the *British Critic* in 1831 of Richard Jones's *Essay on the Distribution of Wealth*. It is difficult to appreciate the absence in the essay itself, or in Mill's correspondence of the early 1830s, or in the retrospective account given in the *Autobiography*, of any reference to the Cambridge 'inductivists' who dealt with matters directly pertinent to the subject of the essay at the time of its composition. Indeed, the issue is more complex still, for on matters of methodological principle and even of detailed formulation there is considerable common ground between the three writers – particularly Herschel and Mill.[110] In effect, Mill stood side by side with the Cambridge 'inductivists' against the championship by Nassau Senior, Richard Whately and James Mill (not to speak of Harriet Martineau) of an axiomatic foundation of 'universal' relevance – an accord that is disguised by the erroneous tradition that Mill's essay makes out an extreme case for a priori procedure in economics designed to yield universally applicable generalizations based on unobservable behavioural motivation.

> those who championed 'fact collection' arguing that 'political economy, although a science which is founded on facts, and which has a practical application in reference to facts . . . yet requires for the establishment of its fundamental principles very little information beyond what is almost unconsciously, and indeed unavoidably, acquired by every one' (225); yet more strongly, 'the prominant part, and that which demands the principal share of our attention, in Political Economy, strictly so called (i.e., considered as to the *principles* of the science), must be the reasoning-process; – the accurate and dexterous application of Logical principles, in combining, and drawing inferences from, those few and simple data from which we set out; – in short, the Logical, not the Physical investigation' (239). And he took issue with Jones for objecting 'to the procedure of *founding our reasonings* [in Political Economy] *on definitions*' (243n).

110 For a full discussion see Hollander (1983).

A major theme running through Herschel's discourse is the requirement for a judicious blend of induction and deduction in scientific method – the former a process of 'reasoning upward' from specific instances to general principle and the latter a process of 'reasoning downward' from general principle to particular application: where 'the inductive and deductive methods of enquiry may be said to go hand in hand, the one verifying the conclusions deduced by the other', we have a 'combination of experiment and theory' which forms 'an engine of discovery infinitely more powerful than either taken separately' (1830, 179–81). For Herschel, the isolation of 'residual' phenomena (or disturbing causes) results by way of verification in applications of theory to specific cases after allowance for the 'compound effect' of known causes, the latter arrived at possibly by way of deductive reasoning (156). But the process of verification may lead not only to the isolation of a modifying cause (a residual) and the conformation of the initial model and its axioms; it may reveal the initial model itself to be defective and to require improvement (164–6). And here Herschel introduces his celebrated warning against casual verification of a theory and his recommendation to place oneself 'in the situation of its antagonists' and even 'perversely' to seek its weaknesses (167).

Herschel had no complaints about contemporary political economy on matters of method. On the contrary, in this branch of social science, he suggested, there was occurring an application of the proven methods of natural science – a blending of experience and reasoning:

The successful results of our experiments and reasonings in natural philosophy, and the incalculable advantages which experience, systematically consulted and dispassionately reasoned on, has conferred in matters purely physical, tend of necessity to impress something of the well weighed and progressive character of science on the more complicated conduct of our social and moral relations. It is thus that legislation and politics become gradually regarded as experimental sciences; and history . . . as the archive of experiments, successful and unsuccessful, gradually accumulating towards the solution of the grand problem – how the advantages of government are to be secured with the least possible inconvenience to the governed. The celebrated apothegm, that nations never profit by experience, becomes yearly more and more untrue. Political economy, at least, is found to have sound principles, founded in the moral and physical nature of man, which, however lost sight of in particular measures – however even temporarily controverted and borne down by clamour – have yet a stronger and stronger testimony borne to them in each succeeding generation, by which they must, sooner or later, prevail (172–3).

Mill's essay of 1836, both its general argument and terminological formulation is, in several major respects, at one with the *Preliminary*

Discourse.[111] No mention is made of it in correspondence of the early 1830s, the essay itself contains no references thereto, and in the *Autobiography* Mill explicitly states that he had originally read (and even reviewed) it 'with little profit' (*CW*, I, 215, 217). Yet this statement is belied by the record. For in a review for *The Examiner* (20 March 1831, 179-80) Mill represented the work as 'an example, and the only example, of a vast body of connected truth, gradually elicited by patient and earnest investigation, and finally recognized and submitted to by a convinced and subdued world'. The prejudice (even of scientists) against the study of methodology, despite the general failure to appreciate what scientific discovery entails, was condemned by Mill; for 'with the exception of the analysis of the syllogism which was performed long ago by the ancients, scarcely any thing has yet been contributed towards an accurate dissection of the mode in which the human understanding arrives at the discovery and verification of truth'. Herschel's work, by contrast, contained 'a clearer and less incomplete view of the nature of philosophical truth, of the evidence on which it rests, and the means of discovering and testing it, than is to be found in any work which has yet been produced . . . We never met with a book so calculated to inspire a high conception of the superiority of science over empiricism under the name of *common sense* – of the advantage of *systematic* investigation, and high general cultivation of the intellect.' The review ends by quoting in full the passage (cited above) relating to political economy, referring to 'the incalculable advantages which experience, systematically consulted and dispassionately reasoned on, has conferred' on physical science and was beginning to confer on the social and moral sciences. And Mill was in fact to use as epigraph for his second volume of the *System of Logic* that passage which champions a *mixed method of induction and deduction* – 'a combination of experiment and theory' forming 'an engine of discovery infinitely more powerful than either taken separately' (*CW*, VII, 284).[112] This passage would have served perfectly as epigraph for the essay of 1836, and Mill's neglect of Herschel is difficult to appreciate considering his approving tone both in 1831 and 1843. We note also that in the manuscript version of the *System of Logic* (and in all editions until the fourth) Mill commended Herschel's formulation in the *Preliminary Discourse* regarding axioms – that they are generalizations from experience (150-1n. regarding Herschel, 1830, 95-6).

111 In the passage just cited, however, Herschel seems to confound the 'science' and 'art' of political economy by conceiving the subject as concerned with the solution to a particular problem in policy. Mill in his essay warned against this type of confusion.
112 Mill also drew in the *System of Logic* upon Herschel's detailed analysis of the four methods of experimental investigation and his cautions regarding the limited scope of 'empirical laws' (cf. 405-6, 414f, 426-8). See also the friendly remarks regarding Herschel's book, in Mill to Herschel (1 May 1843, *CW*, XIII, 583-4).

We must proceed with care as far as concerns the matter of 'precedence', for two of the basic themes upon which the essay on method is predicated and which Mill shared with Herschel – the insistence upon an empirical justification for the axioms and the complementarity of inductive and deductive procedures – had been clearly expressed by Mill himself several years earlier. I refer to the analysis of Richard Whately's *Elements of Logic* for the *Westminster Review* of 1828 where – by and large following Whately on the principle involved[113] – Mill took issue with those 'inductive philosophers of modern times' who (misunderstanding the criticisms of the schoolmen by 'their idol' Bacon) believed that 'it was necessary to discard the syllogism which they thought was one method of reasoning, and confine ourselves to induction, which they imagined was another'.[114] The truth was very different:

That great writer, whom it is now fashionable to style the founder of inductive philosophy, a title which he himself would have been the foremost to disclaim, imputes the errors of Aristotle and the schoolmen, not to their neglecting induction, – for he had read them – but to their performing it ill. They knew that all knowledge must be ultimately derived from the observation of nature; but they were bad observers, and had even (as was remarked by Lord Bacon) fundamentally wrong ideas with respect to the proper mode of directing their observations. They consequently generalized on insufficient evidence, and arrived, by an incorrect induction indeed, but yet by induction, at general principles, which were not true, but which, if they *had* been true, *would have warranted* all the conclusions which they deduced from them. The merit, therefore, of Bacon, did not consist in teaching mankind to employ induction instead of syllogism, but in pointing out to them the insufficiency of the mode of induction which they had hitherto relied on, and communicating some useful hints for the formation of a better . . . The schoolmen erred . . . because they did not perform the *other* and equally necessary part of that process of investigating truth with the same unrivalled skill, with which, by the aid of logic, they performed that part of it with which alone logic is conversant (XI, 12–14).[115]

---

113 Cf. *Elements of Logic*, 207: 'Much has been said by some writers of the superiority of the Inductive to the Syllogistic method of seeking truth, as if the two stood opposed to each other; and of the advantage of substituting the Organon of Bacon for that of Aristotle, &c. &c. which indicates a total misconception of the nature of both.'

114 On Mill's early speculations regarding induction and deduction see Pappé (1979), 295–308; Sparshott (1978).

115 Decades later Mill repeated the charge of 1828 that Bacon's champions totally misread their master; cf. *An Examination of Sir William Hamilton's Philosophy* (1865), IX, 485n: 'It is but just to add, that the English mode of thought has suffered in a different, but almost equally injurious manner, by its exclusive following of what is imagined to be the teaching of Bacon, being in reality a slovenly misconception of him, leaving on one

The question of 'plagiarism' on Mill's part thus does not enter into the story. But how are we to explain the absence of any reference in the essay to the appearance, just at the time of its composition, of Herschel's *Preliminary Discourse*?

The papers on Ricardian theory by Whewell and the review of Richard Jones's *Essay* also insist upon a firm inductive basis for deductive reasoning but charge Ricardian economists for attempting to 'reason downwards' from axioms supposedly universal although at best of local relevance only. In some formulations it is implied that the immediate task at hand in political economy – the construction of generally applicable axioms by careful induction – was so immense that the process of 'downward reasoning' was for the distant future. Whewell also had much to say on the matter of 'disturbing' causes.

The first paper of 1829 is quite clear regarding the positive role for deduction. There is, perhaps, room for debate whether Whewell's readers would have regarded the two items of 1831 as a general attack against deduction as such or merely against premature deduction based on an inadequate axiomatic foundation. While the latter seems to have been Whewell's intention so that the positions of Mill and Whewell have much in common, a degree of ambiguity surrounds some of the formulations. But even if Whewell had been read by Mill as condemning deduction as a matter of principle, in which case there would exist a very substantive difference between their respective positions, his essay still in effect constitutes a defence of orthodoxy against Whewell's charge that it lacked an adequate empirical dimension by use of axioms snatched irresponsibly from narrow empirical observations and applied universally. One would expect some allusion to Whewell – if not of indebtedness or of common position, then at least as presenting an interpretation of orthodoxy which Mill could not accept.

There were, of course, basic differences between Whewell and Mill on the derivation of axioms and the role of hypotheses, which subsequently came to light – matters on which Mill and Herschel were to see eye to eye. Thus in the *System of Logic* Mill expressed his complete agreement with Herschel's criticism of Whewell as developed in the *Quarterly Review* of 1841, which 'maintains, on the subject of axioms, the doctrine advanced in [my] text, that they are generalizations from experience' a view Herschel supported 'by a line of argument strikingly coinciding with mine' (*CW*,

side the whole spirit and scope of his speculations. The philosopher who laboured to construct a canon of scientific Induction, by which the observations of mankind, instead of remaining empirical, might be so combined and marshalled as to be made the foundation of safe general theories, little expected that his name would become the stock authority for disclaiming generalization and enthroning empiricism, under the name of experience, as the only solid foundation of practice.'

VII, 248n).[116] But these differences imply a stricter insistence upon the careful use of empirical evidence on Mill's part than on that of Whewell.[117] The same may be said of the difference between the two on the proof of hypotheses (see above, p.83), namely Mill's objections to Whewell's ('Friedmanesque') position that the verification of predictions flowing from a hypothetically based theory constituted proof of the 'truth' of that theory. And this reinforces the relevance of our question why Mill failed to mention in his essay on political economy Whewell's position expressed at that time which also insisted upon the extrication of the axioms of the subject 'from the mass of facts which observation of the world and of ourselves teaches us . . .' (1831, 2).

In considering the Whewell connection we must take account of the concerted attempt on the part of the Cambridge writers and their allies – though *not* Herschell – to disparage contemporary political economy, a polemical campaign against orthodoxy represented as entailing hasty and illegitimate generalizations derived deductively from axioms supposedly of universal coverage (this in itself a legitimate ideal) but in actuality based on the most casual observation and even introspection. We know (Hollander, 1983, 140–1) that Whewell had growing doubts regarding Jones's prospect to develop an 'inductivist' program. Yet Mill may have been unaware of Whewell's doubts – doubts with which he would certainly have sympathized,[118] since, in that same year, he charged Sedgwick with failing

---

116 Indeed Mill felt it necessary to add that he had arrived at this position quite independently of Herschel. But it will again be noted that he also commended the formulation in Herschel's *Preliminary Discourse*, 95–6 in the manuscript version of the *Logic* and all editions until the fourth (250–1n).

 See too Mill to John Austin, (7 July 1842, *CW*, XIII, 528): '[Herschel's] review of Whewell contains so much that chimes with my comments on the same book that he would probably like to lend a helping hand to a writer on the same side with him' – an allusion to Mill's preference for Herschel as *Quarterly* reviewer of his forthcoming *System of Logic*. On 5 Oct. 1844 (639) Mill wrote to Comte that he found Whewell's (privately circulated) reply to Herschel's review 'très faible'.

117 Whewell represented the axioms of the mathematical and physical sciences – for example, that straight lines cannot enclose a space, that two straight lines which intersect cannot be parallel to a third, the indestructibility of matter, that bodies combine chemically in certain proportions – not as generalizations from experience but as universally and necessarily true 'on evidence' (in Mill's paraphrase in the *Logic*) 'of a higher and more cogent description than any which experience can afford' (*CW*, VII, 238 referring to Whewell, *History of Scientific Ideas* which was the first part of the *Philosophy of the Inductive Sciences*, 3rd ed., 1858).

 On Mill's hostility to the intuitional philosophy of the German metaphysicians and its bearing upon his debate with Whewell see Nagel, xxviif; Strong (1955) Randall (1965), 64f.; McRae, (1973), xxif.; Buchdahl (1971); Ryan (1979), xf.

118 See also Rashid (1980, 289–91), on the Cambridge opposition to 'Ricardian' economics on the grounds that *economics* could not be separated from *theology* or be based on materialistic premises. In the battle against utilitarianism, Ricardian economics was seen as the ally or associate of Benthamism.

to accord history its appropriate place in scientific investigation: 'the usefulness of history depends upon its being kept in second place' (*CW*, X, 45) – a nice summary indeed of the essay on method itself. And taking into account Whewell's participation in the concerted program against Ricardian political economy, and its polemical tone, it is not stretching things far to suggest that, despite the extensive common ground on methodological matters, Mill mistakenly understood Whewell's position to be opposed to deduction as a matter of principle and thus included him amongst those charged in the 1828 review of Whately with failing to appreciate Bacon and wishing 'to discard the syllogism which they thought was one method of reasoning, and confine [themselves] to induction which they imagined was another'.

This is all the easier to appreciate if we have in mind Whewell's weighting of the significance of the various tasks of political economy. The allowances for a deductive economics constructed on a narrowly-based axiomatic foundation appropriate for a local environment emerge only during the course of his detailed analysis of the Ricardian models, whereas his general observations on methodology in the economic pamphlets and the review of Jones give a very different impression – that the task at hand was the building up of general axioms (incorporating the full range of 'causes'), a task of 'far higher philosophical dignity and importance' than mere deduction, and one which might last for generations and even fail because of the peculiar complexity of social data to yield the requisite small number of universal axioms – in which case no use at all for deduction could be envisaged. S. G. Checkland has written of an 'ambiguity' in the public mind regarding Whewell's position: 'Whewell and those who thought with him were not . . . averse to political economy *per se*. Richard Jones provided the escape from the apparent dilemma with his Baconian injunction to "look and see." It was to him that Whewell, Sedgwick, Herschel, and other leading Cambridge scientists gave their blessing. Political economy was to some a long, perhaps indefinite, apprenticeship to the facts. This means, of course, that to the public their attitude appeared ambiguous, involving acceptance of political economy, but rejection of the leading school, at a time when the two appeared to the indiscriminating to be coterminous' (1951a, 61). It may well be that this too was Mill's judgement.

Let us recall also the major theme relating to the derivation of axioms in Whewell's later works on the history of science to which Mill was to object. In the *System of Logic* Mill expressed full agreement with Herschel's criticism of Whewell's a prioristic view of the axioms of the mathematical and physical sciences. But it is also clear that even in the early 1830s, at the time of composition of the methodological essay, Mill recognized much common ground with Herschel's *Preliminary Discourse*; whereas Whewell adopted a critical attitude towards Herschel as far as concerns the social sciences (Hollander, 1983, 137-8). There is no mistaking, also, the notion of an inductive process involving much more than a combination of logic and experience in Whewell's economic papers – the representation of the discovery of the principles which constitute the 'origin of a system', as 'an employment of far higher philosophical dignity and importance than any office to which the Mathematician can aspire', or the reference to 'the higher department of the science of political economy' (1831, 43-4). There is in short a 'metaphysical' dimension to Whewell's championship of induction in the social as well as the physical sciences.

This perspective pervades Whewell's contribution to the celebrated Bridgewater Treatises which appeared precisely during the preparation of Mill's essay (1834, 304, 329). The tone of the chapters on 'Inductive Habits' and 'Deductive Habits' relegates deduction to a lower sphere. Equally significant, Whewell alludes to the 'moral sciences' and warns of the uselessness of mathematical reasoning when prematurely undertaken (336-7). But most important we have here also a clear statement that the task at hand in the social sciences was the seeking for a 'First Cause of the Moral and Material World', a task impeded not by deficiencies of logical habits of mind but deficences of 'the habit of apprehending truth of other kinds' – that of 'ultimate truths and efficient causes'(340). This is precisely the message which Whewell conveys in his *general* statements regarding method in the economic pamphlets.

There are also broader intellectual issues of relevance. The Cambridge 'inductivist' group included leading churchmen and professors opposed to Ricardianism – their conception of Ricardianism – the population and rent principles in particular (Checkland, 52-3; and also 60f. regarding opposition to utilitarianism). To render the tone of their hostility we may refer to Jones's bitter denunciation of the birth control notions said to be implicit in the former doctrine – 'vile', 'degrading' and the work of the devil – and of the notion that the interests of the landlord are opposed to those of society as a whole (1831, xiii-xiv; x-xi; xxxi-xxxii).

Mill himself entered the lists in a review of Adam Sedgwick's discussion of the Cambridge University programme. Here *inter alia* he attacked the old universities as second rate and blamed them for the low cultural state of the community: 'All is right so long as no one speaks of taking away their endowments, or encroaching upon their monopoly' (*CW*, X, 35).[119] Sedgwick's comments on natural philosophy, the utilitarian theory of morals in particular, were described as 'a few trite commonplaces' (39); his conception of the theory of utility was condemned as 'misrepresentation

---

119 Sedgwick's *Discourse* was based on a lecture delivered in the Chapel of Trinity College in December 1832. The published version was dedicated to the 'Master of Trinity' (Whewell). Mill reviewed the 3rd ed. (1834); see *CW*, XII, 235, 238.

in voluntary ignorance' (72; also 51f, 57f); and the whole work was said to be on a 'level with a lower class of capacities' (73).

Whewell referred to Mill's (anonymous) review as the work either of a demagogic scoundrel or an idiot: 'I am somewhat puzzled, not being able to make out whether the reviewer is a scoundrel, who, by bringing together ferocious expressions knowingly, endeavours to excite people's passions against an antagonist, or whether he is a real *bonâ fide* example of that silliness which belongs to Benthamites and the like; and which can see nothing but moral horrors in all persons of opposite opinions' (Todhunter, 1876, II, 212). Mill obviously did not know of this particular comment, but he was perfectly aware of Whewell's connections – that he was enmeshed in the 'Cambridge Network' (Cannon, 1964).[120]

The conflict is quite evident. But nowhere is Mill's hostility towards Whewell better expressed than in a comment of 1852 which envisages Whewell's perspective on induction – his 'necessary truths' – as an invitation to social apologetics:

We do not say the intention, but certainly the tendency, of his efforts, is to shape the whole of philosophy, physical as well as moral, into a form adapted to serve as a support and a justification to any opinions which happen to be established. A writer who has gone beyond all his predecessors in the manufacture of necessary truths, that is, of propositions which, according to him, may be known to be true independently of proof; who ascribes this self-evidence to the larger generalities of all sciences (however little obvious at first) as soon as they have become familiar – was still more certain to regard all moral propositions familiar to him from his early years as self-evident truths. His *Elements of Morality* could be nothing better than a classification and systematizing of the opinions which he found prevailing among those who had been educated according to the approved methods of his own country; or, let us rather say, an apparatus for converting those prevailing opinions, on matters of morality, into reasons for themselves (*CW*, X, 168–9).[121]

120 Whewell had published in defence of mandatory daily attendance at college chapel service (Douglas, 1881, 165, 173). Sedgwick in his *Discourse* had spoken of political philosophy in terms that Mill would probably have found acceptable: 'To what then are we led by considerations such as these? To the belief, that all systems of political philosophy based on the doctrines of utility, and deduced by *a priori* reasoning from assumed simple principles (without comprehending all the great elements of man's mortal nature, and without, perhaps, even regarding his social condition), are either mischievous or impracticable. Universal systems, like universal nostrums, savour more of political quackery than political philosophy' (86). This was, precisely the complaint Mill himself had directed against the Benthamites and Comte years earlier. But on this passage he remains silent in his review, and it may well be because he did not see fit to express agreement with a man – and a community – for whom he had so little respect.

121 Cf. 'in the English Universities no thought can find place, except that which can reconcile itself with orthodoxy' (167).

In the light of all this it cannot be precluded that Mill's neglect of Whewell's economic papers in 1836 is, at least in part, explicable by a sharp division of social attitude, whether or not Mill realized the extent of Whewell's allowance regarding deductive theory in economics within an appropriately limited environment. Common ground on this matter – and it is far from certain that Mill recognized it as such – would have paled in significance compared with the broader issues at stake.

We must allow also for the fact that it was not in Mill's character to advertise differences between political economists. While his essay alludes to certain fallacious perspectives including those of his own father, it constitutes largely a positive defence of the subject as envisaged by the great masters. To emphasize agreement with Whewell and Cambridge with their critical tone, albeit valid when directed at some of the lesser lights amongst the economists, would perhaps have suggested a *Methodenstreit* within economics and was therefore to be avoided.

Mill's silence in his essay regarding Herschel's position cannot, however, be so easily accounted for. Herschel's formulations on scientific method in general were unambiguous; he did not engage in polemic against the Ricardians; he was not apparently linked in Mill's mind with social apologetics; and we have Mill's own very favourable reaction of 1831. The question stands.

There remains a postscript to the Mill-Whewell relation. In a discussion of Mill's *System of Logic* published in 1849 Whewell conceded frankly that deductive procedure took precedence in the research program of political economy; little more could be achieved from further elaboration of the inductive establishment of axioms. We must presume (putting the matter in the terms used in 1831) that Whewell now believed political economy to be a 'much advanced' science. What is striking in all this is his failure to appreciate the great weight Mill placed, in his book on logic (which incorporates the substance of the early essay on definition and method), upon the need for a careful building up of the axioms from the evidence, for the inductive process was envisaged as on-going, the axioms of political economy subject to continual improvement and modification.[122] It is Whewell who in the end emerges as champion of a procedure which he and his inductivist colleagues originally attributed to the economists in rather hostile terms (1860 [1849], 284–5).

---

122 It is doubtful whether Mill would have been pleased by the new ally, considering his comment of 1852 regarding the apologetic implications of Whewell's moral philosophy which reiterates those earlier criticisms of the Cambridge writers discussed above.

Richard Jones himself totally misunderstood the message of Mill's essay. See also Rashid (1979), 168 who falls into the same trap confounding the deductive with the geometrical method.

XII THE UNIVERSALITY OF THE METHOD OF POLITICAL ECONOMY

In the course of his criticisms in 1834 of English political economists who 'attempt to construct a permanent fabric out of transitory materials' and 'presuppose, in every one of their speculations, that the produce of industry is shared among three classes, altogether distinct from one another – namely, labourers, capitalists, and landlords; and that all these are free agents, permitted in law and fact to set upon their labour, their capital, and their land, whatever price they are able to get for it' (*CW*, IV, 225–6) Mill added a qualification which withdraws the barbs as far as concerns the method itself: 'It must not, however, be supposed that the science is so incomplete and unsatisfactory as this might seem to prove. Though many of its conclusions are only locally true, its method of investigation is applicable universally; and as he who has solved a certain number of algebraic equations, can without difficulty solve all others, so he who knows the political economy of England, or even of Yorkshire, knows that of all nations actual or possible: provided he have sense enough not to expect the same conclusions to issue from varying premises.' A similar position is taken in a reaction to Auguste Comte in 1865. It was Comte's view, to which Mill subscribed, that social science constituted the most advanced and complex of all the sciences (X, 282–4). But at this point Mill broke away on a matter of principle – an insistence upon the legitimacy of a quasi-independent science of political economy having in mind the general applicability of its method:

On the whole question [of political economy] he has but one remark of any value, and that he misapplies; namely, that the study of the conditions of national wealth as a detached subject is unphilosophical, because, all the different aspects of social phaenomena acting and reacting on one another, they cannot be rightly understood apart: which by no means proves that the material and industrial phaenomena of society are not, even by themselves, susceptible of useful generalizations, but only that these generalizations must necessarily be relative to a given form of civilization and a given stage of social achievement. This, we apprehend, is what no political economist would deny. None of them pretend that the laws of wages, profits, value, prices, and the like, set down in their treatises, would be strictly true, or many of them true at all, in the savage state (for example), or in a community composed of masters and slaves. But they do think, with good reason, that whoever understands the political economy of a country with the complicated and manifold civilizations of the nations of Europe, can deduce without difficulty the political economy of any other state of society with the particular circumstances of which he is equally well acquainted (305–6).

This claim regarding the universal applicability of the method of political economy has been said to conflict with the position laid down in the *Principles* that 'only through the principle of competition has political economy any pretension to the character of a science' (Winch, 1972, 340). Edgeworth (1910, II, 757), Winch observes, 'drew attention to this inconsistency when he pointed out that it was not really possible for Mill to retain belief in the *à priori* deductive method when "he began to doubt the universality of the principle of self-interest, which he once regarded as the foundation of economic reasoning".' Now the fact is that Mill had 'doubted' the universality of the principle of maximizing behaviour from 1829 onwards. Are we then obliged to charge him with a self-contradiction throughout the essay?

The formulation relating to the applicability of the method of economics to 'all nations actual or possible', (including even 'the savage state') appears indeed to be too strongly stated on Mill's own terms. However, in his paper of 1834 Mill is quite explicit regarding his specific intentions: 'The conclusions of the science,' Mill insisted, 'being all adapted to a society thus constituted, require to be revised whenever they are applied to any other. They are inapplicable where the only capitalists are the landlords, and the labourers are their property; as in the West Indies. They are inapplicable where the universal landlord is the State; as in India. They are inapplicable where the agricultural labourer is generally the owner both of the land itself and of the capital, as in France; or of the capital only, as in Ireland. We might greatly prolongue this enumeration' (IV, 226). In the *System of Logic* Mill pointed out that ' [t] he deductive science of society will not lay down a theorem, asserting in an universal manner the effect of any cause; but will rather teach us how to frame the proper theorem for the circumstances of any given case. It will not give the laws of society in general, but the means of determining the phenomena of any given society from the particular elements or data of that society' (VIII, 899–900). To this he adds that 'whoever has mastered with the degree of precision which is attainable the laws which, under free competition, determine the rent, profits, and wages, received by landlords, capitalists, and labourers, in a state of society in which the three classes are completely separate, will have no difficulty in determining the very different laws which regulate the distribution of the produce among the classes interested in it, in any of the states of cultivation and landed property set forth in the foregoing extract' (904) a reference to a range of specific institutional arrangements listed in Harriet Martineau's *Political Economy* (1834). The argument is referred to with enthusiasm by Torrens (1844, xif) in his defence of Ricardian economics.

The strong statements regarding the applicability of the method of political economy 'to *all* nations and all times' do not, therefore, extend maximization principles to all possible cases including (say) those involving custom or gift or force (see above p.108 and below p.266) which Mill maintained were not amenable to economic analysis and would have to be dealt with by 'some other' science. Mill was pointing rather to the working out of the maximization hypothesis within a wide variety of specific institutional arrangements in addition to the capitalist-exchange system. In all this nothing is said formally of the empirical justification for the hypothesis in the systems in question. But we know from the essay and the *System of Logic* that for economics to be of any great practical service there must be evidence that the maximizing axiom does reflect the 'predominating influence' governing behaviour in the particular sphere of activity and locale under investigation.[123]

---

123 For pertinent comments on the issue of this section cf. Marshall (1925), 156f.

THREE

Transition to the *Principles*

I THE ESSAY ON METHOD AND THE *PRINCIPLES*

Historians of economics frequently allude to an inconsistency between methodological pronouncement and practice on Mill's part, or at least to a marked change between the essay and the *Principles*. In his early article on logical method in economics (see above, p.66–7) Jacob Viner found the one redeeming feature of the record as far as concerned Mill to be the apparent circumstance that he did not live up to the (supposedly) unsatisfactory standards established in the essay when he came to compose his text: 'even those economists who were most decided in their contention that the abstract deductive method was the only one available to the economist made considerable use of inductive methods in their economic researches' (1962 [1917], 117). Similarly: the *Principles* 'has no single methodological character'; like the *Wealth of Nations* 'some portions are predominantly abstract and *a priori*; in others, there is a substantial measure of factual data and of inference from history' (1958, 329).[1]

Nassau Senior remarked early on that 'in the Essays Political Economy is an hypothetical science: in the Principles it is a positive art' (1848, 304). Ingram charged that Mill 'shifted his position' on method and ended up on uncertain ground (1888, 150). John Neville Keynes, who was far more sympathetic to the classics than Ingram, wrote of a 'marked contrast' between Mill's methodological formulations in the essay and his practice

---

[1] In this paper Viner still refers to the *Principles* as a work based on artificial axioms drawn from introspection. But this is qualified in the light of the restriction of that procedure to the 'scientific' part of political economy and its rejection for other established branches of social thought; as well as the according 'application' a wider range of method and the ('platonic') allowance for abstraction from reality in making policy application.

in the *Principles* (1891, 18–19). Marshall observed that 'a change had come over [Mill's] tone of thought and of feeling before he published in 1848 his great economic work'; for Mill

> desired to call attention to the influences which are exerted on human conduct by custom and usage, by the ever-shifting arrangements of society, and by the constant changes in human nature; the pliability of which he agreed with Comte in thinking that earlier economists had underrated. It was this desire which gave the chief impulse to his economic work in the latter half of his life, as distinguished from that in which he wrote his *Essays on Unsettled Questions*; and which induced him to separate distribution from exchange, and to argue that the laws of distribution are dependent on 'particular human institutions', and liable to be perpetually modified as man's habits of feeling, and thought, and action pass from one phase to another (1920, 823–4).

More recent examples of the theme include Schumpeter's criticism of Mill's 'misleading' use of the term 'a priori' in the essay on method and his unnecessary emphasis upon 'deduction', coupled with a concession that such deficiencies never caused any errors in practice, since 'the literal meaning of a methodological profession of faith is of little interest except for the philosopher' (1954, 537).[2] D. P. O'Brien describes Mill's position as 'somewhat of a hybrid mixture. On the one hand Mill defended Ricardian theorems, established by something at least very like the "Geometrical Method" as "tendencies" when they were contradicted by the available evidence as it was gathered by such writers as McCulloch and G. R. Porter. On the other, when one takes the contents of the *Principles* as a whole – in other words, when we see what Mill "did" – it is impossible to deny Schumpeter's view that Mill's formal stress on deduction is excessive' (1975, 73). Some commentators also complain of 'methodological confusion', in that Mill advocated a 'rigorously deductive method for economic science' in the earlier essay and indeed remained 'relatively orthodox in his theory of value and distribution', yet reached 'conclusions on land reform, the social effects of unregulated competition, the protection of infant industries, evolutionary socialism, state sponsored social reform [which] often resembled those of the historical economists' (Koot, 1975, 320–2; 1980, 179).

The foregoing perspectives will not stand up to a close examination. The circumstance of the publication of the essay and the *Logic* at approximately

---

2 Schumpeter estimates that although the proportion of 'factual' material amounted to only one-sixth of the book, when allowance is made for the factual matter implicit in the discussion of applied problems – the presumption that the reader had at hand works such as Babbage's *Economy of Machinery and Manufacture* – then the proportion rises to more than two thirds, the remainder constituting the analytical apparatus (541–2).

the same time that Mill set about the preparation of the *Principles* and the contemporaneous reissues of these works, coupled with Mill's own insistence that the essay and the *Principles* were to be read side by side (*CW*, XV, 517), point away from any likelihood of substantive inconsistency or of a change in position. More positively, if the account of the essay given in the preceding chapter holds good, the entire issue of inconsistency falls away. To the extent that the *Principles* is an eclectic work covering abstract theory, application, and economic and sociological history, it is perfectly consistent with the essay on method which itself insisted upon a subtle blend of the deductive and inductive. Despite some anomalies, on the whole there is close accord between the pronouncements of the essay and the *Principles* so far as concerns the linkages between factual investigation, economic theory and policy recommendation; indeed the *Principles* can only be fully appreciated in the light of the essay.[3]

Contrasting with historians of thought, economic historians tend to play down the historical dimension to the *Principles*. It was the lament of the celebrated British historicist T. E. C. Leslie that Mill had failed to develop a 'doctrine of evolution' or to investigate 'the entire system of human society' with an eye to a better appreciation of 'the origin and growth of human ideas' – a failure for which Mill himself could not fairly be blamed. The historical method was admittedly 'taken up' in the Preliminary Remarks to the *Principles*, but 'only to be laid aside':

it was the fault of [Mill's] age and of his education if the doctrine of evolution found no place in his psychology or his social science; if the historical method was taken up in the preliminary remarks of his treatise only to be laid aside; and if corrections from observation and fact of the inferences from *a priori* reasoning appear . . . only in the form of practical exceptions to abstract theory, or of 'applications' of economic science, when the fault really lay in the original conception of the science itself. It was not possible to weld the abstractions of Ricardo and the actual forces governing economic phenomena into a consistent and scientific system; or to furnish an adequate theory of the origin and growth of human ideas without investigation of the entire system of human society. But if any one individual is especially to be blamed for the shortcomings of his system, it is not John, but James Mill (1888, 57).

This view has been repeated in recent statements by economic historians: 'John Stuart Mill's *Principles of Political Economy* (1848), though concerned to some extent with what he called "applications" as well as with the "principles" themselves, followed Ricardo in treating economics in a

---

3 On the necessity to read the *Principles* in the light of the essay bearing in mind the linkage there championed between theory and 'experience' see de Marchi (1974), 139n.

basically non-historical manner'; the method adopted 'was that of logic and deduction from abstract principles, rather than of empirical investigation and historical inquiry' (Harte, 1971, xiii).

Here again emerges the unjustified notion of a conflict between logical deduction on the one hand and empirical investigation and historical inquiry on the other. It is certainly true, however, that history as a matter of principle, was to be kept in 'second place':

There is not a fact in history which is not susceptible of as many different explanations as there are possible theories of human affairs. Not only is history not the source of political philosophy, but the profoundest political philosophy is requisite to explain history; without it all in history which is worth understanding remains mysterious . . . Mr. Sedgwick mistakes the function of history in political speculation. History is not the foundation, but the verification, of the social science; it corroborates, and often suggests, political truths, but cannot prove them. The proof of them is drawn from the laws of human nature ascertained through the study of ourselves by reflection, and of mankind by actual intercourse with them. That what we know of former ages, like what we know of foreign nations, is, with all its imperfections, of much use, by correcting the narrowness incident to personal experience, is undeniable; but the usefulness of history depends upon its being kept in the second place (1835, *CW*, X, 44-5).

Decades later Mill repeated a charge of 1828 that Bacon's champions totally misread their master:

It is but just to add, that the English mode of thought has suffered . . . by its exclusive following of what is imagined to be the teaching of Bacon, being in reality a slovenly misconception of him, leaving on one side the whole spirit and scope of his speculations. The philosopher who laboured to construct a canon of scientific Induction, by which the observations of mankind, instead of remaining empirical, might be so combined and marshalled as to be made the foundation of safe general theories, little expected that his name would become the stock authority for disclaiming generalization and enthroning empiricism, under the name of experience, as the only solid foundation of practice (1865, *CW*, IX, 485n).

And it is also true that Mill did not attempt the daunting task of developing a 'doctrine of evolution'.

II RICARDIAN THEORY, SMITHIAN FRAMEWORK

Pride of place in Mill's *Principles* is accorded Ricardian theory although this fact is disguised by the form of organization. That Mill intended from the outset to write a work much broader in scope than that of Ricardo or of his father is not in question: 'I believe if I have done any good a large share of it lies in the example of a professed logician & political economist who believes there are other things besides logic & political economy' (23 Dec. 1840, *CW*, XIII, 453). The contrast drawn by Mill (before commencing work) between their 'abstract manner' and his own proposed approach, 'which resembled the practical and popular manner of Adam Smith', is also not at issue. What he envisaged was no small task: 'The whole science requires extremely to be recast, incorporating, of course, Wakefield's and all the other new doctrines and shewing how they do not contradict but *fit into* the others, and such a book if one were able to do it well would at once supersede all the existings treatises, which are, one and all, effete and useless except as matter of history, and would give a right direction to the revived interest which begins to be felt in the study, and which languishes for want of a book at once free from gross error and teaching the applications along with the principles, which it is the beauty of Adam Smith's book that he did' (8 Nov. 1844, 642).[4] Likewise, when reporting to a correspondent in 1847 the completion of his work (bar revision and rewriting) he described it as 'a book to replace Adam Smith, that is, to attempt to do for political economy what A. S. did at the time when he wrote, to make a book which, while embodying all the abstract science in the completest form yet attained, incorporating all important improvements, should at the same time be essentially a book of applications exhibiting the principles of the science in the concrete. I was the more prompted to do this inasmuch as it would enable me to bring in, or rather to bring out, a great number of opinions on incidental matters, moral and social, for which one has not often so good an opportunity, and I have used this privilege as freely as Adam Smith did . . .' (9 March 1847, 708).

In the *Principles* itself Mill elaborated upon the decision to proceed in the Smithian manner:

Political Economy is inseparably intertwined with many other branches of social philosophy. Except on matters of mere detail, there are perhaps no practical questions, even among those which approach nearest to the character of purely economical questions, which admit of being decided on economical premises alone. And it is because Adam Smith never loses sight of this truth; because, in his applications of Political Economy, he perpetually appeals to other and often larger considerations than pure Political Economy affords – that he gives that well-grounded feeling of command over the principles of the subject for purposes of practice, owing to which the 'Wealth of Nations,' alone among treatises of Political

---

4 Cf. letter to John Sterling, 29 May: 'I think my next book will be a systematic treatise on Political Economy, for none of the existing ones are at all up to the present state of speculation' (630).

Economy, has not only been popular with general readers, but has impressed itself strongly on the minds of men of the world and of legislators (II, xci).

The need was great because the *Wealth of Nations* was 'in many parts obsolete, and in all imperfect. Political Economy, properly so called, has grown up almost from infancy, since the time of Adam Smith' (xcii). Even the 'philosophy of society . . . though still in a very early stage of its progress, has advanced many steps beyond the point at which he left it. No attempt, however, has yet been made to combine his practical mode of treating his subject with the increased knowledge since acquired of its theory, or to exhibit the economical phenomena of society in the relation in which they stand to the best social ideas of the present time, as he did, with such admirable success, in reference to the philosophy of his century.'

Mill was also very clear regarding the weighting of his interests: 'I confess that I regard the purely abstract investigations of pol. economy (beyond those elementary ones which are necessary for the correction of mischievous prejudices) as of very minor importance compared with the great practical questions which the progress of democracy & the spread of Socialist opinions are pressing on, & for which both the governing and the governed classes are very far from being in a fit state of mental preparation' (20 March 1852, XIV, 87). In the *Autobiography* the success of the *Principles* was ascribed precisely to its preoccupation with application in addition to abstract theory – and responsibly qualified application at that; to the fact that its conclusions were treated as true only 'conditionally' and subject to correction for what Mill (in the essay) referred to as 'disturbing causes' emanating outside the range of concern of the specialist economist; and to its distinction between the laws of production, common to all industrial societies, and the laws of distribution and exchange, specific to particular social arrangements (I, 243).

Yet for all that the *Principles* contains a core of Ricardian theory – 'embodying all the abstract science in the completest form yet attained'; the broad objective of the *Principles* in no way affects the body of Ricardian theory contained therein as it relates to a capitalist exchange system. The common notion that the 'Smithian' form of Mill's *Principles* itself implies a breakaway from Ricardian theory is an error which has a counterpart in the belief that Smith, merely by dint of his broad concern for application and social arrangement, falls *ipso facto* into a different analytical 'school' from Ricardo. In both instances the misunderstanding flows from a neglect of the intended audience – the fact that, unlike Smith and Mill, Ricardo was addressing professionals. We can go further in Mill's case: his emphasis in the *Principles* upon the 'provisional' character of the conclusions reached by economists concerning practical matters or the need to allow for 'other classes of considerations' constitutes a carefully thought out strategy of using a Smithian form of presentation *in defence of* scientific economics as developed by Ricardo.

III THE COMTE CONNECTION

Mill's critical perspective on contemporary political economy owed something to the Saint-Simonians although, as remarked earlier, the debt was a qualified one (above, pp.86–7), and to Austin. But so far as concerns the *Principles of Political Economy* specifically and its orientation or 'tone', defined by the celebrated contrast between categories of economic laws, Mrs. Taylor was given pride of place in the *Autobiography*:

> The purely scientific part of the Political Economy I did not learn from her; but it was chiefly her influence that gave to the book that general tone by which it is distinguished from all previous expositions of Political Economy that had any pretension to being scientific, and which has made it so useful in conciliating minds which those previous expositions had repelled. This tone consisted chiefly in making the proper distinction between the laws of the Production of Wealth . . . and the modes of its Distribution . . . I had indeed partially learnt this view of things from the thoughts awakened in me by the speculations of the St. Simonians; but it was made a living principle pervading and animating the book by my wife's promptings (*CW*, I, 255, 257).[5]

In this context Mill singled out the chapter on 'The Probable Futurity of the Labouring Classes' – which he described as the most influential on opinion – for special attention. It was

> entirely due to her: in the first draft of the book, that chapter did not exist. She pointed out the need of such a chapter, and the extreme imperfection of the book without it: she was the cause of my writing it; and the more general part of the chapter, the statement and discussion of the two opposite theories respecting the proper condition of the labouring classes, was wholly an exposition of her thoughts, often in words taken from her own lips . . . [She] was much more courageous and far-sighted than without her I should have been, in anticipations of an order of things to come, in which many of the limited generalizations now so often confounded with universal principles will cease to be applicable. Those parts of my writings, and especially of the Political Economy, which contemplate possibilities in the future such as, when affirmed by Socialists, have in general been fiercely denied by political economists, would, but for her, either have been absent, or the

---

5 Some presentation copies of the second edition of the *Principles* had pasted in them a formal dedication to Mrs Taylor (see Hayek (1951), 122, 297).

suggestions would have been made much more timidly and in a more qualified form.[6]

We shall have more to say on the validity of this retroactive view in our discussion of Mill's Socialism.

It was also Mill's belief that the *Principles* – organized within a Smithian framework to emphasize the distinction between the laws of production 'common to all industrial societies' and those of distribution and exchange as manifested in the capitalist-exchange system – would provide powerful propaganda for Comtian 'positivism':

J'ai même encore l'idée, puisque mes méditations éthologiques ne seront pas mûres de longtemps, de faire en attendant ce qui ne serait pour moi qu'un travail de quelques mois, c'est-à-dire un traité spécial d'économie politique, analogue à celui d'Adam Smith qui n'est certainement plus au niveau de ce temps-ci, tandis que sa place n'est pas encore convenablement remplie. Je sais ce que vous pensez de l'économie politique actuelle: j'en ai une meilleure opinion que vous, mais si j'écris quelque chose là-dessus ce sera en ne perdant jamais de vue le caractère purement provisoire de toutes ses conclusions concrètes, et je m'attacherais surtout à séparer les lois générales de la production, nécessairement communes à toutes les sociétés industrielles, des principes de la distribution et de l'échange des richesses, principes qui supposent nécessairement un état de société déterminé, sans préjuger que cet état doive ou même qu'il puisse durer indéfiniment, quoiqu'en revanche il soit impossible de juger les divers états de la société sans prendre en considération les lois économiques qui leur sont propres. Je crois qu'un pareil traité peut avoir, surtout ici, une grande utilité provisoire et qu'il servira puissamment à faire pénétrer l'esprit positif dans les discussions politiques (3 April 1844, XIII, 626).

This is a potentially significant consideration, for Mill went so far as to assert that any hesitation he might have felt regarding the preparation of a general treatise on political economy turned on expectations of Comte's negative reaction. These fears were dissipated in the event by Comte's approval (1 May 1844 in Lévy-Bruhl ed., 1899, 314), which Mill attributed to the proposed emphasis upon the provisional status of those industrial phenomena that abstract from 'the general movement of humanity' and are solely pertinent to specific institutional arrangements – and by implication upon the permanent status of those that do not so abstract:

6   The reference to 'the two opposite theories respecting the proper condition of the labouring classes' is to 'the theory of dependence and protection' and that of 'self-dependence' (*CW*, III, 759).

Je me félicite de l'approbation que vous voulez bien donner à mon projet de faire sur l'économie industrielle des sociétés un traité un peu plus systématique. Je ne me sentais pas auparavant suffisamment assuré de votre adhésion à ce projet, qui pouvait vous paraître essentiellement anti-scientifique, et qui le serait en effet si je n'avais le plus grand soin de bien établir le caractère purement provisoire de toute doctrine sur les phénomènes industriels qui fasse abstraction du mouvement général de l'humanité. Je crois que ce dessein, s'il pouvait être convenablement éxécuté, aurait l'avantage de préparer l'éducation positive de beaucoup d'espirits qui s'occupent plus ou moins sérieusement des questions sociales, et il me semble aussi qu'en prenant pour modèle général le grand et le beau travail d'Adam Smith, j'aurais des occasions importants de répandre directement quelques-uns des principes de la nouvelle philosophie, comme Adam Smith a fait pour la plupart de ceux de la métaphysique négative dans ses applications sociales sans éveiller les défiances onbrageuses en déployant aucun drapeau (3 June 1844, XIII, 630–31).

However, the Comte-Mill link must not be exaggerated. W. J. Ashley has made the very significant point that Mill and Comte were in fact at 'cross purposes' in the foregoing exchange since Comte meant by 'provisional' a legitimate role for political economy '*until a positive Sociology can be created*', whereas Mill meant '*so long as the present system of private property lasts*'. And, Ashley adds correctly, 'until the present social system should be fundamentally changed, Mill clearly regarded the Ricardian economics as so far applicable to existing conditions as to call for no substantial revision in method or conclusions' (Mill, 1917, xxiii). In fact, Mill's differences with Comte are yet more serious, since despite the accommodating letter of 1 May 1844 Comte's opinion of political economy was as low as Mill originally suspected. It is helpful to place all this into suitable perspective.[7]

Mill referred to certain differences with Comte in his very first letter to him, a letter of introduction.[8] They were scarcely dissipated by years of

---

7 For a general account of Comte in the context of the idea of progress, see Pollard, (1968), Ch. III.

8 'Depuis l'heureuse époque où ces deux volumes me sont connus' – a reference to the first two volumes of the *Cours de Philosophie Positive* which Mill first read in 1837 – 'j'attends toujours chaque volume nouveau avec une vive impatience et je le lis et je le relis avec une véritable passion intellectuelle. Je puis dire que j'étais déjà entré dans une voie assez voisine de la vôtre, surtout par l'impulsion que m'avait donnée votre ouvrage précédent' – a reference to the *Traité de Politique Positive* which Mill first read in 1829; 'mais j'avais encore à apprendre de vous bien des choses de la première importance, et j'espère vous donner à quelque temps d'ici la preuve que je les ai bien apprises. Il reste quelques questions d'un ordre secondaire sur lesquelles mes opinions ne sont pas d'accord avex les vôtres . . .' (8 Nov. 1841, *CW*, XIII, 489).

correspondence. Some of the more important were developed formally in his famous review of the *Cours de Philosophie Positive* in 1865,[9] and this document provides us with a convenient summary of Mill's considered position.

The tone of the critique was by and large respectful since Mill adhered consistently to the broad scheme of things expounded in the *Cours*, including Comte's notions regarding the development of science and philosophy of history. Thus he accepted the fundamental proposition – originating, Mill insisted, with Comte – that 'every distinct class of human conceptions passes through all these stages, beginning with the theological, and proceeding through the metaphysical to the positive: the metaphysical being a mere state of transition, but an indispensable one, from the theological mode of thought to the positive, which is destined finally to prevail, by the universal recognition that all phaenomena without exception are governed by invariable laws, with which no volitions, either natural or supernatural, interfere' (X, 269).[10] It was in terms of the latter criterion that Mill defined the 'positivist' standpoint.[11]

This general perspective governed also Mill's warm appraisal in his review of the *Cours* of Comte's philosophy of history: 'We find no fundamental errors in Comte's general conception of history'–which avoided 'the vulgar

---

9 For the *Westminster and Foreign Quarterly Review*, LXXXIII, April 1865, in *CW*, X, 261f. The review was of the second edition of the *Cours* in six volumes (1864). The six volumes of the first edition were published between 1830 and 1842.

10 His agreement extended to matters of detail: 'Combining the doctrines, that every science is in a less advanced state as it occupies a higher place in the ascending scale, and that all the sciences pass through the three stages, theological, metaphysical, and positive, it follows that the more special a science is, the tardier is it in effecting each transition, so that a completely positive state of an earlier science has often coincided with the metaphysical state of the one next to it, and a purely theological state of those further on. This statement correctly represents the general course of the facts, though requiring allowances in the detail' (287–8).

11 'The fundamental doctrine of a true philosophy, according to M. Comte, and the character by which he defines Positive Philosophy, is the following: – We have no knowledge of anything but Phaenomena; and our knowledge of phaenomena is relative not absolute. We know not the essence, nor the real mode of production, of any fact, but only its relations to other facts in the way of succession or of similitude. These relations are constant; that is, always the same in the same circumstances. The constant resemblances which link phaenomena together, and the constant sequences which unite them as antecedent and consequent, are termed their laws. The laws of phaenomena are all we know respecting them. Their essential nature, and their ultimate causes, either efficient or final, are unknown and inscrutable to us' (265–6); cf: 'whoever regards all events as part of a constant order, each one being the invariable consequent of some antecedent condition, or combination of conditions, accepts fully the Positive mode of thought: whether he acknowledges or not a universal antecedent on which the whole system or nature was originally consequent, and whether that universal antecedent is conceived as an Intelligence or not' (270).

mistake of supposing that the course of history has no tendencies of its own' while, at the same time, was 'free from the error of those who ascribe all to general causes' (322). And the insistence upon a 'positive' approach to history in the investigation of social and political phenomena was applauded:

Much has been said and written for centuries past, by the practical or empirical school of politicians, in condemnation of theories founded on principles of human nature, without an historical basis; and the theorists, in their turn, have successfully retaliated on the practicalists. But we know not any thinker who, before M. Comte, had penetrated to the philosophy of the matter, and placed the necessity of historical studies as the foundation of sociological speculation on the true footing. From this time any political thinker who fancies himself able to dispense with a connected view of the great facts of history, as a chain of causes and effects, must be regarded as below the level of the age; while the vulgar mode of using history, by looking in it for parallel cases, as if any cases were parallel, or as if a single instance, or even many instances not compared and analyzed, could reveal a law, will be more than ever, and irrevocably, discredited (308).

Mill also expressed his profound distaste for the lunatic aspects of the man and his work – the appeal to phrenology in place of direct mental observation or psychology, a failing which damaged the attempt by Comte to create a science of sociology (297–8); the educational schemes (312–3); and the conception of a moral authority or spiritual power (314–15, 326f).[12] He commented also on the failure to place the practical recommendations regarding future improvement on a 'scientific' basis:

After so profound and comprehensive a view of the progress of human society in the past, of which the future can only be a prolongation, it is natural to ask, to what use does he put this survey as a basis of practical recommendations? . . . We fail to see any scientific connexion between his theoretical explanation of the past progress of society, and his proposals for future improvement . . . [His recommendations] rest as completely, each on its separate reasons of supposed utility, as with philosophers who, like Bentham, theorize on politics without any historical basis at all (324–5).

There were also criticisms relating to the absence of proof in Comte's

---

12 Mill was clearly concerned at an early stage with the problem of the source of power, although his formulation was muted: 'Vous vous êtes . . . sagement borné, quant à l'ordre temporel, à poser le principe incontestable, que la direction en doit désormais appartenir aux chefs industriels, en laissant indécises bien des questions, destinées à être progressivement résolues par les sociologistes positifs, et sur lesquelles je désirerais

philosophy of science.[13] But for us here the most significant of Mill's reservations relate to Comte's condemnation of contemporary political economy and to his insistence that the subject could not be treated independently as a detached discipline. What Mill had to say about these aspects of Comte's position casts a bright light upon his intentions in the *Principles*.

There was first Mill's insistence on the legitimacy of specialized sciences, including political economy (305–6). Mill did not by this insistence upon the possibility of a detached discipline of political economy deny that its practitioners, like all other scientists, might be narrow minded: 'The principal error of narrowness with which they are frequently chargeable, is that of regarding, not any economical doctrine, but their present experience of mankind, as of universal validity; mistaking temporary or local phases of human character for human nature itself; having no faith in the wonderful pliability of the human mind; deeming it impossible, in spite of the strongest evidence, that the earth can produce human beings of a different type from that which is familiar to them in their own age, or even, perhaps, in their own country.' 'The only security against this narrowness', Mill concluded, 'is a liberal mental cultivation, and all it proves is that a person is not likely to be a good political economist who is nothing else' (306). Interestingly enough, however, Mill – in a different context – found Comte himself guilty of precisely this defect as he made clear in private correspondence of 1848 regarding the *Discours sur l'ensemble du Positivisme* (1848):

You may well call Comte's a strange book. I agree . . . that it is well calculated to stir the mind and create a ferment of thought . . . because it is the first book which has given a coherent picture of a supposed future of humanity with a look of possibility about it, and with enough of *feature* for the reason and imagination to lay hold of it by . . . [But] [w]ith all his science he is characteristically and resolutely ignorant of the laws of the formation of character; and he assumes . . . all

bien entamer déjà avec vous une discussion philosophique' (to Comte, 23 Oct. 1842, XIII, 554).

However, Mill had expressed sharply anti-liberal sentiments in the early 1830s (cf. letter to John Sterling, 20–22 Oct. 1831, XII, 84). Of these it has been said that 'they were only characteristic of a brief phase in [Mill's] intellectual career and were not repeated later. In part, they may be attributed to the continuing dialectic of liberal and authoritarian beliefs in Mill's writings; more specifically, they were the result of his current policy of exaggerating the extent of his agreement with all correspondents, even at the risk of distorting his own views' (Wolfe, 1975, 39).

13 'We are taught the right way of searching for results, but when a result has been reached, how shall we know that is is true? How assure ourselves that the process has been performed correctly, and that our premises, whether consisting of generalities or of particular facts, really prove the conclusion we have grounded on them?' (292).

these [character] differences . . . as ultimate, or at least necessary facts, and he grounds universal principles of sociology on them (Mill to J. P. Nichol, 30 September 1848, *CW*, XIII, 738-9).[14]

This same criticism – on a par with the complaint against his father's procedures–had been made as early as 1829 in the letter to Gustave d'Eichthal (above, p.90).

The insistence upon the investigation of political economy as a distinct specialization was also a central argument of Mill's treatment in his *System of Logic*, composed during the early 1840s and based upon the earlier essay on the methodology of political economy (VIII, 900-4).[15] Mill's criticism of Comte in 1865 is precisely that which he had in mind while preparing the *Principles*. As is clear from the correspondence, Mill certainly found attractive those aspects of Comte's work which were consistent with the 'provisional' nature – as he understood the term – of many of the conclusions of political economy in the light of the temporariness of institutional arrangements. But further than this he did not go.

It seems fair to conclude that nothing in the Comte-Mill nexus of the 1840s implies a breakaway on Mill's part from the orthodox (Ricardian) approach to political economy – subject always to the insistence upon specification of the appropriate institutional framework and recognition of the consequentially restricted scope of its generalizations, which Mill intended to emphasize by the organization of his *Principles*. What comes to the fore in the 1865 review is disappointment in Comte's continued failure to appreciate the possibility of remaining within the orthodox camp, while yet recognizing the historico-institutional dimension of social investigation and the need to avoid 'narrow mindedness'.

The second point of departure from Comte in the review of 1865 that concerns us involves the actual state of social science in general and economics in particular. According to Comte, [t]he most complex of all sciences, the Social, had not . . . become positive at all, but was the subject

---

14 But Mill had kind words for the emphasis on 'the purely *subordinate* role of the intellect as the minister of the higher sentiments' and for recognition of Comte's 'admirable historical views' including his championship of social renovation, and the possibilities for the replacement of religion by the *culte de l'humanité*.

15 In the manuscript version and in the first edition (1843) Mill specifically referred to Comte's position: 'When M. Comte (for of the objections raised by inferior thinkers it is unnecessary here to take account) pronounces the attempt to treat political economy, even provisionally, as a science apart, to be a misapprehension of the scientific method proper to Sociology; I cannot but think that he has overlooked the extensive and practical guidance which may be derived, in any given state of society, from general propositions such as those above indicated; even though the modifying influence of the miscellaneous causes which the theory does not take into account, as well as the effect of the general social changes in progress, be provisionally overlooked' (903).

of an ever-renewed and barren contest between the theological and the metaphysical modes of thought. To make this highest of the sciences positive, and thereby complete the positive character of all human speculations, was the principal aim of his labours and he believed himself to have accomplished it in the last three volumes of his Treatise' (X, 290). The charge against social scientists, including the economists, Mill rejected: 'Montesquieu; even Machiavelli, Turgot, Adam Smith, and the political economists universally, both in France and in England; Bentham, and all thinkers initiated by him, had a full conviction that social phaenomena conform to invariable laws, the discovery and illustration of which was their great object as speculative thinkers. All that can be said is, that those philosophers did not get so far as M. Comte in discovering the methods best adapted to bring these laws to light.'[16]

Interestingly enough, Mill had already made the point in his very first letter to Comte that Benthamism had in its favour precisely this kind of perspective. Although he referred there to his 'sortie définitive de la section benthamiste de l'école révolutionnaire' – in part because of Comte's own influence in the late 1820s – he saw great merit to the mode of thought which Bentham represented from a 'Comtian' point of view: 'Quique le Benthamisme soit resté, sans doute, trés loin du véritable esprit de la méthode positive, cette doctrine me paraît encore à présent la meilleure préparation qui éxiste aujourd'hui à la vraie positivité, appliquée aux doctrines sociales: soit par sa logique serrée, et par le soin qu'elle a de toujours se comprendre elle-même, soit surtout par son opposition systématique à toute tentative d'explication de phénomènes quelconques au moyen des ridicules entités métaphysiques, dont elle m'a appris dès ma première jeunesse à sentir la nullité essentielle' (8 Nov. 1841, XIII, 489).[17]

16 See also: 'All theories in which the ultimate standard of institutions and rules of action was the happiness of mankind, and observation and experience the guides . . . are entitled to the name Positive, whatever, in other respects, their imperfections may be' (299). The special allowance for Comte's contribution was not saying much in the light of his belief that Comte had failed to provide a principle of proof.

17 Comte gave his qualified approval to this formulation: 'Le Benthamisme, où vous avez d'abord vécu, est une preuve sensible de la conformité naturelle de nos tendances intellectuelles, indépendamment de tout contact; car cette doctrine, la plus éminente dérivation de ce qu'on nomme l'économie politique, me semble, comme à vous, surtout pour l'Angleterre, une préparation immédiate à la positivité sociologique; si j'ai moi-même évité cette phrase, cela tient sans doute à des circonstances personelles d'éducation, qui, m'ayant imbu, dés mon enfance, des rudiments de la vraie méthode positive, m'ont permis de sentir à temps combien Bentham avait imparfaitement compris cette méthode, malgré sa tendance évidente à la faire partout prévaloir' (20 Nov. 1841, Lévy-Bruhl ed., 1899, 7).

See also Mill's letter to Theodor Gomperz, 19 Aug. 1854 opposing 'the theory of innate principles'; and the strong attack in the *Autobiography* on the intuitive school of philosophy

In arriving at his dim view of the spectacle provided by political economy, Comte had formally based himself upon two tests of the positive character of that discipline – 'continuity' and 'fecundity':

When the work of the present time, instead of presenting itself as the spontaneous sequel and gradual consummation of former work, takes, in the case of each new author, a character essentially personal, and the most fundamental notions are incessantly brought into question; when the dogmatic constitution of a science, far from engendering any sustained progress, results habitually in the sterile reproduction of illusory controversies, ever renewed, never advancing; when these indications are found, there we may be certain we have to do, not with positive science, but with theological or metaphysical dissertation. Now is not this the spectacle which Political Economy has presented for half a century? If our economists are in reality the scientific successors of Adam Smith, let them show us in what particulars they have effectively improved and completed the doctrine of that immortal master, what discoveries really new they have added to his original felicitous aperçus (Translation by Cairnes, 1873, 285).

To all this Mill protested in strong language: 'Any one acquainted with the writings of political economists need only read his few pages of animadversions on them (Vol. IV, pp. 193–205), to learn how extremely superficial M. Comte can sometimes be. He affirms that they have added nothing really new to the original *aperçus* of Adam Smith; when every one who has read them knows that they have added so much as to have changed the whole aspect of the science, besides rectifying and clearing up in the most essential points the *aperçus* themselves' (X, 305).

There is nothing to suggest any change in Mill's attitude over time to Comte's evaluation of the 'quality' of Ricardian political economy, for twenty years earlier he had alluded to Comte's low opinion – 'Je sais ce que vous pensez de l'économie politique actuelle: j'en ai une meilleure opinion que vous' (above, p.169). One is led to conclude that what influence Comte may have exerted related at most to the 'tone' and organization of the *Principles* – and even this is not certain for there were a variety of other influences pointing in the same direction – but certainly not to substantive content.[18]

(as distinct from that of experience and association) as a hindrance to reform (I, 269). Mueller (1956), 97, makes the point 'that by the time of his association with Comte . . . Mill had given up the exalted notion of the artists' function impressed upon him by Carlyle . . .' (But see the favourable references to Carlyle in letter to Comte, 23 Oct. 1842, XIII, 552).

18 For the limited nature of Comte's influence see also Ekelund and Olsen (1973), 383–416.

IV   THE STRATEGY

The reader of the *Principles of Political Economy* is made aware of the 'provisional' nature of existing British arrangements by an organization that introduces him to widely applicable constraints before he encounters the precise manifestation of those constraints in the capitalist-exchange system. Rough models already existed in the literature for the ordering of topics in the sequence production, distribution and exchange.[19] If we seek for immediate parallels these should perhaps include Thomas Chalmers. In what may be his earliest reference to a project to write a general text Mill observed that 'I think, I shall, some time or other, write a Treatise on the whole Science' noting, with reference to Chalmers' *Political Economy* (1832), that 'I have derived many new ideas from it, and it has even suggested an entirely new view of the order in which the truths of the science ought to be arranged' (30 Aug. 1834, *CW*, XII, 231). Whether Chalmers – including his separation of science and application – indeed left an indelible impression, is an open question; but the possibility cannot be ruled out. Mill also compared aspects of his procedures in the *Principles* with Karl Rau's *Lehrbuch der politischen Oekonomie* (1826–37).[20] But the decision to proceed as he did can be accounted for apart from these precedents.[21]

At the outset of Book III (on exchange), Mill took issue with Richard Whately's celebrated proposal to rename political economy 'Catallactics' or the science of exchanges (above, p. 45, n32): 'If these denominations had appeared to me logically correct, I must have placed the discussion of the elementary laws of value at the commencement of our inquiry, instead of postponing it to the Third Part; and the possibility of so long deferring it is alone a sufficient proof that this view of the nature of Political Economy is too confined.' The reason given for the chosen plan was that 'the conditions

19   Cf. Schwartz, (1972), 18f, for a comparison of the *Principles* with James Mill's *Elements* and Say's *Treatise*. James Mill was apparently influenced by the latter: 'As an elementary book it is much superior to Adam Smith's because the arrangement is much improved, and the principles are stated more completely'; letter to Francis Place, cited Winch (1962), 190n.

20   'Your plan of separating the scientific inquiry into the production & distribution of wealth, as a branch of social science, from the consideration of the economic policy of governments, appears to me both logically and didactically the best, & I have made the same separation in my own treatise' (20 March 1854, XIV, 86). It is not clear that this observation was intended as an acknowledgement of indebtedness.

21   By April 1845 Mill was engaged in general reading for the proposed volume, and by March 1846 he was making rapid progress with the writing (XIII, 664, 682, 698). But six months were lost in devising proposals for Irish land reclamation and peasant ownership (705, 707). Mill apparently wrote the *Principles* in the order of the final product: '[I have] got on well with the *Pol. Econ.* I am on the point of finishing the third book' (Sept. 1846, 704).

and laws of Production would be the same as they are, if the arrangements of society did not depend on Exchange, or did not admit of it', while, even in the contemporary system of production and exchange involving division of labour, 'exchange is not the fundamental law of the distribution of the produce, no more than roads and carriages are the essential laws of motion, but merely a part of the machinery for effecting it' (CW, III, 455).

By the so-called 'laws of production' Mill here intended laws on which are based the need for population control. It is to this order of consideration that he alluded when he stated that the conditions of production are independent of exchange. The main objective of placing the discussion of production at the outset was to emphasize that there are 'physical' laws of this kind at work acting as basic constraints whatever the social arrangements might be, and thus to counter the 'error . . . of not distinguishing between necessities arising from the nature of things, and those created by social arrangements: an error . . . producing two opposite mischiefs; on the one hand, causing political economists to class the merely temporary truths of their subject among its permanent and universal laws; and on the other, leading many persons to mistake the permanent laws of Production (such as those on which the necessity is grounded of restraining population) for temporary accidents arising from the existing constitution of society – which those who would frame a new system of social arrangements, are at liberty to disregard' (455–6).

That Mill had these considerations in mind in according 'production' pride of place is amply confirmed in his correspondence prior to publication. The central issue turns on contemporary attitudes towards the poor, Mill contrasting the 'superficial philanthropy' on the part of the upper classes with the true solution, as he saw it, which involved 'the anti-population doctrine' and 'forethought and self-command on the part of the poor'. Mill expressed himself in very strong language on these matters which play directly on the organization of the *Principles*:

There is a prodigious current setting in every day more strongly, of superficial philanthropy. English benevolence can no longer be accused of confining itself to niggers and other distant folks; on the contrary everybody is all agog to do something for the poor. A great many things have conduced to this, some good, some bad. The anti-poor-law cry; the state of the houses of the poor, and their sanitary condition, as made known by Chadwick's official investigations; the conditions of large masses of people as shown by the enquiries of Commissions about factories, mines, etc., then in another way the speculations of Carlyle, the Puseyites, and others, about the impossibility of any social stability or security if there is not a habitual bond of good offices and sympathy between the ruling classes and the ruled, especially the poor – which speculations would have had no effect whatever if there were no chartism and socialism to frighten the rich. One sees plainly that while the noise is made up by a few sincere people, the bulk of the following has for its motive the desire of preventing revolution, and perhaps still more, the desire of taking the popularis aura out of the sails of the Anti-Corn-law league. In both these things they will fail. The Corn Law *must* go, and very soon, and as for revolution, there has been nothing in our day so calculated to produce it as the talk now in vogue, none of which is lost upon the working class, who do not thank them for it one jot, but whom it greatly strengthens in the faith that it is other people's business to take care of them, that all of the rich have more than they is a wrong to them, and that the rich themselves are partly ashamed of this wrong, and partly afraid of its consequences and desiring to buy them off at the expense of those who are better off, is always asserted; and I never remember a time when any suggestion of anti-population doctrine or of fore-thought and self-command on the part of the poor was so contemptuously scouted as it is now. The 'Times' is at the head of this movement, and has contributed very much to set it going . . . Things never seemed to tend so rapidly to a complete bouleversement of our social system, though whether peaceably or violently, none can tell. I am thinking of saying out my say on all these things in a treatise on political economy . . . (To H. S. Chapman, 8 Nov. 1844, XIII, 640f).

It was in this frame of mind that Mill prepared for the *Edinburgh Review* an article on Arthur Helps' *Claims of Labour* (1844) subtitled 'An Essay on the Duties of the Employers to the Employed'. The following passage regarding the article – alluding to the paternalistic proposals of the 'young Englanders' represented by *The Times* (IV, 371) – provides a splendid illustration of Mill's hostility to philanthropic do-gooders who based themselves on thin air:

It appears to me that along with much of good intention, & something even of sound doctrine, the speculations now afloat are sadly deficient, on the whole, in sobriety & wisdom – forgetful, in general, of the lessons of universal experience, & some of those fundamental principles which one did think had been put for ever out of the reach of controversy by Adam Smith, Malthus, & others. The general tendency is to rivet firmly in the minds of the labouring people the persuasion that it is the business of others to take care of their condition, without any self control on their own part – & that whatever is possessed by other people, more than they possess, is a wrong to them, or at least a kind of stewardship, of which an account is to be rendered to them (To Macvey Napier, 9 Nov. 1844, XIII, 643–4).[22]

22 Two years earlier, Mill had made the same point in even stronger language regarding the misguided enthusiasm of the ruling classes: 'This feeling one can see breaking out in all sort of stupid and frantic forms, as well as influencing silently the opinions and conduct of sensible people. But as to the means of curing or even alleviating great social evils people are as much at sea as they were before' (9 Sept. 1842, 544).

The object of his proposed article on the claims of labour, he explained further to Napier,

> would be to examine & controvert what appears to me an erroneous theory of the condition of the labouring classes. The practical consequences of the theory break out in all sorts of propositions of things to be done for the poor either by the Government, the millowners, the landowners, or the rich in general; some of which propositions have more or less of utility & good sense in them, others are quite chimerical and absurd, but *all* are absurd when looked to as things of great or permanent efficacy. The discussion of the theory will naturally involve a consideration of the real nature of the duties both of Government & of the various classes of society towards the poor; tending mainly to the conclusion, that the greater part of the good they can do is indirect, & consists in stimulating & guiding the energy & prudence of the people themselves: in all which I should wish to use details copiously for purposes of example & illustration, but without laying any particular stress upon them, & still less undertaking to specify with any minuteness what particular things either the Government, or the employers of labour, ought to do or attempt (20 Nov. 1844, 645).

Similarly, to another correspondent he recorded his preoccupation with the 'claims of labour', complaining that 'I never knew a time when so much nonsense, mischievous nonsense, too, was afloat on that subject, and I thought it a most useful thing to enter a protest against the intolerable mass of pseudophilanthrophy now getting into vogue, and to commit the *Edinburgh Review* at the same time (if possible) to strong things in favour of good popular education and just laws' (18 Jan. 1845, 654–5).

It is in the first letter to Napier, that Mill also made a celebrated reference to the need to avoid the 'hard, abstract mode' of treating labour questions: 'I am sure you will agree with me in thinking it very necessary to make a stand against this sort of spirit while it is at the same time highly necessary as well as right, to shew sympathy in all that is good of the new tendencies, & to avoid the hard, abstract mode of treating such questions which has brought discredit upon political economists & has enabled those who are in the wrong to claim, & generally to receive, exclusive credit for high & benevolent feeling.' For misuse of theory had enabled the anti-theorists to claim credit for benevolent feeling.[23]

This defensive strategy entailed in part an emphasis upon the 'provisional' character of the conclusions of orthodox economics – allowing in application for 'other classes of considerations' – as required by the earlier essay. By

---

23 In 1858 Mill complained of those who claimed to show that 'the settled principles of the science are wholly fallacious', writing to one culprit: 'You are not the first, nor the hundredth' (*CW*, XV, 555).

this means, Mill hoped to protect Ricardian economics from unfair charges by candidly conceding its limited scope:

Political Economy, in truth, has never pretended to give advice to mankind with no lights but its own; though people who knew nothing *but* political economy (and therefore knew that ill) have taken upon themselves to advise, and could only do so by such lights as they had. But the numerous sentimental enemies of political economy, and its still more numerous interested enemies in sentimental guise, have been very successful in gaining belief for this among other unmerited imputations against it and the *Principles* having, in spite of the freedom of many of its opinions, become for the present the most popular treatise on the subject, has helped to disarm the enemies of so important a study (Autobiography, I, 243, 245).

But there was also a more positive defence of orthodoxy against charges that it lent itself to extreme *laissez-faire* conclusions and, in general, apologetic class attitudes. Orthodox economists (with the possible exception of J. R. McCulloch) were innocent of these charges. A letter to John Lalor brings out very clearly what concerned him:

My objection to the passage relating to Chalmers did not turn as you seem to suppose, on the word 'baptism'. My remarks did not apply to the phraseology, but to the meaning of the sentence – to the assertion that pol. ec. unless connected with Xtianity is 'a true child of the devil.' Any reader would suppose that by Xtianity was here meant belief in the Bible and on your own interpretation I must still protest against the statement that Chalmers 'began' the baptism in question. I do not know any pol. economist except perhaps McCulloch to whom the accusation you bring against all who preceded Chalmers can be attributed even by the license of caricature – & I especially reject it with respect to A. Smith, Turgot, Say, Ricardo & my father not one of whom was a believer in Xtianity & none of whom regarded pol. ec. as anything but a subordinate though necessary branch of utility or as you prefer to term it 'the doctrine of human welfare.'
 No men ever wrote to whom the charge of seeking in pol. ec. or in anything else a 'justification of universal selfishness' or of any selfishness at all could be applied with less justice, & I cannot, on this point, accept any compliment at their expense. I confess I do not see the good that is to be done by swelling the outcry against pol. economists – or why they should be blamed because people do unjust or selfish things for the sake of money. I do not know what authority you have for saying that the clearing of Irish estates was 'perpetrated in the name of pol. economy' any more than the clearing of English estates from the same motives in the time of the Tudors. But I do know that nearly all the pol. economists supported a poor law in Ireland in order to give the landlords an interest in fighting against the causes of poverty (3 July 1852, XIV, 93).[24]

24 See Checkland (1951a), 51, 55–6, 66f on the Tractarians (including Newman,

We have also to consider the population issue from the perspective of Mill's defensive strategy. Political economy, Mill complained in a variety of contexts, had long had a bad but quite unjustified reputation amongst 'progressive' thinkers. In his *Autobiography*, for example, he recalled a debate in 1825 regarding the question of population, in which he had participated – a *lutte corps-à-corps* between Owenites and political economists, whom the Owenites regarded as their most inveterate opponents: but it was a perfectly friendly dispute. We who represented political economy had the same objects in view as they had, and took pains to shew it' (I, 129). More generally, he wrote of the period of his youth when 'we found all the opinions to which we attached most importance, constantly attacked on the ground of feeling. Utility was denounced as cold calculation; political economy as hard-hearted: anti-population doctrines as repulsive to the natural feelings of mankind. We retorted by the word "sentimentality", which, along with "declamation" and "vague generalities", served us as common terms of opprobrium.' He conceded that the attitude reflected a neglect of 'the cultivation of feeling' yet at the same time insisted that 'we were generally in the right'. The 'Philosophical Radicals', he explained, were Malthusians but in a carefully specified sense: 'This great doctrine, originally brought forward as an argument against the indefinite improvability of human affairs, we took up with ardent zeal in the contrary sense, as indicating the sole means of realizing that improvability by securing full employment at high wages to the whole labouring population through a voluntary restriction of the increase of their numbers' (107).[25] In line with this perspective is Mill's

Pusey and Keble) who maintained that all studies be subservient to theology.
    In his *Money and Morals* (1852) Lalor had asserted of Chalmers that he 'began that baptism, so to speak, of political economy into Christianity, which was the main thing needful to bring about its regneration' (xvii). Of Mill's *Principles* Lalor wrote that 'it has indeed effected, scientifically and conclusively, that subordination of the doctrine of wealth to the doctrine of human welfare, which was the object so earnestly desired by Sismondi and Chalmers' (xxvii–xxviii).

25 The same reformist attitude, Mill explicitly insisted in the *Principles* had been adopted by Malthus himself after the *First Essay*: 'few writers have done more than himself, in the subsequent editions, to promote . . . juster and more hopeful anticipations' (*CW*, III, 753; cf. V, 728). This dimension partly accounts for Mill's qualified enthusiasm for Carlyle. Carlyle's earlier writings, which contributed to a broadening of his hitherto narrow perspective, in themselves had little effect because they demonstrated 'animosity to most of the opinions which were the basis of my mode of thought; religious scepticism, utilitarianism, the doctrine of circumstances, and the attaching any importance to democracy, logic, or political economy' (I, 181). On the influence of Carlyle, see Neff (1926).
    The famous denigration of Coleridge is pertinent from this same perspective, namely Mill's remark made in print in 1840, that 'in political economy especially he writes like an arrant driveller, or worse, and it would have been well for his reputation had he never meddled with the subject. But this department of knowledge can now take care of itself' (X, 155).

dissatisfaction with the published version of his 1845 review for the *Edinburgh Review*. Excisions of materials relating to Wakefield's proposals and the contemporary debate regarding small-holdings, at Napier's insistence, had destroyed an opportunity to emphasize the error on the part of those who envisaged a conflict between orthodox theory and social conscience; for he had hoped to show that 'the doctrines called Malthusian do not as is vulgarly supposed, imply that in one's opinions on social arrangements one looks only to amount of production & not to the producers' (17 Feb. 1845, XIII, 661).[26]

The simple lesson to be transmitted was that 'a greater number of people cannot, in any given state of civilization, be collectively so well provided for as a smaller. The niggardliness of nature, not the injustice of society, is the cause of the penalty attached to over-population' (II, 188). This was positively not a prediction of inevitably low standards. But Mill realized he was fighting an up-hill battle – there existed a 'sentimental horror of Malthus' whose principles had not yet 'gained possession of the general mind' (370). Indeed 'since the population doctrine was first promulgated, nine-tenths of the talk has always been against it, and the remaining tenth only audible at intervals; and . . . it has not yet penetrated far among those who might be expected to be the least willing recipients of it, the labourers themselves' (371).

The reason for the failure of opinion was partly a matter of class bias since 'nearly all who are not labourers themselves, are employers of labour and are not sorry to get the commodity cheap'. The gentry ('charitable people') from whom more could be expected, also bore some of the responsibility, for 'charitable people have human infirmities, and would, very often, be secretly not a little dissatisfied if no one needed their charity; it is from them one oftenest hears the base doctrine, that God has decreed there shall always be poor'. In fact, even the social reformers were an impediment, for each had his pet scheme (corn law abolition, reduced taxation, small-note issue, Chartism, abolition of the Church or aristocracy) 'and looked upon every one as an enemy who thought anything important

---

26 Correspondence with the French novelist Eugène Sue in May 1848 immediately after publication of the *Principles* similarly reveals Mill's insistence upon the consistency of his approach to social questions and the Ricardian analytical framework; indeed hostility to orthodox economic theory came from the apologetic opponents of reform. Thus Mill hoped to prove 'qu'on peut être économiste et même professer un grand nombre des opinions de Malthus et de Ricardo, sans être pour cela un Duriveau, ou un flatteur des Duriveau. Je vous dirai en outre comme fait, que quant aux Duriveau de mon pays si toutefois il y en a, ceux qui se font tous instruments, non seulement, ne professent pas les opinions de ces économistes, mais en général les puent et les conspuent, presqu' autant que vous' (736).

except his object' (370–1). Yet all the discussions of poverty and its solution missed the point:

Discussions on the condition of the [agricultural] labourers, lamentations over its wretchedness, denunciations of all who are supposed to be indifferent to it, projects of one kind or another for improving it, were in no country and in no time of the world so rife as in the present generation; but there is a tacit agreement to ignore totally the law of wages, or to dismiss it in a parenthesis, with such terms as "hard-hearted Malthusianism"; as if it were not a thousand times more hard-hearted to tell human beings that they may, than that they may not, call into existence swarms of creatures who are sure to be miserable, and most likely to be depraved (351–2).

Mill's charge was directed, it will be noted, against radical reformers – the Chartists and opponents of the aristocracy and Church – as well as employers and gentry. This theme is greatly expanded in the first two editions in a bitter denunciation of both the Radicals and Tories:

It is not wonderful that the working classes themselves should cherish error on this subject. They obey a common propensity, in laying the blame of their misfortunes, and the responsibility of providing remedies, on any shoulders but their own. They must be above the average level of humanity if they chose the more disagreeable opinion, when nearly all their professed teachers, both in their own and in every other class, either silently reject or noisily declaim against it. The true theory of the causes of poverty seems to answer nobody's peculiar purpose. Those who share the growing and certainly well-grounded discontent with the place filled and the part performed in society by what are called the higher classes, seem to think that acknowledging the necessary dependence of wages on population is removing some blame from those classes, and acquitting them at the bar of public opinion for doing so little for the people; as if anything they could do, either in their present relation to them or in any other, could be of permanent use to the people in their material interests, unless grounded on a recognition of all the facts on which their condition depends. To this class of opponents, the accidents of personal politics have latterly conjoined nearly the whole effective literary strength of the party who proclaim themselves Conservative of existing social arrangements. Any one with whom the cause of the poor is a principle, and not a pretence, or a mere freak of sensibility, must contemplate with unfeigned bitterness the conduct, during ten important years, of a large portion of the Tory party, including nearly all its popular organs; who have studiously fostered the prejudices and inflamed the passions of the democracy, on the points on which democratic opinion is most liable to be dangerously wrong, for the paltry advantage of turning into a handle of popular declamation against their Whig rivals, an enactment most salutary in principle, in which their own party had concurred, but of which those rivals were almost accidentally the nominal authors (352n).[27]

V SUMMARY AND CONCLUSION

What Mill intended by his *Principles* was the liberation of public opinion from the influence of second-rate economists who gave economic theory a bad name by their apologetic misapplications of theory, and from the false belief that Ricardian doctrine and method itself amounted to social apologetics. The task was an urgent one. At the time of his planning of the work Mill feared that the quality of contemporary texts on political economy was so low as to threaten the public's respect for the subject itself, and leave it open to vulgar empiricism – 'l'empirisme systématique qui nie toute doctrine générale en matière sociologique' (6 June 1844, *CW*, XIII, 631).[28] And the fine line Mill had to tread is apparent: 'On the one hand . . . he wanted to reiterate stern necessities, while avoiding the Scylla of "hard-heartedness" by clearly demonstrating his concern for the working classes. On the other hand, he wanted to show that laissez-faire is inadequate as a precept for all but the preliminary work of pulling down abuses, while avoiding the Charybdis of "sentimentality" by holding firmly to the laws of population and of the production of wealth' (de Marchi, 1974, 136). But in all this it must never be forgotten that the insistence of 'universal' laws of population entails nothing like 'hard-line Malthusianism' with its prediction of inevitably 'low' standards in consequence of land scarcity. On the contrary, it was envisaged as the key to high living standards, a deceleration of the population growth rate entailing favourable consequences from this perspective.

* * *

There was, in addition to the danger of neglecting the 'universally' present phenomenon of land scarcity and its consequences, that of failing to recognize features common to a wide variety of distributional arrangements by paying too close attention to the outward forms of the familiar capitalist-exchange system. Partly to counter this danger Mill placed distribution before exchange.

27 In 1869 Mill repeated the charge against 'the older school of levellers and democrats' (V, 728).
28 Cf. Hayek (1942), vii: '. . . if one may speak of prejudices of so singularly candid a mind, there can be little doubt that Mill had acquired something like prejudice and even contempt not only for English society, which he little knew, but also for contemporary development of English thought and especially of English political economy, which he neglected to a surprising extent'.

It is thus in Book II that we have analysis of alternative land-tenure systems (below, p. 238f). But there is also a Marx-like approach to the sources of profits, which takes issue with the 'popular apprehension' that 'the profits of business depended upon prices' reflecting a failure to look below 'the outside surface of the economical machinery of society' (*CW*, II, 410). Profits, Mill insisted, arose from the circumstance 'that labour produces more than is required for its support'. It was, therefore, possible to conceive of an institutional arrangement involving, for example, no specialization and accordingly no exchange but yet where 'profit' is yielded: 'profit arises, not from the incident of exchange, but from the productive power of labour; and the general profit of the country is always what the productive power of labour makes it, whether any exchange takes place or not' (411). This was added in the fourth edition of 1857. But the same general conception of things also appears during the course of Book I in the context of a discussion of 'fundamental properties respecting capital':

The demand for commodities is a consideration of importance rather in the theory of exchange, than in that of production. Looking at things in the aggregate, and permanently, the remuneration of the producer is derived from the productive power of his own capital. The sale of the produce for money, and the subsequent expenditure of the money in buying other commodities, are a mere exchange of equivalent values for mutual accommodation. It is true that, the division of employments being one of the principal means of increasing the productive power of labour, the power of exchanging gives rise to a great increase of the produce; but even then it is production, not exchange, which remunerates labour and capital. We cannot too strictly represent to ourselves the operation of exchange, whether conducted by barter or through the medium of money, as the mere mechanism by which each person transforms the remuneration of his labour or of his capital into the particular shape in which it is most convenient to him to possess it; but in no wise the source of the remuneration itself (88).

Mill repeatedly referred to money as constituting merely the *machinery* of the exchange process (71–2, 83, 86n; III, 455).

More generally, the laws of distribution were said to be 'regulated by the same principles when paid in money, as they would be if apportioned in kind' under different social arrangements (III, 698). Thus wage-rate determination in terms of the 'ratio between population and capital' would still hold good even 'if all the capital in the world were the property of one association, or if the capitalists among whom it is shared maintained each an establishment for the production of every article consumed in the community, exchange of commodities having no existence' (695–6). Similarly, a broad comprehension of the principle of differential rent did not require a detailed discussion of prices (II, 416f). Even the 'cost of labour', or the real cost of producing the wage basket (as distinct from the commodity wage itself) – which, we shall see, governs the proportionate shares and accordingly constitutes the key variable in the analysis of the profit rate – could, to pedagogic advantage, be discussed in general terms prior to the introduction of exchange (413f).[29] Indeed, Mill referred to the inverse profit-wage relation as 'a law of arithmetic' which necessarily holds true: 'If the labourers really get more, that is, get the produce of more labour, a smaller percentage must remain for profit. From this law of Distribution, resting as it does on a law of arithmetic, there is no escape. The mechanism of Exchange and Price may hide it from us, but is quite powerless to alter it' (III, 479).

All this does not imply a minimization of the significance – empirical and theoretical – of the problem of distribution within the context of a capitalist-exchange system in which specific context distribution is inextricably intertwined with the general pricing system. The point is that Mill wished to make his readers aware that the exchange system constituted one possible form only of social arrangement and this he did by emphasizing at the outset of the work the economic problems of production and distribution common to a wide variety of institutional forms. Nothing more was intended – certainly not the technical divorce of distribution and valuation within the context of an exchange system.

---

29 Cf. 'Of Profits', esp. 413f. See also Mill's letter to H. Fawcett, 17 May 1863: '[T]hat of going at once to money prices, without first discussing the general laws of exchange value, answers very well in the simpler questions . . . I think too (as Ricardo thought) that is is of importance to cultivate in learners the habit of arguing questions at first on the supposition of barter, in order to adjourn the difficulties which arise from the wrong and confused associations which cling to the idea of money' (XV, 859).

FOUR

The sources of increased efficiency

I THE TREATMENT OF FACTUAL MATERIALS

We shall be concerned in this chapter with the factual materials discussed in the 'Preliminary Remarks' to the *Principles* regarding the causes of differences in the origin, magnitude and distribution of wealth over time and space; the more complex account of Book I, Chapter vii ('On What Depends the Degree of Productiveness of Productive Agents') which makes special reference to climate, location, the quality of manpower, and the state of health, knowledge and security; Chapters viii and ix which elaborate further, from the same perspective, the role of industrial organization with particular reference to economies of scale; and Chapter xii on the implications for productivity of land scarcity in its various manifestations. These 'inductive' chapters isolate, in the historical and contemporary record, various 'qualitative' determinants of factor productivity and also the 'quantitative' relationships summarized in the principles of diminishing agricultural returns, the increased productivity of indirect processes and increasing returns to scale in manufacturing.[1] Further empirical exercises elaborating upon the primary causal phenomena acting on wealth – particularly knowledge, security and cooperation – are undertaken in Book IV, Chapter i on the 'General Characteristics of a Progressive State of Wealth'.

To what precise ends were these factual materials devised? There is a

---

1 For a summary of several of these chapters with special reference to their inductive character, see Sidgwick (1883), 33f; Mitchell, (1967), I, 597f. The determinants of factor supply, though dealt with by Mill under production, find a more appropriate location in our chapter on capital, employment and growth since we are now largely limiting the discussion to productivity.

theoretical and an institutional dimension to consider. As to the theoretical dimension we must recall Mill's central methodological proposition in the early essay that deductive theory constitutes a mixed method entailing both logical ratiocination and induction, the latter providing the raw materials for the former – 'the laws of the [individual] causes on which the effect depends must be obtained by an induction from comparatively simple [empirically observed] instances . . .' (above, p. 77). For the construction of any analytical apparatus is designed to overcome the problem of 'composition of causes', or to allow an evaluation of the net effect of combined and perhaps conflicting causes, each of which must in the first place be known at least provisionally; while the accuracy of the chosen axioms and of the ratiocinative exercise remain always subject to 'verification' – the general conclusions derived deductively must 'accord with the results of direct observation wherever it can be had'.

As Mill explained at the outset of Book IV the earlier books had provided 'economical laws of a stationary and unchanging society' (*CW*, III, 705). The factual investigations of the first book provided, it would seem, the empirical justification for the technological relationships presupposed by the 'static' model – particularly the principle of diminishing agricultural returns. Much more ambitious, however, and the apogee of the work as a whole is the 'theory of economic dynamics':

We have still to consider the economical condition of mankind as liable to change, and indeed (in the more advanced portions of the race, and in all regions to which their influence reaches) as at all times undergoing progressive changes. We have to consider what these changes are, and what are their laws, and what their ultimate tendencies; thereby adding a theory of motion to our theory of equilibrium – the Dynamics of political economy to the Statics (705).

For the axioms of the dynamic growth model Mill again drew upon the factual materials – namely diminishing agricultural returns and increasing returns in manufacturing (two technological relationships pertinent quite generally but also to the specific circumstances of contemporary Great Britain); the savings and effort-supply propensities; and in addition three basic causal phenomena shown from the record to be pre-eminently responsible for expanding productivity, namely advances in knowledge, in 'security', and in 'co-operation'.[2]

---

2 J. K. Whitaker, (1975), 1046, singles out the discussion of economies of scale as an instance of the establishment of premises for deduction. But the list can be vastly extended. The behavioural axiom itself, for example, has an 'inductive' underpinning.

   On the basis of his analysis of Book I, Sidgwick (1883), 36f, concluded that Mill's treatment of production is mainly 'inductive' and 'analytical'; and in reply to the question

More closely than anywhere else in the volume, apart from the Preliminary Remarks, Mill touches here upon the problem of 'social progress'. But the constrained nature of the exercise is conspicuous. The deductive model, a Ricardian model of a growing economy, does no more than spell out the consequences of an expanding GNP for productivity (and thus for costs and prices) and for distribution within the limits imposed by a capitalist-exchange system. It does not, in short, extend beyond the traditional borders of specialist political economy, namely the production and distribution of wealth within the contemporary institutional framework, and thus constitutes a 'general theory of the economical progress of society' (752) and not social progress in general.

A second constraint limits the exercise. Mill made no concerted effort to analyse the 'causes' of the observed advance (indeed, the increasing rate of advance) in knowledge, security and co-operation. That such a phenomenon is at play was taken for granted as an empirical tendency characterizing modern 'progressive' society, the rationale for which, in terms of the progressive changes occurring 'in the character of the human race', is largely left an open question. Thus the fundamental problem of social progress, the linkage of empirical laws to the principles of human nature by use of the inverse deductive method as recommended in the essay (see summary below, p. 195) is not broached.[3] On a more mundane level this failure to investigate the sources of increased knowledge implies the absence of any convincing explanation for the observed transition from stagnant or very slowly growing to rapidly expanding 'progessive' economies.

Mill's subscription to a growth model which accords a key role to diminishing agricultural returns despite keen awareness of the empirical significance of advance in knowledge and of innovation, can be appreciated in terms of his rule that each of the individual causal phenomena acting, in the present case, on wealth – and necessarily discernable in the record

---

why, in that case, Mill described political economy as an 'abstract' science Sidgwick suggested that Mill had in mind not the theory of production 'as he himself conceives and expounds it, but the theory of Distribution and Exchange'. There is some truth in this; but Sidgwick did not perhaps sufficiently emphasize that the inductive analysis served Mill in providing the individual axioms of the growth model, while the growth model itself incorporates aspects of the theory of distribution and exchange.

3 Cf. the conclusion by Whitaker (1975), 1046, regarding the absence of a role for the inverse deductive method in the *Principles*. But we shall see that Mill does attempt to rationalize the efficiency of large scale operation in the form of joint-stock companies in terms of behaviour patterns, and examines the efficiency of small-scale (peasant) farms from the same perspective. These analyses involve a form of inverse deduction.

as we have noted – remains operative even when counteracted by an 'antagonistic' cause. Technological progess is thus not envisaged as eradicating the force of land scarcity; it merely disguises its consequences analogously to the operation of conflicting forces in physics.[4] But an additional consideration may be found in Mill's presumption that the problem of the generation and application of knowledge constituted the subject matter of some other science or sciences, to be treated by the economist only as an exogenous disturbance (or 'disturbing cause'). This indeed appears to be the position Mill adopts in justification of his formal distinction between the laws of production and distribution, since this distinction, we shall show, presumes unchanged knowledge or at most allows exogenous changes in knowledge.

This approach to knowledge, though, is not consistently maintained. The distinction between the two classes of laws breaks down as soon as stated since Mill, it turns out, in fact recognized a place for the sources of knowledge within political economy (although he himself did not carry the matter far). As he had explained in the early essay, while scientific political economy presumed maximizing behaviour, some flexibility in drawing disciplinary borders was desirable, as in the case of its incorporation of population growth despite the frequent irrelevance of self-interested motivation in that context (above, p. 124).[5] The conclusion reached in the *Principles* regarding the incorporation of knowledge creation, although largely unrelated to 'standard' behavioural motivation, provides us with a conspicuous instance of the expansion of the axiomatic base of political economy in consequence of changing contemporary experience – in this case a marked increase in the rate of technological progress which rendered it unsatisfactory to treat the issue as a 'disturbing cause'. That Mill himself made no concerted effort to carry the program further requires explanation; we shall suggest that the causal phenomena relevant for the generation of knowledge touched too closely upon the problem of social progress (in the evolutionary sense) for Mill to find the courage to explore the matter far.

* * *

Mill's concern in the *Principles*, as the full title indicates, was not with the 'science' of political economy alone but also with 'application'. The deductive exercises, based upon the empirical materials in the manner described above, served – needless to say, not alone – as a basis for policy recommendations

---

4 See Hollander (1979), 637f, for an analysis of Ricardo's position on the issue.
5 For other examples see below (p. 220).

(such as the repeal of agricultural tariffs) pertinent within the behavioural and institutional framework presupposed by the models. Here a word on the 'use' of theory is in order before proceeding.

It will be remembered that the narrow scope of the specialist social sciences – the constraints imposed by the requirement that the 'state of society' be given – rendered it unlikely that empirical laws pertaining to the given framework could ever be obtained (above, p. 97f). But by this disclaimer Mill referred solely to relationships involving specific policy changes, such as that portraying the effect of corn laws on wealth – Mill's pre-eminent illustration of the theme.[6] Within the constraints of the social and institutional environment relevant to legislation, there were likely to be too few observations to reveal any pattern whatsoever, since various conflicting causes are at play upon wealth in each specific case, some acting positively and some negatively. But, more important for us, even in the (unlikely) event of a large number of observations relating to corn laws, at best all that could ever be yielded is a generalization stating that in a majority of cases such laws were accompanied (let us say) by a positive outcome, and such generalizations would be of little use since it would still remain necessary to interpret them by reducing them to the underlying causal forces at work. It is, as Mill put it, 'from theory' only that the effects on wealth specifically due to corn laws could be appreciated; to base recommendations on apparent empirical regularities might prove disastrous.

The requirement for theory does not, it must be noted, pertain to the technological level of conception. For example, the observed downward trend of agricultural costs over several decades (despite agricultural protection) had to be analyzed by an empirical investigation of the impact of diminishing agricultural and increasing manufacturing returns, innovation, changing work attitudes, labour relations, health, skill, business organization and so forth with the 'abstract' science of economics scarcely entering the picture; on the contrary, the empirical investigation contributes towards a specification of the axioms required by economic theory. The function of the 'abstract' science is rather to explain the data at a more profound level – the level involving human decision making, for which reason precisely

---

6 Mill insisted, it will be recalled, that a priori method professes experience but not *specific* experience of economic facts (above, p. 102). By 'specific experience' I understand such relationships as that between corn laws and wealth. It cannot refer to the individual axioms. Thus Mill described the maximizing behaviour characteristic of the contemporary commercial and industrial sphere as an 'empirical law' albeit of 'the lowest kind' and this we have shown is an axiom based on specific evidence (above, p. 112f).

economic models were restricted to particular behavioural patterns. As far as the Ricardo-Mill growth model is concerned, we have a relation linking falling agricultural productivity to the return on capital (a relationship which entails the inverse wage-profit mechanism and thus the competitive pricing mechanism in labour and commodity markets) and hence to savings decisions; and a reverse relationship brought into play by innovations or rather certain categories of innovations. Theory tells us that the imposition of corn laws sets in motion the former tendency and is thus adverse to the growth of wealth, a matter easily disguised by the statistical data.

There were policy recommendations, however, with the deepest social implications, unrelated to any theoretical model – recommendations regarding institutional change based upon comparisons and generalizations from observed facts which relate differences in productivity and wealth between countries and over time to natural advantages, security, knowledge and co-operation, work habits, savings attitudes and so forth – the inductive materials discussed above. There is danger in relying upon statistical generalizations as the basis for policy recommendations such as corn law legislation because any such generalization is the net outcome of multifold forces and not merely of the change in question. But, as we have reiterated, the individual influences acting on wealth are supposedly discernable in the record, which must be broken down into its component parts by close inductive analysis; and the institutional recommendations relate to measures designed precisely to act appropriately upon the individual components. At the same time, Mill is never satisfied to leave his investigation without attempting wherever possible to provide a firm foundation for the empirical results in 'laws of behaviour'. This concern is well illustrated by his analyses of the consequences for productivity of alternative forms of land tenure, in Book II, Chapters 6–10.

* * *

Our discussion in this chapter will be limited to the empirical treatment of economic growth, with particular reference to increased efficiency. The formal growth model and its 'verification' are taken up in a subsequent chapter which also deals with the determinants of factor supply.

We proceed in Section II to a discussion of the historical dimension to the *Principles* with particular reference to the 'Preliminary Remarks'. We shall demonstrate (bearing in mind the criticisms of the historicists), on the one hand, Mill's relative unconcern with a theory of social progress as such and, on the other, the use to which the historical record is actually put in illustrating geographical, national and temporal differences of wealth, and in suggesting candidates for the determinants of such differences. This latter theme is further developed in Section III with regard to various 'qualitative' determinants of productivity, and in Sections IV and V with regard to the two basic quantitative relationships, economies of scale and diminishing returns.[7]

The great weight placed by Mill upon knowledge creation in the empirical investigation of wealth leads us to consider his suggestion that the subject be recognized as a branch of political economy – a suggestion which sits ill beside the celebrated distinction between the laws of production and distribution – and his failure to get to grips with the underlying determinants of knowledge creation (and other determinants of the advance of wealth). In Section VII we outline Mill's enthusiastic account of contemporary and prospective economic progress, again with particular reference to the role of technological change.

A final section treats the analyses of alternative land-tenure arrangements of the second book. Here we show the intricate inductive-deductive relationship (in this case a form of 'inverse deduction') which figures large in the formal methodological works. The breakdown of the strict dichotomy between laws of production and distribution is further confirmed in this context.[8]

II THE HISTORICAL DIMENSION

To set the stage for our investigation we must recall Mill's modest view, best expressed in the *System of Logic*, of the ability of social scientists to deal with 'social dynamics'. By this he intended the investigation of the laws whereby 'any state of society produces the state which succeeds it and takes its place' – a problem involving no less than 'progressive changes both in the character of the human race' (the problem of 'ethology') 'and in their outward circumstances as far as moulded by themselves' (above, p. 137). Consistently with this position Mill made it clear in the *Principles* that political economy deals with only a very small part of the subject matter covered by 'the theory of human progress', taking as given for its specific purposes various relationships – including (in Leslie's phrase) 'the origin and growth of human ideas':

The opinions and feelings of mankind, doubtless, are not a matter of chance. They

---

7 For a full discussion of Mill's position on economic growth see also Spengler (1960).

8 It is not our intention in this chapter to provide an exhaustive account of Mill's use of history and empirical evidence. We shall encounter more materials in our analysis of Mill's formal work on theory and policy. We shall also take up at a later juncture the opinion that Mill in the last few years of his life came to adopt a more 'historical' perspective under the influence of Cliffe Leslie (below, p. 914).

are consequences of the fundamental laws of human nature, [combined with the existing state of knowledge and experience, and the existing condition of social institutions and intellectual and moral culture]. But the laws of the generation of human opinions are not within our present subject. They are part of the general theory of human progress, a far larger and more difficult subject of inquiry than political economy (*CW*, II, 200).[9]

We will encounter frequently this restriction on the scope of the *Principles* – a deliberate decision not to extend the investigation back beyond a certain stage; in particular, behavioural and technological relationships were to be accepted as 'ultimate' and not further reducible, at least for the constrained purpose at hand.

It is also helpful to recall here Mill's championship of specialization in the social sciences, based upon the view – expressed strongly against Auguste Comte – that only by narrowing the borders of its individual branches could there be hope of effective intellectual advance. Thus the 'scientific' dimension of political economy was constrained to the construction of models pertaining to the production and distribution of wealth assuming a particular behaviour pattern, on the presumption that other 'sciences' would proceed along a parallel course. The objective ideally was a 'synthesis' connecting together theorems relating to the production and distribution of wealth based upon a variety of different behaviour patterns – for example, 'custom' as well as 'competition'; linking the end of wealth creation with other ends of human activity; and developing the whole (which would amount to a theory of 'social statics' or 'the mutual actions and reactions of contemporaneous social phenomena') into a full-fledged theory of social progress.

Approach towards a theory of social progess was not, however, to await the solution to the problem of social statics. The projects could, to some extent, proceed simultaneously. And here let us bear in mind the so-called inverse deductive method: The complexity of the quest for a dynamic science of society, extending as it does over the full range of human history and all human activities, and entailing pre-eminently changes in behaviour patterns, was such as to preclude the direct derivation of laws of progress from the elementary principles of human nature, even in the limited sense of tendencies. What was required were preliminary observations of series of effects portraying regularities, or the discovery of 'empirical laws' of progress, which only at a subsequent stage would be related logically to principles of human nature. (This reverses the sequence characterizing the

9 In place of the bracketed section, the manuscript and first two editions read 'and of the constitution of the planet which we inhabit, modified by local or special peculiarities'. The alteration implies greater emphasis on the malleability of opinion and sentiment.

specialist branches of social science where a ratiocinative analysis can often be undertaken on the basis of assumed behavioural, institutional and technological axioms – or more specifically provisionally assumed axioms, since they are continuously subject to verification.) There were, in fact, already at hand empirical laws relating to 'the progressive change in the condition of the human species', but the problem was to 'connect with the laws of human nature, by deductions showing that such were the derivative laws naturally to be expected as the consequence of those ultimate ones'. Without the latter exercise the regularities in question were 'applicable only within the bounds of actual observation; without any means of determining their real limits, and of judging whether the changes which have hitherto been in progress are destined to continue indefinitely, or to terminate, or even to be reversed' (above, p. 139).

From all this one is led to conclude that there were no substantive differences between Mill and the historicists regarding ultimate objective. Any difference related to strategy reflecting a hard-headed and realistic estimate on Mill's part of the magnitude of the task ahead.[10]

Now there are materials in the *Principles* pertinent to the investigation of social progress in the evolutionary sense discussed above – pre-eminently the 'Introductory Remarks' (which impressed Leslie) and Book IV on the 'Influence of the Progress of Society on Production and Distribution'. But in neither context is the evolutionary theme carried far. In the light of Mill's general views on the scope of political economy and of the textual exegesis below it seems clear that a serious consideration of evolutionary social trends was not a central concern of the *Principles*. To this extent it is a little misleading to speak, as Leslie spoke, of Mill's 'abandonment' of the historical method in the body of the *Principles* (above, p. 164).

* * *

The discussion in Book IV will be examined in a later section. Here we consider only the 'Preliminary Remarks' (II, 3–21).

The general structure of the account turns formally on the celebrated 'four-stages' scheme of socio-economic development – hunting, pasturage (nomadic), agriculture and commerce. According to this doctrine, the mode

10 Mill remarked on Comte's failure to fulfil his promises in this regard: 'After so profound and comprehensive a view of the progress of human society in the past, of which the future can only be a prolongation, it is natural to ask, to what use does he put this survey as a basis of practical recommendation? . . . We fail to see any scientific connection between his theoretical explanation of the past progress of society, and his proposals for future improvement . . . [His recommendations] rest as completely, each on its separate reasons of supposed utility, as with philosophers who, like Bentham, theorize on politics without any historical basis at all' (*CW*, X, 324–5).

of subsistence existing at any time determines the legal and institutional framework of society, a proposition which, coupled with a presumption regarding the uniformity of the human constitution but allowing for an environmental approach to behaviour, was applied by its eighteenth-century originators to an investigation of historical progress – the development of productive organization through the four stages with corresponding changes in civil society (legal and governmental institutions) as well as in customs, manners and morals.[11]

In Mill's hands, however, the scheme was not thus used.[12] No concerted effort was made to account for transitions between the major historical stages – a key consideration had his objective been the construction of a genuine theory of socio-economic development. We should qualify this generalization somewhat: Mill did refer to 'the spontaneous course of events' in alluding to transitions, with particular reference to the consequences of pressure of population on scarce land (12); in these terms he discussed the transition from the nomadic to the agricultural stage and invasions of settled territories by nomadic hordes, and a tendency to the abandonment of agricultural activity in the European case and the adoption of trading. But the argument is casual, little specific evidence is offered and it is difficult to escape the impression that Mill was repeating commonplaces from the literature.[13]

That the pattern of economic organization constitutes a determinant of institutional arrangements, especially those defining governmental authority, is implied at various stages of the account. Thus, for example, in the second stage of nomadic existence, upon the appearance of a surplus and its unequal distribution, the relatively wealthy are enabled to exert authority over the less favoured. This is a general theme extending beyond the nomad stage. But what constitutes the main feature of the four-stages theory is but a passing consideration in Mill's account and inconspicuous at key junctures. Thus the initial success of the system of roughly equal distribution of land is explained by reference to 'the progress of events', 'circumstances of race and climate' and 'favourable accidents'; and no concerted effort is made

11   See the excellent account of the eighteenth-century origins and subsequent fortunes of the doctrine by Meek (1976). On stages of economic growth with particular reference to the German historical school, see Hoselitz (1960).

12   We limit our discussion to the *Principles*. It should not be forgotten that themes consistent with the four-stages approach are touched upon elsewhere; cf. for example Mill's approval of Comte's proposition that the form of government existing at any time is intimately bound up with the contemporary state of civilization, and his own paper 'Coleridge' (1840) regarding government forms and states of society.

13   It is perhaps pertinent that Malthus used the theory to demonstrate the validity of his propositions regarding population at all stages of development. Cf. *An Essay on the Principles of Population* (1798), Chs. III, IV.

to account for the differing forms of government authority in the Asiatic monarchical and the European feudal systems, and this despite their diverse degrees of 'security', landowner status and size of individual holding (15). In any event, personal characteristics play as large a role in determining some of the most important institutional features alluded to: for example, the mode of land tenure adopted following the collapse of the Roman Empire is attributed to the indisposition of the barbarian conquerors to superintend industrial undertakings, ruling out the institution of a full-fledged system of slavery.

It seems fair to conclude with Meek that the four-stages approach was not adopted by Mill in the full technical sense of the term (1976, 223).[14] And this is scarcely surprising when we consider Mill's reflections in the *System of Logic* on the infantile state of 'ethology' and 'progress' – the 'science of human character formation' and the 'general science of society' involving the laws 'according to which any state of society produces the state which succeeds it and takes its place'.

Mill in fact used the four-stages framework merely as a convenient peg to attach a largely descriptive prologue to the *Principles* regarding differences in the form, magnitude, and distribution of wealth over time and from place to place – to portray those 'extraordinary differences in respect of [wealth], which exist between nation and nation, and between different ages of the world; differences both in the quantity of wealth, and in the kind of it; as well as in the manner in which the wealth existing in the community is shared among its members (*CW*, II, 10).

One of the major tasks of political economy was precisely to investigate the 'causes' – or more specifically some of the causes – of the 'remarkable differences in the state of different portions of the human race, with regard to the production and distribution of wealth . . .' (20). This task was not undertaken in any great detail in the introductory section – it was a matter for the treatise itself – but there are brief allusions to 'causal' influences and these must be spelled out.

A central role is accorded the generation of an agricultural surplus over 'absolute necessaries' as the basic pre-condition for the growth of wealth (and accordingly of population), a phenomenon first encountered in the nomadic or pasturage stage (11). Accumulation takes the palms as determinant of growth; the emphasis thereupon is apparent at all stages of the account, but particularly in the discussion of saving by serfs in the

---

14 The historical dimension and the four-stages approach were also a secondary consideration in the *Wealth of Nations*, but what discussion there is was rather more sophisticated in Smith's case, for he attempted to account for stadial transitions and for distortions to the standard pattern. On Smith see Hollander, in O'Driscoll (1979), 71–84 and Moss, *ibid.*, 85–101.

early medieval period, which permitted their transfer to the towns, and by the 'saving class' of burghers during the later Middle Ages, which entailed their transplantation of the feudal aristocrats (18). Some attention is paid invention and knowledge (domestically generated or imported), but security of property seems to be weighted rather more heavily as a necessary condition for accumulation. To a lack of security is attributed what Mill saw as the surprisingly slow progress of the agricultural stage, insecurity dictating a preference for movable riches (jewellery and the like) rather than permanent capital investments. The fixed personal positions of the great landlords of feudal Europe compared with the uncertain positions of those of the Asiatic forms was a 'main reason' for the more favourable progress of Europe, while increasing 'security of person and property' characterized the period of transformation from the feudal to the modern era.

The distribution of wealth is accorded a conspicuous role in determining the use to which the surplus is put. Thus inequality opened the way to the rise of a leisure class and the generation of 'new wants' implying use of the surplus in the support of manufacturing and trading sectors rather than of menials and the like – a pattern repeated at various stages of the historical record. (This provides an example of spontaneous or unplanned change.) Inequality is not, however, represented as a necessary condition for the generation of new wants, since these might also arise (for example) from foreign contacts which tend to break 'the chain of routine' (15–16). Moreover, inequality, carried too far may have debilitating consequences. The distortions of the Roman period led to an ostentatious life style which contributed to the downfall of the Empire – an allusion to the effects of excessive unproductive consumption (17). Similarly, to the smaller individual command over the surplus by European landholders, compared to their Asiatic counterparts, Mill attributed a smaller propensity towards luxury consumption.

It may be noted in passing that unequal distribution of wealth itself is attributed in its initial origins to differential expenditures of effort and thrift on the part of individuals – accidental differentials in character traits – and this as early as the second or nomad stage. Subsequently, inequalities are generated or solidified by institutional arrangements (as where the sovereign, in the Asiatic form, makes decisions immediately relevant to distribution) or in the aftermath of conquest as in the case of the Roman Empire. Also pertinent, although unexplained, is a general observation (drawn from the Roman context) that pressure towards growing inequality is generated by excessive unproductive consumption: 'When inequality of wealth once commences, in a community not constantly engaged in repairing by industry the injuries of fortune, its advances are gigantic; the great masses of wealth swallow the smaller'(16).

Surprisingly little is said of human motivation. What there is suggests the manifestation of maximizing behaviour even at the earliest stages. Thus, for example, references is made to the profit-seeking operations of money dealers and grain dealers in the Asiatic monarchies; the development of new wants and desires are said to have stimulated the free cultivators in the ancient European communities 'to extract from their own soil the utmost which they knew how to make it yield'; and, particularly interesting, there is a contrast drawn between the barbarian conquerors who were not disposed to superintend personally industrial undertakings and were obliged therefore to permit institutions in place of slavery – such as crop sharing – which assured some 'incentive to exertion, some real interest in the soil' on the part of cultivators (16-17).

It thus appears that Mill found evidence in the secular record for at least some of the individual causal influences acting upon the growth of wealth directly and indirectly. But the 'Preliminary Remarks' scarcely constitute an exhaustive and detailed inductive treatment. For elaboration we must turn to subsequent chapters.

III 'QUALITATIVE' DETERMINANTS OF PRODUCTIVITY

It is only in Chapter vii of Book I ('On What Depends the Degree of Productiveness of Productive Agents') that Mill takes up the major issue outlined at the outset of the work – to explain the 'evident' fact that 'productive efficacy varies greatly at various times and places. With the same population and territory, some countries have a much larger amount of production than others, and the same country at one time a greater amount than itself at another' (100).[15] The chapter proceeds to an inductive analysis of the individual causes of productivity differentials on a much broader scale than that undertaken in the 'Preliminary Remarks'. Reference is made to natural advantages incorporating land fertility and climate, the availability of minerals and power, and location; the quality of manpower incorporating energy and manual dexterity, and intelligence; the quality of management; the level of knowledge including that embodied in capital goods; the trustworthiness of the community; and the state of security.

---

15 The status of the first six chapters is rather ambiguous. They include a classificatory and descriptive section defining the meaning of 'wealth', isolating 'the requisites of production' in the generation of wealth, and distinguishing between 'productive' and 'unproductive' labour from the same perspective. But the discussion in Chapter V of 'the fundamental propositions respecting capital' and in Chapter VI of the machinery issue are essentially exercises in deductive theory presuming capitalist-exchange institutions and thus rather out of place in the location accorded them in Book I.

It is from observation of the contemporary as well as the historical records that Mill derives the list of causal influences at play and even the weights to be accorded to them in general and in special circumstances. In all this he relied on his own observations and the literature. A brief summary account of Mill's weighting of the qualitative determinants of productivity is in place here.

As for natural advantages Mill, basing himself on historical illustrations, takes the position that climate is of greater significance even than fertility – sufficiency so to check development at the nomadic stage when unfavourable; conversely, favourable climatic conditions provide scope for the release of resources from the production of necessaries to 'higher' uses provided that 'the character of the inhabitants does not rather induce them to use up these advantages in over-population, or in the indulgence of repose' (102). Most important, however, in this category is a maritime situation, the availability of natural harbours and rivers. Mill draws his evaluation from the observation that at early stages of development (in the Ancient World and Middle Ages) the most prosperous communities were not those with the largest territories and most fertile soils but rather those with excellent maritime facilities permitting savings in resources otherwise absorbed in transportation and encouraging exchange and specialization.

Precedence in the hierarchy is, however, given to quality of manpower over natural advantages: 'But experience testifies that natural advantages scarcely ever do for a community, no more than fortune and station do for an individual, anything like what it lies in their nature, or in their capacity, to do. Neither now nor in former ages have the nations possessing the best climate and soil, been either the richest or the most powerful; but (in so far as regards the mass of the people) generally among the poorest . . .' (102–3). It is precisely in areas with the 'best climate and soil' that there are likely to be few incentives for sustained labour and little concern for 'remote objects' and thus backwardness in the creation of good political institutions and the protection of property, an observation implying that character is a matter of nurture or circumstances;[16] and indeed that institutional change itself reflects character. Mill generalized thus:

Successful production, like most other kinds of success, depends more on the qualities of the human agents, than on the circumstances in which they work: and it is difficulties, not facilities, that nourish bodily mental energy. Accordingly the tribes of mankind who have overrun and conquered others, and compelled them to labour for their benefit, have been mostly reared amidst hardship. They have either been bred in the forests of northern climates, or the deficiency of natural hardships has

16 See also Section VIII below on alternative land-tenure arrangements, particularly the Irish case.

been supplied, as among the Greeks and Romans, by the artificial ones of a rigid military discipline. From the time when the circumstances of modern society permitted the discontinuance of that discipline, the South has no longer produced conquering nations; military vigour, as well as speculative thought and industrial energy, have all had their principal seats in the less favoured North (103).

British industrial predominence it may be noted is, at least in the present context, attributed largely to the energy of its work force – a consequence of climatic conditions and not only original temperament – intending by 'energy' not the efforts people are 'able and willing to make under strong immediate incentives', for in this there was comparatively little distinction between nations, but rather the 'capacity of present exertion for a distant object; and . . . the thoroughness of . . . application to work on ordinary occasions'. In this there were sharp distinctions between the British and (for example) the West Indian labourer with the most significant implications regarding policy. For in underdeveloped areas such as Jamaica the 'stimulus of new wants and desires' was needed as motive to steady effort so that the causal relations pertinent elsewhere are reversed: '. . . in most societies . . . indulgence tends to impoverish rather than enrich; but in the state of mind of the negroes [on emancipation] it might have been the only incentive that could make them voluntarily undergo systematic labour, and so acquire and maintain habits of voluntary industry which may be converted to more valuable ends' (104–5).

As for the general level of intelligence and its implications for productivity Mill (in a later chapter) had the strongest words: 'So low, in some of the most civilized countries, is the present standard of intelligence, that there is hardly any source from which a more indefinite amount of improvement may be looked for in productive power than by endowing with brains those who now have only hands.'[17] The 'carefulness, economy and general trustworthiness of labourers', is weighed on a par with intelligence, as also are friendly relations between employer and employee, although Mill again had harsh words regarding the contemporary state of affairs: 'I know not where any such sentiment of friendly reliance now exists.' Amelioration of the mind and character of the wealthier classes, Mill was hopeful, would release the mental energy, better instruction, feelings of conscience, public spirit, philanthropy – other than self-regarding motives it will be noted – likely to generate 'the most valuable improvements, both in the economical resources of their country, and in its institutions and customs'(183–4).[18]

17 The general level of intelligence plays a role in the stimulus of invention (see below Section VI regarding 128, 129).

18 Mill cites the backwardness of French agriculture as partly due to the devotion of landowners to town pleasures.

Mill describes as 'self evident' the impact on productivity of manual dexterity in routine operations, intelligence where mental operations are involved, and knowledge of the natural powers and properties of objects used in industry. But he does recommend Babbage's *Economy of Machinery and Manufacture* for illustration of the effects upon efficiency of improvements due to invention embodied in tools and machinery (106).[19] As evidence of the scarcity of those fit for superintendence and non-routine positions – a reflection of the level of general intelligence in the community – Mill drew upon the great wage differentials earned in such functions, an empirical procedure characteristic of the *Wealth of Nations* (107).[20]

As for the significance of 'trustworthiness' Mill illustrates from 'countries of Europe, of first-rate industrial capabilities, where the most serious impediments to conducting business concerns on a large scale, is the rarity of persons who are supposed fit to be trusted with the receipt and expenditure of large sums of money'. Here Babbage is cited regarding the case of a branch of export trade stopped because of forgeries and frauds associated with it, and for illustrations of the waste of resources due to the need to counter untrustworthiness (110f).[21]

The matter of 'security' in the sense of protection by the government (and more still from the government) is described as 'secondary', obviously a reference not to its relative significance but to its role as a determinant playing at one remove upon accumulation. The 'acknowledged explanation of the poverty of many fertile tracts of Asia, which were once prosperous and populous' is said to be a lack of security against tyrannical violence permitting the production of little more than subsistence; similar examples are given from periods of Roman history and pre-Revolutionary France (112f). Greece and her colonies in the Ancient World, and Flanders and Italy in the Middle Ages, provide instances of the positive effects of security from government. In the contemporary British case 'manners and opinion' (reflected in fear of exposure) as well as formal law and institutions assured security of property (114–15).

The French revolution is subsequently portrayed as having been 'equivalent to many industrial inventions' – a Ricardian proposition in fact (Hollander, 1979, 114), which Mill illustrates in terms of the strengthening of redress against injury to person and property by people of rank and a reduction of court influence; similarly, 'no improvements operate more directly upon the productiveness of labour, than those in the tenure of farms,

19 On Babbage see Romano (1982).
20 But here we have a case where a theoretical 'model' – that of the wage structure – is being silently put to use.
21 In the manuscript Mill also cited in this context Charles Dunoyer, *De la Liberté du Travail* (1845) on the significance of personal qualities.

and in the laws relating to landed property . . . All these things are as real, and some often as great, improvements in production, as the invention of the spinning jenny or the steam-engine' (*CW*, I, 183).

All of this alludes directly or indirectly to the matter of 'security'. The same issue is again taken up at length in Book IV where along with technological change and 'cooperation' it is represented as a 'tendency' which has 'always hitherto characterized, and will assuredly continue to characterize the progress of civilized society' (706; discussed below, Section VII).

IV QUANTITATIVE TECHNICAL RELATIONSHIPS: SCALE ECONOMIES

The role of 'cooperation' or 'the combined action of numbers' as a determinant of productivity was sufficiently important for Mill to devote an entire chapter to the matter (Book I; viii). His discussion commences with Wakefield's distinction between simple and complex co-operation and turns on very general observation utilizing homely examples. It proceeds to the issue of exchange between town and country, and thence to instances of higher degrees of division of labour, drawing for illustration upon Smith and Say on pins and also Babbage on cards, 'a remarkable proof' of which is seen in the 'trifling sum' at which they can be sold (*CW*, II, 117–18, 121–2, 122–4).[22] Thereafter the precise nature of the advantages of specialization are taken up: Adam Smith's 'increase of dexterity' is accepted as is Babbage's avoidance of waste in learning new arts; but Mill questions Smith's emphasis upon time saving and Babbage's rendition of the theme – 'few workmen change their work and their tools oftener than a gardener; is he usually incapable of vigorous application?' – and suggests counterforces affecting productivity positively due to switching operations (124f). Smith's conception of specialization as a stimulus to invention is accepted but qualified (128), a matter to be discussed in a later section (below, p. 221). But apart from the effects due to increased dexterity Mill was most enthusiastic about savings flowing from classification of workers according to their differential physical capacities – due to Babbage (128–9)[23] – and from the increased utility derived from specialized tools – due to John Rae (130). The limitations to division of labour imposed by the extent of the market and the nature of the particular employment are then taken up, the latter with special reference to agriculture – a well-known Smithian theme: 'Agriculture, for example, is

22 Simple co-operation entails several individuals working together on the same set of operations; complex co-operation entails a division of labour between tasks undertaken simultaneously.

23 This aspect of specialization had been earlier discussed by Josiah Tucker (1757); cf. Hollander (1973a), 66.

not susceptible of as great a division of occupations as many branches of manufactures, because its different operations cannot possibly be simultaneous' (130; cf. 142).

This empirical account continues in Book I; ix on the size of manufacturing establishments: 'The larger the enterprise, the farther the division of labour may be carried. This is one of the principal causes of large manufactories. Even when no additional subdivision of the work would follow an enlargement of the operations, there will be good economy in enlarging them to the point at which every person to whom it is convenient to assign a special occupation, will have full employment in that occupation' – a proposition elaborated by Babbage (131). Other determinants of large scale discussed include the need for expensive machinery (in which context the celebrated Ricardian analysis of machinery is alluded to) and opportunities created for economy of superintendence (133–5). Yet, despite his enthusiasm for the general proposition that '[t]he larger the scale on which manufacturing operations are carried on, the more cheaply they can in general be performed', Mill hesitated to join Senior in describing the phenomenon as a 'law': 'Mr Senior has gone the length of enunciating as an inherent law of manufacturing industry, that in it increased production takes place at a smaller cost, while in agricultural industry increased production takes place at a greater cost. I cannot think, however, that even in manufactures, increased cheapness follows increased production by anything amounting to a law. It is a probable and usual, but not a necessary consequence' (III, 712–13). There were industries where scale economies do not pertain, and '[w]hether or not the advantages obtained by operating on a large scale preponderate in any particular case over the more watchful attention, and greater regard to minor gains and losses, usually found in small establishments, can be ascertained, in a state of free competition, by an unfailing test. Whenever there are large and small establishments in the same business, that one of the two which in existing circumstances carries on the production at greatest advantage will be able to undersell to other' – the preservation of the fittest (II, 133; cf. 140, 141).

Apart from scale economies there is also a particular amenability of manufacturing activity to 'improvement' – in contrast with agricultural activity – yielding a 'tendency' to a decline in costs 'with the progress of society', a matter taken up by Mill in a later chapter of Book I:

the cost of the material forming generally a very small portion of the entire cost of the manufacturer, the agricultural labour concerned in the production of manufactured goods is but a small fraction of the whole labour worked up in the commodity. All the rest of the labour tends constantly and strongly towards diminution, as the amount of production increases. Manufactures are vastly more susceptible than agriculture, of mechanical improvements, and contrivancies for saving labour; and it has already been seen how greatly the division of labour, and its skillful and economical distribution, depend on the extent of the market, and on the possibility of production in large masses. In manufactures, accordingly, the causes tending to increase the productiveness of industry, preponderate greatly over the one cause which tends to diminish it: and the increase of production, called forth by the progress of society, takes place not at an increasing, but at a continually diminishing proportional cost (182).

While Mill was unprepared to designate increased productivity due to scale of operation as a 'law', he was apparently prepared so to describe the upward secular trend in manufacturing productivity incorporating also the effects of innovation proper: 'any tendency which may exist to a progressive increase in that single item' (a reference to the possible secular rise in raw material prices) 'is much overbalanced by the diminution continually taking place in all the other elements; to which diminution it is impossible to assign any limit . . . [I]t may . . . be laid down as a rule, that manufactured articles tend, as society advances, to fall in money prices. The industrial history of modern nations, especially during the last hundred years, fully bears out this assertion' (III, 713). In an alternative rendition, the statistical evidence is said to reveal 'the progressive fall of the prices and values of almost every kind of manufactured goods during the two centuries past; a fall accelerated by the mechanical inventions of the last seventy or eighty years . . .' (II, 182).

We must be cautious in evaluating Mill's generalization regarding increasing manufacturing productivity. In the first place, the term 'progress of improvement' (below, p. 225), frequently includes more than innovation proper, incorporating advances in law, government and education and thus touching upon matters pertaining to the institutional framework itself; the reference to the 'mechanical inventions of the last seventy or eighty years' suggests, however, a narrower dimension. More significant: the 'rule' of increasing manufacturing productivity which allows for the effects of innovation proper, is formally unrelated to scale as such – it is an empirical generalization describing the movement of productivity over time. That there had occurred and was occurring an increase in the average size of operating establishment seems clear and this would account for part of the observed fall in manufacturing costs. But what of innovation proper: Did Mill maintain that innovation, as well as specialization, is related to scale, thus transforming the empirical generalization into a far more potent technical relationship?

Mill indeed suggests as much although casually by maintaining that 'labour tends constantly and strongly towards diminution, as the amount of the production increases' and proceeding immediately to discuss innovation as well as division of labour. But this gives us too little to go on, since nothing is said to rationalize any such relationship. On the other hand, in discussion of the agricultural sector which follows below, we find the observation that experiments 'can seldom be made with advantage except by rich proprietors or capitalists', which perhaps can be generalized. And, what Mill has to say in his chapter on scale economies regarding joint-stock organization – a reflection of an observed trend towards this form by large firms – is also tangentially relevant to the theme. On the basis of common experience and general knowledge of work attitudes such firms are said to be in a position to attract management of a quality particularly suitable for the undertaking of projects 'out of the ordinary routine' (below, p. 229).

* * *

Mill similarly drew upon his own observation, experience, the 'testimony of witnesses' and contemporary literature for the comparative merits of large and small farming – for the most part identified here respectively with capitalist and peasant farming (142f). The weight of evidence, drawn from a wide range of contemporary examples, suggested that small scale was not detrimental to productivity except if carried to extreme lengths as in France.[24]

The issues were, none-the-less, complex: Empirical skill and traditional knowledge were most impressive in the case of the small peasant farms of Flanders and Italy (and even parts of France) (146–7). The standard tools and equipment were not costly and were thus available to small farmers (143). There were, on the other hand, advantages of size in marketing and purchasing (144). Mill also conceded 'an absence of science, or at least of theory; and to some extent a deficiency of the spirit of improvement, so far as relates to the introduction of new processes. There is also a want of means to make experiments, which can seldom be made with advantage except by rich proprietors or capitalists' (147). Major drainage and irrigation works affecting entire areas were also deficient in peasant systems. But some of these disadvantages could be and to some extent had been, overcome by co-operative effort involving major improvements of whole areas, and co-operative ownership of machinery and organization of marketing (143, 146–7); and Mill hoped that 'we have now heard the last of the incompatibility of small properties and small farms with agricultural improvement. The only question which remains open is one of degree; the

24 'The inferiority of French cultivation . . . is probably more owing to the lower general average of industrial skill and energy in that country, than to any special cause [pertinent to small scale activity]; and even if partly the effect of minute subdivision, it does not prove that small farming is disadvantageous, but only (what is undoubtedly the fact) that farms in France are very frequently *too* small . . .' (150).

comparative rapidity of agricultural improvement under the two systems; and it is the general opinion of those who are well acquainted with both, that improvement is greatest under a due admixture between them' (152). It is revealing too that, appealing to the 'testimony of witnesses', Mill dwelt on the impressive force of the self-interested motivation exerted by the owner-cultivator or the metayer, on fixed tenure (148f.), whereas in his discussion of scale of operation in manufacturing he emphasized, by contrast, the possibility of hiring managers of superior intelligence capable of operating more efficiently than self-interested owner-directors.

V QUANTITATIVE TECHNICAL RELATIONSHIPS: DIMINISHING RETURNS

We proceed to the 'law' of diminishing agricultural returns, noting first a standard formulation in terms of average (rather than incremental) product, allowance made (as by Ricardo as well as Carey) for an initial stage of rising productivity: 'After a certain, and not very advanced, stage in the progress of agriculture, it is the law of production from the land, that in any given state of agricultural skill and knowledge, by increasing the labour, the produce is not increased in an equal degree; doubling the labour does not double the produce; or to express the same thing in other words, every increase of produce is obtained by a more than proportional increase in the application of labour to the land' (*CW*, II, 174). Similarly: 'In all countries which have passed *beyond a rather* early stage in the progress of agriculture, every increase in the demand for food, occasioned by increased population, will always, unless there is a simultaneous improvement in production, diminish the share which on a fair division would fall to each individual' (187; the italicized phrase appears in 1865. In earlier editions the reference is to a 'very' early stage).[25] Mill's notion of an optimum population was severely criticized in an early paper by Robbins (1927), on the grounds that it fails to relate the optimum to 'the progress of improvement' but represents it as 'fixed once and for all', whereas at least part of such progress may be the consequence of population increase or at the least can only be realized by a larger population (111–12).

The 'law' manifests itself at the extensive and intensive margins, the former a reflection of land scarcity: 'Land differs from the other elements of production, labour and capital, in not being susceptible of indefinite increase. Its extent is limited, and the extent of the more productive kinds of it more limited still. It is also evident that the quantity of produce capable

25 In a letter to Henry Carey of February 1845 Mill allowed for increasing agricultural productivity 'up to the point of density in population, which is indispensable to the use of costly machinery & to the adequate combination & division of labour', and conceded that 'the limit cannot be fixed precisely'. But there was a limit as was evident to 'any one living in an old country' (*CW*, XIII, 659–60).

of being raised on any given piece of land is not indefinite. This limited quantity of land, and limited productivity of it, are the real limits to the increase of production' (*CW*, II, 173). The limited productivity of given areas of land is not rationalized in the form of 'variable proportions' as in modern formulations, or in an equivalent or alternative formulation.

Casual observation might suggest that 'for the present all limitation of production or population' (deriving from the 'limited quantity of land, and the limited productiveness of it') 'is at an indefinite distance, and that ages must elapse before any practical necessity arises from taking the limiting principle into serious consideration'. But any such impression was merely an optical illusion due to the observation that marginal products in given areas are normally positive – 'there is no country in which all the land, capable of yielding food, is so highly cultivated that a larger produce could not (even without supposing any fresh advance in agricultural knowledge) be obtained from it' – and that there were entirely uncultivated areas on the globe still available. To avoid the error, the phenomenon should be envisaged not as an 'obstacle opposed by a wall' but rather – and here Mill implies a slowly diminishing marginal product – as 'a highly elastic and extensible band, which is hardly ever so violently stretched that it could not possibly be stretched any more, yet the pressure of which is felt long before the final limit is reached, and felt more severely the nearer that limit is approached' (173–4).

As we know, empirical support for the principle of diminishing agricultural returns is essential from the perspective of the position of the essay on method: the individual causes acting upon wealth must be observable for deductive analysis to proceed. Now the 'evidence' Mill provides in this case turns formally on a double-barrelled logical argument, the first part of which constituting the famous 'flower-pot' rationalization – that diminishing returns are encountered at the intensive margin is evident from the fact of extensions from 'good' to 'inferior' land:[26] 'When, for the purpose of raising an increase of produce, recourse is had to inferior land, it is evident that, so far, the produce does not increase [on good land] in the same proportion with the labour' (174). Thus high-farming methods involving intensive cultivation must entail increasing (unit) costs:

[Land] might be ploughed or harrowed twice instead of once, or three times instead of twice; it might be dug instead of being ploughed; after ploughing, it might be gone over with a hoe instead of a harrow, and the soil more completely pulverized; it might be oftener or more thoroughly weeded; the implements used might be of higher finish, or more elaborate construction; a greater quantity or more expensive

26 The physical property distinguishing good and poor lands is primarily thickness of soil which governs the yield for any given application of labour and dressing (174, 177).

kinds of manure might be applied, or when applied, they might be more carefully mixed and incorporated with the soil . . . But, that it is obtained at a more than proportional increase of expense, is evident from the fact that inferior lands are cultivated. Inferior lands, or lands at a greater distance from the market, of course yield an inferior return, and an increasing demand cannot be supplied from them unless at an augmentation of cost, and therefore of price. If the additional demand could continue to be supplied from the superior lands, by applying additional labour and capital, at no greater proportional cost than that at which they yield the quantity first demanded of them, the owners or farmers of those lands could undersell all others, and engross the whole market . . . [I]t never could be the interest of any one to farm [lands of a lower degree of fertility] for profit. That a profit can be made from them, sufficient to attract capital to such an investment, is a proof that cultivation on the more eligible lands has reached a point, beyond which any greater application of labour and capital would yield, at the best, no greater return than can be obtained at the same expense from less fertile or less favourably situated lands (174–5).[27]

In this argument is is taken for granted that 'inferior' land is definable and recognizable, the use of which indicates rising costs at the intensive margin on good land. But Mill then seems to reverse the procedure. He now takes for granted that, with the use of high-farming methods on good land, (unit) costs rise and considers this intelligence to be evidence of the scarcity of good land:

The careful cultivation of a well-farmed district of England or Scotland is a symptom and an effect of the more unfavourable terms which the land has begun to exact for any increase of its fruits. Such elaborate cultivation costs more in proportion and requires a higher price to render it profitable, than farming on a more superficial system; and would not be adopted if access could be had to land of equal fertility, previously unoccupied. Where there is the choice of raising the increasing supply which society requires, from fresh land of as good quality as that already cultivated, no attempt is made to extract from land anything approaching to what it will yield on what are esteemed the best European mode of cultivating. The land is tasked up to the point at which the greatest return is obtained in proportion to the labour employed, but no further: any additional labour is carried elsewhere (175).[28]

27 Some of the high-farming operations listed here, such as pulverizing, are sometimes represented as instances of technical change (as on 107). At the present stage, however, Mill is apparently assuming constant technology.

28 Mill is technically mistaken in his final assertion unless we assume a low 'capacity' output relative to the total market.

    In this context Mill refers critically to one authority's assertion that where, as in the United States, 'land is so plentiful and labour so dear . . . a totally different principle must be pursued to that which prevails in populous countries, and . . . the consequence

There is an obvious circularity of reasoning involved when these two 'proofs' are taken together at face value. It seems probable, however, that Mill accepted as known facts that high-farming and extensions to inferior soil both entail higher average cost,[29] using the formal 'proofs' as additional and, strictly speaking, superfluous evidence. The principle of diminishing returns at both the extensive and intensive margins appears to be supported by knowledge of factual cost observation rather than by specious logic.[30] Relevant cost observations would, of course, be those pertaining to a given point of time or, if historical cost data are utilized, to periods of unchanged technology unless correction is made for changing technology.

Here it is essential to recall the further methodological principle that the individual causal laws are at work continuously even when their effects are counterbalanced or even outweighed by conflicting tendencies. In our context, the law of diminishing returns is operative even when (uncorrected) cost data do not reveal it. This explains Mill's insistence upon the law in the very context of a discussion of the 'countervailing tendency' of technological progress acting in 'habitual antagonism' to it:

The materials of manufacture being all drawn from the land, and many of them from agriculture, which supplies in particular the entire material of clothing; the general law of production from the land, the law of diminishing return, must in the last resort be applicable to manufacturing as well as to agricultural industry. As population increases, and the power of the land to yield increased produce is strained harder and harder, any additional supply of material, as well as of food, must be obtained by a more than proportionally increasing expenditure of labour (182).

Since the context wherein this formulation occurs is the impressive range of new technologies – in agriculture itself, but also in transportation,

will of course be a want of tidiness, as it were, and finish, about everything which requires labour'. Mill objected that only plentifulness of land is relevant, for 'however dear labour may be, when food is wanted, labour will always be applied to producing it in preference to anything else' – although it is true that 'this labour is more effective for its end by being applied to fresh soil, than if it were employed in bringing the soil already occupied into higher cultivation' (176).

29 Thus he writes (176) of Flanders and Tuscany, where (compared with England and the same applied to England compared with the US), 'by the application of a far greater quantity of labour there is obtained a considerably larger gross produce, but on such terms as would never be advantageous to a mere speculator for profit, unless made so by much higher prices of agricultural produce' – a clear statement of observed cost increase with intensive farming.

For cost behaviour at the extensive margin, cf. 179: 'Everyone knows . . . that it is the high lands and their soils which are left to nature, and when the progress of population demands an increase of cultivation, the extension is from the plains to the hills.'

30 Cf. G. J. Stigler (1965) on the so-called 'flower-pot' proof (1952, 119).

engineering, metal working and food processing – which act to reduce agricultural costs, and considering Mill's account of past cost trends (500 years of constant cost) and of contemporary and prospective cost reductions it is difficult to imagine that he was alluding to necessary cost increases of agricultural produce in an historical sense.[31] His references, it is most likely, were to the ever present upward pressure on costs due ultimately to land scarcity irrespective of the fact that such pressure is often disguised by countervailing tendencies.[32] It is in this sense that one must understand the term 'tendency of the returns to a progressive diminution . . . a result of the necessary and inherent conditions of production from the land' (187).[33]

* * *

In a strong passage, with Marxian flavour distinguishing as it does between 'superficial agencies' and 'essence', Mill emphasized the social implications of the denial of the law – itself represented as 'the most important proposition in political economy':

I apprehend this to be not only an error, but the most serious one, to be found in the whole field of political economy. The question is more important and fundamental than any other; it involves the whole subject of the causes of poverty in a rich and industrious community: and unless this one matter be thoroughly understood, it is of no purpose proceeding any further in our inquiry . . . This general law of agricultural industry is the most important proposition in political economy. Were the law different, nearly all the phenomena of the production and distribution of wealth would be other than they are. The most fundamental errors

31 For the general context in which this passage occurs see below (p. 224); for Mill's account of historical cost data and of contemporary and prospective movements, see p. 223f.

32 Thus secular trend observations pertaining to agricultural productivity might pick out positions such as A, B (1700), $A'B'C$ (1800) etc. The equality of productivity at A and B (or A' and C) despite differences in intensity of cultivation reflects the differential quality of the plots.

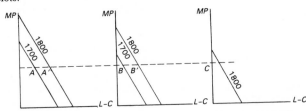

33 Similarly, the law of diminishing returns (supplemented by exhaustibility of the metals) is insisted upon in mining despite emphasis upon improvement as a peculiarly powerful countervailing force (below, p. 216).

which still prevail on our subject, result from not perceiving this law at work underneath the more superficial agencies on which attention fixes itself, but mistaking those agencies for the ultimate causes of effects of which they may influence the form and mode, but of which it alone determines the essence (173–4).

Yet there were some authorities who ignored the 'truth' that 'the produce of land increases, *caeteris paribus*, in a diminished ratio . . .' And there were also those who positively denied it, pre-eminently H. C. Carey in his *Principles of Political Economy* (1837), which formulated a reverse law of increasing agricultural returns based upon a supposed historical trend whereby activity commenced on poor soil and subsequently spread to the richer soils of the valleys – the cultivation of which required preliminary clearing and draining – thus undermining 'the very foundation of what he calls the English political economy, with all its practical consequences, especially the doctrine of free trade' (177–8).[34]

In his defence Mill explained that the law of diminishing returns was not intended to apply to 'the first cultivation in a newly settled country', although some authorities may have been careless in their formulations.[35] At early stages of development land requiring preliminary investments to render it fit for tillage may indeed remain unused for some time despite the higher return to labour on such land, assuming the requisite preparation, than on the 'poor' land initially in use and including in the cost comparison the preparatory investments involved. When circumstances permitted the investments in clearing and draining, there then followed the sequence portrayed by Carey. Yet for all that the standard 'law' set in early enough to merit the conclusions political economy drew from it – conclusions pertinent to 'old countries':

Mr. Carey will hardly assert that in any old country – in England or France, for example – the lands left waste are, or have for centuries been, more naturally fertile than those under tilage . . . [Is] it true that in Engand or France at the present day the uncultivated part of the soil consists of the plains and valleys, and the cultivated, of the hills? Every one knows, on the contrary, that it is the high lands and thin soils which are left to nature, and when the progress of population demands an increase of cultivation, the extension is from the plains to the hills. Once in

34 Before the edition of 1865 Mill had referred generally to those who rejected the principle of diminishing returns and who appealed to experience 'in proof that the returns from land are not less, but greater, in an advanced, than in an early, stage of cultivation – when much capital, than when little, is applied to agriculture'. In 1865 and thereafter he expanded the discussion inserting a specific reference to Carey.
35 Ricardo had always insisted on this same point; cf. Hollander (1979), 397. Until the edition of 1865 Mill himself was sometimes careless, referring to 'the universal law of agricultural industry' (177n).

a century, perhaps, a Bedford level may be drained, or a Lake of Harlem pumped out: but these are slight and transient exceptions to the normal progress of things; and in old countries which are at all advanced in civilization, little of this sort remains to be done (178–9).

Now on a surface view this position has its problems. Only a few paragraphs earlier Mill had raised the issue of imperfections in capital markets which prevented or delayed the adoption of known technology requiring embodiment in investment.[36] Consequently, increased food supplies were 'sometimes raised at an augmenting cost by higher cultivation [more intensive land use], when the means of producing it without increase of cost are known and accessible. A case in point is precisely the potential conversion by soil improvements of 'poor' into 'good' land thereby increasing the supply of the latter and permitting the abandonment of some marginal plots and the less intensive cultivation of land remaining in use:

[P]robably a considerable part of the less productive lands now cultivated, which are not particularly favoured by situation, would go out of culture; or (as the improvements in question are not so much applicable to good land, but operate rather by converting bad land into good) the contraction of cultivation might principally take place by a less high dressing and less elaborate tilling of land generally; a falling back to something nearer the character of American farming; such only of the poor lands being altogether abandoned as were not found susceptible of improvement. And thus the aggregate produce of the whole cultivated land would bear a larger proportion than before to the labour expended on it; and the general law of diminishing return from land would have undergone, to that extent, a temporary suppression (177).

Now this phenomenon is represented as one of great contemporary potential, and by no means an exceptional matter such as that of the Bedford level referred to in the reply to Carey: 'There can be no doubt', Mill asserted, 'that if capital were forthcoming to execute, within the next year, all known and recognized improvements in the land of the United Kingdom which would pay at existing prices, that is, which would increase the produce in as great or a greater ratio than the expense; the result would be such

---

36 In some instances capital had not sought out its most advantageous use because of the delayed return – the capitalist 'must wait longest for its remuneration' – so that 'it is only in a rather advanced state of industrial development that the preference is given to it'. Even in advanced societies 'the laws or usages connected with property in land and the tenure of farms, are often such as to prevent the disposable capital of the country from flowing freely into the channel of agricultural improvement' (176).

(especially if we include Ireland in the supposition) that inferior land would not for a long time require to be brought under tillage . . .'[37]

It is difficult to countenance a blatant self-contradiction within a span of two or three pages. Possibly the resolution of the apparent difficulty turns upon the treatment of Ireland – a note added in 1865 (179n.) treats Ireland as a special case (and even then suggests the possibility that with drainage, the boglands might yet remain 'poor' soils). Yet the original statement regarding productive potential specifically referred to the conversion of poor into good land in the United Kingdom generally – though with special force in the Irish case. The solution seems rather to be that Mill laid down as exceptional in contemporary British circumstances those cases of potential land improvement which embody known technology capable of rendering hitherto infertile or even unusable land *more productive* than land currently cultivated, while allowing for an impressive potential for those land improvements capable of raising the productivity of poor land. The likelihood that this is a legitimate reading is strengthened by a rider attached to the discussion. For Mill questioned whether the distinction between grades of land would ever be entirely eradicated – the potential improvements might raise the productive status of hitherto poor soils but not to the standard of the very best: 'No one, however, can suppose that . . . the whole produce required for the country would be raised exclusively from the best lands, together with those possessing advantages of situation to place them on a par with the best. Much would undoubtedly continue to be produced under less advantageous conditions, and with a smaller proportional return, than that obtained from the best soils and situations' (177).

To summarize: On Carey's analysis there may be no contemporaneous utilization of differential qualities of land if the poor (hillside) lands originally cultivated should be abandoned with the extensions envisaged to the river bottoms.[38] On Mill's allowances by contrast, some poor lands (those unamenable to treatment) might be abandoned by the improvements envisaged which increase the availability of good land, there being a contraction in the intensity of farming on lands remaining in use. But he still insisted upon the likelihood of remaining differentials between plots. It may be added too that diminishing returns operate as always, at the intensive margin. Finally, the phenomenon of improvements embodying known technology can always be envisaged as a shift in the production function on a par with genuine innovation rather than as part of a given

37 Cf. Ricardo's rendition of this same theme; Hollander (1979), 609.

38 Carey's argument from empirical price movements is confusing. Mill observed (179) that Carey's own allusions to secular increase in corn prices – were they accurate – would be evidence for falling agricultural productivity (provided the real cost of producing the monetary metals was stationary or rising over time which he believed was the general case.) For further discussion of the data, see below, (p. 232f).

functional relationship between costs and output; or as Mill put it as 'a temporary suppression' of the regular law which, sooner or later upon expansion of population – though perhaps 'not for a long time' – 'would resume its course, . . . the further augmentation . . . obtained at a more than proportionate expense of labour and capital' (177).

* * *

A brief word now regarding mining which shared with agriculture the characteristic that it 'yields an increase of produce at a more than proportional increase of expense'. Here there is the further problem of exhaustibility so that '[t]he law of diminishing return applies . . . to mining, in a still more unqualified sense than to agriculture', although conversely, 'the antagonizing agency, that of improvements in production, also applies in a still greater degree. Mining operations are more susceptible of mechanical improvements than agriculture: the first great application of the steam-engine was to mining'; and – a reference to sectoral interdependence (below, p. 222) – there were 'unlimited possibilities of improvement in the chemical processes by which the metals are extracted'. In addition, counterbalancing 'the progress of all existing mines towards exhaustion' there was 'the discovery of new ones, equal or superior in richness'(184–5).[39]

VI THE LAWS OF PRODUCTION AND DISTRIBUTION:
POLITICAL ECONOMY AND KNOWLEDGE

At the outset of the second book ('On Distribution') Mill formulates the celebrated distinction between the immutable 'laws of production' and the malleable 'laws of distribution': 'The principles which have been set forth in the first part of this Treatise, are, in certain respects, strongly distinguished from those, on the consideration of which we are now about to enter. The laws and conditions of the production of wealth partake of the character of physical truths. There is nothing optional or arbitrary in them.' Mill had in mind the constraint imposed on industry by capital, the principle of diminishing returns, and the differential effects on wealth of productive and unproductive consumption (all derived from the first book):

Whatever mankind produce, must be produced in the modes, and under the conditions, imposed by the constitution of external things, and by the inherent properties of their own bodily and mental structure. Whether they like it or not, their productions

---

39 On exhaustibility of the mines, cf. *CW*, III, 712. An exceptional reference to exhaustibility of soils will be found (920).

will be limited by the amount of their previous accumulations, and, that, being given, it will be proportional to their energy, their skill, the perfection of their machinery, and their judicious use of the advantages of combined labour. Whether they like it or not, a double quantity of labour will not raise, on the same land, a double quantity of food, unless some improvement takes place in the processes of cultivation. Whether they like it or not, the unproductive expenditure of individuals will *pro tanto* tend to impoverish the community, and only their productive expenditure will enrich it. The opinions, or the wishes, which may exist on these different matters, do not control the things themselves (*CW*, II, 199).

It will be noted that the laws of production here illustrated are explicitly defined with the states of physical energy, organization and knowledge given. What, however, if allowance is made (let us say) for changing knowledge? Mill himself raised this question and arrives at a conclusion which draws much of the substantive content from the original formulation: 'We cannot', he conceded 'forsee to what extent the modes of production may be altered, or the productiveness of labour increased, by future extensions of our knowledge of the laws of nature, suggesting new processes of industry of which we have at present no conception. But however we may succeed in making for ourselves more space within the limits set by the constitution of things, we know that there must be limits. We cannot alter the ultimate properties either of matter or mind, but can only employ those properties more or less successfully, to bring about the events in which we are interested' (199). This conclusion is weak since 'the ultimate properties of matter or mind' are undefined and of no apparent interest to the economist. The only specific illustrations given of the 'immutable' laws of production presume unchanged knowledge.

The Preliminary Remarks are equally revealing. Mill, there, readily conceded the significance of knowledge (as well as religious belief and law) for the production and distribution of wealth, and conversely of changes in production and distribution for knowledge (although he did not elaborate much on this mutual relationship); but he insisted that the determinants of the state of knowledge, religious attitudes, and law are the subject matter of special sciences (each of which may draw upon political economy) and not of political economy itself: The state of knowledge is a *datum* from the perspective of political economy (3, 20). The subject matter of production is then described with this constraint in mind:

The production of wealth; the extraction of the instruments of human subsistence and enjoyment from the materials of the globe, is evidently not an arbitrary thing. It has its necessary conditions. Of these, some are physical, depending on the properties of matter, *and on the amount of knowledge of those properties possessed at the particular time and place.* These Political Economy does not investigate, but assumes;

referring for the grounds, to physical science or common experience. Combining with these facts of outward nature other truths relating to human nature, it attempts to trace the secondary or derivative laws by which the production of wealth is determined; in which must lie the explanation of the diversities of riches and poverty in the present and past, and other grounds of whatever increase is reserved for the future (21).

The italicized phrase relating to the state of knowledge was added to the text in 1852, and places knowledge on a par with the 'properties of nature'. It is indeed only thus that the distinction in question can be made at all since once the state of knowledge is recognized as an endogenous variable from the perspective of political economy the so-called 'physical' laws of production become inevitably very much diluted.

The practical outcome of the formal categorization of laws is thus that political economy by itself can generate not even a tentative explanation of wealth differentials over space and trends over time, since it considers only the narrowest range of determinants – allocative efficiency and changing supplies of labour and capital in the presence of decreasing and increasing returns – relegating the 'qualitative' determinants, which include the state of knowledge, to the *ceteris paribus* pound as the subject matter of other sciences. This is on reflection a quite extraordinary position to take for one sensitive to the impact of new technology who sets out to write on the wealth of nations and who strenuously warns against over-specialization in the social sciences.

It turns out, however, that Mill by no means proposed so narrow a perspective as the categorization itself implies. We revert again to the Preliminary Remarks and the following cryptic statement on the scope of political economy:

In so far as the economical condition of nations turns upon the state of physical knowledge, it is a subject for the physical sciences, and the arts founded on them. But in so far as the causes are moral or psychological, dependent on institutions and social relations, or on the principles of human nature, their investigation belongs not to physical, but to moral and social science, and is the object of what is called Political Economy (20–1).

It is not made explicit what is intended by the dependence of the 'economical condition of nations' upon 'institutions and social relations', but it is not difficult to imagine, considering Mill's subsequent preoccupation with the notion that productive organization and thus productivity and the 'economical condition of nations' is in very large part an institutional and social matter. I refer to the analysis of the impact on productivity of the joint-stock arrangement, of different systems of land tenure, of laws relating to inheritance and poor relief, of civil protection. Here we have obvious examples of the fact that production *is* 'malleable'.[40] And Mill also is clear that its 'malleability' as far as it turns upon organization, does in fact fall within the domain of the economist.

The same can be said of knowledge, as is apparent from the enumeration of 'Modes of Labour in Production' (Book I, ii). I refer to the implicit widening of the scope of political economy to cover activities not motivated by maximizing behaviour. Consider the statement that while the mental labour of invention is sometimes 'undergone . . . in the prospect of a remuneration from the produce', yet in the case of speculative activity in general, the 'material fruits [of thought], though the result, are seldom the direct purpose of the persuit of savants, nor is their remuneration in general derived from the increased production which may be caused incidentally, and mostly after a long interval, by their discoveries', so that 'this ultimate influence does not, for most of the purposes of political economy, require to be taken into consideration'. Mill, however, proceeds to qualify himself: 'But when (as in political economy one should always be prepared to do) we shift our point of view, and consider not individual acts, and the motives by which they are determined, but national and universal results, intellectual speculation must be looked upon as a most influential part of the productive labour of society, and the portion of its resources employed in carrying on and remunerating such labour, is a highly productive part of its expenditures.' Accordingly, 'In a national or universal point of view, the labour of the savant, or speculative thinker, is as much a part of production in the very narrowest sense, as that of the inventor of a practical art; many such inventions having been the direct consequences of theoretic discoveries, and every extension of knowledge of the powers of nature being fruitful of applications to the purposes of outward life . . .' (42–3)[41]

The same distinction between the 'individual' and the 'national' points of view is made regarding elementary education and health expenditures. From the former perspective such outlays are not a matter for political economy whereas from the latter they positively are: 'to the community at large, the labour and expense of rearing its infant population form a part of the outlay which is a condition of production, and which is to be replaced with increase from the future produce of their labour. By the individuals, this labour and expense are usually incurred from other motives

---

40 On this issue, cf. Sidgwick (1883), 30f; Bladen (1965), xlviii.
41 A contemporary example is the 'electromagnetic telegraph', an 'unexpected consequence of the experiments of Oersted and the mathematical investigations of Ampère'. Cf. III, 706: 'The most marvellous of modern inventions . . . the electromagnetic telegraph – sprang into existence but a few years after the establishment of the scientific theory which it realizes and exemplifies.'
    On Marx's rather more potent analysis of the growth of scientific knowledge, see Rosenberg (1984).

than to obtain such ultimate return, and, for most purposes of political economy, need not to be taken into account as expenses of production.' Similarly, the doctor's labour 'when made use of by persons engaged in industry, must be regarded in the economy of society as a sacrifice incurred, to preserve from perishing, death or injury that portion of the productive resources of society which is fixed in the lives and bodily or mental powers of its productive members'. But by the individual's paying for treatment, it is 'not principally from economical motives . . . This is, therefore, one of the cases of labour and outlay which, though conducive to production, yet not being incurred for that end, or the sake of the returns arising from it, are out of the sphere of most of the general propositions which political economy has occasion to assert respecting productive labour; though, when society and not individuals are considered, this labour and outlay must be regarded as part of the advance by which society effects its productive operations, and for which it is indemnified by the produce' (41).

In the course of his formal work on method, it will be recalled, Mill had allowed that where a 'disturbing cause' is encountered which acts through the same law of human nature as that reflected in the general behavioural axioms it might be 'brought within the pale of the abstract science [of political economy]' (above, p. 124). What we apparently have now, however, is the proposed incorporation of 'disturbing causes' governed specifically by other than wealth-maximizing motivation. For this there was a precedent even in the early essay of 1836, when Mill had allowed the analysis of population growth within the subject despite the absence of the standard motive force.

* * *

The presumption that wealth-maximizing motivation is generally not a play in matters pertaining to 'intellectual speculation' (theoretical science) is potentially of the first importance. It is regrettable that Mill himself failed to carry out an investigation of the topic as a whole – an analysis of the fraction of the community's resources devoted to technological advance, health and education; clearly the incorporation of the determinants of pure science was merely a 'proposal' flowing from a realization that, though maximization motives may be irrelevant the subject was too close to home to be safely relegated to 'another science'. What we have are scattered observations and these not always consistent with the initial presumption itself. A brief note on what can be gleaned is appropriate.

In the context of his generally favourable discussion of small-scale farming (above, p. 207) Mill conceded 'an absence of science, or at least of theory; and to some extent a deficiency of the spirit of improvement, so far as relates to the introduction of new processes. There is also a want of means to make experiments, which can seldom be made with advantage except by rich proprietors or capitalists.' Theoretical science (in addition to experimentation and innovation proper) is thus to some extent undertaken in commercial undertakings, the expenditures evidently governed by profit calculations.

The relatively slow advance of agricultural compared with manufacturing technology has been referred to earlier and this provides relevant material.[42] To the extent that specialization is limited in agriculture (130, 142) we would have a rationalization for the supposed contrast in terms of Adam Smith's discussion of the extensions to knowledge deriving from division of labour. Mill, however, accepted Smith's position only with qualification, warning of the dangers to 'general intelligence and habitual activity of mind' to which he also looked for the source of inventiveness:

The third advantage attributed by Adam Smith to the division of labour, is, to a certain extent, real. Inventions tending to save labour in a particular operation, are more likely to occur to any one in proportion as his thoughts are intensely directed to that ocupation, and continually employed upon it. A person is not so likely to make practical improvements in one department of things, whose attention is very much diverted to others. But, in this, much more depends on general intelligence and habitual activity of mind, than on exclusiveness of occupation; and if that exclusiveness is carried to a degree unfavourable to the cultivation of intelligence, there will be more lost in this kind of advantage, than gained. We may add, that whatever may be the cause of making inventions, when they are once made, the increased efficiency of labour is owing to the invention itself, and not to the division of labour (128).[43]

The emphasis upon the 'general intelligence and habitual activity of mind' as a determinant of productivity advance is, of course, a pervasive theme in the *Principles*. We shall see, too, that when ventures 'out of the ordinary routine' are at stake – and this might apply to invention as well as innovation – Mill had little confidence in the self-interested calculation of capitalist owners placing his faith in hired managers of above-average intelligence (below, p. 229). But Mill's various *obiter dicta* give us very little of substance to go on.

* * *

Let us turn to distribution, which in the context of the formal dichotomy

42 We shall see that the weight of evidence apparently obliged Mill to rethink his minimization of the relative significance of agricultural improvement (below, pp. 225-6).
43 Smith himself had warned of the negative effects on the inventiveness of plant operators when division of labour is carried too far, but was confident in the effect on the average level of intelligence, having in mind specialized scientific workers. See Hollander (1973a), 214-15; also Rosenberg (1965).

of Book II is said to be 'a matter of human institution solely. The things once there, mankind, individually or collectively, can do with them as they like . . . The distribution of wealth, therefore, depends on the laws and customs of society' (199–200).[44] These 'rules by which it is determined, are what the opinions and feelings of the ruling portion of the community make them, and are very different in different ages and countries . . .' That the state of opinion is itself not an arbitrary matter in the literal sense of the term, is recognized, but said to be irrelevant since 'the laws of the generation of human opinions are not within our present subject' but rather part of the 'general theory of human progress, a far larger and more difficult subject of inquiry than political economy'.[45]

This formulation places the state of opinion on a par with the state of knowledge, in that both are said to be exogenous from the perspective of political economy. But if that is indeed the case there seems no reason to treat the laws of distribution any differently from the laws of production: Given the state of opinion – and the corresponding distributive system – there will be certain results which follow and which are as much 'immutable' as are the laws of production. This is indeed precisely what Mill proceeds to say: 'We have here to consider, not the causes, but the consequences, of the rules according to which wealth may be distributed. Those, at least, are as little arbitrary, and have as much the character of physical laws, as the laws of production' (200).[46]

* * *

The inevitable conclusion follows that the distinction between the laws of production and distribution, understood literally, proved a brittle one, first because productive organization and knowledge creation are, like

44 A more qualified statement given in the Preliminary Remarks asserts that 'governments or nations *can in some measure determine* what institutions shall be established' (21). The italicized phrase appears only in the manuscript version and first two editions. Thereafter it was replaced by 'have the power of deciding' more in line with the stronger statement given in our text. Yet Mill failed to alter the rendition only a few lines earlier reading in all editions: 'Unlike the laws of Production, those of Distribution are partly of human institution: since the manner in which wealth is distributed in any given society, depends on the statutes or usages therein obtaining'.
45 The qualifying 'of the ruling portion' of the community was added in 1852.
46 Cf. 21: 'But though governments or nations have the power of deciding what institutions shall exist, they cannot arbitrarily determine how those institutions shall work. The conditions on which the power they possess over the distribution of wealth is dependent, and the manner in which the distribution is effected by the various modes of conduct which society may think fit to adopt, are as much a subject for scientific enquiry as any of the physical laws of nature' (ms. 1848, 1849: '. . . determined by laws as rigid, and as independent of human control, as those of Production itself').

distribution, subject to human decision, and indeed in some contexts are said to fall within the domain of political economy, and because the consequences of distributive change are subject to causal analysis just as are productive relationships. Mill himself pointed all this out; it cannot be said that he was 'unaware that production and distribution are but two sides of the same coin and that either both must be considered to be governed by inflexible laws, or both be seen as susceptible of institutional modification' (Schwartz, 1972, 59). The 'inflexibility' of the laws of production applies only when key determinants of productivity are held constant; the 'modifiability' of the laws of distribution does not apply to the consequences flowing from choice of particular distributive forms. But the prime purpose of Mill's distinction did not need more than this diluted statement. A strong, mutually exclusive, formulation was not called for to convey his message regarding, on the one hand, the constraints of a technical order that must be faced under all forms of social organization – the implications flowing from land scarcity, the limits imposed on industry by capital, the differential impact of productive and unproductive consumption – and, on the other, the dependence of such organization itself upon human will. Taken in this light we are close to the more mundane distinction between problems bearing only on efficiency and problems involving value judgement. Thus, for example, Book I, ix is devoted to large- and small-scale farming 'as a question of production, and of the efficiency of labour'; examination of the issue 'as affecting the distribution of the produce, and the physical and social well-being of the cultivators themselves' (152) was postponed. Similarly, Mill in Book I applauded large-scale industrial marketing '[w]ith a view merely to production, and to the greater efficiency of labour', conceding 'drawbacks, rather social than economical . . .' (141).[47]

VII ON CONTEMPORARY AND PROSPECTIVE ECONOMIC PROGRESS

Mill takes for granted as a phenomenon characteristic of 'civilized' countries, indisputably at work and not itself requiring demonstration, the 'progress in wealth' in the sense of growing 'material prosperity' – an allusion to aggregate wealth (CW, III, 705–6). Drawing entirely upon observation of the record he spells out early in Book IV three 'tendencies' which 'characterize', 'attend', are 'incident to' or 'accompany' 'this progressive economical movement of civilized nations', namely advances in technology, in security and in 'cooperation' treating them as causal phenomena

---

47 Léon Walras (1954) took the position that 'The will of man is free to influence the production, as well as the distribution, of social wealth. The only difference is that in distribution, man's will is guided by consideration of justice, whereas in production his will is guided by considerations of material well-being' (75). This is close to Mill's 'weak' formulation.

responsible (directly or indirectly) for expanding aggregate output (706, 707, 708, 711). Mill provides in effect, both a summary and an elaboration of the inductive or factual materials relating to the primary determinants of wealth given in Book I, with particular reference to contemporary British conditions. We shall consider Mill's vision of his own world and its prospects, and see what, if anything, can be gleaned in this context regarding the fundamental tendencies, particularly the advance of knowledge.

The general long-term record presented a picture of stagnation or at best very slow growth: 'In some, perhaps in most states of society (looking at the whole surface of the earth,) both agricultural skill and population are either stationary, or increase very slowly . . .' (713). Similarly: 'Agricultural skill and knowledge are of slow growth, and still slower diffusion. Inventions and discoveries, too, occur only occasionally' (729). By contrast, there was the contemporary 'progressive' state characterized by the apparently unlimited extensions of knowledge and application of knowledge to productive processes – a phenomenon which (we shall see presently) applied throughout the economy including agriculture:

Of the features which characterize this progressive economical movement of civilized nations, that which first excites attention, through its intimate connexion with the phenomenon of Production, is the perpetual, and so far as human foresight can extend, the unlimited, growth of man's power over nature. Our knowledge of the properties and laws of physical objects shows no signs of approaching its ultimate boundaries: it is advancing more rapidly, and in a greater number of directions at once, than in any previous age or generation, and affording such frequent glimpses of unexplored fields beyond, as to justify the belief that our acquaintance with nature is still almost in its infancy. This increasing physical knowledge is now, too, more rapidly than at any former period, converted by practical ingenuity, into physical power (706).

There is vague reference here to the 'ultimate boundaries' to knowledge, but what stands out is the empirical statement regarding an increasing contemporary rate of knowledge creation. Indeed, Mill in Book I was prepared to commit himself to a trend 'susceptible of being prolonged and extended beyond any limit which it would be safe to specify' (II, 182), with apparently boundless potential for applied knowledge: 'We cannot . . . forsee to what extent the modes of production may be altered, or the productiveness of labour increased, by future extensions of our knowledge of the laws of nature, suggesting new processes of industry of which at present we have no conception' (199). Similarly, there were 'unlimited possibilities of improvement in the chemical processes . . .' (185).

These references to science and invention may be read in conjunction with an observation regarding the absence of constraints imposed by the supplies of skilled labour called for by innovatory applications of new knowledge: 'the manual part of [the] great scientific operations is never wanting to the intellectual; there is no difficulty in finding or forming, in a sufficient number of the working hands of the community, the skill requisite for executing the most delicate processes of the application of sciences to practical uses'. The implications were clear: 'From this union of conditions, it is impossible not to look forward to a vast multiplication and long succession of contrivances for economizing labour and increasing its produce; and to an even wider diffusion of the use and benefit of these contrivances' (III, 706).

It may appear that these observations relate solely to industrial activity. But we must not forget Mill's emphasis elsewhere on sectoral inter-dependence. Thus agricultural technology itself is categorized as either land- or labour-saving – both types counteracting 'the tendency in the cost of production of agricultural produce, to rise in the progress of population and demand' (II, 180) – but productivity advance in agriculture hinged also upon technology developed externally, innovations in transportation, metal working, engineering and food processing acting in the manner of labour-saving agricultural inventions to 'counteract or retard the diminution of the proportional return to labour from the soil . . .' (181–2; cf. III, 724, 744, 751 on technical interdependence). Also relevant (and not only to the agricultural sector although that was Mill's particular concern in the context) is the effect on productivity of the whole range of improvements in law and government affecting industrial organization and security in their manifold manifestations, as well as advance in the general level of intelligence of the work force and in the mind and character of the wealthier classes (183f).

Mill's recognition of the impact of technology on the agricultural sector considerably deepened with the passage of time; there are several indications of a change in perspective. Thus, for example, in the first edition (and with inconsequential variation until 1857) Mill had maintained that agriculture 'so far as present foresight can extend, does not seem to be susceptible of improved processes to so great a degree as some branches of manufacture; but inventions may be in reserve for the future, which may invert this relation'. But in 1862 this passage was deleted, the remaining text stating without qualification that (assuming population unchanged so that the problem of diminishing returns is set aside) there would be nothing to distinguish the two sectors: 'Mankind would . . . have the full benefit of all improvements in agriculture, or in the arts subsidiary to it, and there would be no difference, in this respect, between the products of agriculture and those of manufactures' (III, 712).[48] In the manuscript and the first

48 Mining would be distinguished by the special problem of exhaustibility.

two editions of the *Principles*, Mill wrote of machinery having 'done little' in farming, the great agricultural inventions involving treatment of the soil and plants – and rotation and pulverizing of soil, cultivation of subsoil, manures, conversion of bogs, pruning and so forth; but in the edition of 1852 he allowed that '[i]n agriculture and horticulture, machinery is only now beginning to show that it can do anything of importance, beyond the invention and progressive improvement of the plough and a few other simple instruments' (II, 107). In his discussion of the merits of scale in farming Mill observed that since the principal agricultural implements (such as the threshing machine) were inexpensive, it would be practicable for small farmers to own such equipment if not individually then co-operatively and thus take advantage of one of the advantages usually relegated to large size. In 1852 this was qualified: 'The observations in the text may hereafter require some degree of modification from invention such as the steam plough and the reaping machine' (143n).

Considering Europe as a whole and particularly France, where land was subdivided into extremely small plots such that prospects for improvement were impaired, the contemporary evidence pointed to remarkable agricultural advance – namely an unprecedented urban growth rate compared with that of the general population with no indication of falling real wages of town workers, and in addition, higher food consumption by the farm population itself (151). In 1862 Mill added that he had hitherto moderated his account of French agricultural advance not to be accused of exaggeration; but 'I little knew', he continued, 'how much stronger my language might have been without exceeding the truth, and how much the actual progress of French agriculture surpassed anything which I had at that time sufficient grounds to affirm', drawing upon newly published evidence demonstrating 'that since the Revolution of 1789, the total produce of French agriculture [had] doubled; profits and wages having both increased in about the same, and rent in a still greater ratio' (152).[49]

* * *

Despite his recognition of the increasing contemporary significance of extensions of knowledge both pure and applied (and despite the formal allowance of a place for the topic in political economy) Mill himself made little effort to explore the problem of determinants of resources devoted

49 In some formulations given in the Book on progress Mill seems to emphasize French agricultural innovation only in a relative sense: 'In some other countries, and particularly in France, the improvement of agriculture gains ground still more decidedly upon population, because though agriculture, except in a few provinces, advances slowly, population advances still more slowly, and even with increasing slowness; its growth being kept down, not by poverty, which is diminishing, but by prudence' (714).

to knowledge creation, and failed even to direct his readers to relevant literature. There is indeed an allusion to '[t]he changes which the progress of industry causes or presupposes in the circumstances of production' (III, 710) with reference to technological advances, implying a mutual cause-effect relation, restating an observation made to that effect in the Preliminary Remarks:

These remarkable differences in the state of different portions of the human race, with regard to the production and distribution of wealth, must, like all other phenomena, depend on causes. And it is not a sufficient explanation to ascribe them exclusively to the degrees of knowledge, possessed at different times and places, of the laws of nature and the physical arts of life. Many other causes co-operate; and that very progess and unequal distribution of physical knowledge are partly the effects, as well as partly the causes, of the state of the production and distribution of wealth (II, 20).

But neither here nor in Book IV are the details of this mutuality of relationship entered into. Although one can never be sure about such matters, Mill's reticence may reflect a belief that the topic approached too closely the 'progress of society' in the broadest sense of this term, entailing pre-eminently the issue of ethology, or human character formation which Mill felt would be a premature exercise. But whatever the reasons, his reflections regarding knowledge in the context of 'economical' or 'industrial' progress remain almost entirely at the level of observation.[50]

The second change, 'which has always hitherto characterized, and will assuredly continue to characterize the progress of civilized society', namely 'a continual increase of the security of person and property' is also not explained in any causal sense. All we have is a panagyric to the phenomenon with optimistic observations regarding the future:

The people of every country in Europe, the most backward as well as the most advanced, are, in each generation, better protected against the violence and rapacity of one another, both by a more efficient judicature and police for the suppression of private crime, and by the decay and destruction of those mischievous privileges which enabled certain classes of the community to prey with impunity upon the rest. They are also, in every generation, better protected, either by institutions or by manners and opinion, against arbitrary exercise of the power of government. Even in semi-barbarous Russia, acts of spoliation directed against individuals, who have not made themselves politically obnoxious, are not supposed to be now so

---

50 Innovation is more fully treated than the generation of scientific knowledge. See in particular the role of competition (pp. 758–9, 762, 776, 780–1); of the falling profit rate (p. 467); and – in our present chapter – of business organization (p. 229).

frequent as much to affect any person's feelings of security. Taxation, in all European countries, grows less arbitrary and oppressive, both in itself and in the manner of levying it. Wars, and the destruction they cause, are now usually confined, in almost every country, to those distant and outlying possessions at which it comes into contact with savages. Even the vicissitudes of fortune which arise from inevitable natural calamities, are more and more softened to those on whom they fall, by the continual extension of the salutary practice of insurance (III, 706–7).

Once again 'progress' – in this case social progress in the large – is not only encouraged by increased security but is itself responsible for improvements in that regard: '. . . one of the acknowledged effects of that progress is an increase of general security. Destruction by wars, and spoliation by private or public violence, are less and less to be apprehended: and the improvements which may be looked for in education and in the administration of justice, or, in their default, increased regard for opinion, afford a growing protection against fraud and reckless mismanagement' (737).

The third tendency refers to industrial co-operation and in this case the mutual linkage with 'progress' is better justified. A most conspicuous form of co-operation is joint-stock organization, and to this we turn referring first to Mill's empirical analysis in Book I of the efficacy of hired managers.

We find there a negative evaluation of hired workers in general: 'experience shows, and proverbs, the expression of popular experience, attest, how inferior is the quality of hired servants, compared with the ministration of those personally interested in the work, and how indispensable, when hired service must be employed, is "the master's eye" to watch over it' (II, 137). The 'fidelity and zeal of hired managers was assured (by fear of dismissal and by conscience) in the case of routine tasks; but 'to carry on a great business successfully, requires a hundred things which, as they cannot be defined beforehand, it is impossible to convert into distinct and positive obligations'. There were two main issues: 'First and principally, it requires that the directing mind should be incessantly occupied with the subject; should be continually laying schemes by which greater profit may be obtained, or expense saved' – an 'intensity of interest in the subject [which] it is seldom to be expected that any one should feel, who is conducting a business as a hired servant and for the profit of another', for evidence of which Mill drew on 'experiments in human affairs' in government and in the market for hired labour. A second consideration is the disregard of 'small gains and small savings' by hired managers in contrast to the large capitalist who 'arrange[s] his business on a *system*, which if enforced by a sufficiently vigilant superintendence, precludes the possibility of the habitual waste, otherwise incident to a great business' (137–8).

Despite all this Mill maintained that Adam Smith's negative evaluation on similar grounds of the joint-stock ogranization constituted 'one of those overstatements of a true principle often met with' in the *Wealth of Nations*. For Smith had 'fixed his observation too exclusively on the superior energy and more unremitting attention brought to a business in which the whole stake and the whole gain belong to the persons conducting it; and he overlooked various countervailing considerations which go a great way towards neutralizing even that great point in superiority'.[51] The disadvantages alluded to earlier might be reduced by resort to some form of profit-sharing relating the 'interest of the employés with the pecuniary success of the concern',[52] and by attraction of 'a class of candidates superior to the common average intelligence':

Of these one of the most important is that which relates to the intellectual and active qualifications of the directing head. The stimulus of individual interest is some security for exertion, but exertion is of little avail if the intelligence exerted is of an inferior order, which it must necessarily be in the majority of concerns carried on by the persons chiefly interested in them. Where the concern is large, and can afford a remuneration sufficient to attract a class of candidates superior to the common average, it is possible to select for the general management, and for all the skilled employments of a subordinate kind, persons of a degree of acquirement and cultivated intelligence which more than compensates for their inferior interest in the result. Their greater perspicacity enables them, with even a part of their minds, to see probabilities of advantage which never occur to the ordinary run of men by the continued exertion of the whole of theirs, and their superior knowledge, and habitual rectitude of perception and of judgment, guard them against blunders, the fear of which would prevent the others from hazarding their interests in any attempt out of the ordinary routine (139).

The net result of such devices and the careful selection of candidates 'often raises the quality of the service much above that which the generality of masters are capable of rendering to themselves' (140).[53]

51 See also Mill's reference to 'the "human *too much*" in [Smith's] premises; the portion of them that is over and above what was either required or is actually used for the establishment of his conclusions' (162n).
52 Mill adds a reference to the unskilled: 'there is such a thing as task-work, or working by the piece: and the superior efficiency of this is so well known, that judicious employers always resort to it when the work admits of being put out in definite portions . . .'
53 Whitaker (1975), 1046, observes that 'often, especially in the discussion of alternative methods of organizing production and cultivation, Mill may be construed as merely searching for empirical regularities and judgments, with only a remote hope of rationalizing these deductively – in other words of verifying any findings by the indirect deductive route'. There seems to be some truth to this; yet in the present instance regarding joint-stock organization Mill makes an effort to rationalize the observations. See also below (p. 238f) regarding land tenure.

It will be noted that Mill emphasizes the negative effects of self-interest on the part of 'the ordinary run of men'; self-interest, it appears, is a potent driving force only in the more mundane task of assuring small gains and savings, but mitigates against undertaking ventures 'out of the ordinary routine'. This proposition is of significance in the context of our discussion of innovation (above, p. 206) for it implies a rationalization of the relationship between scale (subject to altered organization) and innovation.

Mill envisaged good prospects for a great extension of the joint-stock and other forms of complex organization determined by the progress of wealth and reinforcing such progress. There is an objective argument pertinent to our theme, that the introduction of large establishments is facilitated in a growing economy: '[T]his as well as other changes in the system of production is greatly favoured by a progressive condition of capital. It is chiefly when the capital of a country is receiving a great annual increase, that there is a large amount of capital seeking for investment: and a new enterprise is much sooner and more easily entered upon by new capital, than by withdrawing capital from existing employments' (140). More generally: 'In the countries in which there is the largest markets, the widest diffusion of commercial confidence and enterprise, the greatest annual increase of capital, and the greatest number of large capitals owned by individuals, there is a tendency to substitute more and more, in one branch after another, large establishments for small ones' (141).[54] Allusion is made to the 'preservation of the fittest' in competitive conditions for external evidence of a manifestation of the tendency in the British case including the retail sector.

Apart from this objective phenomenon there is some brief discussion of the influences exerted by the 'progress of modern society' upon the capacity for industrial co-operation of a rather subjective order, involving altered behavioural patterns. Thus Mill suggests, without however justifying the suggestion, that while 'economical progress' had a debilitating effect upon intelligence and efficiency of the individual – his ability to adapt means to ends – yet collective intelligence and efficiency increases: 'What is lost in the separate efficiency of each, is far more than made up by the greater capacity of united action', an increased capacity for co-operation reflected in better industrial discipline, adherence to plan, subordination of individual caprice and so forth (III, 708).[55] Now such capacity reinforces itself, for 'this, like other faculties, tends to improve by practice, and becomes capable of assuming a constantly wider sphere of action'. This same set of

54 The same positive effect of unequal income distribution can, however, be obtained in an advanced society by the opportunities available for the accumulation of many small capitals (140).

55 See also Mill's discussion (719–32) of the effects of economical progress on capital population and capital production.

issues had been raised by Adam Smith in the context of invention, and it is not certain whether Mill wished to limit the discussion specifically to organization or whether his observations were illustrative of a more general theme.[56]

* * *

The general conclusion thus far is that '[f]rom the causes enumerated . . . unless counteracted by others, the progress of things enables a country to obtain at less and less of real cost, not only its own productions, but those of foreign countries' (711). But Mill asks, '[a]re no causes of an opposite character brought into operation by the same progress, sufficient in some cases not only to neutralize, but to overcome the former, and convert the descending movement of cost of production into an ascending movement?' The answer is that in the case of 'the most important classes of commodities, food and materials, there is a tendency diametrically opposite to those of which we have been speaking':

When, however, population increases, as it has never failed to do when the increase of industry and the means of subsistence made room for it, the demand for most of the productions of the earth, and particularly for food, increases in a corresponding proportion. And then comes into effect that fundamental law of production from the soil, on which we have so frequently had occasion to expatiate; the law that increased labour, in any given state of agricultural skill, is attended with less than proportional increase of produce. The cost of production of the fruits of the earth increases, *caeteris paribus*, with every increase of the demand (712).

On a first glance this perspective will appear entirely artificial: Mill appears to assume technological and other forms of improvement stimulating the expansion of wealth and at one remove that of population, which latter force will, given technology (and apparently a cessation of 'improvement' in its multifarious forms), imply upward pressure on agricultural costs. It is, however, clear that this is not Mill's intention at all. He is attempting to get across the point that even if 'improvement' is ongoing there is also an ongoing trend acting against it – namely the ever-present consequences of land scarcity. We must at all times bear in mind the methodological principle of 'composition of causes'.[57] (See also de Marchi, 1974, 142–3).

---

56 It should be noted too that 'the progress of wealth and industry' is said to encourage not only the three major tendencies discussed here, but also the propensities towards freer trade and emigration (711–12).

57 At one point, however, this conception is slightly modified, for Mill closed his account of five hundred years of European cost movements (730) by the suggestion that '[a]gricultural improvements may . . . be considered to be not so much a counterforce

The 'progress of improvement' in the broad sense of the term (which involves more than technological change narrowly defined) is thus represented as acting in 'habitual antagonism' to diminishing agricultural returns and is effectively placed on a par with it: 'I do not assert that the cost of production, and consequently the price, of agricultural produce, always and necessarily rises as population increases. It tends to do so; but the tendency may be, and sometimes is, even during long periods, held in check. The effect does not depend on a single principle, but on two antagonizing principles' (II, 179). Similarly 'all natural agents which are limited in quantity, are not only limited in their ultimate productive power, but, long before that power is stretched to the utmost, they yield any additional demands on progressively harder terms. This law may however be suspended, or temporarily controlled, by whatever adds to the general power of mankind over nature; and especially by any extension of their knowledge, and their consequent command of the properties and powers of natural agents' (185). So central indeed is the role accorded 'improvement' that the 'law' of diminishing returns is itself modified at one point to read 'the law . . . of diminishing return to industry, wherever population makes a more rapid progress than improvement' (193), the historical record indicating that 'at some periods the progress of population has been the more rapid of the two, at others that of improvement' (189).

What now of the actual historical record of cost movements. We have already referred (above) to the downward movement in manufacturing costs during the course of 'progress' – 'a fall accelerated by the mechanical inventions of the last seventy or eighty years'. There remains the agricultural sector.

The secular data indicated that since 'both agricultural skill and population are either stationary, or increase very slowly . . . the cost of production of food, therefore, is nearly stationary' (III, 713). This statement is amplified within a broader context:

Agricultural skill and knowledge are of slow growth, and still slower diffusion. Inventions and discoveries, too, occur only occasionally, while the increase of population and capital are continuous agencies. It therefore seldom happens that improvement, even during a short time, has so much the start of population and capital as actually to lower rent, or raise the rate of profits. There are many countries in which the growth of population and capital is not rapid, but in these agricultural

conflicting with increase of population, as a partial relaxation of the bonds which confine that increase', implying that the notion of conflicting forces playing upon productivity does not sufficiently bring to light that the rate of population growth had itself been governed by that of agricultural innovation and other manifestations of progress. No substantive issues seem to be involved, however, in the modified formulation.

improvements is less active still. Population almost everywhere treads close on the heels of agricultural improvement, and effaces its effects as fast as they are produced.

The reason why agricultural improvement seldom lowers rent is that it seldom cheapens food, but only prevents it from growing dearer; and seldom, if ever, throws land out of cultivation, but only enables worse and worse land to be taken in for the supply of an increasing demand (729–30).

This pattern of roughly unchanged agricultural productivity is said to describe half a millenium of European experience. For five hundred years the production function had slowly shifted outwards providing scope for a slowly expanding population, constant productivity reflecting the net outcome:

[I]n Europe five hundred years ago, though so thinly peopled in comparison to the present population, it is probable that the worst land under the plough was, from the rude state of agriculture, quite as unproductive as the worst land now cultivated; and that cultivation had approached as near to the ultimate limit of profitable tillage, in those times as in the present. What the agricultural improvements since made have really done is, by increasing the capacity of production of land in general, to enable tillage to extend downwards to a much worse natural quality of land than the worst which at that time would have admitted of cultivation by a capitalist for profit; thus rendering a much greater increase of capital and population possible, and removing always a little and a little further off, the barrier which restrains them; population meanwhile always pressing so hard against the barrier, that there is never any visible margin left for it to seize, every inch of ground made vacant for it by improvement being at once filled up by its advancing columns (730).

According to this account, European economies, during the extensive period in question, would have been (in the absence of new technology) in a stationary state. By contrast there was the United States where 'the worst land in cultivation is of a high quality . . . and even if no further improvements were made in agriculture or locomotion, cultivation would have many steps yet to descend, before the increase of population and capital would be brought to a stand . . .'

In the British case the price of corn had apparently increased during the eighteenth century at least until the French Revolution; and since the value of money had not fallen, and Britain by the close of a century had become a net importer, it seemed that 'improvements' had lagged behind even the slow growth of population. But these data proved little because of the misleading impression of a trend created by a series of extraordinarily abundant harvests early in the century and not recurring subsequently

(II, 189).[58] In any event, since the late 1820s 'so rapid has been the extension of improved processes of agriculture, that even the land yields a greater produce in proportion to the labour employed; the average price of corn had become decidely lower, even before the repeal of the corn laws had so materially lightened, for the time being, the pressure of population upon production' (190). We have then an allusion to some two decades of positive agricultural productivity increase, despite rapidly rising population and demand for food, and this during a period of agricultural protection.[59] We also have a later observation that '[i]n England and Scotland, agricultural skill has of late increased considerably faster than population, insomuch that food and other agricultural produce, notwithstanding the increase of people, can be grown at less cost than they were thirty years ago' (III, 713).[60]

Mill also alluded to a 'tendency' towards an increase in the *relative* prices of agricultural to manufactured products:

The tendency, then, being to a perpetual increase of the productive power of labour in manufactures, while in agriculture and mining there is a conflict between two tendencies, the one towards an increase of productive power, the other towards a diminution of it, the cost of production being lessened by every improvement

58 Cf. 179 where Mill observed that the cost of production (and accordingly the general purchasing power) of the metals had fallen in only two historical periods – following the opening of the Mexican and Peruvian mines and in the contemporary period. At all other times a secular increase in the money price of corn would be an accurate indication of rising (real) cost of agricultural production were such a trend discernable, which (against Carey) he denied.

See also III, 714 for a reference to Tooke's demonstration that even half a century of price movements are subject to seasonal bias.

59 The precise period of remarkable productivity improvement was occasionally amended (189–90): in the manuscript and in the first two editions (1848, 1849) Mill referred to the past 15 or 20 years; in 1852, to 20 or 25 years; in 1857, to 20 or 30 years, suggesting the onset of rapid agricultural change from the late 1820s. The years 1827 or 1828 are evidently significant ones: 'the average price of corn had become decidely lower, and the country more nearly feeds its own population without foreign aid than it did in 1827 [ms; 1828, in eds. of 1848, 1849]'. That Mill did not amend the years in later editions is accounted for by a wish to emphasize the effect on corn prices of 'improvement' apart from that of a free corn trade, so that only the period until corn law repeal was strictly relevant.

60 All editions. We note that Mill had just referred to 'a strong impulse . . . towards agricultural improvement . . . in Great Britain during the last twenty or thirty years'. But here differences between editions cloud the period intended, the reference to 20 or 30 years in the final edition (1871) replacing not only an earlier reference to 20 or 25 years (in the 1865 edition) but one to 15 or 20 years (in the first edition of 1848). The amendations suggest that from 1865 Mill no longer dated the onset of striking 'agricultural improvement' in the late 1820s but only in the early 1840s. It is not certain, however, that Mill intended to convey this impression.

in the processes, and augmented by every addition to population; it follows that the exchange values of manufactured articles compared with the products of agriculture and of mines, have, as population and industry advance, a certain and decided tendency to fall (713).

No evidence is brought to demonstrate the actual pattern of relative prices (although an index in this case would not have been open to criticism on the grounds of distortion flowing from changes in general purchasing power). And Mill's formulation is misleading considering his own recognition of extensive technological interdependencies.

* * *

We have traced Mill's account of an historical trend pattern extending for centuries of roughly constant agricultural costs reflecting the combined effects of sluggish agricultural improvement and the consequential slow growth of population; and recognition in the British case of two or even three decades of falling agricultural prices in real terms – 'decided' reductions achieved, quite apart from corn law repeal, during a period of rapid population growth. Did the facts pertaining to the nineteenth century indicate an entirely new pattern, or was nothing more involved than a temporary respite to be followed in all likelihood by a reversal to the norm? In brief, does 'an incapacity to understand the dynamicism of the capitalist system' characterize Mill's thought as has been maintained (Schwartz, 1972, 231; cf. also Spengler, 1960, 135)?

The evidence is complex. But it appears that as far as concerns technological advance as such there was, in Mill's view, every prospect of continued rapid advance. We have his statement that 'inventions may be in reserve' to 'invert' the relative unsusceptibility to improvement of agricultural activity revealed in the record, followed by the apparent suggestion that such 'inversion' had indeed occurred. There is also his keen appreciation of the sources of productivity improvement external to agriculture itself which take on particular significance considering recent chemical and mechanical inventions and prospects for further change 'of which at present we have no conception'. All of this suggests that the contemporary period ushered in a new era on all fronts. This is indeed what Mill concludes: 'The progress which is to be expected in the physical sciences and arts, combined with the greater security of property, and the greater freedom in disposing of it, which are obvious features in the civilization of modern nations, and the more extensive and more skillful employment of the joint-stock principle, afford space and hope for an indefinite increase of capital and production, and for the increase of population which is its ordinary accompaniment' (708–9).

Now in Mill's opinion, at least in this context, there was little danger that the growth of population would surpass the increase of production, but he insisted that '[w]e must . . . in considering the effects of the progress of industry, admit as a supposition . . . an increase of population, as long continued, as indefinite, and possibly even as rapid, as the increase of production and accumulation':

That the growth of population will overpass the increase of production, there is not much reason to apprehend and that it should even keep pace with it, is inconsistent with the supposition of any real improvement in the poorest classes of the people. It is, however, quite possible that there might be a great progress in industrial improvement, and in the signs of what is commonly called national prosperity; a great increase of aggregate wealth, and even, in some respects, a better distribution of it; that not only the rich might grow richer, but many of the poor might grow rich, that the intermediate classes might become more numerous and powerful, and the means of enjoyable existence be more and more largely diffused, while yet the great class at the base of the whole might increase in numbers only, and not in comfort and in cultivation. We must, therefore, in considering the effects of the progress of industry, admit as a supposition, however greatly we deprecate as a fact, an increase of population as long continued, as indefinite, and possibly even as rapid, as the increase of production and accumulation (709).

That Mill should express his fears regarding population growth is not in itself inconsistent with recognition of continued rapid technological progress.

We also find statements, however, that are difficult to reconcile with what has been said so far. There is an unsupported assertion (713) that '[i]n a society which is advancing in wealth, population generally increases faster than agricultural skill, and food consequently tends to become more costly', apart from 'times when a strong impulse sets in towards agricultural improvement', which implies that the contemporary experience of rapid technological progress constituted a deviation from the norm of falling agricultural productivity. This position is stated even more strongly subsequently in Book V with references to a prospective downward trend in agricultural productivity even in an open economy, the onset of which was merely postponed by the contemporary advance in agricultural invention and innovation:

The imposition of corn laws raises rents, but retards that progress of accumulation which would in no long period have raised them fully as much. The repeal of corn laws tends to lower rents, but it unchains a force which, in a progressive state of capital and population, restores and even increases the former amount. There is every reason to expect that under the virtually free importation of agricultural produce, at last extorted from the ruling powers of this country, the price of food, if population goes on increasing, will gradually but steadily rise; though this effect may for a time be postponed by the strong current which in this country has set in (and the impulse is extending itself to other countries) towards the improvement of agricultural science, and its increased application to practice (849–50).[61]

We have too the statement in Book I regarding the contemporary technological advance which generated lower corn prices 'even before the repeal of the corn laws had so materially lightened, for the time being, the pressure of population upon production' (II, 190).

It is regrettable that Mill failed to delve deeper into the sources of new technology to allow us better to understand his representation in these extracts of sharply rising agricultural productivity as a temporary deviation from a falling trend.[62] This is especially troublesome considering a further cryptic assertion that 'the abolition of the Corn Laws has given an additional stimulus to the spirit of improvement' (III, 713–14), and having in mind his enthusiasm for the limitless prospects for new technology and other positive influences upon productivity characterizing a progressive state. It will be remembered too that the cost data referred to by Mill himself portrayed several centuries of constant agricultural productivity, whereas he now refers to a prospective reversion to a pattern of declining productivity.

It appears that Mill was on occasion carried away by his concern with population growth and stated his case more strongly than either the historical cost data or the analytical model and its methodological status or his own appreciation of the contemporary impact of new technology merited. It is, in short, perhaps fair to say that he sometimes engages in 'Malthus-mongering which can only be excused as rhetorical' (Whitaker, 1975, 1047).[63]

61 In early editions until 1857 Mill wrote only conditionally that the 'current' towards agricultural invention and innovation 'may extend' abroad.

62 In his letter to Carey 1845 (above, n25) Mill proceeds in terms that confirm the principle of composition of causes but seems to 'predict' that the force of technical change will be swamped by that of land scarcity: '[W]hen this point is attained' – a reference to optimum population size – 'the natural law by which (in any given state of knowledge) a double expenditure on land yields less than a double return, prevails, on the average, over the antagonistic influence of new inventions & discoveries' (CW, XIII, 660). But in this context the 'new inventions and discoveries' allowed for are assumed to occur within 'a given state of knowledge'. Apparently, the 'prediction' does not (necessarily) apply once the possibility of a transformation in the state of pure scientific knowledge is introduced.

63 One small indication of this is seen in a slight amendation made to the passage from CW, III, 849–50, cited above. In the first two editions Mill had written of the prospect of a steadily rising price of food (following the impact effect of repeal of the corn laws) 'if population goes on increasing at its present rate'. Thereafter this qualification was removed rendering the danger more potent, notwithstanding the fact that by 1852 Mill was *more* optimistic not less so.

VIII   LAND TENURE AND MOTIVATION

In the chapter summarizing the theory of value Mill examines the qualifications required when cultivation is undertaken, not for profit by the capitalist but for subsistence by the labourer – peasant ownership, metayage and allotment holding – who seeks 'not an investment for his little capital, but an advantageous employment for his time and labour' (*CW*, III, 499). Here many of the tools or theorems of competitive analysis prove indispensable. Mill closes by the hope that 'further adaptations of the theory of value to the varieties of existing or possible industrial systems' would be undertaken by 'the intelligent reader' (501). He himself had worked out illustrations in Book II; and his analyses of various contemporary forms of land tenure indicate his preoccupation with the empirical validity of the maximization axiom in a wide variety of alternative institutional contexts.[64] These same analyses further demonstrate the untenability of the distinction between laws of production and distribution in any strict sense – for the distributional arrangements are shown to play strategically upon efficiency.

Mill proceeds in his treatment of peasant proprietorship by first establishing the 'facts' of the case before 'entering into the theory of it' (II, 254), drawing on Sismondi for Switzerland, Laing for Norway, Howitt and Kay for Germany and McCulloch for Belgium. This empirical evidence indicated a striking and rapid improvement in per capita income reflecting highly motivated labour and productive methods of cultivation, and effective population control, which conflicted with the standard 'dogma of our land

---

64   Jacob Viner (1958) emphasizes the influence on Mill's *Principles* of Richard Jones regarding the importance of custom especially in connection with land tenure and landlord-tenant relations (328). Mill himself, while noting a debt, disparaged him: 'Jones I believe to be quite incapable of having a fundamentally new, and at the same time true, idea in Political Economy. His merit was that he called attention to the great variety in the tenures of land as affecting the laws of distribution' (May 12, 1860, *CW*, XV, 699). For a critical evaluation of Jones in this context see Miller (1977). On Mill's 'debt' to Jones (and to W. T. Thornton) see Dewey (1974), Martin (1976).

   Whewell, writing to Jones, criticized Mill in the *Principles* – much as he criticized McCulloch – for having 'spoilt your great normal division of tenants by confounding Cottiers with Ryots, and not having the grace to acknowledge where he got what he spoilt' (23 April 1849; Todhunter, II, 353).

   Marshall regarded appeal to 'custom' as an admission of our ignorance (1925, 169-70; cf. also Viner, 1958, 329). Yet there is surprisingly little on 'custom' in the analysis of land tenure and its impact on productivity and population; as we shall see, the principles of 'competition' based on maximization axioms predominate.

   For recent theoretical analyses of Mill on proprietorship, productivity and population see Bradley (1983), Bush (1973).

doctors at home, that large farms worked by hired labour and great capital can alone bring about the greatest productiveness of the soil and furnish the greatest supply of the necessaries and conveniences of life to the inhabitants of a country' (262). The task he set himself was to explain the facts. This he does in terms of the results which can be shown to flow 'logically' from a model based upon self-interested (maximization) behaviour within the pertinent institutional framework. There is no direct appeal to evidence for such behaviour, but there is *indirect* appeal in so far as the analysis accounts for the facts. This procedure is in line with the formal pronouncements regarding the inverse deductive method (above, p. 140). Mill attempts, therefore, to satisfy the requirement for an inductive foundation for the model, which might then be used in policy applications, subject always to caution considering the possible intervention of 'disturbing causes'.[65]

The specific argument shows how property rights governing the pattern of distribution affect productivity and population growth. The rationale for the impressive record regarding adoption of advanced agricultural processes is found in Laing's overview of British farming compared with European peasant proprietorships:

large capital applied to farming is . . . only applied to the very best of the soils of a country. It cannot touch the small unproductive spots which require more time and labour to fertilize them than is consistent with a quick return of capital. But although hired time and labour cannot be applied beneficially to such cultivation, the owner's own time and energy may. He is working for no higher terms at first from his land than a bare living. But in the course of generations fertility and value are produced; a better living, and even very improved processes of husbandry are attained. Furrow draining, stall feeding all summer, liquid manures, are universal in the husbandry of the small farms of Flanders, Lombardy, Switzerland. Our most improving districts under large farms are but beginning to adopt them (261–2).[66]

---

65 Mill did not desist, however, from recommending technical exercises for their own sake as it were. Thus he recommended analysis based upon a profit-maximizing capitalist slave holder (III, 500–1) although he believed (following Richard Jones) that 'great landowners are everywhere an idle class' (II, 249).

66 Cf. 279: 'the far larger gross produce which, with anything like parity of agricultural knowledge, is obtained from the same quality of soil on small farms, at least when they are the property of the cultivator', Mill explains by the preponderance of higher motivation over an 'inferiority of resources, imperfection of implements, and ignorance of scientific theories'. For while peasant farming entails 'an amount of labour which, if paid for by an employer, would make the cost to him more than equivalent to the benefit . . . to the peasant it is no cost, it is the devotion of time which he can spare, to a favourite pursuit, if we should not rather say a ruling passion'.

Has this 'indefatigable industry . . . of which so large a portion is expended not in the mere cultivation, but in the improvement, for a distant return of the soil itself', Mill asks rhetorically, 'no connexion with *not* paying rent? Could it exist, without presupposing either a virtually permanent tenure, *or the certain prospect, by labour and economy on hired land, of becoming one day a landed proprietor?'* (270).[67]

Mill was highly conscious that facts do not speak for themselves. He warned that the 'economy' exercised by Flemish labourers had been mistaken by some observers, with narrow experience, as an index of their poverty: 'So little are English labourers accustomed to consider it possible that a labourer should not spend all he earns, that they habitually mistake the signs of economy for those of poverty' (270). To establish his case Mill therefore drew upon supplementary evidence showing that Flemish workers 'eagerly seize every opportunity of purchasing a small farm, and the price is so raised by competition, that land pays little more than two per cent interest for the purchase money. Large properties gradually disappear, and are divided into small portions, which sell at a high rate.' Similarly: 'the tendency of peasant proprietors, and of those who hope to become proprietors, is to the contrary extreme' – compared to day-labourers – 'to take even too much thought for the morrow' (282).[68]

There will also be found in Mill's account an argument that smacks of *post hoc ergo propter hoc* reasoning which he himself condemned in his formal methodological work (above, p. 97). For he asserts at one point that '[t]hree times during the course of ages the peasantry have been purchasers of land; and these times immediately preceded the three principal eras of French agricultural prosperity . . . [C]ompare these historical periods, characterized by the dismemberment of large and the construction of small properties, with the wide-spread national suffering which accompanied, and the permanent deterioration of the condition of the labouring classes which followed, the "clearing" away of small yeomen to make room for large grazing farms, which was the grand economical event of English history during the sixteenth century' (296n). Taken by itself, this form of reasoning would not do at all, on Mill's own terms; but as supplement to the main argument which traces back an empirical reality to behavioural axioms (and it appears first in a note in 1862) little damage could be done.

---

67 Italicized section added in 1871. But in earlier editions Mill had also insisted that 'the idea of property does not . . . necessarily imply that there should be no rent, any more than that there should be no taxes. It merely implies that the rent should be a fixed charge, not liable to be raised against the possessor by his own improvement, or by the will of a landlord . . .' (278).

68 Mill alludes also to the training of intelligence by peasant proprietorship in consequence of the wide variety of tasks to be undertaken (280–1). This is an allusion to the 'alienation problem' that exercised Adam Smith and Karl Marx.

There is next the matter of population growth. On a casual glance there appears to be a strong '*à priori*' dimension to Mill's argumentation – indeed Mill himself formally phrases the issue in precisely those terms: '[S]upposing a peasantry to possess land not insufficient but sufficient for their comfortable support, are they more, or less likely, to fall from this state of comfort through improvident multiplication, than if they were living in an equally comfortable manner as hired labourers? All *à priori* considerations are in favour of their being less likely' (283–4).[69] The rationale offered is that 'every peasant can satisfy himself from evidence which he can fully appreciate, whether his piece of land can be made to support several families in the same comfort as it supports one', evidence not available to the day labourer.

However, the à priori dimension must not be exaggerated. Mill was directly addressing the unsubstantiated opinion of Richard Jones that peasant proprietorship encouraged population growth.[70] With this in mind he protested: 'I am totally unable to conjecture from what theory of human nature, and of the motives which influence human conduct, he would have derived [his opinions]' (283), showing by way of contrast how the opposite conclusions followed from 'the motives which influence human conduct' as Mill understood them. But those motives were not drawn from a hat. For he was presuming the successful establishment of the positive impact of peasant proprietorship upon productivity and accumulation, and therefore, a presumption in favour of the behavioural axiom selected: 'Is it likely that a state of economical relations so conducive to frugality and prudence in every other respect, should be prejudicial to it in the cardinal point of increase of population?' Secondly, there was direct empirical evidence (due to Sismondi) of the impact of peasant proprietorship on population growth from Switzerland: 'It is not found that in the Swiss Cantons, the patrimonies of the peasants are ever so divided as to reduce them below an honourable competence'; and similarly from Norway due to Laing (285). Some explanation was required and 'it is reasonable to give part of the credit of this prudential restraint to the peculiar adaptation of the peasant-proprietary system for fostering it'. In the French case the evidence was more indirect; a positive impact of peasant proprietorship on population must have pushed wages down generally as people were forced off the land, whereas in fact wages had risen since the revolution (286f).

69   Cf. 277: 'The benefits of peasant properties are conditional on their not being too much subdivided; that is, on their not being required to maintain too many persons, in proportion to the produce that can be raised from them by those persons. The question resolves itself, like most questions respecting the condition of the labouring classes, into one of population. Are small properties a stimulus to undue multiplication, or a check to it?'

70   Mill refers to Jones' *Distribution of Wealth* (1831) and the promise there to explain his position in a second volume that never appeared.

'I am not aware', Mill concluded, 'of a single authentic instance which supports the assertion that rapid multiplication is promoted by peasant properties' (291). The facts of the case thus vindicated the axiomatic basis of the model. Even so he warns against positive prediction in the application of the model to future cases considering the possible intervention of disturbing causes.[71]

The theme (citing de Lavergne) that '[c]ultivation spontaneously finds out that organization that suits it best' (293) justifies Mill's argument that should there exist (in special circumstances) excessive numbers under a peasant proprietorship system it cannot be presumed that the technical farming unit will be too small: 'a subdivision of occupancy is not an inevitable consequence of even undue multiplication among peasant proprietors' (292). Thus in Flanders there were cases where the sons, rather than divide up the patrimony, sell the unit; and in France there was indirect evidence of the general theme from the great number of land sales, combined with observations of big landowners who distribute units among several tenants and individual farmers who rent land from several landowners to permit *grande culture*.

* * *

Output sharing between landlord and labourer (or peasant) comprises two subcategories, the metayer system where the division is determined by 'custom' and the cottier system where the proportionate division is a matter of 'competition' (297). In the former case there are 'no laws of distribution to investigate' (299), the issues being the condition and efficiency of the peasantry as in the case of peasant proprietorship. Indeed, what had been said of that case applied though to a somewhat lesser degree, for metayage. The validity of evidence by Arthur Young, McCulloch and Jones is denied, as based upon special circumstances peculiar to pre-Revolutionary France; the evidence brought by Châteauvieux, and especially Sismondi, for Tuscany, showed a very different picture:

I do not offer these quotations as evidence of the intrinsic excellence of the metayer system; but they surely suffice to prove that neither 'land miserably cultivated' nor a people in 'the most abject poverty' have any necessary connexion with it, and that the unmeasured vituperation lavished upon the system by English writers, is grounded on an extremely narrow view of the subject. I look upon the rural economy of Italy as simply so much additional evidence in favour of small occupations with

---

71 In the Belgian case, for example, there was the impact of the Church to consider (292). We must beware of presuming that Mill championed peasant proprietorship as the best of all possible worlds as some commentators appear to do (Dewey, 1974, 34). On preferable alternatives, see Chapter 11.

permanent tenure. It is an example of what can be accomplished by those two elements, even under the disadvantage of the peculiar nature of the metayer contract, in which the motives to exertion on the part of the tenant are only half as strong as if he farmed the land on the same footing of perpetuity as a money-rent, either fixed, or varying according to some rule which would leave to the tenant the whole benefit of his own exertion (310).

Sismondi is also drawn upon for evidence of population control (311).

The cottier system applies where, as above, 'the labourer makes his contract for land without the intervention of a capitalist farmer', but 'in which the conditions of the contract, especially the amount of rent, are determined not by custom but by competition' (313). Here wages are treated as a residual: 'The effect, therefore, of this tenure, is to bring the principle of population to act directly on the land, and not, as in England, on capital. Rent, in this state of things, depends on the proportion between population and land. As the land is a fixed quantity, while population has an unlimited power to increase; unless something checks that increase, the competition for land soon forces rent to the highest point consistent with keeping the population alive. The effects, therefore, of cottier tenure depend on the extent to which the capacity of population to increase is contolled, either by custom, by individual prudence, or by starvation and disease' (314). Some of Mill's finest writing is reserved for a condemnation of the system as it existed in Ireland.[72]

The essential defects of the system include the lack of motive to effort; on the contrary, any increase in value given to the land by the tenant would raise the rent against himself, since the landlord can always hire it out at a better rate to competitors. And while the landlord might desist from taking full advantage of such opportunities, no tenant would rely on his benevolence: 'it is never safe to expect that a class or a body of men will act in opposition to their immediate pecuniary interest; and even a doubt on the subject would be almost as fatal as a certainty, for when a person is considering whether or not to undergo a present exertion or sacrifice for a comparatively remote future, the scale is turned by a very small probability that the fruits of the exertion or of the sacrifice will be taken away from him' (315). Secondly there is a weakening of the prudential motives on the part of each individual, since notwithstanding his own responsible behaviour the market would turn against him in consequence of imprudence by others (316).

In a sense, these rationalizations are academic; for the degradation that results after the system had been long in existence was such that rent agreements become 'merely nominal; the competition for land makes the

---

72 'Until very lately,' inserted 1862 (313).

tenants undertake to pay more than it is possible they should pay [and stay alive], and when they have paid all they can, more almost always remains due' (317). The landlord, therefore, does not in fact exact the full competitive rent, and allows the tenant to get into debt:

In such a condition, what can a tenant gain by any amount of industry or prudence, and what lose by any recklessness? If the landlord at any time exerted his full legal rights, the cottier would not be able even to live. If by extra exertion he doubled the produce of his bit of land, or if he prudently abstained from producing mouths to eat it up, his own gain would be to have more left to pay to his landlord; while, if he had twenty children, they would still be fed first and the landlord could only take what was left (318).

At this point Mill – in the Smithian tradition – bitterly denounced those who blamed the backwardness of Irish agriculture on the '*insouciance* in the Celtic race' (319). He pointed to the energy of Irishmen in England and America as evidence that national character is not the culprit, but rather faulty institutions encouraging 'listlessness and indifference'. This was not however, to deny all national differences under identical conditions: 'It is very natural that a pleasure-loving and sensitively organised people like the Irish, should be less addicted to steady routine labour than the English, because life has more excitements to them independent of it; but they are not less fitted for it than their Celtic brethren the French, nor less than the Tuscans, or the ancient Greeks'.

FIVE

Allocation, trade and distribution

I INTRODUCTION

In earlier work I have demonstrated that Ricardo's exposition of his fundamental theorem on distribution (the inverse profit–wage relationship based upon the standard measure of value) had a positive and lasting effect on professional thought well after his death in 1823 (Hollander, 1977b). I have also argued that the fundamental theorem was not divorced from that corpus of analysis involving the principles of allocation economics (Hollander, 1979, Ch. 6; 1982). It would appear that Ricardianism and neo-classicism – while not sharing identical procedures and certainly not identical preoccupations – have in common a similar 'central core' amounting to allocation theory based upon the mechanisms of demand and supply. The evidence tends to confirm the position adopted by Alfred Marshall who insisted that he himself wrote in the Ricardian tradition, and objected to the strictures against Ricardo (and J. S. Mill) by Jevons and others writing from a 'marginalist' perspective.

In the present chapter the story is carried further. A primary theme is the profound extent of J. S. Mill's 'Ricardianism'. By this I again mean the inverse profit–wage relationship based upon the standard measure, and those same linkages between distribution and the competitive allocation mechanisms which were isolated in the investigation of Ricardo. On my reading of the record, the contributions to theory by Mill fit into the existing structure as 'improvements', as Mill himself always insisted, and do not constitute elements of a different general model. The demonstration is all the more important when we consider the repeated assertions in the literature that Mill effected, or wished to effect, or ought to have effected, a breakaway from Ricardian analysis – positions which imply the need for 'explanations' to account for his unwillingness to recognize the innovatory character of his own work.

There is a further reading of the record that it will be helpful to have in mind during the course of our argument. In his famous critique of the Ricardian theory of production and distribution, Frank Knight described distribution theory as having to do with the pricing of productive services. 'These prices', he explained, 'are significant in connection with the division of the product, but their prior and even more significant function is the dual one of apportioning productive capacity in the various forms among industries and among financial and technical productive units within each industry' (1956, 42). It is inviting to make much of Mill's formal arrangement of the *Principles* – the discussion of 'distribution' prior to 'exchange' – as indicating a failure on his part to envisage any such relationship between value theory and distribution, particularly the notion that distribution entails the pricing of scarce factor services; and to see in the discussion of 'production' prior to 'distribution' a failure to appreciate the relation between the valuation of scarce services and the technical conditions of production on the one hand, and their allocation between different uses on the other. This conclusion is in fact formulated clearly in J. A. Schumpeter's account:

The central theory of value, which should come first on logical grounds (and does come first with Ricardo and Marx) is presented in [Mill's] Book III as if it had to do only with the 'circulation' of goods and as if production and distribution could be understood without it. This is worth mentioning because it points to a funda-mental weakness of the 'classic' construction . . . There is some truth in Professor Knight's indictment . . . that the problem of *distribution* . . . was not approached as a problem of valuation at all. To this extent, we must qualify our recognition of Ricardo's chief merit. He and all the 'classics', including Mill, did indeed make progress toward the acquisition of an analytic apparatus that would unify all purely economic problems; but, partly owing to the shortcomings of their groundwork, they never realized its possibilities to the full. They still divorced production from distribution – J. S. Mill even took credit for doing so – as if they were governed by different 'laws' (1954, 543; but see the more qualified position cited below).

Similarly, we have it on high authority that 'by treating the problem of value in Book III *after* discussing production and distribution, [Mill] more or less suggests that distribution has nothing to do with valuation, being a product of historical accident' (Blaug, 1978, 188).

This is an unconvincing perspective. The organizational practice was standard as far as concerns the precedence accorded production; it was followed by so-called embryonic neo-classicists – Say (1821) and Senior (1836) – as well as so-called Ricardians – McCulloch (1864) and James

Mill (1821). As for J. S. Mill specifically, his intentions in organizing the *Principles* as he did must be explained on grounds of broad objective which do not touch at all upon purely analytical considerations. His objective, above all, was to emphasize the 'provisional' status of many of the generalizations in political economy while at the same time insisting upon the 'permanent' constraints applicable in a wide range of social organization. But nothing that he had to say regarding these matters implied a rejection – for the case of a capitalist-exchange system – of distribution as a problem in factor pricing, or more generally a network of prices which may be conceptualized in terms of general equilibrium. For when Mill limited himself to discussion of such a system he was faced by precisely the same dilemma as Ricardo had been, namely that distribution could not be adequately appreciated without the *prior* discussion of exchange. In a system involving specialization and exchange the relation between profits and wages could not be appreciated except in value terms. He followed Ricardo also in the resolution of the problem.

It is generally recognized that the classical theory of international trade accords allocation principles based on demand a high profile. But this is all too frequently regarded as the exception that proves the rule – the absence of a general 'neo-classical' allocative perspective (cf. Myint, 1948, 13, 68–9; Hicks, 1985). This kind of reading simply does not ring true; it is inconceivable that the tools of allocation theory would have been used in one specific application by an economist who is ignorant of their foundation in a more general system of demand–supply analysis. No such presumption is required on our reading. It is appropriate, therefore, to devote a section in the present chapter to the issue of international values.

In what follows, Mill's conception of national income is first taken up to clear away some of the underbrush; we shall attend to the dichotomy between productive and unproductive labour and its implications for the role of utility and scarcity in Section II. Mill's recognition of the principle of derived demand confirms the role accorded final demand in employment decisions by firms (Section III). Section IV deals with consumer behaviour where a fundamentally non-Jevonian perspective emerges in which consumers are not treated as maximizers. The next five sections deal with price formation in the short and long run, under various cost and market conditions. Particular attention is paid to price adjustment to costs, following the introduction of a new technology into a competitive (constant-cost) industry, *prior to* supply variation – apparently a new perspective on the competitive process. Section X deals with the wage- and profit-rate structures with special reference to the implications flowing from the allowance for 'non-competing groups' and for 'entrepreneurship'. International value is taken up in Section XI.

The remainder of the chapter considers major aspects of the theory of distribution, with particular attention paid to various linkages with the theory of value and allocation. Following Mill, the emphasis will be almost entirely on what Cannan refers to as 'pseudo-distribution', or wages per man, profits per cent and rent per acre, as distinct from 'distribution proper', or aggregate wages, profits and rent as shares in the total national product (1917, 180–1, 282). (But there is a presumption that the rental share rises with progress *ceteris paribus* : 'The ordinary progress of a society which increases in wealth, is at all times tending to augment the incomes of landlords; to give them both a greater amount and a greater proportion of the wealth of the community . . .' *CW*, III, 819). Mill's requirement for the investigation of exchange prior to a full analysis of distribution is developed in Section XII; here the fundamental Ricardian theorem on distribution based upon the standard measure of value is shown to constitute a central theme of Mill's economics, quite consistently with a profound appreciation of the relativity of exchange value. Thereafter, we show that Mill, again in the Ricardian manner, also maintained the fundamental theorem in terms of the quantity theory of money independently of the measuring device (XIII). The logic of his economic reasoning (as in Ricardo's case), however, turned upon the principles of competitive allocation theory although he seriously weakened his argument by presuming (as did McCulloch) that capital goods can be entirely reduced to labour costs (XIV).

II ASPECTS OF NATIONAL ACCOUNTING

In a formal statement regarding distribution Mill includes in national income the wages of 'productive' labourers alone; the wages paid for 'unproductive services' are treated as a transfer payment:

The three requisites of production . . . are labour, capital and land . . . Since each of these elements of production may be separately appropriated, the industrial community may be considered as divided into landowners, capitalists and productive labourers. Each of these classes, as such, obtains a share of the produce: no other person or class obtains anything except by concession from them. The remainder of the community is, in fact, supported at their expense, giving, if any equivalent, one consisting of unproductive services (*CW*, II, 235).

The exclusion from national income of the earnings of service labour – we shall see presently that this exclusion is subject to some fundamental qualifications – reflects the limitation of the national product to material goods yielded, in classical terminology, by the application of 'productive' labour. Appreciation of Mill's national income accounting requires, therefore, some attention to the productive–unproductive dichotomy.

A problem faced by the reader of Mill's *Principles* is the too easy transfer between 'wealth' as a flow of income and as capital stock; the 'Preliminary

Remarks', designed partly to clarify what is intended by the term, is notoriously ambiguous. The ambiguity arises from a more basic distinction between commodities capable of storage for however brief a time, a condition – though, we shall see, positively not a sufficient condition – for designation as 'wealth', and services; as Lord Robbins has phrased the matter, the distinction is between 'the production of capital in some form or other, and pure services leaving directly or indirectly no lasting sources of enjoyment behind' (1967, xi). According to this classical usage, the annual output of 'wealth' (or capital) said to result from the application of 'productive' labour, extends beyond producer goods (machinery and structures) to include those consumer goods which satisfy the storage condition.[1] Currently produced wealth can, in short, be used for either investment or consumption in our sense; and needless to say, the stock of wealth at any moment may include items which were produced during a variety of periods. Myint (1948, 11, 82, 125) goes too far in maintaining that Mill turned his back on Adam Smith's notion of wealth as a flow of annual 'produce' by defining wealth in terms of susceptibility of accumulation. Mill indeed focussed on this dimension but he too was preoccupied with the magnitude and distribution of the flow of annual produce, choosing (much in the manner of Smith, 1937, 314f.) to include in 'produce' only those things amenable to accumulation and to exclude –subject, however, to qualifications not made by Smith – immaterial services.

We proceed to the details of Mill's formulation. It must be understood at the outset that the productive–unproductive distinction in no way involved a rejection of the role of utility in price formation.[2] Mill did not accept the view of J. B. Say and J. R. McCulloch that all labour be regarded as productive that is 'useful', or engaged in producing 'a benefit or a pleasure worth the cost';[3] but he insisted, none the less, that 'what we produce, or desire to produce, is always, as M. Say rightly terms it, an utility. Labour is not creative of objects, but of utilities' (*CW*, II, 45–6).[4] By 'productive

1 Thus though, for Mill, the process of 'production' in a capitalist-exchange system, ends with the final consumer, the national stock of wealth, at any time, includes not only items in the pipe-line – capital goods proper and manufacturers' and traders' stocks of all kinds – but also stocks in the hands of final purchasers.

2 The key point has been made by Lewis (1954), 147, in discussing Adam Smith's categories: 'The distinction between productive and unproductive workers had nothing to do with whether the work yielded utility as some neo-classicists have scornfully but erroneously asserted'.

3 Surprisingly, Mill does not refer to Nassau Senior's rejection of the productive-unproductive categories (Senior, 1938 [1836], 50f).

4 Myint (1948), 10–11, 71, 82, 125f, 174, 206, makes much of what he calls Mill's 'materialist bias' (see also below, p. 251). He commends Lauderdale and J. B. Say for their criticism of the productive–unproductive distinction, citing Say's statement that

labour' he intended labour productive of 'wealth', defined such that utility is a necessary but insufficient condition to include material products: 'things which cannot, after being produced, be kept for some time before being used, are never . . . regarded as wealth, since however much of them may be produced and enjoyed, the person benefited by them is no richer, is nowise improved in circumstances' (48). 'For some time' is left undefined, and quite deliberately, for any degree of 'susceptib[ility] to accumulation', however small, suffices to distinguish wealth from 'a mere service . . . a pleasure given, an inconvenience or a pain averted, during a longer or a shorter time . . . ', that is from the yield of 'unproductive' labour 'employed in producing an utility directly, not . . . in fitting some other thing to afford a utility' (47); similarly, 'pleasures which only exist while being enjoyed, and services which only exist while being performed, cannot be spoken of as wealth, except by an acknowledged metaphor' (48).

The same categories had been similarly drawn up in an early essay 'On Productive and Unproductive Labour' (1830) published in the *Unsettled Questions* (1844): 'The end to which all labour and all expenditure are directed, is twofold. Sometimes it is *enjoyment* immediately; the fulfilment of those desires, the gratification of which is wished for on its own account. Whenever labour or expense is not incurred *immediately* for the sake of enjoyment, and is yet not absolutely wasted, it must be incurred for the purpose of enjoyment *indirectly* or mediately; by either repairing and perpetuating, or adding, to the *permanent sources* of enjoyment' (IV, 284).

If we are fully to appreciate the subsequent transfer to 'neo-classicism' it is essential to recognize that materiality as such was, for Mill, irrelevant. It is the flow of future utilities embodied in a (scarce) material product that renders it an item of wealth – wealth constitutes a store of utilities. This fact governs the rationalization given for an extension of the productive-labour category to include transport workers, merchants and dealers, namely that their labour 'adds the property of being in the place where they are wanted . . . which is a very useful property, and the utility it confers is embodied in the things themselves, which now actually are in the place where they are required for use, and in consequence of that increased utility could be sold at an increased price, proportioned to the labour expended in conferring it' (II, 47–8). There is, thus, no conflict between a utility and a labour theory of value.

That materiality in and of itself is insufficient for a physical item to be counted as 'wealth' and thus be included in the national dividend is also

'production is the creation, not of matter, but of utility' and continues: 'But the doctrine continued to be the centre of interest up to the time of J. S. Mill, partly because the classical school was, as a rule, impervious to the influence of the early demand economists . . . ' (71). Myint neglects the fact that Mill himself firmly insisted upon Say's statement.

apparent in the fact that '[p]roductive labour may render a nation poorer, if the wealth it produces, that is, the increase it makes in the stock of useful and agreeable things, be of a kind not immediately wanted: as when a commodity is unsaleable, because produced in a quantity beyond the present demand; or when speculators build docks and warehouses before there is any trade' (52).[5] Ricardo, it will be recalled, had also rejected the charge that political economists wished 'to heap up what we call valuable commodities, without any regard to quantity . . . ' (1951, IX, 248-9). More generally, unproductive labour may be wastefully applied to satisfy given ends but so too may productive labour 'if more of it is expended than really conduces to production' and 'if defect of skill in labourers, or of judgment in those who direct them, causes a misapplication of productive industry' (*CW*, II, 51).

Needless to say, on this view, the condition that the productive labourer 'produces more than he consumes' (50) must be interpreted in value not physical terms. In a capitalist-exchange system it is the market process that can normally be relied upon to assure against 'wastages' of various kinds. But the process might function poorly, as an observation regarding some canal and railway ventures in the US and England suggests – an observation with broad implications for development policy to the present day: 'Labour sunk in expectation of a distant return, when the great exigencies or limited resources of the community require that the return be rapid, may leave the country not only poorer in the meanwhile, by all which those labourers consume, but less rich even ultimately than if immediate returns had been sought in the first instance, and enterprises for distant profit postponed' (52).

In the light of all this it is difficult fully to accept Hla Myint's qualified observation that 'none of the major classical economists, with the possible exception of J. S. Mill, seems to be guilty of the "confusion between the technical and the economic problem" . . . [They] were not bemused by the purely technological or engineering possibilities of increasing output divorced from the economic calculus based on the relation between cost and output' (1948, 13). Considering the foregoing evidence there is no reason to exclude Mill from this otherwise valid generalization.[6]

5 Presumably Mill would have agreed that 'the argument for *laisser faire* does not tend to show that the spontaneous combination of individuals pursuing self-interest will lead to the production of a maximum of *material* wealth, except so far as the individuals in question prefer material wealth to utilities not embodied in matter' (Sidgwick, 1883, 408). But on this matter compare Myint (1948), 127.

6 Myint (1948), bases his case partly on the organization of the *Principles* which places distribution and exchange after production, suggesting to him that 'the size of the national dividend is determined entirely by technology and the laws of changes in the supply of factors, almost independently of the equilibrium process of the market . . . [Distribution and exchange] merely played a secondary role of parcelling out this predetermined bloc

It is sometimes claimed that the productive–unproductive dichotomy reflects for Mill, as for Smith, a concern with economic development. Thus Vincent Bladen: 'This is perhaps clearer in Adam Smith than in Mill, but . . . the continued use of the distinction between productive and unproductive labour indicates a continued concern for the liquidation of the primitive sector of the economy in which menial servants were maintained in idleness on a more or less feudal basis, and for the development of "industry", the advanced sector of the economy in which workers, well managed, well disciplined, would probably be employed at wages considerably higher than those prevailing in the primitive sector' (1965, xlii). There may be something to this, but a concern for 'development' cannot alone or even primarily explain the classification. Physical commodities are distinguished from services because of an insistence that satisfaction as such be divorced from its source. Thus the annual product measures sales of producer and consumer goods envisaged as the source of utilities and not utilities as such, while services are excluded since in their case the source and the utilities cannot be separated. This contrast differentiates between a pleasure enjoyed which (at least beyond the instant) is a mere memory, and pleasures stored up. It appears truistic to declare that labour which does not terminate in the creation of material wealth . . . does not render the community, and the world at large, richer in material products, but poorer by all that is consumed by the labourers while so employed' (CW, II, 50) but this assertion reflects the perfectly legitimate view of wealth as a store of future utilities. This position is better expressed in the essay: expenditure for 'mere pleasure . . . renders a country poorer in the permanent sources of enjoyment' (IV, 287).

Here we should note Mill's observation that the individual unproductive worker may grow 'richer' – in the acquisition of permanent means of enjoyment – but if so it is at the expense of the consumer of (final) services:

An unproductive labourer may receive for his labour, from those who derive pleasure or benefit from it, a remuneration which may be to him a considerable source of wealth; but his gain is balanced by their loss; they may have received a full equivalent for their expenditure, but they are so much poorer by it. When a tailor makes a coat and sells it, there is a transfer of the price from the customer to the tailor, and a coat besides which did not previously exist; but what is gained by an actor

of wealth, national dividend, among different individuals according to the prevailing system of economic organization. It was not thought that distribution and exchange could affect the size of the national dividend.' For our view of the implications deriving from the organization of the *Principles*, and from the distinction between the laws of production and distribution which also plays a part in Myint's case, see above, pp. 164–84 esp. 174, 175.

The view that the Ricardo–Mill concept of maintaining capital intact 'belongs essentially to the physical level of analysis' (Myint, 1948, 174) also needs modification.

is a mere transfer from the spectator's funds to his, leaving no article of wealth for the spectator's indemnification (II, 50–1).

That a transfer of '*wealth*' is involved is not in question – this is implied by the definition of national income. Yet clearly Mill allows both for the satisfaction of the consumer of final services and of the service worker who is supposed to acquire wealth. He avoids Smith's error in the *Theory of Moral Sentiments* (1966, 264), asserting that distribution is more egalitarian than appears since the landowner pays out means of subsistence to his servants, an assertion which neglects that the master is enjoying his income in the services enjoyed and the servants their incomes in the form of wage goods. Distribution is more equal only so far as concerns material goods (the utilities embodied therein) not final utilities.[7] Cannan has remarked of the kind of error alluded to above that 'few or none of the economists who have expressed themselves in favour of excluding the produce of "unproductive" labour from the annual produce, have attempted to adhere consistently to the exclusion' (1917, 24–5), thereby implying that to exclude services logically entails the error in question. But this is not so; we merely have to recall that classical national income does not catch in its net – is not designed to catch – currently enjoyed utility.

We turn now to Mill's treatment of human capital. The non-accumulativeness of final services posed some severe strategic problems for Mill, who was much concerned not to confuse readers by unfamiliar usage of familiar terms. He himself maintained that 'wealth' extends beyond physical objects 'serviceable to human beings' (*CW*, II, 46), and wished to include knowledge and the state of health along with tools and machinery within the community's stock of producer goods. From this perspective, the labour engaged 'in creating permanent utilities, whether embodied in human beings, or in any other animate or inanimate objects' should ideally be categorized as productive (48). This indeed had been a leading theme of the early paper to which he now refers the reader: 'Sources of enjoyment may be accumulated and stored up; enjoyment itself cannot. The wealth of a country consists of the sum total of the permanent sources of enjoyment, whether material or immaterial, contained in it: the labour or expenditure which tends to augment or to keep up these permanent sources, should, we conceive, be termed productive' (IV, 284). Unproductive labour is that exerted 'directly and exclusively for the purpose of enjoyment, and not calling

---

7 At times, however, Mill seems rather close to the borderline: e.g. letter of Dec. 21, 1861 to Cliffe Leslie (*CW*, XV, 757): 'You say that a duke's family does not consume very much more "of certain things" than an artisan's or a clerk's. Not nearly so much in proportion to their means; but much more absolutely, since they pay for all that is consumed by their servants & dependants.' Yet even this need not be read as denying the duplicate benefit of utility.

into existence anything, whether substance or quality, but such as begins and perishes with the enjoyment' (287).

On this early formulation, wealth – envisaged as a source of enjoyment – may be 'material or immaterial', 'substance or quality'. Mill includes not only capital goods proper and stocks of consumer goods, opera houses and musical instruments – since they constitute 'a permanent source of enjoyment, which does not begin and end with the enjoying, and therefore admits of being accumulated' (284–5) – but also, for the same reason, operatic and musical skill (and the skill of the instrument maker). Skill 'must be considered as wealth; and the labour and funds employed in acquiring skill in anything tending to the advantage or pleasure of mankind must be considered to be productively employed and expended'. Similarly, 'the architect who built the theatre was a productive labourer; so were the producers of the perishable [theatrical] articles; so were those who constructed the musical instruments; and so . . . were those who instructed the musicians, and all persons who by the instructions which they may have given to Madame Pasta, contributed to the formation of her talent. All the persons contributed to the enjoyment of the audience in the same way, and that a remote way viz., by the production of a *permanent source of enjoyment*' (285–6).

In the *Principles* Mill reiterates his preference for a distinction that would 'turn upon the permanence rather than upon the materiality of the product' (II, 49) – a formulation incorporating human capital in general no matter that the ultimate yield may be an immaterial service. But usage disallowed it: 'in applying the term wealth to the industrial capacities of human beings, there seems always, in popular apprehension, to be a tacit reference to material products. The skill of an artisan is accounted wealth, only as being the means of acquiring wealth in a material sense; and any qualities not tending visibly to that object are scarcely so regarded at all.' Mill chose, therefore, to limit the term wealth to indicate 'only what is called material wealth, and by productive labour only those kinds of exertion which produce utilities embodied in material objects'.

Yet even on this view, it was still possible to consider as productive 'labour which yields no material product as its direct result, provided that an increase of material products is its ultimate consequence'; thus 'labour expended in the acquisition of manufacturing skill' is covered, though 'not in virtue of the skill itself, but of the manufactured products created by the skills, and to the creation of which the labour of learning the trade is essentially condusive'.[8] Similarly, police protection of industry entails productive

8 In the 'Preliminary Remarks' Mill dismisses the whole issue of the treatment of skill as 'a question, not of very great importance' (II, 10). But he evidently did not really mean this. He stood firmly for the inclusion of skill within the stock of national capital, as is clear from an added note in the edition of 1865, which rejects the argument that

activity 'because without it, material wealth, in anything like its present abundance, could not exist'. So too (as we have earlier emphasized) do transportation and other distributive services.[9] Excluded from wealth are theatrical and musical skills however valuable they might be.[10] It would seem to follow logically that the theatre, props, musical instruments and so forth must also be excluded from capital wealth (and their producers denied the 'productive' appelation) since they too are not conducive to material products as ultimate consequence. But it is doubtful whether Mill went so far. All the indications are that 'real' or 'physical' capital goods relating to the (unproductive) service sector are still regarded as wealth and their producers as 'productive' labourers; Mill bowed to public usage only concerning human capital, adopting a narrower perspective than he would have preferred.

A second important qualification, apart from that for human capital, arises in an open economy: 'A country would hardly be said to be richer, except by a metaphor, however precious a possession it might have in the genius, the virtues, or the accomplishments of its inhabitants; unless indeed these were looked upon as marketable articles, by which it could attract the material wealth of other countries, as the Greeks of old, and several modern nations have done' (49). Similarly: 'A community . . . may add to its wealth by unproductive labour, at the expense of other communities, as an individual may at the expense of other individuals. The gains of Italian opera singers, German governesses, French ballet dancers, &c., are a source of wealth, as far as they go, to their respective countries, if they return thither' (51).

At this point it is most important to emphasize, considering the topicality of the issue (particularly due to the work of Johnston, 1975, and Bacon

skill be excluded on the grounds of non-transferability: 'the skill of an artisan (for instance) being both a desirable possession, and one of a certain durability (not to say productive even of national wealth), there is no better reason for refusing it the title of wealth because it is attached to a man, than to a coalpit or manufactory because they are attached to a place' (48n). This argument he further supplemented by adding that though 'the skill itself cannot be parted with to a purchaser, the use of it may; if it cannot be sold, it can be hired'. Accordingly, while the individual could not be classed as wealth, 'his acquired capacities, which exist as means, and which have been called into existence by labour, fall rightly, as it seems to me, within that designation.'

9   In this regard Mill differed from Smith for whom all service workers earn derivative rather than original incomes (Hollander, 1973a, 146–7).

10 Nassau Senior (1848), in his famous review of the *Principles* noticed that 'the Treatise . . . limits the epithet productive to the exertion which directly or indirectly produces utilities embodied in material objects. This definition excludes what [the essay] admitted, labour employed in conferring permanent benefit unconnected with the increase of material objects' (308). As Senior points out, Adam Smith seems to have excluded from his productive group even those teachers of vocational skills in the narrowest sense – skills of use in the production of material goods.

and Eltis, 1978) that the marketability criterion for the division of activities is not utilized by Mill, although marketability is necessarily referred to in the foregoing context of foreign trade in services. This fact emerges in a contrast between the ms. of the *Principles* and the published versions – namely, that in the former, utilities are presumed to be 'fixed and embodied in human beings [by] parents, so far as they concern themselves in the education of their children', whereas in the published version education is presumed to be undertaken by 'governments, so far as they aim successfully at the improvement of the people' (47). Conversely, theatrical performers organized by a capitalist-entrepreneur who engages his players with an eye to a return on his investment derived through the market mechanism still falls within the unproductive category. Consistently with this position, Mill (following McCulloch) took issue in the essay with those who insisted that labour can only be classified as 'productive' if the return is paid to the same person or authority that makes the outlay (IV, 282). This position yielded the unsatisfactory implication that labour employed by government in building and maintaining toll-free roads are engaged in unproductive activity, which becomes productive when a toll is charged. Similarly, labour employed by government in the protection of private industry would have to be classified as unproductive, whereas equivalent private outlays fall into the productive category. This kind of outcome Mill rejected.

Government financed services are therefore not excluded out of hand as they were by Smith. Indeed in the essay, the judge, legislator, police officer and soldier – and menials too – are classified as 'partly productive and partly unproductive' (286).[11] For they protect physical and human capital thereby indirectly increasing production 'in a degree far more than equivalent to the expense which is necessary for their maintenance . . .'; however, 'they protect mankind, not merely in the possession of their permanent resources, but also in their actual enjoyments; and so far, although highly useful, they cannot . . . be considered productive labourers'. This position still allows for quite a range of productive government services. In the *Principles*, however, Mill seems to take a somewhat narrower view. For national defence is entirely excluded from the productive category, on the grounds that the military 'at the best, prevent a country from being conquered, or from being injured or insulted, which is a service, but in all other respects leave the country, neither improved nor deteriorated' (II, 47). This is also the view taken of 'the legislator, the judge, the officer of

---

11 Menial servants 'are entertained mainly as subservient to mere enjoyment; but most of them occasionally, and some habitually, render services which must be considered as of a productive nature; such as that of cookery, the last stage in the manufacture of food; or gardening, a brand of agriculture' (286-7).

justice, and all other agents of government, in their ordinary function . . . '. Mill did concede, on the other hand, that some government services have an impact on 'the national mind', remarking in the same context that theatrical services too might produce 'some good . . . beyond the moment, upon the feelings and disposition' of the spectators, rendering the final outcome a 'permanent' source of enjoyment. The latter extension is, however, dismissed – even from the 'ideal' categorization – on the grounds that the permanent effect is accidental since 'nothing but the immediate pleasure' is intended by actor and audience. The rationale implies, though, that a service designed to impinge on 'feelings and dispositions' (a course, let us say, in the history of art whether taught privately or by a government employee) would be included within the productive category, at least if Mill had stood by his own preferences.

* * *

We now must consider the breakdown of national income between the gross and net components. Mill's further elaboration of the productive and unproductive dichotomy is pertinent to this issue.

The formal categorization based upon materiality is, of course, subject to notorious anomalies; a loaf of bread is classed with a coat as 'wealth', though in terms of materiality – its incorporation of use value for some finite period after purchase – it is much closer along the spectrum to a theatrical service. In the manuscript Mill himself provisionally suggested that labour engaged in the production of material 'luxury' goods – 'gold lace, pine apples, or champagne' (53) – be excluded 'since the temporary addition which it makes to the wealth of the community, unavoidably perishes in the very next stage', and although he concluded that in fact 'society has been richer' while the objects lasted, the initial suggestion is removed from the published version, perhaps because of a realization that some 'luxury' goods last a very long time while some 'necessaries' are rapidly consumed.

Whether or not it is from a consideration of this fact, Mill proceeded in the essay and the *Principles* to offer an alternative classification focussing more upon function than materiality (though the materiality dimension is not abandoned), with particular reference to capital maintenance and accumulation. The modified classification turns on the distinction between productive and unproductive *consumption*:

What [production labourers] consume in keeping up or improving their health, strength, and capacities of work or in rearing other productive labourers to succeed them, is productive consumption. But consumption on pleasures or luxuries, whether by the idle or by the industrious, since production is neither its object, nor in any way advanced by it, must be reckoned unproductive: with the reservation perhaps of a certain quantum of enjoyment which may be classed among necessaries, since anything short of it would not be consistent with the greatest efficiency of labour. That alone is productive consumption, which goes to maintain and increase the productive powers of the community; either those residing in its soil, in its materials, in the number and efficiency of its instruments of production, or in its people (II, 52–3).[12]

This distinction – between intermediate use and final use – leads to a classification of labour which Mill considered 'more important to the wealth of a community than even that between productive and unproductive labour', namely that between 'labour for the supply of productive, and for the supply of unproductive, consumption; between labour employed in keeping up or in adding to the productive resources of the country, and that which is employed otherwise':

Of the produce of a country, a part only is destined to be consumed productively; the remainder supplies the unproductive consumption of producers, and the entire consumption of the unproductive classes. Suppose that the proportion of the annual produce applied to the first purpose amounts to half; then one-half of the productive labourers of the country are all that are employed in the operations on which the permanent wealth of the country depends. The other half are occupied year by year and generation by generation in producing things which are consumed and disappear without return; and whatever this half consumes is as completely lost, as to any permanent effect on the national resources, as if it were consumed unproductively (54).

In the foregoing extracts Mill refers to 'productive labour', 'production', 'annual produce'. We are therefore still faced with the problem of defining these categories – who are the producers? Presumably Mill had not abandoned the earlier discussion but rather introduced subcategories of 'productive' labour. All the earlier problems therefore remain – and some new ones.[13] But it had at least emerged that transport and distributive

12 Cf. IV, 288: productive consumption is 'what is necessary to keep the productive labourer in perfect health and fitness for his employment' supplemented by 'what he expends in rearing children to the age at which they become capable of productive industry'.

13 From his silence on the matter it appears that Mill still included within capital wealth the real capital goods and stocks-in-process from which flows the annual stream of luxury goods. If this is so, the maintenance of labour engaged on their upkeep – productive labour on the materiality criterion – involves productive consumption. For earlier versions of this distinction see James Mill (1844, 220f), J. B. Say (1964 [1819], 105f) and Nassau Senior (1938 [1836], 54f). It is pertinent that in Say and Senior we have authors who objected to the productive–unproductive labour division as such, based on materiality, yet found helpful the distinction between consumption involving intermediate use and that involving final use. Thus Senior: 'So far as a man's consumption is essential to his production, he belongs to the first class; so far as it is not essential to the second.' By 'products' Senior, however, includes menial service.

personnel and some security, educational and health workers, whether in the private or government sector, fall within the productive category so that their maintenance (at least part of it) entails 'productive consumption'. Conversely, part of the annual flow of material products purchased by labour unambiguously classified as 'productive' in terms of materiality turns out to be unproductively consumed. In these respects materiality is played down.[14]

The principle is clear even if the practical designation of particular items is not: Part only of the 'national income' (however we choose to define it) is absorbed in the maintenance of human or physical 'capital' (again however we choose to fill the box). This implies the classical contrast between net and gross national income. The chapter on productive labour in the *Principles* closes with an unfortunate passage implying a clear-cut distinction between that part of the annual produce constituting the community's 'necessities', and that part available for its 'pleasures and for all higher uses', the latter constituting 'the fund from which all the wants of the community, other than that of mere living, are provided for' (54). This formulation smacks of unambiguous, physiological, 'subsistence'. But of course, as his references to their 'unproductive consumption' indicate, Mill presumed that workers earned wages exceeding that minimum. The essay closes with a firm insistence that wages exceeding subsistence – defined as that level assuring constancy of population (the 'productive' population) – be included in net national income. The passage is important for it emphasizes saving, not only unproductive consumption, out of wages:

Political economists generally define the 'net produce' to be that portion of the gross annual produce of a country which remains after replacing the capital annually consumed. This, as they proceed to explain, consists of profits and rent; wages

---

14 Logically if national capital ('wealth') includes the skills of the opera singer – as in the essay – then his maintenance will have to be designated productive consumption. Yet Mill was loath to draw this conclusion; in fact 'he ties himself in an extraordinary knot while trying to work backward from the unproductiveness or productiveness of a final good to the character of its necessary inputs' (Boss, 1981). The performer's maintenance during the time he learns his skill is designated 'productive' since the labour then exerted tends 'to increase the sum of the permanent sources of enjoyment in the country, by effecting a new creation of those sources, more than equal to the amount of the consumption' (*CW*, IV, 285). Yet his maintenance whilst actually exercising his profession is classified as unproductive since 'the accumulated total of the sources of enjoyment which the nation possess is diminished by the amount he has consumed' (284). (The same position seems to be taken of the maintenance of the theatre: 'The building and decorations are consumed unproductively; . . . the epithet unproductive must be equally applied to the gradual wearing out of the bricks and mortar, the nightly consumption of the more perishable "properties" of the theatre . . . ', 285.) Thus, even in the early essay Mill failed to face up to the implications of his own allowance within the country's capital stock for skill with no potential to yield final material product.

being included in the other portion of the gross produce, that which goes to replace capital. After this definition, they usually proceed to tell us that the net produce, and that alone, constitutes the fund from which a nation can accumulate, and add to its capital, as also that which it can, without retrograding in wealth, expend unproductively, or for enjoyment. Now, it is impossible that both the above propositions can be true. If the net produce is that which remains after replacing capital, then net produce is not the only fund out of which accumulation may be made: for accumulation may be made from wages; this is in all countries one of the great sources, and in countries like America perhaps the greatest source of accumulation. If, on the other hand, it is desirable to reserve the name of net produce to denote the fund available for accumulation or for unproductive consumption, we must define net produce differently. The definition which appears the best adapted to render the ordinary doctrines relating to net produce true, would be this: The net produce of a country is whatever is annually produced beyond what is necessary for maintaining the stock of materials and implements unimpaired, for keeping all productive labourers alive and in condition for work, and for just keeping up their numbers without increase. What is required for these purposes, or, in other words, for keeping up the productive resources of the country, cannot be diverted from its destination without rendering the nation as a whole poorer. But all which is produced beyond this, whether it be in the hands of the labourer, of the capitalist, or of any of the numerous varieties of rent-owners, may be taken for immediate enjoyment, without prejudice to the productive resources of the community; and whatever part of it is not so taken, constitutes a clear addition to the national capital, or to the permanent sources of enjoyment (IV, 288–9).

Not to take this position would be to argue 'as if all profit and loss to the nation were to be seen in the capitalist's account-book'.

In all this Mill followed Ricardo (*Works*, I, 348n.) who treated the 'subsistence' element of wages as an intermediate cost (along with the maintenance of 'fixed' capital) counting only the excess within net income. Smith by contrast – like Kuznets in our day (1941, 1948) – included wages (though not of service labour) in their entirety.

* * *

We return to the basic classification. Throughout our account we have encountered references to the surplus-yielding requirement of productive labour, supplementing the accumulatability requirement. This too has Smithian pedigree; in the *Wealth of Nations* only labour bestowed on the production of material goods is said to be capable of 'adding value' – service labour 'adds to the value of nothing' (1937, 314). For Mill productive labour is not restricted to the private sector, yet his primary concern was with the capitalist, not the government sector; and if we can set aside the problem of the capitalistic employment of opera singers and also make the appropriate allowances for distributive personnel and the like we are left with the broad classical distinction between labour engaged in the capitalist sector with an eye to profit and service labour engaged for purposes of final consumption.

Nassau Senior in his famous review of the *Principles* (1848, 309) maintained that Mill made no practical use of the classification.[15] But this is not quite so; we shall see that one of the *ceteris paribus* conditions in the analysis of the declining profit-rate trend is the distribution of the work force between the unproductive and productive classes – any transfer from the former to the latter counteracting the decline (below, p. 459). We shall also find that the production function differs between the two sectors with important analytical consequences (below, p. 407).

III DERIVED DEMAND

At the close of a long footnote towards the end of his mammoth history of analysis J. A. Schumpeter has tucked away the remark that 'Walras really reformulated the theories of production of A. Smith, J. B. Say, and J. S. Mill. The latter's theory of production, of course, must not be looked for exclusively in his Book I' (1954, 1010n.). Apart from the absence here of Ricardo's name, this is a justified assertion of the first importance for any correct perspective on historiographical trends.

Ricardo's subscription to J. B. Say's famous embryonic notion of 'imputation' has been dealt with elsewhere (Hollander, 1979, Ch. 6; 1982).[16] Here I shall demonstrate the presence in Mill's work of an appreciation of derived demand in the sense that the ultimate source of factor remuneration is said to be in sales proceeds and the motive for factor employment to be the added revenue-product expected. The argument is obviously not technically watertight because of the absence of a clear marginal principle;[17] but it provides evidence of the close connection envisaged by Mill between value and distribution.

15 Cf. Stigler (1982), 151: 'Senior and McCulloch denied the distinction, and John Mill refined it almost out of existence.'

16 The term 'imputation' is not used in Menger's specific technical sense, namely that of a marginal utility theory of imputation, but in the more general sense of derived factor demand, which does not require necessarily a utility underpinning. (But see Mill's pronouncement (*CW*, II, 47–8), cited above, p. 250, which is consistent with utility analysis.)

17 There is already an indication of this at the outset of the *Principles* in a discussion of the relative contributions to output of land and labour: '[w]hen two conditions are equally necessary for producing the effect at all, it is unmeaning to say that so much of it is produced by one and so much by the other; it is like attempting to decide which half a pair of scissors has most to do in the act of cutting' (II, 28–9).

We encounter a brief suggestion of the relationship in question in Mill's reference to 'the present system of industrial life, in which employments are minutely subdivided, and all concerned in production depend for their remuneration on the price of a particular commodity . . . ' (*CW*, III, 455). The principle was further elaborated at the very outset of the *Principles* in a chapter dealing with indirect inputs of labour in lengthy processes of production: 'All these persons ultimately derive the remuneration of their labour from the bread, or its price: the ploughmaker as much as the rest; for since ploughs are of no use except for tilling the soil, no one would make or use ploughs for any other reason than because the increased returns, thereby obtained from the ground, afforded a source from which an adequate equivalent could be assigned for the labour of the ploughmaker. If the produce is to be used or consumed in the form of bread, it is from the bread that this equivalent must come' (II, 31). It is presumably the expectation of future yield that provides the motive for the use of the input.

In the case of materials which are 'destroyed as such by being once used, the whole of the labour required for their production, as well as the abstinence of the person who supplied the means of carrying it on, must be remunerated from the fruits of a single use'. By contrast, 'implements . . . being susceptible of repeated employments, the whole of the products which they are instrumental in bringing into existence are a fund which can be drawn upon to remunerate the labour of their construction, and the abstinence of those by whose accumulations that labour was supported. It is enough if each product contributes a fraction . . . towards the remuneration of that labour and abstinence, or towards indemnifying the immediate producer for advancing that remuneration to the person who produced the tools' (37). Similarly, labourers in the capital-goods sector 'do not depend for their remuneration upon the bread made from the produce of a single harvest, but upon that made from the produce of all the harvests which are successively gathered until the plough, or the buildings and fences, are worn out' (32).

The general principle also covers workers involved in transportation of a product 'from the place of its production to the place of its destined use . . . its final consumption'; they too derive their remuneration from the value of the ultimate product. The same applies to the wholesale and retail functions of the 'Distributing Class, whose agency is supplementary to that of the Producing Class' – 'the produce so distributed, or its price, is the source from which the distributors are remunerated for their exertions, and for the abstinence which enabled them to advance the funds needful for the business of distribution' (40). We have already noted that it is precisely because of the 'increased utility' afforded by these functions that the product 'could be sold at an increased price, proportioned to the labour expended in conferring it' (see above, p. 250).

Although the 'distributive' functions are formally separated from the strictly 'productive', it is clear that the process of production in the capitalist-exchange system was envisaged as ending with sale to the final consumer. This applies also to wage goods. For Mill distinguished labourers' accommodation from industrial structures which have what he termed a 'protective' function in production – 'factories, warehouses, docks, granaries, barns, farm buildings devoted to cattle, or to the operations of agricultural labour' – on the grounds that the housing of workers 'is destined for their personal accommodation: these, like their food, supply actual wants, and must be counted in the remuneration of their labour' (38). Coal may be employed 'not only in the process of industry, but in directly warming human beings. When so used, it is not a material of production, but is itself the ultimate product' (35). That 'the finished products of many branches of industry are the materials of others' (36) was an irrelevant consideration in the case of workers' consumables which were treated on a par with all other final goods. The point at stake is an important one since in some contexts Mill envisaged wage goods (or at least that part constituting 'subsistence') as intermediate products, implying a sort of 'production of commodities by means of commodities'.

There is also to be found in the *Principles* a passage of potential significance for Mill's intentions by his 'recantation' in 1869 of the wages-fund doctrine. It contains an observation drawn from Thomas De Quincey focussing upon the implications of the fact that input use is characterized by the properties of derived demand and joint demand. Again the perspective is one of microeconomics involving particular industries (III, 474; discussed below, pp. 412–13).

Mill's notion that 'demand for commodities is not demand for labour' has often been misunderstood as a denial of the role of final demand in employment decisions. This was the view of Jevons, who read Mill as asserting that capitalists 'maintain and pay for labour whether or not there is a demand for the commodities produced' and that 'production goes on independently of the use to which the produce is to be put' (1905, 127).[18] But as we shall show later (pp. 371–3) the fourth proposition is, in fact, unrelated to employment decisions by individual firms and refers to aggregate employment or earnings.[19] There is certainly no conflict with the principle of derived demand.

18 For a recent statement, cf. Thompson (1975), 188: Mill 'was oblivious to the fact that without consumption there would be no demand for resource inputs, labour included'.

19 Cf. Hayek (1941), 434: 'That . . . the doctrine has suffered a marked eclipse is mainly due to the fact that the modern subjective theory of value was erroneously thought to have provided an effective refutation' (see also for this same view, Marshall, 1920, 827, Viner, 1958, 402–3).

IV CONSUMER BEHAVIOUR

The closest Mill comes to the notion of diminishing marginal utility is in the context of income distribution, specifically his proposal to limit the right of bequest, where the distribution of generalized purchasing power is at stake: 'I do not conceive that the degree of limitation . . . would be felt as a burthensome restraint by any testator who estimated a large fortune at its true value, that of the pleasures and advantages that can be purchased with it . . . [It] must be apparent . . . that the difference to the happiness of the possessor between a moderate independence and five times as much, is insignificant when weighed against the enjoyment that might be given . . . by some other disposal of the four-fifths' (*CW*, II, 225). This seems to imply the principle of diminishing utility applied to income although the incremental dimension is not necessarily intended.

In the first two editions Mill observed that his position went counter to the view that the legator does best to the legatee if he should 'heap on him to satiety all the external goods of life' which from a marginalist perspective is promising. Therefore this latter clause is altered to read: ' . . . those intrinsically worthless things on which large fortunes are expended', Mill now engaging the kind of value judgement to be found in *Utilitarianism* (1860): 'It is quite compatible with the principle of utility to recognise the fact, that some *kinds* of pleasure are more desirable and more valuable than others. It would be absurd . . . [if] [the] estimation of pleasures should be supposed to depend on quantity alone' (*CW*, X, 211). It may, however, perhaps be said that Mill had in mind something like Carl Menger's blueprint of household planning which also involves classes of utility (Menger, 1871, 122f; cf. Kauder, 1965, 74). But in any event there is even here no explicit reference to incremental variation. (But see the opinion of Edgeworth, 1910, 762).

The discussion of progressive taxation suggests even more strongly the absence of that dimension. Even the looser principle of diminishing utility is not, however, applied to income taxation. Mill was wary of the general case for progression according to which 'a tenth part taken from a small income is a heavier burthen than the same fraction deducted from a much larger one . . . the percentage [rising] with the amount of the income' (*CW*, III, 808–9). He allowed only that degree of progression implicit in a tax system that exempts some minimum income, which represents 'necessaries', and imposes a flat proportionate rate on the remainder:

To take a thousand a year from the possessor of ten thousand, would not deprive him of anything really conducive either to the support or to the comfort of existence; and if such *would* be the effect of taking five pounds from one whose income is fifty, the sacrifice required from the last is not only greater than, but entirely incommensurable with, that imposed upon the first. The mode of adjusting these inequalities of pressure, which seems to be the most equitable, is that recommended by Bentham, of leaving a certain minimum of income, sufficient to provide the necessaries of life, untaxed.

* * *

In Mill's case a number of discernable impediments were at play which detract from a marginal utility conception. First, the materiality dimension of wealth which focusses upon the acquisition of a stream of future utilities rather than immediate utility. In the central chapter 'Of Value', Mill deals with a case of bilateral exchange involving a durable consumer good, namely Thomas De Quincey's famous musical snuff-box, the object of negotiations between two passengers on a steamboat on Lake Superior (III, 457; cf. 462-3; V, 397f). Utility, or the capacity 'to satisfy a desire or serve a purpose' (456) is here recognized as a necessary condition of exchange value, but the absence of an incremental dimension is very apparent and is scarcely surprising given the exercise: 'Value or use, or as Mr. De Quincey calls it, *teleologic* value, is the extreme limit of value in exchange. The exchange value of a thing may fall short, to any amount, of its value in use; but that it can ever exceed the value in use, implies a contradiction; it supposes that all persons will give, to possess a thing, more than the utmost value which they themselves put upon it as a means of gratifying their inclinations' (457).

There is a further impediment to arrival at incremental utility. I have in mind the famous declaration that 'only through the principle of competition has political economy any pretension to the character of a science' (II, 239), repeated in a 'warning' at the outset of Book III ('On Exchange') that the analysis of price formation to follow presumes competition, for 'only so far as [prices] are thus determined, can they be reduced to any assignable law' (III,460). On these grounds the analysis – which turns on 'the axiom . . . that there cannot be for the same article, of the same quality, two prices in the same market' – is limited to the wholesale sector, for Mill represents individual consumers as typically failing to act in maximizing fashion, the 'axiom' actually constituting a second stage deduction from more basic behaviour postulates which do not apply in the retail sector:

The values and prices . . . to which our conclusions apply, are mercantile values and prices; such prices are as quoted in price-currents; prices in the wholesale markets, in which buying as well as selling is a matter of business; in which the buyers take pains to know, and generally do know, the lowest price at which an article of a given quality can be obtained . . . Our propositions will be true in a much more qualified sense, of retail prices; the prices paid in shops for articles of personal consumption. For such things there often are not merely two, but many prices, in different shops, or even in the same shop; habit and accident having as much to do in the matter as general causes. Purchases for private use, even by people in business, are not always made on business principles: the feelings which come into play in the operation of getting, and in that of spending their income, are often extremely different. Either from indolence, or carelessness, or because people think it fine to pay and ask no questions, three-fourths of those who can afford it give much higher prices than necessary for the things they consume; while the poor often do the same from ignorance and defect of judgment, want of time for searching and making inquiry, and not unfrequently, from coercion, open or disguised.

In this passage we have a perfect instance of the counteracting-forces to 'wealth-maximization' of which so much is made in the early essay on definition and method.

The same perspective had been taken in the discussion of 'Competition and Custom' (Book II, iv) where it was also emphasized that the single-price axiom turning upon wealth maximization in the sense of pecuniary interest does not apply at the retail level because of the failure of consumers to act as maximizers: 'Not only are there in every large town, and in almost every trade, cheap shops and dear shops, but the same shop often sells the same article at different prices to different customers: and, as a general rule, each retailer adapts his scale of prices to the class of customers whom he expects. The wholesale trade, in the great articles of commerce, is really under the domination of competition. There, the buyers as well as the sellers are traders or manufacturers, and their purchases are not influenced by indolence or vulgar finery, nor depend on the smaller motives of personal convenience, but are business transactions' (242–3).[20] Strictly speaking the retailer may be said to act in maximizing fashion by resorting to price discrimination, but our concern here is the consumer whose failure to act 'on business principles' makes such discrimination possible.[21]

Not only is maximization ruled out for the typical consumer, but some kinds of behaviour lead to what from a marginal utility perspective seem perverse. The reference to 'vulgar finery' in the foregoing passage recalls the term 'costly indulgencies' in the statement of the basic behavioural

20 The reference to 'the smaller motives of personal convenience' was added in 1865. Mill here seems to distinguish rational types of non-pecuniary interest from irrational types.

21 The notion of carelessness in budgeting made famous in our day in the context of Leibenstein's X-efficiency is said to be particularly striking in the traditional sector: 'Large accumulations are continually made by the agents, stewards, and even domestic servants, of improvident persons of fortune; and they pay much higher prices for all purchases than people of careful habits . . . ' (72n.; for Adam Smith's version, see Hollander 1973a, 118–19.)

postulates given in the essay on definition (IV, 321). There is every indication that Mill here had in mind, at least in part, the issue of conspicuous consumption, as implied in the discussion of differential qualities of utility (above, p. 264), and emerging in a variety of other contexts either explicitly or implicitly. Thus, we have the pronouncement that 'I know not why it should be a matter of congratulation that persons who are already richer than any one needs to be, should have doubled their means of consuming things which give little or no pleasure except as representatives of wealth' (III, 755).[22] A related proposition emerges in a striking application to the taxation of luxuries:

they operate in some cases as a useful, and the only useful, kind of sumptuary law. I disclaim all asceticism, and by no means wish to see discouraged, either by law or opinion, any indulgence (consistent with the means and obligations of the person using it) which is sought from a genuine inclination for, and enjoyment of, the thing itself; but a great portion of the expenses of the higher and middle classes in most countries, and the greatest in this, is not incurred for the pleasure afforded by the things on which the money is spent, but from regard to opinion, and an idea that certain expenses are expected from them, as an appendage of station; and I cannot but think that expenditure of this sort is a most desirable subject of taxation. If taxation discourages it, some good is done, and if not, no harm; for in so far as taxes are levied on things which are desired and possessed from motives of this description, nobody is the worse for them. When a thing is bought not for use but for its costliness, cheapness is no recommendation. As Sismondi remarks, the consequence of cheapening articles of vanity, is not that less is expended on such things, but that the buyers substitute for the cheapened article some other which is more costly, or a more elaborate quality of the same thing; and as the inferior quality answered the purpose of vanity equally well when it was really expensive, a tax on the article is really paid by nobody: it is a creation of public revenue by which nobody loses (869).[23]

22 Pigou (1960), 89–90, cites Mill, ('Posthumous Essay on Social Freedom', *Oxford and Cambridge Review*, Jan. 1907) to the effect that 'Men do not desire to be *rich*, but to be richer than other men. The avaricious or covetous man would find little or no satisfaction in the possession of any amount of wealth, if he were the poorest amongst all his neighbours or fellow-countrymen.' I wish to thank Professor J. Robson for confirming that this is not in fact a Mill item. (The true author was probably Ebenezer R. Edger; on Edger and the work in question see editorial notes, *CW*, XV, 792).
23 Mill cites John Rae (1834), 169–71, to similar effect. Rae in his detailed discussion 'Of Luxury' 265f., refers to Smith, Storch, Say, Mandeville and McCulloch. On Smith see *Wealth of Nations* (1937), 172, regarding 'true' utility and 'status-oriented' utility; for any given degree of the former, desirability rises with expense: 'To the rich the merit of an object which is in any degree either useful or beautiful, is greatly enhanced by its scarcity, or by the great labour which it requires to collect any considerable quantity of it, a labour which nobody can afford to pay but themselves. Such objects they

We have here a very clear case of an upward sloping demand curve.

The extensive implications of the view that utility turns on relative not absolute income has long been appreciated.[24] Stigler (1949, 3) adds the point that Nassau Senior – because of concern with the 'possession of *superior* wealth' – 'reached and stated the law of diminishing marginal utility only to dismiss it'.

Mill's statement in the essay on method that '[w]e know not of any *laws* of the *consumption* of wealth as the subject of a distinct science: they can be no other than the laws of human enjoyment' (*CW*, IV, 318n) has attracted much attention as indicating an unconcern with demand theory.[25] The 'laws of human enjoyment', he maintained, are not the concern of economists whose interest in consumption is constrained to three areas all of which relate to production and distribution: the first involves 'capital' maintenance and growth ('productive' and 'unproductive' consumption), the second, the possibility of excess accumulation (an obvious allusion to the Malthus–Sismondi perspective and the opposing law of markets), and the third, taxation incidence and the impact of taxes on factor supply. If we take Mill's comments on scientific economics seriously, it will indeed follow that 'the laws of human enjoyment' as such fall outside the economists' domain, since (as we have seen) Mill does not consider the behaviour of the typical consumer to be amenable to generalization except in the negative sense that he fails to act in maximizing fashion.

However, matters are not so simple. The fact is that Mill does refer to 'general principles of demand' as in a discussion of the precious metals which are said (at least in their non-monetary role) to be no exception to those principles: 'So far as they are wanted for purposes of luxury or the arts, the demand increases with the cheapness, in the same irregular way as the

are willing to purchase at a higher price than things much more beautiful and useful, but more common . . . With the greater part of rich people, the chief enjoyment of riches consists in the parade of riches, which in their eyes is never so compleat as when they appear to possess those decisive marks of opulence which nobody can possess but themselves' (on conspicuous consumption, see also 224–5, 329, 390, 414, 794–6, 861; also 'Of the Imitative Arts', 1980, 182f, on the dating of this work and the specific argument see the editors' introduction, 172f).

24 Cf. letter of Marshall to Pigou, March 1903, restating a theme by Pigou: 'Though we may pass from the utility curve of an individual to the demand curve of a nation (or other group) as regards bread or milk or any other commodity which is valued only for its direct benefit to us, yet we cannot do that for commodities which we value partly because they impart social distinction. For a large change in the supply all round of such a commodity alters the conditions which we have assumed to be practically constant when making out the curve of an individual' (Pigou, ed., 1925, 433).

25 Mill was here opposing the 'usual' categorization of topics in texts on political economy into sections on production, distribution and consumption. He probably had in mind J. B. Say's *Treatise* in particular.

demand for any other commodity. So far as they are required for money, the demand increases with the cheapness in a perfectly regular way, the quantity needed being always in inverse proportion to the value' (III, 619). Similarly, 'every tax on a commodity tends to raise its price, and consequently to lessen the demand for it in the market in which it is sold' (850).[26] Now considering Mill's position on consumer behaviour one might be justified in supposing that the negative slope of the demand curve is attributed solely to wholesale purchasers. Certainly that seems to be the case when demand relations arise in the context of international trade (below, p. 322f). Yet we also find in Mill's *Principles* rationalization of this property that seems to apply to final consumers. Mill apparently applies the methodological distinction between 'scientific' economics – based on the standard maximization axiom – and 'applied' economics which allows for qualifications, the subject matter of 'other sciences'; it is with the former that he is now concerned. Equally, however, we must bear in mind that though the majority of consumers 'give much higher prices than necessary for the things they consume' a price *increase* yet has some restrictive impact on consumption that requires analysis, the imperfections merely precluding a single price for all consumers.

The argument for a negative slope runs in terms of the 'income effect' generated by price variation. Depending upon the kind of commodity involved more or less of any increase in purchasing power generated by a fall in price will be devoted to the commodity in question rather than to others. From this perspective 'absolute necessaries' were demand inelastic: 'In the case of food, as those who had already enough do not require more on account of its cheapness, but rather expend in other things what they save in food, the increased consumption, occasioned by cheapness, carries off, as experience shows, only a small part of the extra supply caused by an abundant harvest' (467). (Similarly, in the polar case of 'peculiar luxuries, the taste for which is confined to a small class'.) Conversely, in the event of price increases, 'it is other things rather than food that are diminished in quantity by them, since, those who pay more for food not having so much to expend otherwise, the production of other things contracts itself to the limits of a smaller demand' (475).[27]

26 Cf. 854: '[T]he imposition of a tax on a commodity almost always diminishes the demand more or less; and it can never, or scarcely ever, increase the demand'. What Mill must have meant by the possible exception we have indicated above (pp. 267–8). See also: 'if the thing is cheap, there is usually a demand for more of it than when it is dear' (466); 'the demand . . . varies with the value, being generally greater when a thing is cheap than when it is dear' (497).
27 Cf. *CW*, V, 448: 'In the case of an article of necessity like food, it might easily happen that as much might be demanded and as much consequently produced after the rise of price as before. The inconvenience to the consumers would then consist in the privation of something else, a greater part than before of their means of expenditure being required for food.'

All this is important, although already well dealt with by Ricardo (Hollander, 1979, 276–7). But Mill's later correspondence shows how easy it would have been for him to have incorporated into the *Principles* a wider range of considerations relating to demand theory had he wished to devote more attention to the rationalization of the price–quantity relationship at the level of the individual.[28] I have in mind Mill's debate in 1864–5 with Cairnes regarding consumption theory, arising from Cairnes' critique of the section in the *Principles* on the effects produced of duties on international exchange (*CW*, III, 850f.; see below, p. 333f). It is a line concentrating on the logic of choice.

In an initial communication proposing changes to Mill's fifth edition Cairnes asserted that changes in the relative prices of commodities will leave unchanged the pattern of consumption by an individual, thus implying, in our terms, the absence of a substitution effect:

A man has £1000 a year, and with this sum he obtains annually necessaries comforts and luxuries in certain proportions. His power of commanding these things is curtailed to a certain extent by taxation; but the amount thus deducted from his income being given, I contend that the character of his expenditure will not be affected by the mode in which the deduction is made. If £50 a year be taken in the form of remitted taxes from the price of necessaries, and placed in the form of new taxes on the price of luxuries; or if both necessaries and luxuries are relieved at the expense of a direct deduction from his income – *so long as the total amount taken from him is the same*, I cannot see (apart from objections to particular taxes on *other* grounds) why this shd affect the proportions in which he consumes commodities. His means of commanding commodities remains in all cases the same, and if his tastes remain the same, why shd the mode of taxation affect the quality of his demand? (29 Nov. 1864, III, 1053).[29]

28 Even so, as just indicated, his explanation of the law of demand in the *Principles* was not as 'rudimentary' as is sometimes believed (cf. Smith, 1951, 248). Mill focusses generally upon quantity demanded as a function of price; by contrast Mountifort Longfield's (1834) deservedly famous exposition of the law of demand, involves an approach in terms of (marginal) demand price (114–15). Longfield's account relates market to individual demand, and provides an excellent explication of market clearing, but the relation between marginal demand price and the individual's subjective feelings is left implicit in the argument. Nassau Senior (1836), concentrated heavily on the subjective dimension of demand (e.g. 1938, 15), but was rather weak on the relation of market demand to individual utility. The best formulated contemporary statement of the principle of marginal utility – and one explicitly shown to be consistent with a labour theory – is by Lloyd (1834) (on these issues, cf. Smith, 1951; Bowley, 1972; Moss, 1974; 1976; Hollander, 1977b; Block, 1983).

29 A qualification is allowed to cover the immediate response to a change in price: 'It is conceivable indeed that for a time, the expenditure of people on particular commodities having been regulated with reference to a certain scale of prices, any sudden change

Mill admitted in reply that in his analysis of duties he had 'omitted one of the elements of the question, viz. the competing demands of other commodities on the purse of the consumer', but he stood by his conclusions on the grounds that he had recognized the negative slope of the demand curve: 'Suppose that I have a given sum, say £10 a year, the expenditure of which I am determined, whatever happens, to divide between two commodities, A and B. I conceive that even then, if A rises in price and B falls, the effect in the average of cases will be that I shall buy more of B and less of A' (12 Dec. 1864, III, 1072; also XV, 976).

Cairnes reiterated his insistence upon the absence of substitutability in consumption, explicitly specifying in his next formulation the assumption of constant income in the sense of unchanged purchasing power: 'Substitute for A & B, beer & tobacco. Suppose a man has £10 to spend on these luxuries, & that after the transference of the tax from one commodity to the other, his money will enable him to consume them in the same quantities & in the same proportion as before, is it conceivable that he will continue permanently to regulate the proportion of his smoking and drinking not by his tastes – his means being by hypothesis sufficient – but by the relative prices? I conceive that he might do so for a time under the influence of association; but this influence would be constantly diminishing, while his tastes & means would remain constant forces' (25 Dec. 1864, III, 1087). It was in reply to this position that Mill formulated a truly beautiful statement of the substitution effect in consumption resulting from changes in relative price:

Of the two or three points which we differ about, I will only touch upon one – the influence of price on demand. You say, if a tax is taken off beer and laid on tobacco in such a manner that the consumer can still, at the same total cost as before, purchase his usual quantity of both, his tastes being supposed unaltered, he will do so. Does not this assume that his taste for each is a fixed quantity? or at all events that his comparative desire for the two is not affected by their comparative prices. But I apprehend the case to be otherwise. Very often the consumer cannot afford to have as much as he would like of either: and if so, the ratio in which he will share his demand between the two may depend very much on their price. If beer grows cheaper and tobacco dearer, he will be able to increase his beer more, by a smaller sacrifice of his tobacco, than he could have done at the previous prices: and in such circumstances it is surely probable that some will do so. His apportionment of self-denial between his two tastes is likely to be modified, when

in relative prices might induce them to alter the character of their expenditure; but I imagine they wd very soon ascertain what their most urgent wants were, and find also the means of distributing their expenditure in such a way as most effectually to satisfy them.'

the obstacle that confined them is in the one case brought nearer, in the other thrown farther off (5 Jan. 1865, III, 1089; also XVI, 986 and O'Brien, 1948, 281).

Cairnes in reply restated his position that while there might occur, on first impact, some such a reaction as Mill proposed, with time the normal consumption pattern would reassert itself; we have here a clearly expressed description of the basket chosen, as it were technologically by 'physical conditions' rather than by 'mental impressions' as Mill would have it:

Touching the taxation question, after weighing carefully what you say I am still inclined to think that the position is *substantially* sound that 'a man's comparative desire for two commodities is not affected by their comparative prices'. The animal propensity towards beer and tobacco in certain proportions to each other depends on physical conditions: I can conceive that these may be overborne in some degree by the force of mental impressions; but then I think the mental impressions depending for their force on the principle of association are liable to become weak, while the force of the former is a constant quantity (9 Jan. 1865, III, 1089–90).[30]

In the Preface to the sixth edition of the *Principles* (II, xciv) Mill refers to the introduction of new matter and minor improvements due to suggestions and criticisms by Cairnes, but he did not mention the effects on quantity demanded of relative price changes either in the Preface, or in the text upon which Cairnes had commented. Yet nothing in the correspondence clashes with what is in the *Principles* ; the new materials would merely have provided a further rationalization for the negative slope of the demand curve.

V   SHORT-RUN PRICE FORMATION

If we heed Mill's 'warning' (above, pp. 265–6) we shall have to presume that the central analysis of price determination in the *Principles* applies to the wholesale trade. This analysis runs in terms of the equation of quantity demanded and supplied, and is applied in the first instance to commodities the supply of which is 'not susceptible of being multiplied at pleasure' (*CW*, III, 468). The exposition concentrates largely (though not entirely) upon the role played by the negative slope of the 'demand curve' in the process of correction of excess demand and supply:

30 Interestingly enough, Cairnes added that the whole matter might be settled experimentally: 'At all events we have, I think, brought the question to a point at which it can only be decided by experiment, which, next to agreement, is the most satisfactory issue of an economic argument.'

Meaning by the word demand, the quantity demanded, and remembering that this is not a fixed quantity, but in general varies according to the value, let us suppose that the demand at some particular time exceeds the supply [by one-third], that is, there are persons ready to buy, at the market value, a greater quantity than is offered for sale. Competition takes place on the side of the buyers, and the value rises . . . At what point . . . will the rise be arrested? At the point whatever it be, which equalizes the demand and the supply: at the price which cuts off the extra third from the demand, or brings forward additional sellers sufficient to supply it. When, in either of these ways, or by a combination of both, the demand becomes equal and no more than equal to the supply, the rise of value will stop.

The converse case is equally simple. Instead of a demand beyond the supply, let us suppose a supply exceeding the demand. The competition will now be on the side of the sellers: the extra quantity can only find a market by calling forth an additional demand equal to itself. This is accomplished by way of cheapness; the value falls, and brings the article within the reach of more numerous customers, or induces those who were already consumers to make increased purchases (466–7).[31]

Allowance is also made for the withdrawal of supply into stocks upon price reductions as part of the adjustment mechanism: in some cases of excess supply 'the fall [of price] is practically arrested only when the farmers withdraw their corn, and hold it back in hopes of a higher price; or by the operations of speculators who buy corn when it is cheap, and store it up to be brought out when more urgently wanted. Whether the demand and supply are equalized by an increased demand, the result of cheapness, or by withdrawing a part of the supply, equalized they are in either case.'

Mill's analysis provided a resolution of the apparent 'paradox, of two things each depending upon the other' – that while 'demand . . . partly depends on the value' at the same time 'value depends on the demand' (466). The solution, Mill conceded, 'must have been frequently given', but he added, 'I cannot call to mind any one who had given it before myself, except the eminently clear thinker and skilful expositor, J. B. Say.' The solution, of course, turns upon the distinction between 'demand' in the sense of quantity demanded and demand in the sense of the entire schedule, the

31 The stability of equilibrium also arises conspicuously in the context of the interest rate: 'The rate of interest will be such as to equalize the demand for loans with the supply of them. It will be such, that exactly as much as some people are desirous to borrow at that rate, others shall be willing to lend. If there be more offered than demanded, interest will fall; if more is demanded than offered, it will rise; and in both cases, to the point at which the equation of supply and demand is re-established' (III, 647). The term 'stable equilibrium' is used in the context of pricing (476; cf. also the international monetary context, 630).

equilibrium price being the solution to the *equation* of quantity demanded and quantity supplied:

Thus we see that the idea of a *ratio*, as between demand and supply, is out of place, and has no concern in the matter: the proper mathematical analogy is that of an *equation*. Demand and supply, the quantity demanded and the quantity supplied will be made equal. If unequal at any moment, competition equalizes them, and the manner in which this is done is by an adjustment of the value. If the demand increases, the value rises; if the demand diminishes, the value falls: again, if the supply falls off, the value rises; and falls if the supply is increased. The rise or the fall continues until the demand and supply are again equal to one another: and the value which a commodity will bring in any market, is no other than the value which, in that market, gives a demand just sufficient to carry off the existing or expected supply (467–8).[32]

It will be recalled that Ricardo too had objected to the notion of the 'proportion' of demand to supply in price determination (1951, I, 382; see Hollander, 1979, 282–3).

The negatively sloped demand curve and its rationalization, the appreciation of the equilibrating function of price in the short run, were part and parcel of Ricardian analysis. In some respects Mill's formulation clearly constituted an improvement in rigour – particularly the formal conception of an equation of demand and supply and the distinction between displacements of the demand schedules and movements from one position to another on the same schedule. But their merit lies less in substantive content than in their location at a conspicuous juncture amongst the basic theoretical principles. It may, however, be that Mill was unaware of the sophistication of Ricardo's appreciation of price determination, for Ricardo made many of his statements regarding the theory of allocation in various informal contexts relating to applied problems. But there is nothing to suggest that he believed the principles elucidated to be incompatible with Ricardian doctrine. On the contrary, what we already know of Mill's general reaction to Ricardianism suggests that he quite rightly regarded his own analysis

32 Mill refers to various other notions of demand – as 'a desire for the commodity, or as the desire combined with the power of purchase' – as 'infantine' (*CW*, V, 635). The approach to demand and price formation along the lines of the *Principles* appears in marginal notes made by Mill on a copy of Senior's *Outline of the Science of Political Economy* (1836) (see Mill, 1945, 134).

In an editorial note attached to a letter from Mill to Cairnes (15 May 1872, *CW*, XVII, 1894), reference is made to Cairnes' objections to Mill's definition of 'demand' as 'quantity demanded' in *Principles*, Book III, Ch. 2. But a reading of these objections as expressed in the *Leading Principles* (1874, 36f) reveals dissatisfaction with the definition at the aggregative level, *not* at the level of the individual commodity.

as a clarification of sometimes obscure, or ambiguous, or incomplete formulations in the original statements of 1817. The same applies to contributions by others, as may be illustrated by the response made in 1869 to W. T. Thornton's criticisms of orthodox demand–supply theory.

* * *

W. T. Thornton (1869) prefaced his famous criticisms of orthodox labour economics by more general objections to standard price theory. (For a general account of Thornton's position see Breit, 1967, 517–20.) Mill's defence throws further light on his demand-supply analysis.

All of Thornton's strictures, Mill was able to demonstrate, turned on special cases, which did not (as Thornton believed) conflict with received – that is, Ricardian – doctrine: 'The doctrine he controverts, though true, is not the whole truth. It is not the entire law of the phenomenon; for he has shown, and has been the first to show, that there are cases which it does not reach. And he has, if not fully defined, at least indicated, the causes which govern the effect in those exceptional cases.' His 'fault' was shared by 'all those improvers of political economy by whom new and just views "have been promulgated as contradictions of the doctrines previously received as fundamental, instead of being, what they almost always are, developments of them" ' – citing his earlier review of De Quincey (1845) – 'the almost invariable error of those political economists, for example, who have set themselves in opposition to Ricardo'(1869, *CW*, V, 641–2; cf. 634). As Mill phrased the matter in correspondence with J. E. Cairnes: 'My object in the Fortnightly was to shew that the cases supposed by Thornton do not contradict and invalidate, as he thinks they do, the equation of supply and demand' (23 June 1869, XVII, 1616).

Thus Mill demonstrated in his review that one of Thornton's main illustrations, designed to show that different equilibrium prices may emerge under identical market conditions, amounts simply to a special case involving a demand curve containing a vertical section which coincides with the given supply: 'He has not proved that the law [of supply–demand] is not strictly conformed to in that case . . . [but] that the law is, in this particular case, consistent with two different prices, and is equally and completely fulfilled by either of them . . . If there is a part of the scale through which the price may vary without increasing or diminishing the demand, the whole of that portion of the scale may fulfil the condition of equality of supply and demand' (V, 637). Here the actual price 'becomes simply a question whether buyers or sellers hold out longest; and depends on their comparative patience, or on the degree of inconvenience they are respectively put to by delay' (642). We must recall here that the classical approach to competition does not assume atomistic (price-taking) buyers and sellers; individuals are allowed some price-setting role.[33] Mill found Thornton's suggestions regarding this role constructive, agreeing that 'as a general rule, the initiative of price does rest with the dealers, and the competition which modifies it is the competition of dealers'.[34] In the event that 'several prices are consistent with carrying off the whole supply', dealers were likely to set relatively high prices because there was 'no motive to compete amongst each other in cheapness . . . '. There was an initial presumption that the advantage would lie with the price-setting party. Buyers, however, were 'not compelled by each other's competition to pay that higher price; for (since, by supposition the case is one in which a fall of price does not call forth an additional demand) if the buyers hold out for a lower price and get it, their gain may be permanent. The price, in this case, becomes simply a question whether sellers or buyers hold out longest; and depends on their comparative patience, or on the degree of inconvenience they are respectively put to by delay.'

However, the main point stands. Such cases of indeterminacy were the exception, and ruled out by assuming large numbers, when a negatively sloped demand curve is the rule: 'But how many such cases really exist? Among a few chafferers on the beach of a small fishing port, such a case, though even there improbable, is not totally out of the question. But where buyers are counted by thousands, or hundreds, or even scores; in any considerable market – and, far more, in the general market of the world – it is the next thing to impossible that more of the commodity should not be asked for at every reduction of price' (637).

A second case devised by Thornton was shown by Mill to involve the special situation where 'the scale of the demand is broken by [an] extraordinary jump'. In this event – where supply intersects the flat section of a step in the demand schedule – 'the law fails in its application' not, once again, 'from any fault in the law, but because the conditions on which its applicability depends do not exist' – there being no price that fulfils the condition that quantity demanded and supplied are equated. Such cases, while 'just possible in a very small market', were 'practically impossible in the great market of the community' (638-9).

Thornton's objections extended further. Even if formally valid the theory of price determination in terms of the equalization of supply and demand would amount to very little; for only a small proportion of goods offered for sale would exchange at such a price, a dealer disposing of much of his

---

33 Cf. Stigler (1965), 235 regarding Adam Smith: competition entails 'rivalry in a race – a race to get limited supplies or a race to get rid of excess supplies. Competition is a process of responding to a new force and a method of reaching a new equilibrium' (see also Hollander 1973a, 126).

34 Thornton preferred to consider competition of customers as acting indirectly by showing dealers 'that a higher price than they previously supposed is attainable, and to induce them to consequently to relax their own competition so as to attain it'.

stock at a higher price 'before he will lower the price in order to get rid of the remainder' (cited by Mill, 639). This criticism, amounting to recognition of exchanges at 'non-equilibrium' prices, Mill also refuted as involving a misunderstanding of standard doctrine. The law (like all laws of economics) operates, he explained, 'not suddenly but, gradually' so that '[t]hough a dealer may keep up his price until buyers actually fall off, or until he is met by the competition of rival dealers, still if there is a larger supply in the market than can be sold on these terms, his price will go down until it reaches the point which will call forth buyers for his entire stock'.

This latter defence is further elaborated in response to Thornton's contention regarding received doctrine that it presumed goods to be 'offered for sale unreservedly, and that dealers are always content to let them go for what they will fetch' (cited, 639–40). To this Mill objected that economists were well aware that in conditions of excess supply part of the stock might be withheld – '[r]eserving a price is, to all intents and purposes, withholding a supply'. J. E. Cairnes understood Mill's statement regarding withholding to imply that at a particular price there may correspond not a single quantity but a range of quantities supplied, and objected that this qualification rendered the whole doctrine of market-price determination in terms of the equality of quantities demanded and supplied 'an identical proposition' (To Mill 23 May 1869, in O'Brien, 1943, 283; also Cairnes 1874, 101f). This is not accurate. While Mill may have allowed for a horizontal section of the market supply curve he did not here actually presume the intersections of the demand curve to occur at some point along the step:

When no more than forty shillings a head can be obtained for sheep, all sheep whose owners are determined not to sell them for less than fifty shillings are out of the market, and form no part at all of the supply which is now determining price. They may have been offered for sale, but they have been withdrawn. They are held back, waiting for some future time, which their owner hopes may be more advantageous to him; and they will be an element in determining the price when that time comes, or when, ceasing to expect it, or obliged by his necessities, he consents to sell his sheep for what he can get. In the meanwhile, the price has been determined without any reference to his withheld stock, and determined in such a manner that the demand at that price shall (if possible) be equal to the supply which the dealers are willing to part with at that price. The economists who say that market price is determined by demand and supply do not mean that it is determined by the whole supply which would be forthcoming at an unattainable price, any more than by the whole demand that would be called forth if the article could be had for an old song. They mean that, whatever the price turns out to be, it will be such that the demand at that price, and the supply at that price, will be equal to one another (*CW*, V, 640–1).[35]

One of the situations envisaged by Thornton, Mill continued, involved an extreme form of the 'withholding' of supplies – a case where there existed no equilibrium price at all since '[a]t the actual price the supply exceeds the demand; at a farthing less the whole supply would be withheld', a further instance of Thornton's predilection for exceptional cases involving 'either the demand or the supply advancing or receding by such violent skips, that there is no halting point at which it just equals the other element'.[36]

Since Cairnes misunderstood Mill it is scarcely surprising that he should have been troubled by the assertion that supplies withdrawn from the market do not affect the market price. He enumerated a wide range of considerations including estimates on the part of prospective buyers and sellers of current stocks, productive capacity and foreign-trade opportunities which he maintained were relevant to market-price determination: 'These and similar circumstances, it seems to me, are not less than the commodities actually sold, real conditions determining the fluctuations of prices from month to month and from day to day; for they form equally with them, the data – or, to put the point more distinctly – they are the constituent elements

35 This case may be represented diagramatically thus:

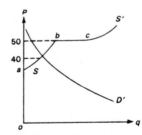

36 Such a case entails an intersection along the step and a vertical section to the *S*-curve coinciding with the *y*-axis. Doubtless Mill's reaction to Thornton would have been the same regarding the stretch *bc* in the preceding note.

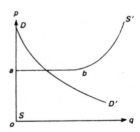

of the opinions of vendors and purchasers, which opinions, coming into contact in the higgling of the market, are the proximate causes of market price . . . But the doctrine of demand and supply, as expounded in the essay, not only does not embody them, but pointedly excludes them from consideration' (in O'Brien, 283–4). To this Mill replied that such considerations, involving expectations regarding the future, certainly influence the position of the current market curves, but that market price is determined solely by the latter:

The numerous considerations which you notice as influencing the minds of sellers, are, all of them, considerations of probable future demand and supply, modifying the effect which would take place if nothing but present facts were considered. Now it appears to me important to point out that these prospective considerations operate by inducing the sellers either to convert a possible present supply into an actual one, or to withdraw an actual present supply into the region of merely possible ones; and that in either case the relation of the price to the actual supply and demand is constant, i.e. the price is that which will make them equal (XVII, 1616–17).[37]

The same point had been made clearly enough in the review itself: 'If . . . the supply, being in excess of the demand, cannot be all disposed of at the existing price, either a part will be withdrawn to wait for a better market or a sale will be forced by offering it at such a reduction of price as will bring forward new buyers, or tempt the old ones to increase their purchases' (V, 636).[38]

Mill put up a sound defence as far as it went. But it is unfortunate that he did not delete the foolhardy statement in the *Principles* that '[h]appily, there is nothing in the laws of Value which remains for the present or any future writer to clear up' (III, 456). Even in his review of Thornton he only scratched the surface of what was later realized to be a notoriously complex matter, that involving small numbers. On the other hand, his statement is rather more justified if we understand him as referring specifically to general satisfaction with the theory of *competitive* price – as our discussion of his reaction to Thornton indeed suggests.

37 In brief, the precise position of sections such as *ab* (note 36) will depend upon expectations. On 'withholding' see also Mill's discussion in the course of his review of De Quincey's *Logic of Political Economy* (IV, 401).

38 Mill seems to be hiding his uncertainties from his readers. There is an intriguing allusion to some of his doubts in a letter of May 1869 where he writes to Cliffe Leslie 'that the doctrine respecting value & cost of production is true within wider limits of error – is true much more roughly & only in the gross, than is often supposed by political economists' (*CW*, XVII, 1600). But see the preface to the 1871 edition where Mill does publicly refer to the recent 'instructive discussion on the theory of Demand and Supply . . . not yet ripe for incorporation in a general treatise on Political Economy (II, xciv).

* * *

That notwithstanding the absence of any notion of incremental utility the so-called 'paradox of value' posed no problem (II, 30) is scarcely surprising considering the price mechanism outlined above. But Mill was unfair in his complaint that Smith by his formulation had introduced an irrelevant moralistic dimension: 'Political economy has nothing to do with the comparative estimation of different uses in the judgment of a philosopher or moralist. The use of a thing, in political economy, means its capacity to satisfy a desire to serve a purpose. Diamonds have this capacity in a high degree, and unless they had it, would bear no price' (III, 456–7; on Smith, cf. Hollander, 1973a, 137–8; also Levy 1982 for a recent discussion of the paradox).

One of the most interesting applications of demand–supply analysis is to the joint-production case in the celebrated chapter 'Of Some Peculiar Cases of Value' (Book III, xvi): 'Since cost of production here fails us, we must revert to a law of value anterior to cost of production, and more fundamental, the law of demand and supply. The law is, that the demand for a commodity varies with its value, and that the value adjusts itself so that the demand shall equal the supply' (583). In the case of joint products, competition reduces the total income generated or sum of prices to joint cost and assures the going profit rate, but the relative equilibrium prices (say of coke and gas) will be such as to assure sales in the same relative proportions as those generated by the joint process, i.e. equality between quantity supplied and quantity demanded of each product:

Equilibrium will be attained when the demand for each article fits so well with the demand for the other, that the quantity required of each is exactly as much as is generated in producing the quantity required of the other. If there is any surplus or deficiency on either side; if there is a demand for coke, and not a demand for all the gas produced along with it, or *vice versâ*; the values and prices of the two things will so adjust themselves that both shall find a market.

When, therefore, two or more commodities take a joint cost of production, their natural values relatively to each other are those which create a demand for each, in the ratios of the quantities in which they are sent forth by the productive process (584).

This specific theorem is said not to be itself important, 'but the illustration it affords of the law of demand, and of the mode in which, when cost of production fails to be applicable, the other principle steps in to supply the vacancy, is worthy of particular attention'. (For Smith's analysis of supply and demand in conditions of joint product, cf. 1937, 233.)

* * *

It has been suggested that Thomas De Quincey (1844) corrected Ricardian value theory by stressing the mutual determination of exchange value by 'intrinsic utility' and 'difficulty of attainment', and that this 'correction' – in the Hutcheson–Smith tradition of value theory – 'greatly influenced' J. S. Mill's treatment of value in the *Principles* (Groenewegen, 1974, 193). This is not a convincing evaluation. The 'mutual determination' of exchange value by demand and cost considerations was a central aspect of Ricardian doctrine although we shall show in the next section that Mill extended the account of adjustment to equilibrium. As for the utility dimension De Quincey may have believed he was 'correcting' Ricardo but it is unlikely that Mill was convinced. On various occasions he denied De Quincey's claim to originality: 'the larger half [of the *Logic of Political Economy*] is occupied with the theory of Value; which [De Quincey] rightly esteems the master key to the principal difficulties of the science', but the analysis of the relation between value-in-use and value-in-exchange lacked 'all the originality which [De Quincey] ascribes to them', the author having merely 'brought out into full theoretical explicitness what was known to all clear thinkers . . . ' (*CW*, IV, 395–6). He complained, similarly, that De Quincey had not given due credit to received doctrine relating to prices diverging from cost: 'Have not all political economists distinguished between articles which can be multiplied to an indefinite extent by labour, and articles naturally or artificially limited to a quantity short of the demand [quantity demanded at cost price]; and have they not all, from Ricardo downwards, affirmed that in the former, and more common case, the value conforms on an average to the cost of production, while in the latter there are no limits to the value except the necessities or desires of the purchaser?' (398). Mill, furthermore, took De Quincey to task in the *Principles* for failing to appreciate the distinction between 'demand' and 'quantity demanded' (III, 466).

VI COST PRICE AND PROFIT RATE EQUALIZATION

Mill's appreciation of the relativity of exchange value and the allocative underpinnings of the process of equilibration is nowhere expressed more clearly than in his discussion of costs of production. ' [V]alue is a relative term', he wrote in his chapter on the 'Ultimate Analysis of Cost of Production', 'not a name for an inherent and substantive quality of the thing itself . . .' (*CW*, III, 479). Accordingly, he defended the emphasis upon labour in Ricardo's treatment of value – despite the fact that the primary costs to be met by the capitalist-employer are wage costs – on the grounds that ' [i]n considering . . . the causes of *variations* in value, quantity of labour is the thing of chief importance; for when that varies, it is generally in one or a few commodities at a time, but the variations of wages (except passing fluctuations) are usually general, and have no considerable effect on value' (481). None the less, wage differentials will be reflected in the price structure (as well as relative labour inputs) and changes thereof will generate changes in the price structure: 'Although, however, *general* wages, whether high or low, do not affect values, yet if wages are higher in one employment than another, or if they rise and fall permanently in one employment without doing so in others, these inequalities do really operate on values' (480; cf. also 692). The same principles apply to profits: 'Value . . . being purely relative, cannot depend upon absolute profits, no more than upon absolute wages, but upon relative profits only . . . In so far as profits enter into the cost of production of all things [equally], they cannot affect the value of any. It is only by entering in a greater degree into the cost of production of some things than of others, that they can have any influence on value' (482).

The foregoing qualification referred not merely to the consequence for the price structure of profit-rate differentials reflecting unequal risk and so forth but, more significantly, covered compensation for differential time periods of production from industry to industry: 'one commodity may be called upon to yield profit during a longer period than the other'. In consequence of differential factor proportions it followed that 'commodities do not exchange in the ratio simply of the quantities of labour required to produce them' (484). Even general wage changes will then influence the structure of prices; while Mill gave more weight to labour quantity than to the factor returns in his analysis, he certainly made full allowance for the latter (477, 499). This is in line with the early rejection of the labour theory in the article on Malthus's criticism of the Ricardians (1825). Ricardo, Mill insisted, had fully allowed 'that commodities which have the same quantity of labour bestowed on their production, will differ in exchangeable value, if they cannot be brought to market at the same time', as was clear from his proposition that 'the difference in value arises . . . from the profits being accumulated as capital, and is only a just compensation for the time that the profits are withheld [1951, I, 37]', and from his allowances (*ibid.*, 30) for the potential impact upon exchangeable values of general wage-rate variations (IV, 31-2).[39] But in his formulation in the *Principles* Mill did not express himself with complete technical accuracy. He identified a rise in the general wage rate with an inverse change in the general return on capital and then applied the lower profit rate in the calculation of the new set of cost prices: 'even a general rise of wages . . . does in some degree influence values. It does not affect them in the manner vulgarly supposed,

---

39 Professor Schwartz (1972) seems to have missed this aspect of Mill's review, for he argues that only in 'On Profits and Interest', written about 1830, did Mill concede that 'into cost of production there enters another element besides labour' (26).

by raising them universally. But an increase in the cost of labour lowers profits; and therefore lowers in natural value the things into which profits enter in a greater proportion than the average, and raises those into which they enter in a less proportion than the average' (III, 485).[40] This was also the frequent practice of Ricardo and Marx (cf. Hollander, 1979, 303; 1981, 138-9). It is not strictly speaking legitimate, however, to presume the validity of the inverse profit-wage relationship before the calculation of the new price structure since the lower equilibrium return on capital emerges along with the new equilibrium prices (below, p. 359). Yet read as a statement involving comparative statics little is lost by the formulation.

These general principles had profound implications for taxation, most important of which is the distinction between a universal and a partial tax. For example:

A tax on profits, like a tax on rent, must, at least in its immediate operation, fall wholly on the payer. All profits being alike affected, no relief can be obtained by a change of employment. If a tax were laid on the profits of any one branch of productive employment, the tax would be virtually an increase of the cost of production, and the value and price of the article would rise accordingly; by which the tax would be thrown upon the consumers of the commodity, and would not affect profits. But a general and equal tax on all profits would not affect general prices, and would fall, at least in the first instance, on capitalists alone (*CW*, III, 826).

This result was a central theme of Mill's until the very end emerging unqualified in correspondence of 1872.[41] Evidently the generalization must be modified for differential factor proportions as Mill indeed insists in some contexts:

When the cost of production is increased, artificially by a tax, the effect is the same as when it is increased by natural causes. If only one or a few commodities are affected, their value and price rise, so as to compensate the producer or dealer for the peculiar burden; but if there were a tax on all commodities, exactly proportioned to their value, no such compensation would be obtained: there would neither be a general rise of values, which is an absurdity, nor of prices, which depend on causes entirely different. There would, however, as Mr. McCulloch has pointed out, be

40 Schwartz (1972) is in error when he writes that 'Mill was quite unaware of the effects of a general rise in wages when different capital-labour ratios in the several industries were allowed in the model' (276).

41 '. . . vous ne semblez pas peutêtre distinguer suffisamment entre un impôt général sur les profits de tout capital productif et un impôt qui frappe seulement ceux de quelques branches de production. D'après les principes généraux de la politique la possibilité de faire retomber l'impôt sur les consommateurs me paraît manquer dès que l'impôt, frappe toutes les industries sans distinction' (to C. Baer, 20 May 1872, *CW*, XVII, 1901).

a disturbance of values, some falling, other rising, owing to . . . the different durability of the capital employed in different occupations (838).

To summarize the essence of the matter: 'under the régime of competition, things are, on the average, exchanged for each other at such values, and sold at such prices, as afford equal expectation of advantage to all classes of producers; which can only be when things exchange for one another in the ratio of the cost of production' (582). The point at issue is of the first importance involving a conception of 'necessary' price as one satisfying the condition that the return to capital in each sector meets the alternative available elsewhere, a condition hinging upon the capital-mobility axiom: 'The value at any particular time is the result of supply and demand' but 'unless that value is sufficient to repay the Cost of Production, and to afford, besides, the ordinary expectation of profit, the commodity will not continue to be produced'; for 'necessary price' includes a return on capital 'as great . . . as can be hoped for in any other occupation at that time and place', and in the event of a return in excess of the going rate 'capital rushes to share in this extra gain, and by increasing the supply of the article, reduces its value'; conversely in the reverse case, output is restricted (471–2).[42]

Mill, although optimistic, was clearly conscious that the adjustment process was not instantaneous and partly for this reason emphasized expectation of return in stating the profit-rate equalization theorem: 'On an average (whatever may be the occasional fluctuations) the various employments of capital are on such a footing as to hold out, not equal profits, but equal expectations of profit, to persons of average abilities and advantages' (II, 406).[43] Allowance is in fact made for quite sluggish adjustments although extreme cases are said to be 'distinctly exceptional, and even in these [cases] the equalization is at last effected':

In the case of an altogether declining trade, in which it is necessary that the production should be, not occasionally varied, but greatly and permanently diminished, or perhaps stopped altogether, the process of extricating the capital is, no doubt, tardy and difficult, and almost always attended with considerable loss; much of the capital fixed in machinery, buildings, permanent works, &c. being either not applicable to any other purpose, or only applicable after expensive alterations;

42 In fact more than an axiom is involved in capital mobility: 'This is not a mere supposition or surmise, but a fact familiar to those conversant with commercial operations', an important instance of appeal to experience.

Mill frequently identified the natural price with the average of market prices (e.g. 470, 473, 490, 582) (on this faulty practice see Blaug, 1978, 205).

43 The term 'equal expectations of profit' was changed in 1865 from 'equal chances of profit' (on this see Knight, 1921, 363). The text remains unchanged, however, in the remainder of the paragraph; Mill seems to use the terms interchangeably.

and time being seldom given for effecting the change in the mode in which it would be affected with least loss, namely, by not replacing the fixed capital as it wears out. There is besides, in totally changing the destination of a capital, so great a sacrifice of established connexion, and of acquired skill and experience, that people are always very slow in resolving upon it, and hardly ever do so until long after a change of fortune has become hopeless. These, however, are distinctly exceptional cases, and even in these the equalization is at last effected (408).

The importance of the allowance lies in its implication that before the capital can be extricated from a declining trade (or increased in the case of expansion) the returns thereon are not in the nature of profit, but rather that of Marshallian 'quasi-rent' a notion already firmly established by Ricardo (Hollander, 1982).[44] Profit equalization occurs, so to speak at the margin of investment. A striking instance of this fact will be found in the account given of taxes on house rents. Such taxes, 'for some time after the tax was first imposed' fall on the owners, reducing net profits on their investment below the going rate so that 'houses would not be built'; '[b]y degrees, however, as the existing houses wore out, or as increase of population demanded a greater supply, rents would again rise; until it became profitable to recommence building, which would not be until the tax was wholly transferred to the occupier' (III, 834).[45]

* * *

In his *Leading Principles* J. E. Cairnes criticized Mill's inclusion of wages and profits within costs, although that position was 'generally accepted by economists'. Wages and profits, he argued, were not 'costs' in the legitimate sense of that term – namely, 'sacrifices incurred by man in productive industry' – but, on the contrary, they constituted 'the return made by nature to man upon that sacrifice'. 'Labour' and 'abstinence' were the true costs of production (1874, 48-9). Mill (who already had some idea of the nature of the criticism) observed to his friend that '[t]he two modes . . . of expressing cost of production, are, I imagine, both of them admissible, and both of them useful, as presenting different points of view'. The difference was a matter of 'the most convenient or most scientific mode of expressing the same doctrine' according to whether the context involved the economic system as a whole or the motives of the individual participants in activity: 'Of course, when we go down to the fundamentals of the matter, the cost to society, as a whole, of any production, consists in the labour and abstinence required for it. But, as concerns individuals and their mutual

---

44 For the extreme case of permanent land improvements see below (pp. 369-70).
45 Cf. 72, 83, 86n for an application to the fourth proposition on capital.

transactions, wages and profits are the measure of that labour and abstinence, and constitute the motives by which the exchange of commodities against one another is immediately determined' (15 May 1872, *CW*, XVII, 1894–5).[46]

Here we have a conscious distinction between aggregative and micro-economic theorizing. It is a distinction which, we shall see, constituted for Mill (as it had done for Ricardo) the foundation for the fundamental theorem on distribution. The matter is raised here for its confirmation of the fact that the theory of costs was treated by Mill from a micro-economic perspective involving relative value and the motives underlying allocation.

We may consider Mill's position on rent from this perspective. Rent was distinguished from the other factor returns on the basis of the conditions of land supply in the aggregate. Rent defined as 'the price paid for the use of an appropriated natural agent' is 'as indispensable (and even more so) as any implement: but the having to pay a price for it, is not. In the case of the implement (a thing produced by labour) a price of some sort is the necessary condition for its existence: but the land exists by nature. The payment for it, therefore, is not one of the expenses of production . . . ' (II, 58). It is purely a matter of scarcity in the case of an 'agent' subject to 'engrossment and appropriation', as Mill's very clear statement of the basic technical failing of the Physiocrats, shows: 'The rent of land being a price paid for a natural agency, and no such price being paid in manufactures, these writers imagined that since a price was paid, it was because there was a greater amount of service to be paid for: whereas a better consideration of the subject would have shown that the reason why the use of land bears a price is simply the limitation of its quantity, and that if air, heat, electricity, chemical agencies, and the other powers of nature employed by manufacturers, were sparingly supplied, and could like land, be engrossed and appropriated, a rent could be extracted for them also' (29; cf. Ricardo, 1951, I, 286–7).

Thus far the aggregative perspective. When Mill focussed upon individual sectors–and in this he went a step further than Ricardo or rather he reverted to Smith – the picture was a very different one:

The question . . . respecting the influence which the appropriation of natural agents produces on values, is often stated in this form: Does Rent enter into Cost of production? and the answer of the best political economists is in the negative. The temptation is strong to the adoption of these sweeping expressions, even by those who are aware of the restrictions with which they must be taken; for there is no

---

46 For Cairnes' reply see O'Brien (1943), 285. In his 1869 review of Thornton, however, Mill erroneously referred to 'labour' and 'abstinence' as costs, although the context involved micro-economics (V, 635).

denying that they stamp a general principle more firmly on the mind, than if they were hedged round in theory with all its practical limitations. But they also puzzle and mislead, and create an impression unfavourable to political economy, as if it disregarded the evidence of facts. No one can deny that rent sometimes enters into cost of production. If I buy or rent a piece of ground, and build a cloth manufactory on it, the ground-rent forms legitimately a part of my expenses of production . . . which must be repaid by the product. And since all factories are built on ground, the rent for it must, on the average be compensated in the values of all things made in factories (*CW*, III, 487).

The implication of multi-use land for cost pricing is laid out clearly in the 'summary of the theory of value', namely that 'when land capable of yielding rent in agriculture is applied to some other purpose, the rent which it would have yielded is an element in the cost of production of the commodity which it is employed to produce' (498). Similarly, '[t]he ground rent of a building, and the rent of a garden or park attached to it, will not be less than the rent which the same land would afford in agriculture' (494).[47]

VII DEMAND–SUPPLY AND COST PRICE:
THE ADJUSTMENT MECHANISM

We turn now to the precise relationship conceived by Mill between his cost price and demand–supply analyses. In dealing with this issue it is helpful to keep in mind the opinion, expressed by Malthus in 1824 (and repeated ever since) that Ricardo had limited demand–supply analysis solely to the market period and cases of monopoly, treating long-run cost price quite independently – that the two theories were mutually exclusive (1963, 188). Mill rejected Malthus's attribution as soon as it appeared, insisting that in the opinion of the Ricardo school long-run cost prices were arrived at by way of supply variation (*CW*, IV, 33–4). It is worth noting that Robert Torrens in 1822, who was critical of the details of Ricardo's cost theory, accepted this perspective as received doctrine: 'Political Economists seem on all hands agreed, that the quantity in which commodities exchange for one another depends, in any given instance, upon the proportion of demand and supply. It is also on all hands agreed, that with respect to all commodities which industry can indefinitely increase, the cost of production is the circumstance which, by limiting the quantity of them brought to market,

---

47 For some complex cases of multi-use land see III, xvi: 'Of some peculiar cases of value' (584). Allowance is also made for the use of particular inputs 'which cannot be increased *ad libitum* in quantity, and which, therefore, if the demand goes beyond a certain amount, command a scarcity value'. In these cases, scarcity value indirectly 'enters into the cost of production, and consequently into the value of the finished article' (486).

regulates the proportion of supply to demand, and ultimately determines the exchangeable value' (1935 [1822], 9).[48]

It would be wrong, however, to deny the presence in Mill's work of formulations which leave a different impression. Thus a letter of 1845 to the American economist Henry Carey implies that a law other than that of demand–supply is at work in the establishment of long-run cost price:

[I]t will perhaps be preferable to state in few words what I conceive to be the principal difference between your scheme of Political Economy, as very clearly set forth in your book [*Principles of Political Economy*, 1837–40], & my own. Most of what appears to me erroneous in your opinions, flows by direct consequence from your refusing to admit (as of any influence in practice) any other source of exchangeable value than labour.

In the case of all useful articles which admit of indefinite increase by means of labour, the labour required for producing them is, no doubt, the regulator & measure of their value; & I look upon it as one of the chief merits of Ricardo to have established in its full generality this law, which is still far from being admitted by the common herd of political economists. It therefore gave me pleasure to see so fundamental a principle thoroughly recognized & enforced by an American political economist; but it seems to me that you overlook half the truth when you deny that a different law governs all cases of either natural or artificial monopoly . . . (*CW*, XIII, 659).

Similar examples can be drawn from the *Principles* itself, including one from an important summary statement of the general laws of value given at the outset of the chapter 'Of Some Peculiar Cases of Value': 'From this it appears that demand and supply govern the fluctuations of value and prices in all cases, and the permanent values and prices of all things of which the supply is determined by any agency than that of free competition: but that, under the régime of competition, things are, on the average, exchanged for each other at such values, and sold at such prices, as afford equal expectation of advantage to all classes of producers; which can only be when things exchange for one another in the ratio of their cost of production' (III, 582). At the close of his discussion of the 'joint-cost' case Mill similarly wrote of 'the mode in which, when cost of production fails to be applicable, the other principle steps in to supply the vacancy . . .' (584). As late as 1869, we find the observation that '[s]upply and demand determine the perturbations of price; but (when the article admits of unlimited increase) not the permanent, or average price' (XVII, 1596).

Such statements imply that permanent or long-run cost prices are determined independently of demand and supply. But these, we shall argue,

48 On Mill's own position in 1822 (see below, n114).

are misleading formulations. In fact in the key chapter 'Of Demand and Supply, in their Relation to Value', there will be found an apparently unambiguous reference to 'another law' than that of demand and supply – namely cost price – 'for that much larger class of things, which admit of indefinite multiplication', which is immediately followed by the caution that in dealing with production costs 'it is not less necessary to conceive distinctly and grasp firmly the theory of this exceptional case' – that involving 'commodities not susceptible of being multiplied at pleasure' – which 'will be found to be of great assistance in rendering the more common case intelligible (III, 468). Indeed, 'the principle of the exception stretches wider, and embraces more cases than might at first be supposed'. As is well clarified in a letter of 1869, a major objective was to deter analysts from 'stopping short at demand & supply as the final regulators of price, without going on to that which in the last resort, adjusts the demand and supply to one another, viz. Costs of production (including all cost necessary for bringing the article to the place of sale)' (XVII, 1596).

Much of the ambiguity surrounding the formulations of cost price in its relation to demand and supply dissipates on a close inspection of the texts. For it becomes apparent that it arises from an impressive attempt to get a grip on the notoriously complex issue of long-run price and output adjustments in unconventional terms from the perspective of the modern reader.[49]

We take as the standard case the adjustment to a new equilibrium following a technological change that reduces natural price. Mill's approach allows a price-setting role to individual (competitive) entrepreneurs who, aware of the likelihood of entry into the industry by firms in response to supernormal profit act to forestall them. It is, therefore, not increase in supply that works to reduce price to the lower cost level, but price that is lowered directly, at a rate depending upon the estimate of the immediate danger of entry made by existing entrepreneurs who calculate the risks of entry as viewed by prospective entrants, a calculation which turns partly upon demand elasticity.

The latent influence by which the value of things are made to conform in the long run to the cost of production, is the variation that would otherwise take place in the supply of the commodity. The supply would be increased if the things continued to sell above the ratio of its cost of production . . . The value of the thing would in a little time, if not immediately, fall . . . because if it did not, the supply would

49 Through the researches of Evelyn Forget I was first made aware of the full complexity of Mill's approach concerning the adjustment mechanism with reference to the relation between firm and industry, expectations by the firm regarding price and market shares, and the role of dealers.

. . . be increased, until the price fell . . . If, indeed, the supply *could* not be increased, no diminution in the cost of production would lower the value: but there is by no means any necessity that it *should*. The mere possibility often suffices; the dealers are aware of what would happen, and their mutual competition makes them anticipate the result by lowering the price . . . Nobody doubts . . . that the price and value of all these things [even things in inelastic demand] would be eventually lowered by any diminution of their cost of production; and lowered through the apprehension entertained of new competitors, and an increased supply; though the great hazard to which a new competitor would expose himself, in an article not susceptible of any considerable extension of its market, would enable the established dealers to maintain their original prices much longer than they could do in an article offering more encouragement to competition (III, 473–4).

It is expectation of increased supply that brings about the price reduction to cost, for which reason precisely '[i]f . . . the supply *could* not be increased, no diminution in the cost of production would lower the value'.[50] Next to be noted is the actual output expansion, in the usual case of non-zero elasticity of demand, by existing firms who will be faced by larger markets which they can serve at the going (economy wide) profit rate. In the limiting case of zero demand elasticity (characterizing basic foodstuffs) there will, of course, be no such opportunities at all; this possibility Mill emphasized in order to counter the view that there necessarily occurs an output expansion to assure that price falls to cost:

We must not . . . suppose it to be necessary that the supply should actually be . . . increased . . . [Many] persons suppose that . . . the value cannot fall through a diminution of the cost of production, unless the supply is permanently increased . . . But this is not the fact: there is no need that there should be any actual alteration of supply; and when there is, the alteration, if permanent, is not the cause, but the consequence of the alteration of value . . . Whether there will be a greater permanent supply of the commodity after its production has been cheapened, depends on . . . whether a greater quantity is wanted at a reduced value. Most commonly a greater quantity is wanted, but not necessarily.

The main proposition – the active role of entrepreneurs in reducing price to cost as a means of dissuading potential entrants and, in the usual case, of assuring larger markets for themselves at the ruling profit rate – relates

---

50 Mill does not seem to consider the possibility that potential entrants will contemplate displacing existing firms even if there should be some technical inability to expand global industry supply, and that those in the industry will protect themselves by cutting prices.

to the long run. In the short run the standard market-clearing process is at play. It is scarcely surprising that readers have been misled into believing that Mill's cost price is divorced entirely from demand and supply: 'It is . . . strictly correct to say, that the value of things which can be increased in quantity at pleasure, does not depend (except accidentally, and during the time necessary for production to adjust itself), upon demand and supply; on the contrary, demand and supply depend upon it. There is a demand for a certain quantity of the commodity at its natural or cost value, and to that the supply in the long run endeavours to conform' (475). As explained, it is not a matter of the supply curve shifting outwards with price sliding down the demand curve in our case of cost-reducing technological improvement; rather, output of each firm expands to meet the quantity demanded at cost price.

There arise a number of complications. Before these expansions have been accomplished there will be excess demand at cost price and one must suppose that consumers or perhaps retailers, for their part, will put upward pressure on the price. This kind of complexity is actually raised by Mill in his treatment of a tax: 'Again, reverse the case, and suppose the cost of production increased . . . The value would rise; and that probably immediately. Would the supply be diminished? Only if the increase of value diminished the demand. Whether this effect followed, would soon appear, and if it did, the value would recede somewhat, from excess of supply, until the production was reduced, and would then rise again' (474).[51] In this case, as before, price changes 'immediately' (although the counterpart to the logic of fear of entry is not clarified). At the higher price, in the first instance, supply exceeds demand in the event that demand contracts, generating some downward pressure on the price although it seems that the market does not fully clear until the adjustment is completed; only after output is actually reduced can the higher price level be permanently maintained.[52] The implication is that in our case of technological change

---

51 Cf. correspondence of 1869 where Mill similarly observed that price might rise following an increase in production costs 'even if there were no diminution of supply. Whether the supply would be finally diminished or not would depend on whether the rise of price caused a falling off in consumption' (XVII, 1596). See also review of Thornton (1869), V, 661–2.

52 Mill may have had this in mind by his opaque observation in the review of Thornton (1869) that 'the influence even of cost of production depends on supply; for the only thing which compels price, on the average, to conform to cost of production, is that if the price is either above or below that standard, it is brought back to it either by an increase or by a diminution of the supply; though, after this has been effected, the supply adjusts itself to the demand which exists for the commodity at the remunerating price' (V, 635).

In a criticism of Newman's *Lectures on Political Economy* (1851): Mill allows implicitly for a miscalculation by firms who reduce output unnecessarily: 'A doctrine respecting

prices will rise somewhat from the initial (lower cost) level although firms would be reluctant to accept improved offers for fear of attracting new entrants and would be engaged in attempting to expand capacity to meet that demand probable at the cost price even if current price should be held somewhat above it. But again, the market does not actually clear until the adjustment process is completed.[53] (That a degree of trial and error is involved is also clarified by the remark that 'the permanent tendency of supply is to conform itself to the demand which is found by experience to exist for the commodity when selling at its natural value'.)

Mill's firms are supposed to engage in a passive form of collusion to satisfy what amounts to an implicit market division arrangement. A characteristic of the analysis is its failure to indicate the long-run optimum size of the firms, each of which is apparently presumed to have the technical ability to expand at constant cost without limit. What constraint exists, is self-imposed. That more generally in his work Mill fails to specify the profit-maximizing output is further suggested by certain remarks made in the context of the wage–profit relation: 'A rise of wages . . . confined to particular employments . . . raises the value and price of the particular article, and falls on the consumer; the capitalist who produces the commodity being only injured in so far as the high price tends to narrow the market; and not even then, unless it does so in a greater ratio than that of the rise of price: for though, at higher wages, he employs, with a given capital, fewer work people, and obtains less of the commodity, yet if he can sell the whole of this diminished quantity at the higher price, his profits are as great as before' (930).[54]

In adopting the 'internal adjustment' mechanism Mill diverged from tradition; but key features of the standard approach are unaffected. Whatever

---

price continually recurs in the book . . . that price (the price of food, for instance) can only be raised by diminishing the supply. We apprehend it is quite possible that the supply may be as great at a high as at a low price. We grant, that if there were no *power* of diminishing the supply, the price would not rise; but it is not necessary that the power should be exercised; and even if it is exercised, the diminution of supply will not necessarily be more than temporary. As much will be produced at the increased price as can find a market at that price: there will be no permanent diminution of quantity, unless the heightened price has placed the article beyond the means or the inclination of some of the consumers' (V, 448).

53 Mill remarks that following a change in data such as a cost reduction, '[t]he real law of demand and supply, the equation between them, still holds good: if a value different from the natural value be necessary to make the demand equal to the supply, the market value will deviate from the natural value; but only for a time . . . ' (III, 475). This must refer to the situation before output increase is technically feasible or at least before full adjustment is achieved.

54 Cf. also Mill's treatment of a particular wage increase in Part II of his 1869 review of Thornton (below, pp. 416–17).

the precise mechanism of adjustment it remains true that in comparing long-run equilibria following the adoption of new technology a lower price will usually be accompanied by a higher output. Secondly, it would certainly be going too far to attribute to Mill sole reliance on internal adjustment. That mechanism applies where the disturbance is amenable to treatment in partial equilibrium terms, where the economy-wide profit rate can be taken by firms in a particular industry as given. It would not apply to a disturbance such as a general wage change. It is also difficult to see how it would operate effectively alone in variable-cost industries where there is no unambiguous level of costs to which firms can attempt to adjust price. Even in the standard case of adjustment to technological change in a constant-cost industry the smaller the success of existing firms in maintaining price near the new cost level the greater will be the attraction for newcomers.[55] The mechanism will be totally irrelevant in the case of newly established industries.

Mill apparently chose to emphasize internal adjustment in order to weigh heavily a neglected component and emphasize that actual supply variation as the initiator of price change was not always essential. It is not surprising, therefore, to find the standard classical representation of the process of equilibration as a 'tendency': 'where there is no monopoly, the profits of a trade are likely to range sometimes above and sometimes below the general level but tending always to return to it like the oscillations of the pendulum' (408). Typical also is the notion of the 'gravitation' to natural price, or to that value which 'will afford the ordinary profit' (570). And the detailed account of the role of credit in the profit-rate equalization process is compatible with both kinds of adjustment – expansion and contraction in response to profit-rate differentials, and expansion and contraction to satisfy excess demand at cost price:

This equalizing process, commonly described as the transfer of capital from one employment to another, is not necessarily the onerous, slow, and almost impracticable operation which it is very often represented to be. In the first place, it does not always imply the actual removal of capital already embarked in an employment. In a rapidly progressive state of capital, the adjustment often takes place by means of the new accumulations of each year, which direct themselves in preference towards the more thriving trades. Even when a real transfer of capital is necessary, it is by no means implied that any of those who are engaged in the unprofitable employment, relinquish business and break up their establishments. The numerous and multifarious channels of credit, through which, in commercial

---

55 Though even at the lower level outsiders might be inclined to enter since the going profit rate can be obtained on net investment in the industry, the signal will be less obvious than if price exceeds the cost level.

nations, unemployed capital diffuses itself over the field of employment, flowing over in greater abundance to the lower levels, are the means by which the equalization is accomplished. The process consists in a limitation by one class of dealers or producers, and an extension by the other, of that portion of their business which is carried on with borrowed capital. There is scarcely any dealer or producer on a considerable scale, who confines his business to what can be carried on by his own funds. When trade is good, he not only uses to the utmost his own capital, but employs, in addition, much of the credit which that capital obtains for him. When, either from over-supply or from some slackening in the demand for his commodity, he finds that it sells more slowly or obtains a lower price, he contracts his operations, and does not apply to bankers or other money dealers for a renewal of their advances to the same extent as before. A business which is increasing holds out, on the contrary, a prospect of profitable employment for a larger amount of this floating capital than previously, and those engaged in it become applicants to the money dealers for larger advances, which, from their improving circumstances, they have no difficulty in obtaining. A different distribution of floating capital between two employments has as much effect in restoring their profits to an equilibrium, as if the owners of an equal amount of capital were to abandon the one trade and carry their capital into the other. This easy, and as it were spontaneous, method of accommodating production to demand, is quite sufficient to correct any inequalities arising from the fluctuations of trade, or other causes of ordinary occurrence (II, 407).[56]

* * *

We can deepen our appreciation of Mill's special adjustment mechanism by following a contrast drawn between 'regular' commodities and money. Mining is described as an industry where adjustments of market to natural price occur by way of actual supply variation: 'Suppose that [the average value of gold] were selling above its natural values . . . A part of the mass of floating capital which is on the look out for investment, would take the direction of mining enterprise; the supply would thus be increased, and the value would fall' (III, 520). The producer is 'stimulated to increase his production' by the deviation of market from natural price which implies a higher than average return on capital. Why this should be the mechanism

56 It is, however, pertinent that precisely in the context of cost-price analysis Mill should refer back to this passage with an eye specifically to internal expansion: 'As already intimated, these variations in the quantity produced do not presuppose or require that any person should change his employment. Those whose business is thriving, increase their produce by availing themselves more largely of their credit, while those who are not making the ordinary profit restrict their operations, and (in manufacturing phrase) work shorttime. In this mode is surely and speedily effected the equalization, not of profits perhaps but of the expectations of profit in different occupations' (III, 472).

rather than that outlined above where production is expanded to satisfy the excess demand at cost price is not clarified, but possibly it relates to the peculiarly long adjustment period involved in consequence of the durability of the metal. This is particularly true of a contraction, but an increase 'must be very great before it can make itself much felt over such a mass of the precious metals as exists in the whole commercial world' (521).

It is true that an important parallel is drawn between money and regular commodities in so far as both are subject to one law: 'It is simply the law of demand and supply, which is acknowledged to be applicable to all commodities, and which, in the case of money as of most other things, is controlled, but not set aside by the law of cost of production, since cost of production would have no effect on value if it could have none on supply.' But this is a general parallel. The specific adjustment differs: [T]here really is, in one respect, a closer connexion between the value of money and its quantity, than between the values of other things and their quantity. The value of other things conforms to the changes in the cost of production, without requiring, as a condition, that there should be any actual alteration of the supply: the potential alteration is sufficient; and if there even be an actual alteration, it is but a temporary one, except in so far as the altered value may make a difference in the demand, and so require an increase or diminution of supply, as a consequence, not a cause, of the alteration in value' (521–2).

Part of Mill's intention in all this was to contrast the (necessarily) unitary elastic demand for money with the (possibly) zero elastic demand for ordinary commodities. In fact what was said of commodities other than money applies to the metals too in their capacity as plate or jewellery: 'If the permanent cost of production of gold were reduced one-fourth, it might happen that there would not be more of it bought for plate, gilding or jewellery, than before; and if so, though the value would fall, the quantity extracted from the mines for these purposes would be no greater than previously. Not so with the portion used as money; that portion could not fall in value one-fourth, unless actually increased one-fourth; for, at prices one-fourth higher, one-fourth more money would be required to make the accustomed purchases; and if this were not forthcoming, some of the commodities would be without purchasers, and prices would not be kept up' (522; cf. 509, 512). But it does seem clear that more was involved – the actual mechanism of price adjustment to a cost reduction in the case of regular commodities does not involve actual supply variation as the necessary 'cause' of price reduction.[57]

---

57 Unfortunately Mill immediately proceeds to a statement regarding the case of money difficult to reconcile with the general trend of his argument. After repeating that cost of production regulates quantity 'in the long run', he proceeds to assert that 'precisely

VIII   VARIABLE-COST CONDITIONS

Thus far we have followed Mill in assuming constant-cost industries. In the event of increasing costs, variations in conditions of demand lead to changes in long-run cost price as well as output. Agriculture, of course, provides the main example:

We have already made abundant reference to the fundamental truth, that in agriculture, the state of the art being given, doubling the labour does not double the produce; that if an increased quantity of produce is required, the additional supply is obtained at a greater cost than the first . . .
If the first hundred quarters were all raised at the same expense (only the best land being cultivated); and if that expense would be remunerated with the ordinary profit by a price of 20s the quarter; the natural price of wheat, so long as no more than that quantity was required, would be 20s . . . But if the population of the district advanced, a time would arrive when more than a hundred quarters would be necessary to feed it . . . Suppose that it will not answer to cultivate the second best land, or land of the second degree of remoteness, for a less return than 25s the quarter; and that this price is also necessary to remunerate the expensive operations by which an increased produce might be raised from land of the first quality. If so, the price will rise, through the increased demand, until it reaches 25s. That will now be the natural price; being the price without which the quantity, for which society has a demand at that price, will not be produced (III, 488–9; cf. 464–5).

The 'law of value', in this case, is that natural price 'determined by the cost of that portion of the supply' – 'even the smallest' – 'which is produced and brought to market at the greatest expense' (490). Elsewhere, the increasing cost case is put thus: 'the permanent value is determined by the greatest cost which it is necessary to incur in order to obtain the required supply' (582). In some formulations equal weight is placed on the so-called intensive margin:'[t]he real expenses of production are those incurred on the worst land, or by the capital employed in the least favourable circumstances' (II, 429).[58]

because the quantity cannot be prevented from affecting the value, the quantity itself will (by a sort of self-acting machinery) be kept at the amount consistent with that standard of prices – at the amount necessary for performing, at those prices, all the business required of it' (522). This implies a very rapid response indeed of metallic output to deviations of the 'market' from the 'natural' price.

58   Cf. Mill's early defence of Ricardian rent doctrine (1828) where he insists that 'what the theory requires is, that of the whole *capital* employed in agriculture, there should always be one portion which yields no rent; one portion which barely replaces itself,

In the case of increasing costs the adjustment mechanism outlined above might in principle be applied, firms setting lower prices to dissuade entrants otherwise attracted by supernormal profits in order themselves to satisfy the higher demand at cost price. But there is now the severe complexity that cost price is a variable even given technology. Understandably Mill did not himself formally apply his internal-adjustment mechanism to this case.

Increasing returns is a yet more complex matter. At the enterprise level, opportunities for specialization hinge on size; but even when specialization has reached a technical maximum, output expansion may lead to higher productivity as the labour force is used to capacity (131, citing Babbage, 1832). Various other overhead costs are also taken into account including that of direction; and the adoption of machine-intensive process usually turns on size (133).[59] It is recognized that an increase in industry output is compatible with a larger number of small firms, but a high industry output is said to encourage the establishment of large firms – concentration increases with industry size. In contemporary England with its growing market there is said to be 'a perpetual growth not only of large manufacturing establishments, but also, wherever a sufficient number of purchasers are assembled, of shops and warehouses for conducting retail business on a large scale' (141).

In this context we encounter a fundamentally important test of scale economies – that provided by the market:

Whether or not the advantages obtained by operating on a large scale preponderate in any particular case over the more watchful attention, and greater regard to minor gains and losses, usually found in small establishments, can be ascertained, in a state of free competition, by an unfailing test. Wherever there are large and small establishments in the same business, that one of the two which in existing circumstances carries on the production at greatest advantage will be able to undersell the other (133; cf. 140).

It is also appropriate to note Mill's important observation that taxation, by restricting demand, may prevent increasing returns from coming into effect: 'The higher price necessitated by the tax, almost always checks the

---

with the ordinary profits of stock. *This* principle is the real foundation of the theory of rent' (IV, 171). It was not essential that no-rent land exist as some critics believed. Moreover, the objection that 'best' land is characteristically not taken up first is rejected. For it neglected that fertility must be interpreted not in terms of absolute produce but produce relative to costs and that highly productive land may need much preparatory investment (173). In this paper Mill asserts that Ricardo knew of the rent principle before his essay on the profits of stock (1815) and informed his friends of it (180).

59 See further the discussion of 'machinery' (below, p. 376f).

demand for the commodity; and since there are many improvements in production which, to make them practicable, require a certain extent of demand, such improvements are obstructed, and many of them prevented altogether' (III, 840).[60]

Bowley has complained that by his organization and weight of emphasis Mill unjustifiably played down the role of demand in price formation. Thus he made 'no attempt to point out the difference in importance of demand in determining price in [the increasing-costs] case compared to the constant-cost case' with the result that the theory of value is made to 'look simple'. In this, she argues, he followed Ricardo rather than Senior (1836) and Bailey (1825), whose 'efforts to explain the varied relationships between demand, cost of production and value certainly had none of that "appearance of lucidity" which Marshall stigmatised as often "mischievous," nor did either of them advance "a plain and simple doctrine"'. In comparison, J. S. Mill's treatment seems deliberately to evade the complex issues, in what might be described as a resolution "to find it easy" ' (1973, 168-9).

There is some justification for this evaluation;[61] but we cannot accept it as a fair statement of Mill's overall approach. That Mill did not emphasize the operation on equal terms of demand and supply in the case of products produced under conditions of increasing cost is easily understandable: 'Their value is not, correctly speaking, a scarcity value, for it is determined by the circumstances of the production of the commodity, and not by the degree of dearness necessary for keeping down the demand to the level of a limited supply' (*CW*, III, 490). This is not to play down demand, but a matter of definition, Mill intending by 'scarcity value' a totally inelastic supply – and clarifying that where such conditions do prevail 'both the land and its produce would really rise to a monopoly or scarcity price' (491).[62] In

---

60   For the same proposition see also Senior (1938 [1836]), 120.
     In one account Mill asserts that 'a tax of one shilling per pound on tea, or of two shillings per bottle on wine, raises the price of each pound of tea and bottle of wine . . . by that and more than that amount' (III, 864). This need not imply increasing-cost industries; he had earlier alluded to a variety of indirect costs on firms subject to taxation (839–40).

61   Cf. Mill's reference to 'the operation of demand' in the general competitive case, as the occasional disturber of value' (III, 585), which neglects entirely the variable cost case.
     Mill is ambiguous as to the relative significance of the constant-cost category. Bowley (167) has cited his belief that it 'embrac[ed] the majority of all things bought and sold'. But Mill wrote to Carey in 1845 that 'cases partaking more or less of [the character of natural or artificial monopoly] are so far from being exceptions, that they constitute a full half of all the cases which arise; land & its produce being the more conspicuous, but very far from the sole, example' (XIII, 659).

62   The term monopoly is here used in classical fashion to represent the case of given aggregate output irrespective of numbers: 'A thing which is limited in quantity, even though its possessors do not act in concert, is still a monopolized article' (II, 417). In

brief, Mill realized that rent might be generated even in the absence of differentials in the event of an absolute constraint on land supply: 'It is also distinctly a portion of Ricardo's doctrine, that . . . the land of a country supposed to be of uniform fertility would, all of it, on a certain supposition, pay rent: namely, if the demand of the community required that it should all be cultivated, and cultivated beyond the point at which a further application of capital begins to be attended with a smaller proportionate return' (428).[63]

Mill, indeed, had himself protested at the sort of reading we are now considering. As early as 1828 in 'The Nature, Origin, and Progress of Rent' he had defended the differential-rent theory against the strictures of Senior (1821) and others 'who affect to suppose that Sir Edward West, Mr. Malthus, and Mr. Ricardo, considered the cultivation of inferior land as the *cause* of a high price of corn'. Rather the reverse: That 'the cultivation of inferior soils' is the *effect* of high price 'itself the effect of demand' was a doctrine 'explicitly laid down by the distinguished authors previously referred to, and particularly by Mr. Ricardo' (IV, 174). Similarly in the *Principles*: 'Mr. Ricardo does not say that it is the *cultivation* of inferior land' [that is the 'cause of rent on the superior'], 'but the *necessity of cultivating* it from the insufficiency of the superior land to feed a growing population (III, 428). This is a perfectly justified defence of the Ricardian position (cf. Hollander, 1979, 665). The analysis of a tax imposed on a decreasing-cost industry confirms vividly the emphasis on the role of demand in long-run price formation.

We summarize the analysis of this and the preceding sections on competitive pricing in terms of Marshall's judicious account: 'Some persons fail to see that [Mill's] "law of Cost of Production" is regarded by him as operative only as a result of, or corollary from, the law according to which the action of the producers of a commodity is governed by their calculations of the circumstances of the future supply and demand in the market.' In this Mill followed Ricardo although 'the true nature of this doctrine would have been more manifest had not Mill, after Ricardo, judged it important

---

fact, Mill even uses the term 'monopoly price' to describe the case of a competitive industry wherein demand increases more rapidly than supply (408).

    Knight (1921) 184, has complained of this usage for confusing 'control' with 'limitation' of supply. In fact, nothing but a peculiar verbal usage is involved with no substantive implications. Mill fully realized that prices of commodities and factor services emerge in consequence of scarcity, no matter what the market structure and cost conditions.

63 But this was regarded as an exceptional case: 'this state of things never can have really existed anywhere, unless possibly in some small island cut off from the rest of the world . . . ' (III, 491); also: 'Rent is not an element in the cost of production which yields it; except in the cases (rather conceivable than actually existing) in which it results from, and represents, a scarcity value' (498).

to use terms that should bring into prominence the properties which distinguished rather than the properties which united the various propositions of the theory of value' (1925 [1876], 127–8; for recent examples of the faulty perspective of which Marshall complained, see Bharadwaj, 1978, de Vivo, 1981).

IX IMPERFECT COMPETITION

Mill recognized that increasing returns may be incompatible with competition.[64] In the context of increasing returns he observed that 'where competitors are so few, they always end up agreeing not to compete. They may run a race of cheapness to ruin a new candidate, but as soon as he has established his footing they come to terms with him' (II, 142). Public utilities provide a case in point; it is to gas and water companies that Mill alludes in referring to a trade which 'from the nature of the case, [is] confined to so few hands, that profits may admit of being kept up by a combination among the dealers' (405). Mill, of course, justified government regulation of public utilities. But governments were also charged with unjustifiable encouragement of barriers-to-entry:

Governments . . . are oftener chargeable with having attempted, too successfully, to make things dear . . . The usual instrument for producing artificial dearness is monopoly. To confer a monopoly upon a producer or dealer, or upon a set of producers or dealers not too numerous to combine, is to give them the power of levying any amount of taxation on the public, for their individual benefit, which will not make the public forego the use of the commodity. When the sharers in the monopoly are so numerous and so widely scattered that they are prevented from combining, the evil is considerably less: but even then the competition is not so active among a limited as among an unlimited number. Those who feel assured of a fair average proportion in the general business, are seldom eager to get a larger share by foregoing a portion of their profits (III, 927–8).

The implication is clear that (*ceteris paribus*) price will be higher the smaller the number of firms in the industry.

A 'strict or absolute' monopoly – a single seller – was easily dealt with as the limiting case, Mill providing a nice statement of the total revenue function, which implies revenue rather than profit maximization as the objective: 'The monopolist can fix the value as high as he pleases, short of what the consumer either could not or would not pay; but he can do so only by limiting the supply. The Dutch East India Company obtained a monopoly price for the produce of the Spice Islands, but to do so they were obliged, in good seasons,

64 For this insight Bowley (1973), 172, gives priority to Senior and Cournot.

to destroy a portion of the crop. Had they persisted in selling all that they produced, they must have forced a market by reducing the price, so low, perhaps,
that they would have received for the larger quantity less total return than for
the smaller: at least they showed that such was their opinion by destroying the
surplus' (468). Cost of production determines the minimum. In all of this
Mill was following closely along Adam Smith's path (Hollander, 1973a, 125).

Here there emerges a fascinating insight into Mill's practice. Mill formally
restricted the scientific treatment of pricing to the competitive case (above,
p. 265). Monopoly, whether natural or artificial, is in fact designated as
a 'disturbing cause'. It is, however, one which had 'always been allowed
for by political economists' and, as we have just seen, Mill himself does
make use of the tools of economic analysis – specifically the total revenue
function. It is presumably in markets characterized by 'small numbers' that
problems arise which Mill did not perceive to be subject to 'assignable law'.

* * *

The practical application of analytical tools emerges also in what amounts
to 'monopolistic competition' in the modern sense of the term. This is so
notwithstanding the role accorded to 'custom'. I refer here to Mill's
discussion of markets subject to unrestrained freedom of entry, 'yet in which
the result is not determined by competition, but by custom or usage;
competition either not taking place at all, or producing its effect in quite
a different manner from that which is ordinarily assumed to be natural to
it', namely price competition (II, 243).

The dual situation characterized the retail trade: 'retail price, the price
paid by the actual consumer, seems to feel very slowly and imperfectly the
effect of competition; and when competition does exist, it often, instead
of lowering prices, merely divides the gains of the high price among a greater
number of dealers' (243). More specifically, 'custom' indicated a particular
mark-up over the wholesale price which was adhered to, though loosely,
and what competition there was avoided price cutting:

[E]ven in countries of most active competition, custom also has a considerable
share in determining the profits of trade. There is sometimes an idea afloat as to
what the profit of an employment should be, which though not adhered to by all
the dealers, nor perhaps rigidly by any, still exercises a certain influence over their
operations. There has been ['is', till 1862] in England a kind of notion, how widely
prevailing I know not, that fifty per cent is a proper and suitable rate of profit in
retail transactions; understand, not fifty per cent on the whole capital, but an
advance of fifty per cent on the wholesale prices; from which have to be defrayed
. . . all the expenses of the retail business. If this custom were universal, and strictly
adhered to, competition indeed would still operate, but the customer would not

derive any benefit from it, at least as to price [this qualification added 1862]; the way in which it would diminish the advantages of those engaged in the retail trade, would be by a greater sub-division of the business (409–10).

Even before the change from present to past tense in his account of the practice of standard mark-ups, Mill had begun to note the 'intensity of modern competition' which encouraged a policy of 'a great business at low prices, rather than high prices and few transactions' (410).[65] This trend, however, was limited to the major cities which offered ' a sufficient inducement to large capitalists to engage in retail operations' such that it was 'generally found a better speculation to attract a large business by underselling others, than merely to divide the field of employment with them' (243).

From 1852 on, Mill continued to complain that price competition had 'a limited dominion over retail prices' with consequential losses in efficiency (410). Mill was here touching on efficiency losses in retailing now identified with 'monopolistic competition': 'the share of the whole produce of land and labour which is absorbed in the remuneration of mere distributors, continues exhorbitant; and there is no function in the economy of society which supports a number of persons so disproportioned to the amount of work to be performed', a formulation that strongly suggests excess capacity.[66] There can certainly be no doubt of his deep concern with the inefficiencies due to excessive numbers for this is a pervasive theme of his critique of contemporary society: ' . . . the class of mere distributors, who are not producers but auxiliaries of production, and whose inordinate number, far more than the gains of capitalists, are the cause why so great a portion of the wealth produced does not reach the producers' (III, 791; cf. 983).[67]

There can also be no doubt that a major source of the problem (in addition to the degree of rigidity imposed by 'custom') was seen to lie in the ability to differentiate – by location if not by product – an ability already weakened as noted above in the 'great emporia of trade' (II, 410) and likely to be further undermined by the transport revolution which breaks down the dependency of consumers on local dealers (243). Viewing Mill's position as a whole and recalling what was said earlier regarding consumer behaviour (above, p. 266) it is clear that allowance is made both for the establishment of prices by 'custom' with entry accorded the primary role of assuring normal profits by acting on market shares – the monopolistic competition

---

65 By turning over their capital more rapidly, and adding to it by borrowed capital when needed, the dealers often obtain individually higher profits; though they necessarily lower the profits of those among their competitors, who do not adopt the same principle.

66 Retail costs are said to include shop rent, and pay of clerks, shopmen and agents of all kinds (409).

67 For Mill's extensive discussions of the issue, cf. the posthumous 'Chapters on Socialism' (1879), *CW*, V, 721, 730, 732, 735, 747; also *Utilitarianism* (1861), *CW*, X, 253–4.

model – and for local areas of monopoly which provide scope for price discrimination.

Apart from the 'static' efficiency losses in retailing there is also the failure of prices to reflect new technology, although where a major change occurs the effect is likely to filter through belatedly. Here again the precise properties of demand elasticity emerge – for prices are more easily maintained in the case of a 'rich' clientele:

retail prices do not follow with all the regularity which might be expected, the action of the causes which determine wholesale prices. The influence of those causes is ultimately felt in the retail markets, and is the real source of such variations in retail prices as are of a general and permanent character. But there is no regular or exact correspondence. Shoes of equally good quality are sold in different shops at prices which differ considerably; and the price of leather may fall without causing the richer class of buyers to pay less for shoes. Nevertheless, shoes do sometimes fall in price; and when they do the cause is always some such general circumstance as the cheapening of leather: and when leather is cheapened, even if no difference shows itself in shops frequented by rich people, the artizan and the labourer generally get their shoes cheaper (III, 460).

The 'monopolistic competition' model emerges also in the discussion of professional remuneration where 'competition operates by diminishing each competitor's chance of fees, not by lowering the fees themselves' (II, 243).[68] Banking too is said to fall into the category at least to some extent, competition acting to keep the return on capital on a par with opportunities elsewhere, in part by reducing market shares:

Whether competition operates by lowering the rate of interest, or by dividing the business among a larger number, it is difficult to decide. Probably it operates in both ways; but it is by no means impossible that it may operate in the latter way alone; just as an increase in the number of physicians does not lower the fees, though it diminishes an average competitor's chance of obtaining them . . . [A] fall in the rate of interest, sufficient to enable the money market to absorb the whole of the paper-loans, may not be sufficient to reduce the profits of the lender who lends what costs him nothing, to the ordinary rate of profit upon his capital. Here, therefore, competition will operate chiefly by dividing the business (IV, 306–8).

Again one is led to conclude that Bowley's strictures are rather too harsh (above, p. 298). For she alludes to Mill's emphasis on the fixed- and elastic-supply classification of monopoly and competition respectively as less

---

68 Mill here alludes to coercive practices whereby competitors are kept in line, as in bookselling.

sophisticated than the barriers-to-entry classification of Bailey and Senior (1973, 169). Yet (setting aside the particular classical terminological usage) the fact is that Mill's analysis both of price and non-price competition turns strategically on the absence of barriers, whereas that of 'strict' monopoly and of imperfect competition requires them; in fact Mill took a major step forward by recognizing the implications for market structure of the 'natural' barriers generated by scale economies. And all this apart from the adjustment of prices to cost in the standard (constant-cost) case where potential entry plays a key role and, from the analysis of the wage-rate structure where barriers are of the essence.

X THE WAGE AND PROFIT STRUCTURES

The recognition of heterogeneous labour led both Adam Smith and Ricardo to abandon the pure labour theory even prior to allowance for capital and land and have recourse to a labour cost theory.[69] Karl Marx writhed in circular reasoning to avoid the conclusion that reducing 'skilled' to 'unskilled' labour by use of an index of wage rates implies the rejection of a pure labour theory; in effect he too abandoned the theory even before introducing the problems flowing from differential 'organic composition of capital' (cf. 1965, 43-4). Mill faced no problems in this regard since he had no particular stake in a strict labour theory.

Our concern in this section is to outline Mill's criticisms of Adam Smith's analysis of the wage structure and bring out the novel perspective adopted. The policy implications of his analysis we take up elsewhere.

Adam Smith recognized certain institutional impediments which in the Britain of his day tended to hinder the transfer of labour from one occupation to another. Apart from these artificial constraints he was confident in the

---

69 Cf. Smith (1937), 47. That Smith continued to speak of 'labour' immediately after allowing for heterogeneity can be explained by the fact that he proceeds to introduce capital and land and for the exposition of the general theory of cost pricing did not require to enter into the complexity of the wage-rate structure; he could assume for his purposes that when wages change they change for all labour proportionately. This is true also of Ricardo whose predominant concern was always with the impact of changes in data with particular reference to the impact of across-the-board alterations in the wage. Ricardo in fact took James Mill to task for neglecting the implications for the labour theory of a wage change not common to all types of labour: 'If a watch and a common jack altered in relative value without any more or less labour being required for the production of either of them, could we say [as Mill did say] that the proposition "that quantity of labour determines exchangeable value" was universally true? What I call exceptions and modifications of the general rule you appear to me to say come under the general rule itself' (1951, IX, 127). Samuel Bailey's charge that Ricardo as well as James Mill refused to draw the implications of their recognition of heterogeneity is unjustified (1825, 209-10).

market process. He envisaged a mechanism which assured that earnings in different occupations would tend to keep in line, so that different monetary returns in long-run equilibrium merely reflected the varying degrees of attractiveness attached to each occupation. Thus, free competition would not assure an equality of monetary returns but would assure that monetary and non-monetary returns together were equalized between occupations. Smith distinguished five main characteristics which, as he put it, 'make up for a small pecuniary gain in some employments, and counterbalance a great one in others':

first, the agreeableness or disagreeableness of the employments themselves; secondly, the easiness and cheapness, or the difficulty and expense of learning them; thirdly, the constancy or inconstancy of employment in them; fourthly, the small or great trust which must be reposed in those who exercise them; and fifthly, the probability or improbability of success in them (1937, 100).

An implicit assumption throughout is that labour can transfer from one occupation to another to assure that in the long run relatively high money earnings in fact exactly counterbalance non-monetary disadvantages. Smith further insisted that the logic of the argument only applied when the different occupations were 'well known and long established', for equalization of net advantages cannot be expected if knowledge of available opportunities is limited, and when the occupations are 'the sole or principal employments of those who occupy them' since workers operating on a part-time basis are likely to accept particularly low monetary returns (114).

There are certain features of his analysis in the famous chapter on 'Differences of Wages in Different Employments' (Book II, xiv) that Mill shared with Smith. Heterogeneity of labour can refer to productivity differentials which reflect natural or learned qualities; or to attitudes to work. In Smith's central account the emphasis is upon the characteristics of the job and the attitude towards them by labour – the 'five circumstances' relate to 'the employments themselves'. There is nothing said of natural differences of 'talent or genius' (though elsewhere this is recognized (e.g. 1937, 14)); indeed there is no formal inclusion of skill and productivity at all though they are implied in the discussion of learning. These characteristics are true also for Mill who lays out the general principle thus: 'There is no difficulty in understanding the operative principle in all these cases. If, with complete freedom of competition, labour of different degrees of desirableness were paid alike, competitors would crowd into the more attractive employments, and desert the less eligible, thus lowering wages in the first, and raising them in the second, until there would be such a difference of reward as to balance in common estimation the differences of eligibility. Under the unobstructed influence of competition, wages tend to adjust themselves in such a manner, that the situation and prospects of the labourers in all employments shall be, in the general estimation as nearly as possible on a par' (*CW*, II, 381-2n).

Mill was particularly impressed by Smith's discussion of the implications for the wage structure of uncertainty and expectation. The fifth feature refers to the argument that in equilibrium the monetary return should allow for the differing degrees of chance of success in the occupation: 'In a perfectly fair lottery, those who draw the prizes ought to gain all that is lost by those who draw the blanks' (Smith, 1937, 106). In fact, Smith argues, people tend to overestimate their chance of outstanding success and crowd into professions which promise a small likelihood of very great profits, so that what may appear to be very large monetary payments are in fact lower than they would be were people's calculations more realistic. Mill, in his account, designates the latter as 'another principle of human nature, [whereby] if the reward comes in the shape of a few great prizes, it usually attracts competitors in such numbers, that the average remuneration may be reduced not only to zero, but even to a negative quantity' (*CW*, II, 383-4); thus for Mill, for whom the phenomenon illustrates 'the equalizing effect of free competition' (385), even a negative return might satisfy the conditions of competitive equilibrium.[70]

The low returns in the professions are said to reflect inducement to entry generated by the 'lottery' calculation in addition to various non-monetary attractions – honour and the like (384). That a negative return may result is further illustrated by the interpretation of the low earnings derived from 'scholarly and bookish occupations', which runs partly in terms of the competition from 'amateurs' working in a part-time capacity whose training costs are covered in the course of preparing for other activities but also by the quest for 'personal celebrity' which implies a miscalculation of the prospective prizes (391-2).[71] The net outcome is that 'although the highest pecuniary prizes of successful authorship are incomparably greater than at any former period, yet on any rational calculation of the chances, in the existing competition, scarcely any writer can hope to gain a living

---

70 Smith was concerned to prove that in equilibrium monetary differentials are zero per unit of 'disutility'. The overestimation of the chances of outstanding success creates a problem since it implies that risk is not a pain cost (Hollander, 1973a, 131-2). This is not a problem for Mill who avoided Smithian welfare economics.

71 Smith too recognized some of the implications of part-time work. But in his account of scholarly occupations, he referred rather to those following from the 'provisions for eleemosynary education' – scholarships, exhibitions, bursaries and the like (1937, 129-30).

The low return of some cottage workers is also explained by Mill partly by reference to their part-time activity. He recognizes that a sufficient demand for the product to allow the entry of full-time labour would generate an economic rent for the original workers (393).

by books, and to do so by magazines and reviews becomes daily more difficult'.

There are various other interesting applications of the basic Smithian analysis appropriately modified for changing conditions. For example, in contrast to Smith who had been impressed by the steadiness of manufacturing activity, Mill traces the implications for relative wages of manufacturing instability – the 'interruptions of business arising from overstocked markets, or from a suspension of demand, or from a commercial crisis' (382). Similarly, the low earnings of hand-loom weavers he explains partly in terms of the importance attached by domestic workers to freedom from those controls characterizing factory work (381). But these variations are minor compared to two alterations touching on matters of principle to which we now turn.

The summary of 'the operative principle' at work (above, p. 305) was withdrawn in the third edition apparently because Mill had grown increasingly dissatisfied with the Smithian analysis and feared his readers might understand him as emphasizing the broad empirical validity of the account. Quite apart from the absence of artificial impediments to mobility – and Smith too had recognized immobility due to 'policy' – the pure logic of the competitive structure, Mill now insisted, presumed the labour market, as a whole, to be in equilibrium; in the event of general unemployment the differentials will become totally distorted:

These inequalities of remuneration, which are supposed to compensate for the disagreeable circumstances of particular employments, would, under certain conditions, be natural consequences of perfectly free competition: and as between employments of about the same grade, and filled by nearly the same description of people, they are, no doubt, for the most part, realized in practice. But it is altogether a false view of the state of facts, to present this as the relation which generally exists between agreeable and disagreeable employments. The really exhausting and the really repulsive labours, instead of being better paid than others, are almost invariably paid the worst of all, because performed by those who have no choice. It would be otherwise in a favourable state of the general labour market. If the labourers in the aggregate, instead of exceeding, fell short of the amount of employment, work which was generally disliked would not be undertaken, except for more than ordinary wages. But when the supply of labour so far exceeds the demand that to find employment at all is an uncertainty, and to be offered it on any terms a favour, the case is totally the reverse. Desirable labourers, those whom every one is anxious to have, can still exercise a choice. The undesirable must take what they can get. The more revolting the occupation, the more certain it is to receive the minimum of remuneration, because it devolves on the most helpless and degraded, on those who from squalid poverty, or from want of skill and education are rejected from all other employments. Partly from this cause, and partly from the natural and artificial monopolies . . . the inequalities of wages are generally in an opposite direction to the equitable principle of compensation erroneously represented by Adam Smith as the general law of the remuneration of labour. The hardships and the earnings, instead of being directly proportional, as in any just arrangements of society they would be, are generally in an inverse ratio to one another (383).

This constitutes an aspect of the linkage of macro- and micro-economics of the very first importance.

The reference to 'natural and artificial monopolies' involves a second major breakaway. Here we note Mill's observation that the first, third and fifth of Smith's 'circumstances' involve 'cases in which inequality of remuneration is necessary to produce equality of attractiveness, and are examples of the equalizing effect of free competition' (385). As for 'trust' Mill believed Smith had erred by implying that the adoption of responsibility was burdensome; he urged rather that a monetary differential paid because of the high degree of trust involved must be seen as a monopoly not a compensatory payment: 'The superiority of reward is not here the consequence of competition, but of its absence; not a compensation for disadvantages inherent in the employment, but an extra advantage; a kind of monopoly price, the effect not of a legal, but of what has been termed a natural monopoly. If all labourers were trustworthy, it would not be necessary to give extra pay to working goldsmiths on account of the trust. The degree of integrity required being supposed to be uncommon, those who can make it appear that they possess it are able to take advantage of the peculiarity, and obtain higher pay in proportion to its rarity' (385).

The case in point, Mill proceeds, illustrated a wider principle which 'Smith, and most other political economists, have taken into far too little account, and from inattention to which, he has given a most imperfect exposition of the wide difference between the remuneration of common labour and that of skilled employments' (385–6). Smith's allowances for educational costs went only some way to account for these differentials in practice: the skilled worker must indeed 'have a prospect of at last earning enough to pay the wages of all this past labour [training], with compensation for the delay of payment, and an indemnity for the expenses of his education. His wages, consequently, must yield, over and above the ordinary amount, an annuity sufficient to repay these sums, with the common rate of profit, within the number of years he can expect to live and be in working condition' (386).[72] But this

---

72 Smith's allowance for an interest element within the wage played havoc with his desire to reduce differentials to disutility payments (Hollander, 1973a, 131). This allowance did not create problems for Mill, first, because interest for him was a return to the 'pain cost' of abstinence; but more importantly because anyway he had no objective of reducing labour units to quantities of disutility by use of a wage index.

consideration (even when supplemented by allowance for legal restrictions on mobility) did not suffice to explain ruling differentials, for the costs even of a minimal education and of maintenance during the training period 'exclude the greater body of the labouring people from the possibility of any such competition' as would reduce the 'monopoly' return of the skilled.

Here we touch on the celebrated concept of 'non-competing industrial groups' a term coined by Cairnes (1874, 96) reflecting social and financial obstacles to upward mobility. In his account Mill makes some allowance for contemporary improvements due to broader educational opportunities: 'Until lately, all employments which required even the humble education of reading and writing, could be recruited only from a select class, the majority have had no opportunity of acquiring those attainments. All such employments, accordingly, were immensely overpaid, as measured by the ordinary remuneration of labour. Since reading and writing have been brought within the reach of a multitude, the monopoly price of the lower grade of educated employments has greatly fallen, the competition for them having increased in an almost incredible degree. There is still, however, a much greater disparity than can be accounted for on the principle of competition' (CW, II, 386–7).[73] Allusion is made also to various 'customary' expenses attaching to some trades, such as the cost of maintaining clothes and appearance, which hinder entry (387).[74] The outcome (at least historically) is strikingly described:

So complete, indeed, has hitherto been the separation, so strongly marked the line of demarcation, between the different grades of labourers, as to be almost equivalent to an hereditary distinction of caste; each employment being chiefly recruited from the children of those already employed in it, or in employments of the same rank with it in social estimation, or from the children of persons who, if originally of a lower rank, have succeeded in raising themselves by their exertions. The liberal professions are mostly supplied by the sons of either the professional, or the idle classes: the more highly skilled manual employments are filled up from the sons of skilled artizans, or the class of tradesmen who rank with them: the lower classes of skilled employments are in a similar case; and unskilled labourers, with occasional exceptions, remain from father to son in their pristine condition. Consequently the wages of each class have hitherto been regulated by the increase of

---

73 Mill recognizes that some skills 'can only be acquired by long practice' (cf. Smith, 1937, 47).

74 Cf. also: 'Similar considerations apply in a still greater degree to employments which it is attempted to confine to persons of a certain social rank, such as what are called the liberal professions; into which a person of what is considered too low a class of society, is not easily admitted, and if admitted, does not easily succeed.'

its own population, rather than of the general population of the country (387–8).[75]

Two lines of development follow from the foregoing analysis, one relating to the population mechanism and the other to the theory of price. We shall consider them in turn in appropriate context.

Mill's chapter 'Of Wages' (Book II, Ch. xi) commences with the caution that the analysis would 'proceed in the first instance as if there were no other kind of labour than common unskilled labour, of the average degree of hardness and disagreeableness' (II, 337). The chapter on differentials which now concerns us reiterates the formal contrast between the wages of 'ordinary or average labour' determined by wage fund and population principles and the wage structure 'depending in some degrees on different laws' (380). Mill failed to fulfil a promise given here to relate formally and explicitly the two analyses, but it is necessary for us to do so.

Characteristically for Mill, and contrasting with modern analysis, the discussion of the wage-rate structure takes into account the population variable and is not limited to the allocation of a given work force. Population growth in any sector if relatively excessive will force down the wage below the due competitive level; conversely, a relatively restrained growth is required to maintain a 'monopoly' return: 'If the professions are overstocked, it is because the class of society from which they have always mainly been supplied, has greatly increased in number, and because most of that class have numerous families, and bring up some at least of their sons to professions. If the wages of artizans remain so much higher than those of common labourers, it is because artizans are a more prudent class, and do not marry so early or so inconsiderately' (388).[76]

Thus the maintenance of monopoly returns requires control of supply *internal* to the group even in the absence of upward mobility and increase of supply from outside. Conversely, prudence on the part of the skilled will not suffice to maintain their 'monopoly' returns in the event of a disintegration of barriers. Here arises a pervasive consideration exercising Mill throughout his career – the danger that, in the absence of prudential

75 Cf. Cairnes (1874), 66: 'What we find, in effect, is, not a whole population competing indiscriminately for all occupations, but a series of industrial layers, superposed on one another, within each of which the various candidates for employment possess a real power of selection, while those occupying the several strata, are, for all purposes of effective competition, practically isolated from each other.'
     Stigler describes Mill on non-competing groups as 'the first major advance beyond [Smith's] theory in its 'recognition of the barriers to mobility created by the costs of education' (1965, 7–8) (but see the earlier account by Longfield, 1834, 83–5).

76 Cf. 372 regarding the extension of the population issue beyond 'the labouring classes' narrowly defined.

restraint on the part of the unskilled, there would emerge a 'competitive' wage structure based upon the lowest category:

The changes, however, now so rapidly taking place in usages and ideas, are undermining all these distinctions; the habits or disabilities which chained people to their hereditary condition are fast wearing away, and every class is exposed to increased and increasing competition from at least the class immediately below it. The general relaxation of conventional barriers, and the increased facilities of education which already are, and will be in a much greater degree, brought within the reach of all, tend to produce, among many excellent effects, one which is the reverse; they tend to bring down the wages of skilled labour. The inequality of remuneration between the skilled and the unskilled is, without doubt, very much greater than is justifiable; but it is desirable that this should be corrected by raising the unskilled, not by lowering the skilled. If, however, the other changes taking place in society are not accompanied by a strengthening of the checks to population on the part of labourers generally, there will be a tendency to bring the lower grades of skilled labourers under the influence of a rate of increase regulated by a lower standard of living than their own, and thus to deteriorate their condition without raising that of the general mass; the stimulus given to the multiplication of the lowest class being sufficient to fill up without difficulty the additional space gained by them from those immediately above (388).

Mill's population and wage-structure analyses together imply that in the absence of non-competing groups, population growth rates across sectors cannot get out of line and that a stable wage structure emerges. In a state of full stationarity, and assuming full mobility, there will be groups of workers earning more than the 'base' subsistence wage, which excess will not stimulate expansion on their part. This dual condition of equalizing money-wage differentials yet zero population growth in each sector must, of course, be modified for a growing system, a competitive wage structure then emerging around a 'base' wage appropriate for a particular positive growth rate of population. It is, therefore, in principle possible to apply the analysis involving capital and population to the determination of the average wage throughout the system rather than to the wages of 'common unskilled labour'.[77]

When allowance is made for non-competing groups, wage differentials are maintained by independently operating population constraints, and it is no longer clear how meaningful the notion of an 'average' wage is. Yet

77 But now the distribution of activity between sectors must be taken into account since there is no longer a one-to-one relation between capital and demand for labour – the same problem precisely as in the case of differential factor proportions.

on at least one occasion – the treatment of 'customary' payments in excess of competitive rates – Mill apparently retained the concept. Thus the earnings of some professionals are said to be subject to a minimum demand price governed by 'the prevalence of an opinion that such persons are more trustworthy, if paid highly in proportion to the work they perform' (398); the same is said of the pay of domestics:

most persons who can afford it, pay to their domestic servants higher wages than would purchase in the market the labour of persons fully as competent to the work required. They do this, not merely from ostentation, but also from more reasonable motives; either because they desire that those they employ should serve them cheerfully, and be anxious to remain in their service; or because they do not like to drive a hard bargain with people whom they are in constant intercourse with; or because they dislike to have near their persons, and continually in their sight, people with the appearance and habits which are the usual accompaniments of a mean remuneration. Similar feelings operate in the minds of persons in business, with respect to their clerks, and other employees (398–9).

It is in generalizing this observation that Mill makes at least a formal application of the capital–population analysis as applied to 'average' wages rather than the wages of unskilled labour and this in a non-competitive world: 'Liberality, generosity, and the credit of the employer, are motives which, to whatever extent they operate, preclude taking the utmost advantage of competition: and doubtless such motives might, and even now do, operate on employers of labour in all the great departments of industry; and most desirable is it that they should. But they can never raise the average wages of labour beyond the ratio of population to capital. By giving more to each person employed, they limit the power of giving employment to numbers'. It is not clear, however, that anything more than a formality is involved by this 'application' once movement between sectors are disallowed, for the 'average' wage then is but a statistic without economic significance.

Mill's continued references to the wages of the unskilled as an adequate first approximation to the labour problem, or even to an 'average' wage, notwithstanding recognition of the non-competing group phenomenon, are troublesome. But we must not forget that Mill considered barriers to be in the course of disintegration thus widening the potential scope of the standard analysis rather than rendering it obsolete. It was not merely the increasing availability of education for the masses. Mill justified the application of wage-fund analysis by the empirical proposition that though '[i]n this country there are few kinds of labour of which the remuneration would not be lower than it is, if the employer took the full advantage of competition' yet competition 'must be regarded, in the present state of society as the principal regulator of wages, and custom or individual character only as a modifying circumstance, and that in a comparatively slight degree' (337).[78]

There are, on the other hand, problematic features of Mill's account. On one conspicuous occasion the population mechanism is said to apply uniquely to the lowest group: 'Even in a progressive stage of capital, in old countries, a conscientious or prudential restraint on population is indispensable, to prevent the increase of numbers from outstripping the increase of capital, and the condition of classes who are at the bottom of society from being deteriorated. When there is not, in the people, or in some very large proportion of them, a resolute resistance to this deterioration – a determination to preserve an established standard of comfort – the condition of the poorest class sinks, even in a progressive state, to the lowest point which they will consent to endure' (III, 753). This does not sit well with the general analysis of the wage structure which requires the extension of the population principle to all sectors of the work force. Mill may not have intended to generalize, having specifically in mind contemporary circumstances involving non-competing groups where the skilled groups were, unlike the unskilled, actually exercising prudential control. But even so, as Mill himself noted, once allowance is made for a breakdown of barriers at least 'the lower grades of skilled labourers' will come 'under the influence of a rate of increase regulated by a lower standard than their own'.

Similarly troublesome is the treatment of taxes on wages later in the *Principles*. For Mill there asserts that the burden falls on employers in the case of the unskilled but on 'skilled' or 'privileged' workers whose 'monopoly' returns can be reduced without consequence:

I have already remarked, that in the present low state of popular education, all the higher grades of mental or educated labour are at a monopoly price; exceeding the wages of common workmen in a degree far beyond that which is due to the expense, trouble, and loss of time required in qualifying for the employment. Any tax levied on these gains, which still leave them above (or not below) their just proportion, falls on those who pay it; they have no means of relieving themselves at the expense of any other class (828–9).

---

78 In the first edition Mill provides a justification for the weighting: 'A custom on the subject [of wages], even if established could not easily maintain itself unaltered in any other than a stationary state of society. An increase or falling off in the demand for labour, an increase or diminution of the labouring population, could hardly fail to engender a competition which would break down any custom respecting wages, by giving either to one side or the other a strong direct interest in infringing it.' It is not at all evident why this important proposition, reminiscent of Adam Smith (cf. Hollander, 1973a, 184–5) was deleted.

Taking account, however, of the prudential control of the skilled alluded to in the formal discussion of the wage structure, one must suppose that (assuming no change in habitual standards) any reduction in the net earnings of the skilled will discourage population (or the population growth rate) thus transferring the burden to employers. But Mill reasoned as if the supply of skilled labour is zero elastic (or in a growing system that the growth rate is exogenous with respect to earnings).

The principle of population proves important in a variety of other applications involving the wage structure. For example, while the earnings of part-time labour (per unit of effort) are relatively low there is a reverse impact upon the major source of income deriving from the possibilities open for part-time work: 'The habits of the people . . . everywhere require some particular scale of living and no more, as the condition without which they will not bring up a family. Whether the income which maintains them in this condition comes from one source or from two, makes no difference: if there is a second source of income, they require less from the first; and multiply (at least this has always hitherto been the case) to a point which leaves them no more from both employments, than they would probably have had from either if it had been their sole occupation' (II, 393). Similarly, opportunities for female labour will (*ceteris paribus*) imply a lower wage for male heads of family: 'For the same reason it is found that, *cæteris paribus*, those trades are generally the worst paid, in which the wife and children of the artizan aid in the work. The income which the habits of the class demand, and down to which they are almost sure to multiply, is made up, in those trades, by the earnings of the whole family, while in others the same income must be obtained by the labour of the man alone.' It is possible even that the net family wage will ultimately be reduced following entry into the work force of the wife or child should the consequence of the immediate improvement be a weakening of the 'prudential restraint' (394). And the principle contributes to explain the low wages of women engaged in the relatively few occupations open to them by 'law and usage', compared with those of men in occupations of 'equal skill' in the sense of efficiency and 'equal disagreeableness'. The low earnings indicate 'overstocking' in the presence of a low long-run supply price:

The wages, at least of single women must be equal to their support, but need not be more than equal to it; the minimum, in their case, is the pittance absolutely requisite for the sustenance of one human being. Now the lowest point to which the most superabundant competition can permanently depress the wages of a man, is always somewhat more than this. Where the wife of a labouring man does not by general custom contribute to his earnings, the man's wages must be at least sufficient to support himself, a wife, and a number of children adequate to keep up the population, since if it were less the population would not be kept up. And even if the wife earns something, their joint wages must be sufficient to support not only themselves, but (at least for some years) their children also. The *ne plus ultra* of low wages, therefore (except during some transitory crisis, or in some decaying employment), can hardly occur in any occupation which the person employed has to live by, except the occupations of women (395–6).

In all these cases Mill analyses the wage structure assuming a base 'subsistence' wage in the technical sense. He leaves to the reader the appropriate modification required for a growing economy.

* * *

We proceed to the implications of the analysis of non-competing groups for the theory of value. Mill is clear that the monopoly return to skill is 'limited by the price which purchasers are willing to give for the commodity they produce', whereas entry sufficing to reduce the wage to the competitive level would entail a price which reflects (*inter alia*) wage costs (387) – a nice illustration of 'cost' envisaged in terms of alternative opportunities (above, p. 284). The 'monopoly' issue emerges also in a discussion of restrictions on entry due to apprenticeship laws or unions, in which case 'wages of that employment may be kept much above their natural proportion to the wages of common labour . . . without any assignable limit, were it not that wages which exceed the usual rate require corresponding prices, and that there is a limit to the price at which even a restricted number of producers can dispose of all they produce' (396).

Now, as in the case of population, so too here regarding value Mill played down the implications of his allowance for non-competing groups rather than carry the argument through energetically. This cautious tone pervades the appropriate principles listed in the 'Summary of the Theory of Value (Book III, Ch. vi):

IV. The natural value of some things is a scarcity value; but most things naturally exchange for one another in the ratio of their cost of production, or at what may be termed their Cost Value.

V. The things which are naturally and permanently at a scarcity value are those of which the supply cannot be increased at all, or not sufficiently to satisfy the whole of the demand which would exist for them at their cost value.

VI. A monopoly value means a scarcity value. Monopoly cannot give a value to anything except through a limitation of the supply . . .

VIII. Cost of Production consists of several elements, some of which are constant and universal, others occasional. The universal elements of cost of production are, the wages of the labour, and the profits of the capital. The occasional elements are taxes, and any extra cost occasioned by a scarcity value of some of the requisites.

IX. Rent is not an element in the cost of production of the commodity which yields it; except in the cases (rather conceivable than actually existing) in which it results from, and represents, a scarcity value . . .

X. Omitting the occasional element; things which admit of indefinite increase, naturally and permanently exchange for each other according to the comparative amount of wages which must be paid for producing them, and the comparative amount of profits which must be obtained by the capitalists who pay those wages (III, 497–8).

Item IX refers to (one-use) land, demand for the product of which is sufficiently high to generate rent even at the 'margin'. Even if one accepts that such cases are 'rather conceivable than actually existing', non-competing groups entail the same principle of 'scarcity' or 'monopoly' value. Yet Mill chose to include the phenomenon among 'occasional elements' and focussed upon the propositions regarding cost turning upon labour and capital which have alternative uses and are mobile between them. As in the discussion of capital and population he conspicuously refused to allow his modification to strike at the roots of orthodoxy.[79] Again, it may well be that this was because his confidence in the breakdown of barriers promised to reinforce the relevance of the orthodox approach.

* * *

We turn now to the profit-rate structure. Within gross profits Mill included interest or 'the remuneration of abstinence', wages of superintendence or the return to the 'assiduity and skill' of management, and compensation for risk or insurance (II, 400–1).[80] The profit rate varies according to the nature of the trade – mainly differences in risk but also (and to a greater degree than differential wages) in 'the circumstances which render one employment more attractive, or more repulsive, than another' (403). Allowance must also be made for 'artificial' and for 'natural monopoly', including that reflecting small numbers under decreasing-cost conditions (405). Allowing for these variations, and setting aside monopoly, profit rates tend to an equality.

79 Cairnes was preoccupied by the implications for cost theory of barriers between groups; exchange between occupants of each group would have to be explained in terms of Mill's 'reciprocal demand' principle (1874, 65–7, 89–94).

80 When risks are 'commuted for a fixed payment' the insurance premium becomes a regular production cost. Mill refers to 'the difference between the interest and the gross profits' as remuneration for 'the exertions and risks of the undertaker', a word he takes from French economists who 'enjoy a great advantage in being able to speak currently of *les profits de l'entrepreneur*' (401n). See Schumpeter (1954), 556n., for objections to the inclusion of risk-bearing as an entrepreneurial function.

There are several qualifications to the generalization regarding profit-rate equalization (or rather tendency to an equality). First and foremost it must be understood as referring to employments not individuals; for (excluding pure interest) profit which (in equilibrium) varies little between employments will still vary greatly between individuals, depending on 'the knowledge, talents, economy, and energy of the capitalist himself, or of the agents whom he employs; on the accidents of personal connexion; and even on chance' (406). The full sense of equalization is then explained with particular emphasis upon 'prospect or estimated probability' (Knight, 1921, 363):

But though profits thus vary, the parity on the whole, of different modes of employing capital (in the absence of any natural or artificial monopoly) is, in a certain, and a very important sense, maintained. On an average (whatever may be the occasional fluctuations) the various employments of capital are on such a footing as to hold out, not equal profits, but equal expectations of profit, to persons of average abilities and advantages. By equal, I mean after making compensation for any inferiority in the agreeableness or safety of an employment. If the case were not so; if there were, evidently, and to common experience, more favourable chances of pecuniary success in one business than in others, more persons would engage their capital in the business, or would bring up their sons to it; which in fact always happens when a business, like that of an engineer at present, or like any newly established and prosperous manufacture, is seen to be a growing and thriving one. If, on the contrary, a business is not considered thriving; if the chances of profit in it are thought to be inferior to those in other employments; capital gradually leaves it, or at least new capital is not attracted to it; and by this change in the distribution of capital between the less profitable and the more profitable employments, a sort of balance is restored. The expectations of profit, therefore, in different employments, cannot long continue very different: they tend to a common average, though they are generally oscillating from one side to the other side of the medium (406–7).

Or again, 'although profits are very different to different inviduals, and to the same individual in different years, there cannot be much diversity at the same time and place in the average profits of different employments . . . ' (408).[81]

81 It is not at all apparent why Mill did not treat differential managerial ability on a par with differential land quality and focus upon the marginal character rather than the average. A second qualification relates to the possibility of a negative (equilibrium) rate of return in a particular venture. This arises (as in the Canadian timber trade) where 'the chance of great prizes operates with a greater degree of strength than arithmetic will warrant, in attracting competitors' (409). And thirdly, 'even in the countries of most active competition, custom has a considerable share in determining the profits of trade'

Let us look more closely at the 'wages' of management. These are said to be determined in a different manner from ordinary labour (cf. Knight, 1921, 24). In some instances the return necessarily derives from the price of the product, there being, it is implied, no labour market for the relevant skill (*CW*, II, 404). In the early essay 'On Profits and Interest' Mill was very explicit: 'The wages of superintendence . . . are not paid in advance out of capital, like the wages of all other labourers, but merge in the profit, and are not realized until the production is completed. This takes them entirely out of the ordinary law of wages. The wages of labourers are paid in advance, are regulated by the number of competitors compared with the amount of capital; the labourers can consume no more than what has been previously accumulated. But there is no such limit to the remuneration of a kind of labour which is not paid out of wealth previously accumulated, but out of that produce which it is itself employed in calling into existence' (IV, 301).

It is not clear that adequate formal allowance was ever made for hired management with an eye to competing employment opportunities. This characteristic reflects in part the empirical significance of the self-employed capitalist, Mill taking the view that 'the control of the operations of industry usually belongs to the person who supplies the whole or the greatest part of the funds by which they are carried on, and who, according to the ordinary arrangement, is either alone interested, or is the person most interested (at least directly), in the result' (II, 401). But even where the manager is hired 'prudence' requires that he be somehow controlled by the capitalist or that he be stimulated by a share in profits; for '[m]anagement . . . by hired servants, who have no interest in the result but that of preserving their salaries, is proverbially inefficient, unless they act under the inspecting eye, if not the controlling hand, of the person chiefly interested . . . ' (401–2). All this is important since it confirms that Mill perceived the so-called 'manager' (at least in some instances) not as earning a contractual wage but as the recipient of an income out of the 'residual' income from sales proceeds, according him entrepreneurial status (cf. Rainelli, 1983, 800, 802).

Throughout this discussion there also emerge suggestions of a significant 'monopoly' element in the return to management at least in large-scale enterprises: 'Some occupations require a considerable amount of scientific or technical education, and can only be carried on by persons who combine with that education a considerable capital. Such is the business of the

---

as in the British retail sector. This latter is an unhelpful example for the reference to a fifty per cent profit as a 'proper and suitable rate' refers not to the return on capital but to a mark-up over wholesale prices which includes a variety of costs. Competition acts here to reduce market size.

engineer, both in the original sense of the term, a machine-maker, and in its popular and derivative sense, an undertaker of public works. These are always the most profitable employments' (404). Aspects of this perspective are touched upon in evidence of 1850, where Mill maintained that 'the wages of the labour of superintendence are not regulated like other wages by demand and supply, but are in reality the subject of a sort of monopoly; because the management of capital is a thing which no person can command except the person who has capital of his own, and therefore he is able, if he has a large capital, to obtain, in addition to interest, often a very large profit, for one-tenth part of which he could, and very often does, engage the services of some competent person to transact the whole of the labour of management, which would otherwise devolve upon himself' (*CW*, V, 411).

This latter observation is not entirely clear, but Mill seems to have in mind monopoly power over sales which maintains revenues above competitive levels, part of which excess is formally counted as the wages of management, although retained by the capitalist and exceeding the wages of 'some competent person . . . '. However in the *Principles* Mill also allowed implicitly for a 'monopoly' return even to managers hired on contract rather than paid from residual surplus, who are treated as a 'non-competing' group: '[w]here the concern is large, and can afford a remuneration sufficient to attract a class of candidates superior to the common average, it is possible to select for the general management, and for all the skilled employments of a subordinate kind, persons of a degree of acquirement and cultivated intelligence which more than compensates for their inferior interest in the result' (II, 139). Such managers may do a better job than the capitalists themselves – including the task of undertaking ventures 'out of the ordinary routine' (above p. 230). In our present context dealing with profits Mill also spells out that the efficient exercise of managerial 'control', 'if the concern is large and complicated, requires great assiduity, and often, no ordinary skill [which] must be remunerated' (401) – again pointing to a 'monopoly' return to salaried management.

With this in mind we can approach the view that the British classical economists all but 'accomplished the impossible feat of overlooking the most colourful figure in the capitalist process' – the entrepreneur (Schumpeter, 554). Thus Smith's merchant or master 'accumulates "capital" – that is really his essential function – and with this "capital" he hires "industrious people", that is, workmen who do the rest. In doing so he exposes these means of production to risk of loss; but beyond this, all he does is to supervise his concern in order to make sure that the profits find his way to his pocket' (see the almost identical terminology in Hébert and Link, 1982, 44). There was a promising potential development with J. B. Say who – following along Cantillon's path – referred to the entrepreneur's function in combining productive factors; but even Say failed to realize that 'the task of combining factors becomes a distinctive one only when applied not to the current administration of a going concern but to the organization of a new one'. J. S. Mill, runs this argument, took a further step in going from ' "superintendence" to "control" and even to "direction", which he admitted, required often "no ordinary risk" ' but even he did not proceed to a function 'distinct from mere administration'; and so 'Say's suggestion came to very little' (556–7).

On our reading this perspective requires some qualification. In the case of large-scale enterprises Mill certainly attributed non-routine decision making even to salaried managers treating such ability as a resource sufficiently scarce in contemporary circumstances to attract a 'monopoly' return. If this is indeed so then the notion of a tendency to profit-rate equality would be under threat – just as the notion of an average wage is unhelpful once the concept of non-competing groups is recognized – unless Mill's remarks were deliberately constrained to cases of imperfect competition. But we have seen that he allowed for differential managerial 'knowledge, talent, economy and energy' under 'competitive' conditions, emphasizing a trend to equality of (expected) profits on the part of *average* personnel. Now this suggests non-routine decision making even in the most usual market structure, a fact confirmed by Mill's account of adjustment by competitive firms to a change in cost data which requires complex calculations regarding prospective entry, demand elasticity and market shares. There is also the appreciation of the calculations required before committing funds to long-lived projects (p. 385, n29). Schumpeter's heavy emphasis upon the organization of a totally new concern as providing the key to true entrepreneurial activity doubtless governed his rather narrow reading of Mill's position.

XI INTERNATIONAL VALUES

We shall keep in mind in this section the recently published opinion that in his essay on trade written in 1830, though not published until 1844, Mill 'made what was then the epoch-making discovery that the relative prices of internationally traded goods could not be determined in terms of costs (as Ricardo had supposed all prices to be determined) but that recourse must be had to "a principle antecedent to that of cost of production, and from which that law holds as a consequence – the principle of demand and supply" (Hicks, 1983, 61). The proposition that international values depend on reciprocal demands 'was the first breakthrough to post-classical economics'.

The brilliance and originality of Mill's elaborations regarding international values is not at all in question (Chipman, 1965a, 483f).[82] In question is the assertion regarding Ricardo that he had supposed all prices to be determined in terms of cost – for this is untrue of the general theory of value and *a fortiori* untrue of the theory of international values. Because this view of Ricardo is unacceptable it is impossible to conceive of Mill as effecting a breakthrough to post-classical economics. The general theory of scarcity value, in fact, has its roots deep in the *Wealth of Nations*, and was greatly refined by Ricardo in the context of foreign trade but also more generally, having in mind the dependence of cost price upon the theory of demand and supply. It is within this framework that Mill's further contributions to international trade must be viewed.

The decision to publish the essay on 'International Value' (as the first of the essays in *On Some Unsettled Questions*) turned partly on the appearance of Torrens's *The Budget* (1844). For the essay maintained 'opinions identical in principle with those promulgated by Colonel Torrens' – on matters of theory if not of application – which 'have been held by the writer for more than fifteen years: although he cannot claim to himself the original conception, but only the elaboration, of the fundamental doctrine of the Essay', namely, the comparative-cost doctrine (*CW*, IV, 231).

On Mill's estimate, in the essay itself, Ricardo deserved the credit for replacing 'the former vague and unscientific, if not positively false, conceptions with regard to the advantage of specialization and trade', the vent-for-surplus doctrine in particular (233). The Ricardian 'philosophical exposition' turned upon the potential for increased productivity of (fully-employed) labour and capital in the trading partners, wherever the terms of commodity transformation by way of production (governed by internal cost conditions) differ between countries, irrespective of the absolute coefficients of production – the gains emerging when each country specializes in the production of that commodity for which it possesses the greatest comparative advantage. Ricardo had further shown that the introduction of money (in a gold standard system) did not distort the trading pattern, 'the precious metals always tending to distribute themselves in such a manner throughout the commercial world, that every country shall import all that it would have imported, and export all that it would have exported, if exchanges had taken place . . . by barter' (234).

---

82 It is probably fair to state, as Hicks states, that from the essay of 1830 to the elaborations in the *Principles* to Marshall's *Pure Theory of International Trade* (1879) (and his *Money, Credit and Commerce*, 1923) there is a 'clear pedigree'.

   According to Viner (1937), 447, James Pennington, in *A Letter . . . on the Importation of Foreign Corn* (1840) 'seems to have been the first explicitly to point out in print that the comparative costs set maximum and minimum rates for the terms of trade, and that within these limits the operation of reciprocal demand could fix the terms of trade at any point'. Mill makes no reference to Pennington's work.

Mill's own account proceeds on the assumption of a two-country, two-commodity world. By merging capital and labour into labour alone ('for shortness') he precluded any elaboration along the later lines of Hecksher-Ohlin (235; cf. also *Principles*, *CW*, III, 588). Moreover, by presuming throughout constant-cost industries, he precluded the possibility of partial specialization. His major concern, however, was not to elaborate upon the comparative-cost theory itself, but rather to specify, within Ricardo's basic model, the terms of trade that emerge and thus the division of the total (world) 'gain' between the trading partners. We note Mill's characteristic respect for Ricardo in his representation of his own contribution as a mere spelling out of the principles already laid down:

This question [the distribution of the gains from trade] was not entered into by Mr. Ricardo, whose attention was engrossed by far more important questions, and who, having a science to create, had not time, or room, to occupy himself with much more than the leading principles. When he had done enough to enable any one who came after him, and who took the necessary pains, to do all the rest, he was satisfied. He very rarely followed out the principles of the science into the ramifications of their consequences. But we believe that to no one, who has thoroughly entered into the spirit of his discoveries, will even the minutiae of the science offer any difficulty but that which is constituted by the necessity of patience and circumspection in tracing principles to their results (IV, 235).

There is but one (gentle) note of criticism, an allusion to an 'unguarded' expression by Ricardo implying that the 'gain' – in the sense of the total saving in resources for a given output, or increase in output for given resources – is enjoyed by each of the parties rather than shared between them, an error first corrected by James Mill (1844 [1926], 122–4). Even so, James Mill had not proceeded to analyse the division of the gain.

The analysis runs in terms of the 'law of exchangeable value' applied to the determination of the terms of trade between German linen and English cloth (the standard illustration), where international factor mobility is precluded: 'The principle, that value is proportional to cost of production being consequently inapplicable, we must revert to a principle anterior to that of cost of production, and from which this last flows as a consequence – namely, the principle of demand and supply' (237).[83] Here Mill

83 In the *Principles* Mill asserts a labour theory of value for the closed economy; implicitly assuming uniform factor ratios: 'When two articles are produced in the immediate vicinity of one another, so that, without expatriating himself, or moving to a distance, a capitalist has the choice of producing one or the other, the quantities of the two articles which will exchange for each other will be, on the average, those which are produced by equal quantities of labour. But this cannot be applied to the case where the two articles are produced in two different countries; because men do not usually leave their country,

formulates, in a statement of general applicability, the negative slope to the demand curve and the process of adjustment whereby excess supplies (or demands) are corrected by price variation:

In order to apply this principle [of demand and supply], with any advantage, to the solution of the question which now occupies us, the principle itself, and the idea attached to the term demand, must be conceived with a precision, which the loose manner in which the words are used generally prevents.

It is well known that the quantity of any commodity which can be disposed of, varies with the price. The higher the price, the fewer will be the purchasers, and the smaller the quantity sold. The lower the price, the greater will in general be the number of purchasers, and the greater the quantity disposed of. This is true of almost all commodities whatever: though of some commodities, to diminish the consumption in any given degree would require a much greater rise of price than of others.

Whatever be the commodity – the supply in any market being given – there is some price at which the whole of the supply exactly will find purchasers, and no more. That, whatever it be, is the price at which, by the effect of competition, the commodity will be sold. If the price be higher, the whole of the supply will not be disposed of, and the sellers, by their competition, will bring down the price. If the price be lower, there will be found purchasers for a larger supply, and the competition of these purchasers will raise the price.

This, then, is what we mean, when we say that price, or exchangeable value, depends on demand and supply. We should express the principle more accurately, if we were to say, the price so regulates itself that the demand shall be exactly sufficient to carry off the supply (237–8).

It is particularly noteworthy that the negative slope to the demand curve is here represented as a well-known property. Whatever may be said of Mill's minimization of the novelty of his own specific application – universally regarded as a brilliant achievement – there is every reason to accept his representation of received doctrine regarding the theory of demand itself.[84]

A direct application is then made to the case at hand involving the German demand for English cloth at various 'linen' prices and the English demand

---

or even send their capital abroad, for the sake of those small differences of profit which are sufficient to determine their choice of a business, or of an investment, in their own country and neighbourhood' (III, 588).

84 It will be noted that in the foregoing passage the increase of quantity demanded with price reduction occurs by way of new entrants into the market. (This was to be Marshall's procedure also in some of his analyses of trade, e.g. 1923, 161.) In the *Principles* Mill took a more general approach allowing also for demand responsiveness by individual purchasers (see below, p. 328).

for German linen at the equivalent 'cloth' prices (the celebrated analysis of reciprocal demand), the illustrative data yielding a unique equilibrium linen: cloth rate of exchange (lying of course between the limits imposed by the autarchic cost ratios): 'It may be considered . . . as established', Mill concluded, 'that when two countries trade together in two commodities, the exchangeable value of these commodities relatively to each other will adjust itself to the inclinations and circumstances of the consumers on both sides, in such manner that the quantities required by each country, of the article which it imports from its neighbour, shall be exactly sufficient to pay for one another' (239–40).

In the course of his general account Mill declared that 'the inclinations and circumstances of consumers cannot be reduced to any rule . . . The circumstances on which the proportionate share [of the gain from trade] of each country more remotely depends, admit only of a very general indication' (240). By this he cannot have meant to deny the rule of a negative slope to the demand curve; he intended the impossibility of specifying a priori where precisely the equilibrium rate of exchange will lie between the limits considering the range of possible positions and shapes of the respective demand curves.[85] But Mill proceeds to establish the direction to follow in seeking a general formula by showing that in the limiting case of zero demand elasticity – for example of German demand for English cloth – the 'gain' falls entirely to Germany. In general, however, while 'there is no absurdity in the hypothesis, that of some given commodity a certain quantity is all that is wanted at any price, and that when that quantity is obtained, no fall in the exchangeable value would induce other consumers to come forward, or those who are already supplied to take more . . . there will not be this extreme inequality in the degree in which the demand in the two countries varies with variations in the price (240–1).

Allowance for transport costs is made in the essay only after the introduction of an international monetary medium (243f). But reducing the formulation to barter terms, the linen:cloth exchange rate will differ in equilibrium between England and Germany by the cost of carriage of both products. This differential will not generally be divided equally between the parties. The extent to which the linen price of cloth rises in Germany and falls in England compared with the initial equilibrium must be such as to assure equality of reciprocal demand net of transport costs. Again Mill emphasizes the wide range of possible elasticity values – the fact that

85 In the *Principles*, the argument is phrased to specify formally both the position ('the extent') and the elasticity ('the intensity') of the demand curves: 'If . . . it be asked what country draws to itself the greatest share of the advantage of any trade it carries on, the answer is, the country for whose productions there is in other countries the greatest demand, and a demand the most susceptible of increase from additional cheapness . . . It also gets its imports cheaper, the less the extent and intensity of its own demand for them' (III, 602, cf. 614, 616). But at some places (e.g. 609) elasticity alone is referred to.

'the increase of the demand for a commodity as its price falls, is not governed by any fixed law' (245). As Mill rephrased the matter in 1848: 'No absolute rule . . . can be laid down for the division of the cost, no more than for the division of the advantage: and it does not follow that in whatever ratio the one is divided, the other will be divided in the same. It is impossible to say, if the cost of carriage could be annihilated, whether the producing or the importing country would be most benefited. All would depend on the play of international demand' (III, 601).[86]

In the fifth edition of the *Principles* (1862), Mill paid tribute to Torrens for having 'at least a joint claim' with Ricardo to the doctrine regarding the nature and measure of the benefit from trade and 'an exclusive one to its earliest publication' – an allusion to *The Economists Refuted* (1808) and Torrens's claim for priority made in 1857 (III, 589). The vent-for-surplus alternative is rejected in the *Principles* as a 'mercantilist error' responsibility for which lay partly with Adam Smith, who had implicitly assumed specificity of national resources to particular products and a low (even zero) domestic demand elasticity for those products such that in the absence of trading opportunities resources must lie idle – an approach that clashes with the law of markets (592–3). Mill's extended account emphatically underscores the consumers' perspective: 'When commerce is spoken of as a source of national wealth, the imagination fixes itself upon the large fortunes acquired by merchants, rather than upon the saving of price to consumers . . . [whereas] [c]ommerce is virtually a mode of cheapening production; and in all such cases the consumer is the person ultimately benefited.'[87] But here we should note Edgeworth's criticism (and before him that of Cournot) that by identifying the 'gains' from trade with the improvement in the terms

---

86 Mill here notes that 'in consequence of cost of carriage there are many things, especially bulky articles, which every, or almost every country produces within itself' and are not traded: 'This is the case with numerous commodities of common consumption; including the coarser qualities of many articles of food and manufacture, of which the finer kinds are the subject of extensive international traffic.' He was thus not rigidly limited in his applications by the restrictive two-commodity axiom.

87 This at least is the 'direct economical advantage' of trade (593). See also Mill's charges against 'mercantilism' which constituted the 'political economy of the selling classes, as distinguished from the buyers or consumers' (686).

There are in addition various indirect consequences, including the impact of market size on technology and scale economies, and the stimulus to effort and savings generated by the availability of new and cheap products. In addition, there are various desirable intellectual and moral effects of an open world, which Mill held to be the most significant of all. 'Commerce', he declared, is rapidly rendering war obsolete' (594). Similarly, '[t]he feelings of rival tradesmen, prevailing among nations, overruled for centuries all sense of the general community of advantage which commercial countries derive from the prosperity of one another: and that commercial spirit which is now one of the strongest obstacles to wars, was during a certain period of European history their principal cause' (686).

of trade, Mill was implicitly 'confound[ing] "final" with integral utility; ignoring the principle of "consumer's rent" ' (cf. Edgeworth, 1925, II, 22).

The basic account given in the *Principles* draws *verbatim* upon the essay as far as that goes (596f.), although Mill deleted the formal preliminary statement of the principle of demand and supply involving the negative slope of the demand curve. Why he did so is not clear, but his decision obviously reflects editorial rather than doctrinal considerations. For the account is introduced by the appeal to the 'antecedent law' to that of cost of production, namely that of supply and demand (597; cf. 469–70); it proceeds to emphasize the position of the demand curve and its elasticity (above, n85); and (in the course of explaining the term 'reciprocal demand') clarifies, as in the essay, that the problem of international value presents but one instance of the general pricing problem:

The law which we have now illustrated, may be appropriately named, the Equation of International Demand. It may be concisely stated as follows. The produce of a country exchanges for the produce of other countries, at such values as are required in order that the whole of her exports may exactly pay for the whole of her imports. This law of International Values is but an extension of the more general law of Value, which we called the Equation of Supply and Demand [466–8]. We have seen that the value of a commodity always so adjusts itself as to bring the demand to the exact level of the supply. But all trade, either between nations or individuals, is an interchange of commodities, in which the things that they respectively have to sell, constitute also their means of purchase: the supply brought by the one constitutes his demand for what is brought by the other. So that supply and demand are but another expression for reciprocal demand: and to say that value will adjust itself so as to equalize demand with supply, is in fact to say that it will adjust itself so as to equalize the demand on one side with the demand on the other (604).

In the central chapter on 'International Values' Mill attempts the extension of the analysis to several countries and several traded commodities, observing that '[t]hose who are accustomed to any kind of scientific investigation will probably see, without formal proof, that the introduction of these circumstances cannot alter the theory of the subject' since no essential principles are affected (600). But the bulk of the chapter proceeds to ring various changes on the basic two-country, two-commodity model, the most important of which involves the impact of technical change.[88]

In the illustration of this problem (605–6) the real cost of German linen

88 A further issue treated in Book III, Ch. xxv, involves the conditions under which third countries can displace German linen in the English market or English cloth in the German market (686–8; cf. the criticisms by Edgeworth, 1925, II, 23–4). The chapter also deals with the problem of undercutting by 'low wage' economies, which we take up elsewhere (pp. 356–7).

is assumed to fall by one third; Mill inquires whether and to what extent England will 'benefit' from the mere fact that the 'theoretical limits' – the limiting terms of trade – have been altered (assuming that the original equilibrium exchange rate lies between the limits). It is taken for granted that the initial impact of the change is to increase the cloth:linen rate of exchange by 50% from 10:17 to 10:25½: 'It is evident that in the outset, the improvement will lower the value of linen in Germany, in relation to all other commodities in the German market, including, among the rest, even the imported commodity, cloth.'[89] Given this assumption it is shown that only if English demand for linen in terms of cloth is of unitary elasticity will this rate represent the new equilibrium, England obtaining linen one third cheaper (as in Germany); in the case of an elastic demand the equilibrium terms of trade will fall somewhat against cloth 'so that England will not have the full benefit of the improvement in the production of linen, while Germany, in addition to that benefit, will also pay less for cloth'. In the inelastic case, 'linen will be cheapened in England in a still greater degree than in Germany; while Germany will obtain cloth on more unfavourable terms; and at a higher exchange value than before'. This result is subject to a 'modification':

In the case supposed the consumers of Germany have had a part of their incomes set at liberty by the increased cheapness of linen, which they may indeed expend in increasing their consumption of that article, but which they may likewise expend in other articles, and among others, in cloth or other imported commodities. This would be an additional element in the international demand, and would modify more or less the terms of interchange (607; cf. 613).

The general result that an increase in productivity in the export sector will usually be attended with a non-proportional variation in the terms of international trade is certainly of much importance (cf. Edgeworth, 1925, II, 21). The 'modification' regarding the impact of the innovation on German demand conditions is revealing for its allusion to the income effect of a price reduction. Mill elaborated upon this matter in a rather full discussion of price elasticity extending in its implications far beyond the specific context in question:

Of the three possible varieties in the influence of cheapness on demand, which is the more probable – that the demand would be increased more than the cheapness,

89 Mill is assuming two commodities in the sense of two traded commodities, since the argument implies the existence of various 'home' commodities. Relative (labour) costs retain their role as far as concerns the price structure of these commodities and the ratios between their prices and that of the exported product. This is also true of Ricardo's analysis (cf. Hollander, 1979, 469–70).

as much as the cheapness, or less than the cheapness? This depends on the nature of the particular commodity, and on the tastes of purchasers. When the commodity is one in general request, and the fall of its price brings it within reach of a much larger class of incomes than before, the demand is often increased in a greater ratio than the fall of price, and a larger sum of money is on the whole expended in the article. Such was the case with coffee, when its price was lowered by successive reductions of taxation; and such would probably be the case with sugar, wine, and a large class of commodities, which, though not necessaries, are largely consumed, and in which many consumers indulge when the articles are cheap and economize when they are dear. But it more frequently happens that when a commodity falls in price, less money is spent on it than before: a greater quantity is consumed, but not so great a value. The consumer who saves money by the cheapness of the article, will be likely to expend part of the saving in increasing his consumption of other things: and unless the low price attracts a large class of new purchasers who were either not consumers of the article at all, or only in small quantity and occasionally, a less aggregate sum will be expended on it. Speaking generally, therefore, the third of our cases is likely to be the most probable . . . (607).

These notions regarding elasticity all have Ricardian pedigree. That the elasticity of the market-demand curve will be higher the higher the opportunities to attract additional purchasers by further price reductions was recognized by Ricardo, who had actually gone further by noting that the degree of attraction varies with the range of prices in question. Secondly, there is Mill's very clear understanding that price elasticity (for the individual) turns on the implicit income elasticity of demand – a fall in price generating an effective increase in purchasing power which may be expended on products other than the one whose price had altered; this too had been spelled out by Ricardo who rationalized the zero demand elasticity for corn in these terms (on Ricardo, see Hollander, 1979, 275f).

A further extension by Mill, appearing in 1852, is to multiple equilibria, or the possibility that various exchange rates all satisfy the equilibrium condition; indeed '[i]t is conceivable that the conditions might be satisfied by every numerical rate which could be supposed' (608) – the case of neutral equilibrium in Edgeworth's terminology (1925, II, 23). Mill attributes this complication to 'intelligent criticisms' (especially by Thornton)[90] of the original statement and subsequent debate (608), but promises to show how

90 The reference to Thornton appears first in 1865 and not in the original statement of 1852. Chipman (1965, 492–3; 1979, 489) suggests that Mill must have had in mind unpublished criticisms by Thornton (whose published critique of demand-supply analysis had not yet appeared) and suggests also that Mill could not possibly have neglected Whewell (1850). Chipman attributes Mill's failure to make specific mention of Whewell partly to his deep antagonism towards him.

the range of indeterminateness can be removed by reference to the extent that resources are released in consequence of the opening of trade. This latter analysis proceeds by the standard arithmetical examples, supplemented by a modest attempt to generalize in algebraic terms, leading purportedly to a 'general law' – that the terms of trade depend on two variables: 'first, on the amount and extensibility of their demand for its commodities, compared with its demand for theirs; and secondly, on the capital which it has to spare, from the production of domestic commodities for its own consumption. The more the foreign demand for its commodities exceeds its demand for foreign commodities, and the less capital it can spare to produce for foreign markets, compared with what foreigners spare to produce for its markets, the more favourable to it will be the terms of interchange . . . ' (614). The lengthy analysis closes, however, with the admission that his second variable is but a mirror image of the first, so that it remains true 'that the countries which carry on their foreign trade on the most advantageous terms, are those whose commodities are most in demand by foreign countries, and which have themselves the least demand for foreign commodities' (615).

As Edgeworth pointed out, Mill in fact did nothing by his extended analysis 'to remove that sort of indeterminateness which does occur in the actual case of plural, though definite, positions of equilibrium – not to speak of that sort of indeterminateness which would occur in the case of that neutral equilibrium which he imagines'. In any event, this problem does not fit well with Mill's usual presumption of unique and stable equilibrium in competitive markets – to be found not only in the *Principles* but in his review of Thornton in 1869 (above, p. 276–7). As for the extended analysis itself Edgeworth had rather harsh words: 'The splendid edifice of theory constructed in the first five sections is not improved by the superstructure of later date which forms the latter part of the chapter.'[91]

* * *

Mill's extension of the analysis of international values from a barter to a money system concludes in terms of Ricardo's famous proposition: 'Gold and silver having been chosen for the general medium of circulation, they are, by the competition of commerce, distributed in such proportions amongst the different countries of the world as to accommodate themselves to the natural traffic which would take if no such metals existed, and the

---

91 For an account of the massive literature that has developed around the added sections see Appleyard and Ingram (1979), 459–78, 500–4. The authors conclude that Edgeworth's criticisms are valid. But see for a different view Chipman (1979) who maintains that Mill provided a rigorous proof of the existence of equilibrium.

trade between countries were purely a trade of barter' (636; citing Ricardo, 1951, I, 137). For this proposition Mill had the highest praise: 'Of this principle, so fertile in consequences, previous to which the theory of foreign trade was an unintelligible chaos, Mr. Ricardo, though he did not pursue it into its ramifications, was the real originator.'

The demonstration that the extension makes no substantive difference to the conclusions already reached in terms of barter proceeds by distinguishing the state of 'stable equilibrium' from that of the 'process' of equilibration. In equilibrium commodity exports pay for imports without money flows so that exchanges occur as if a barter system pertained. Here, in effect, we have an application of that version of the law of markets according to which aggregate demand and supply are equal in equilibrium – not, that is, identically equal – extended to the international plane: 'All interchange is, in substance and effect, barter: whoever sells commodities for money, and with that money buys other goods, really buys those goods with his own commodities. And so of nations' (630). As for the 'process' of equilibration from an initial state of trade imbalance, it is the mechanism not the result that differs between a barter and money system (631).[92]

On this matter Mill cites almost *verbatim* the formulation of the essay involving a two-country, two-commodity world and, abstracting from transport costs and importers' profits, Germany and England having a comparative advantage in linen and cloth respectively (632n.–634n. citing IV, 241–3). Assuming (for illustration) initial equality in the money prices of cloth per unit in each country, English cloth exports are precluded whereas Germany exports linen, which sells in the English market (at the domestic German price).[93] The English deficit is met in gold, the outflow lowering the British cloth price relative to that in Germany, rendering feasible cloth sales in Germany (at the English price).[94] But '[a]s cloth fell in price and

---

92 Under the barter system, 'the country which wants more imports than its exports will pay for, must offer its exports at a cheaper rate, as the sole means of creating a demand for them sufficient to re-establish the equilibrium. When money is used, the country seem to do a thing totally different. She takes the additional imports at the same price as before, and she exports no equivalent, the balance of payments turns against her; the exchange becomes unfavourable, and the difference has to be paid in money.'

93 Cf. Viner (1937), 483–4: 'For the classical school, the immediate determinant of whether a particular commodity will be obtained abroad or at home, or exported, is the *absolute* difference in the prices at which domestic and foreign producers are willing to furnish it. Actual market prices being assumed to be everywhere equal in the absence of transportation costs, commodities will be exported or imported according as their domestic *supply prices* or money costs of production are *absolutely* lower or higher than their foreign supply prices or money costs of production expressed in the same currency.'

94 Mill proceeds on the basic assumption that gold flows do affect prices, namely cases of trade imbalance such that at 'ordinary' French and English prices the English demand for French goods is not covered by the French demand for English goods and price changes

linen rose, there would be some particular price of both articles at which the cloth exported, and the linen imported, would exactly pay for each other. At this point prices would remain, because money would cease to move out of England into Germany' (633n.). Correction of the initial imbalance in Germany's favour thus hinges on the responsiveness of German demand to price reductions of cloth and of English demand to price increases of linen – the less elastic the demands the more the required gold flows, and the greater the decline in P_C/P_L, England gaining relatively little from the opening of trade.

'We have thus arrived', Mill declared, 'at precisely the same conclusion, in supposing the employment of money, which we have found to hold under the supposition of barter' (633n.).[95] In the expanded formulation of the *Principles* the identity of the barter and the money systems is conveniently summarized thus:

The country whose exports are not sufficient to pay for her imports, offers them on cheaper terms, until she succeeds in forcing the necessary demand: in other words, the Equation of International Demand, under a money system as well as under a barter system, is the law of international trade. Every country exports and imports the very same things, and in the very same quantity, under the one system as under the other. In a barter system, the trade gravitates to the point at which the sum of the imports exactly exchanges for the sum of the exports: in a money system, it gravitates to the point at which the sum of the imports and the sum of the exports exchange for the same quantity of money. And since things which are equal to the same thing are equal to one another, the exports and imports which are equal in money price, would, if money were not used, precisely exchange for one another (632).

Mill's advance over Ricardo, it must be emphasized, does not consist

are required to effect equilibrium. Mill allows that there are cases where gold flows are derived from hoards and bankers' reserves – cases such as bad harvests where a once-and-for-all movement is required (*CW*, III, 633).

95 There remains to note Mill's allusion to the gains from trade, which in the case at hand entails a higher level of money income in Germany coinciding with an absolute fall in the money price of cloth; while in England, money incomes have fallen and the price of linen in greater proportion: 'The greater the efflux of money required to restore the equilibrium, the greater will be the gain of Germany, both by the fall of cloth, and by the rise of her general prices. The less the efflux of money requisite, the greater will be the gain of England; because the price of linen will continue lower, and her general prices will not be reduced so much. It must not, however, be imagined that high money-prices are a good, and low money-prices an evil, in themselves. But the higher the general money-prices in any country, the greater will be that country's means of purchasing those commodities which, being imported from abroad, are independent of the causes which keep prices high at home' (634n.).

in the foregoing proposition that the equation of international demand is satisfied in full equilibrium. Ricardo was perfectly aware of this condition and, by implication, of the general conception of reciprocal demand (cf. Hollander, 1979, 468). His 'failure', as Mill pointed out, was to cut the analysis short by neglecting to investigate the precise conditions of demand upon which depended the terms of trade satisfying the equation of international demand. This lacuna is reflected in his presumption (both for a monetary and a barter regime) of a 1:1 equilibrium exchange rate (462–3, 467).

Apart from Mill's amplification in this latter respect we find in his account all the key features of the Ricardian analysis. These include the identity in both trading communities of the prices of common tradable goods, under all circumstances including *disequilibrium* situations. This result, Mill noted, holds good only in the absence of transport costs and allowance for importers' profits: 'In practice, cloth and the linen would not, as here supposed, be at the same price in England and in Germany: each would be dearer in money-price in the country which imported than in that which produced it, by the amount of the cost of carriage, together with the ordinary profit on the importer's capital for the average length of time which elapsed before the commodity could be disposed of (*CW*, III, 634n; cf. 595 regarding the appropriate qualification in the case of barter). This qualification too has been made by Ricardo, subject to the proviso referred to already in the analysis of barter – that the incidence of transport costs depends upon demand conditions: 'it does not follow that each country pays the cost of carriage of the commodity it imports; for the addition of this item to the price may operate as a greater check to demand on one side than on the other; and the equation of international demand, and consequent equilibrium of payments, may not be maintained. Money would then flow out of one country into the other, until . . . the equilibrium [is] restored: and, when this [is] effected, one country would be paying more than its own cost of carriage, and the other less'.

* * *

A further Ricardian theme, the impact on the pattern of trade of technical progress which affects export industries, is considered for the monetary regime in both the essay and the *Principles*. An improvement which opens a new export branch is shown to benefit the country 'not only by the cheapness of the article in which the improvement has taken place, but by a general cheapening of all imported products' (III, 635; IV, 251). An application of the analysis to the exportation of machinery is made in the essay, yielding the reverse results (IV, 252).

In the case of an improvement affecting an established trade (say English cloth), the initial impact is to reduce the price thereof (at home and in foreign markets), but the further consequences will depend upon foreign demand elasticity. Assuming an inelastic foreign demand, an imbalance is created against England, generating a monetary outflow and fall in general prices, including the price of cloth. It follows that the cloth price falls in Germany by more than the improvement has reduced it in England. At this point Mill pays tribute to Ricardo's general analysis of the distribution of the precious metals (III, 636). But Ricardo had in fact argued 'that the improvement of a manufacture in any country tends to alter the distribution of the precious metals amongst the nations of the world: it tends to increase the quantity of commodities, at the same time that it raises general prices in the country where the improvement takes place' (1951, I, 141). Mill's analysis, with its formal allowance for elasticity of demand, thus suggested the limits of that theorem.

The impact on the equation of international demand of export taxes and import duties constitutes another of Mill's celebrated applications of demand-supply theory. The problem, as stated in the essay, is 'whether a country can engross to itself a larger share of the benefits of foreign commerce than would fall to it in the natural or spontaneous course of trade' (IV, 245).

A major outcome of the analysis is that the incidence of a tax on exports does not necessarily fall entirely on foreigners. This early analysis which Mill reproduces in the *Principles*, turns on the foreign (German) demand elasticity for the domestic product (English cloth) the price of which is presumed to rise initially by the full amount of the tax, on the implicit assumption that cloth is a constant-cost industry. Notwithstanding the wide range of possible outcomes, casual empiricism regarding elasticity values leads Mill to conclude that '[i]n general . . . there could be little doubt that a country which imposed such taxes would succeed in making foreign countries constitute something to its revenue; but unless the taxed article be one for which their demand is extremely urgent' – an allusion to an inelastic foreign demand – 'they will seldom pay the whole amount, which the tax brings in' (III, 853).[96] It is conceivable that the burden falls

---

96 In the case of an elastic German demand – which Mill supposes to be the most likely case – total expenditure on English cloth falls following the imposition of the tax; money flows, therefore, from England to Germany to correct the payments imbalance thus reducing the price of cloth in England somewhat below the amount of the tax and raising the price of linen in Germany (and thus in England). Readjustments of demand to the price changes assure renewed balance. In the final analysis 'Germany has paid only a part of the tax, and the remainder of what has been received into our treasury has come indirectly out of the pockets of our own consumers of linen, who pay a higher price for that imported commodity in consequence of the tax on our exports, while at the same time they, in consequence of the efflux of money and the fall of prices, have smaller money incomes, wherewith to pay for the linen at that advanced price' (852).

In the case of an inelastic German demand for English cloth, the tax generates a money flow from Germany to England and an increase in the cloth price in England (and thus

entirely on domestic consumers. Such will be the case if an elastic German demand for English cloth is accompanied by an inelastic English demand for German linen.[97]

Import duties, Mill proceeds to show, need not fall on the home consumer. He assumes initially that the price of linen rises in the English market by the full amount of the tax, the German exporter receiving the original price. However inelastic the English demand may be – provided it is not zero elastic – total expenditure on German linen necessarily falls, generating a gold flow from Germany to England which lowers the linen price in Germany (and thus England) – the linen price in England still exceeding the original price but by less than the tax – and raises the cloth price in England (and thus Germany). Consequently Germans pay a higher price for cloth but with a reduced money income; whereas English consumers pay a lower price for linen with a higher money income: 'It may, therefore, be laid down as a principle' – since zero elasticity is the exception – 'that a tax on imported commodities, when it really operates as a tax, and not as prohibition either total or partial, almost always falls in part upon the foreigners who consume our goods; and that this is a mode in which a nation may appropriate to itself, at the expense of foreigners, a larger share than would otherwise belong to it of the increase in the general productiveness of the labour and capital of the world, which results from the interchange of commodities among nations' (854).

In the foregoing case the burden on foreigners takes the form of the higher prices paid for imported goods; foreign producers are supposedly unaffected. Thus far the essay. In the *Principles* Mill considers two cases where duties on German linen would fall partly on linen producers. First, that of zero

in Germany) and a fall in the linen price in Germany (and thus in England) with appropriate demand adjustments, cloth exports falling and linen imports rising till equilibrium is reestablished: 'It thus appears (what is at first sight somewhat remarkable) that by taxing her exports, England would, in some conceivable circumstances, not only gain from her foreign customers the whole amount of the tax, but would also get her imports cheaper. She would get them cheaper in two ways; for she would obtain them for less money, and would have more money to purchase them with. Germany on the other hand, would suffer doubly: she would have to pay for her cloth a price increased not only by the duty, but by the influx of money into England, while the same change in the distribution of the circulating medium would leave her less money to purchase it with' (851).

97 For now the money outflow from England which raises the price of German linen itself is not diminished but worsens as English expenditure on the commodity rises. Equilibrium thus requires a fall in the cloth price (including the tax), possibly even below the initial level which was tax free. (This will be so if at the original price German consumers choose to spend their (higher) money incomes on a variety of commodities other than cloth.) In this case the German gain is 'at the expense of the English consumers of linen, who, in addition, will be the real payers of the whole of what is received at their own custom-house under the name of duties on the export of cloth' (852).

supply elasticity where cost-pricing is ruled out and price is 'only limited by the desires of the buyer'. Here 'the sum obtained from the restricted [i.e. given] supply being the utmost which the buyers would consent to give rather than go without it, if the treasury [intercepts] a part of this, the price cannot be further raised to compensate for the tax, and it must be paid from the monopoly profits' (854).[98] (Evidently, Mill assumes in the main case that German producers of linen escape the burden entirely by transfer to other sectors – infinite elasticity of supply.) A less extreme case involves diminishing returns; here in consequence of reduced demand, 'the consumers, both in the country itself [Germany] and in those which dealt with it [England], would obtain the produce at smaller cost; and a part only, instead of the whole, of the duty would fall on the purchaser, who would be indemnified chiefly at the expense of the landowners or mine-owners in the producing country' (855).[99]

Mill's application of demand theory to the problem of export taxation is original, but in its use of the range of demand elasticity bears a close family resemblance to Ricardo's discussion at the very outset of his chapter 'On Foreign Trade' (1951, I, 128f). The analysis of import duties is also original but the tools both of demand and supply elasticity – and their importance in tracing through the impact of disturbances to the data – were readily at hand throughout the full range of Ricardo's writings.

XII   THE FUNDAMENTAL THEOREM ON
DISTRIBUTION AND THE MEASURE OF VALUE

Although, in his 1825 defence of McCulloch against Malthus, Mill spelled out the fundamental Ricardian theorem that 'profits depend upon wages' (*CW*, IV, 38) he did not emphasize the role of value theory in its derivation. From this perspective the review was rather superficial, neglecting as it does McCulloch's 'Marxian' analysis of deviations of price from labour value as a preliminary step in the demonstration of the inverse wage–profit relation (McCulloch, 1851, 165–6); and this despite his eulogy of McCulloch (25n). But the neglect of McCulloch's analysis does not amount to its rejection;

98   Mill refers to this as a case of 'monopoly' and Edgeworth (1925), II, 25, reads the term as implying a literal monopoly whereas Mill, in classical fashion, intends competition under conditions of zero supply elasticity.

99   Edgeworth (1925) who is critical of Mill for assuming constant cost industries in Book III: xviii ('International Values') and xxv ('Competition of Different Countries in the Same Market'), concedes that he makes up for it by the analysis of duties which fall on products subject to diminishing returns (II, 21–2). He also cites approvingly Mill's discussion of a tax which by checking demand prevents the law of increasing returns from coming into action (above, p. 297).

Mill's reply was partly dictated by Malthus's selection of topics for criticism.[100]

The formulation in Mill's article 'The Corn Laws' (1825) is not only consistent with basic Ricardianism, but also implies an underlying foundation in value theory. (See the discussion below.) The better known essay 'On Profits and Interest' (1830) presents the inverse profit-wage relationship in a favourable light although with some qualification. In his *Autobiography*, Mill describes the formulation as a 'modified form of Ricardo's theory of profit . . . ' (I, 123).[101] In the *Principles* even the qualifications are withdrawn. These matters will be taken up in the present section.

The rate of profit (the ratio of current surplus to capital) is said, in the essay of 1830, to be given simply by 'the ratio between the wages of labour and the produce of that labour', on the grounds that produced means of production may be reduced entirely to labour: 'In the ultimate analysis . . . labour appears to be the only essential of production. To replace capital, is to replace nothing but the wages of the labour employed.' Thus, 'the whole of the surplus, after replacing wages, is profits', so that 'we arrive at Mr. Ricardo's principle [regarding the rate of profit], that profits depend upon wages; rising as wages fall, and falling as wages rise' (IV, 293). The rate of profit is here calculated upon the capitalists' 'outlay' rather than some measure of the capital stock strictly defined.[102]

---

100 In the early paper on rent (1828) Mill refers to the rent and population principles as the 'great' discoveries after 1776 (IV, 179). The neglect here of the inverse wage–profit relation is noteworthy but may reflect the particular topic under investigation.

101 The status of this paper is difficult to pin down. In April 1834 he wrote to J. P. Nichol regarding what later was to be the *Unsettled Questions*, 'I believe almost all I have written in the fourth essay concerning Interest is erroneous . . . ' (XII, 222). Later in the year he elaborated further: 'I am fearful that the Essay on Wages and Profits . . . is little better than elaborate trifling, and that the doctrine that profits depend on wages, though scientifically correct, does not present the important aspect of the law of profits, perhaps not the ultimate law at all, and is, therefore, of little use in philosophy' (XII, 231). Yet Mill records in his *Autobiography* that in '1830 and 1831 I wrote the five Essays since published as *Essays on Some Unsettled Questions* almost as they now stand, except that in 1833 I partially rewrote the fifth essay' (I, 189). It seems fair to conclude that in 1844 'he was in agreement with the views expressed earlier despite his volte-face in the interim' (Hunter, 1959, 65).

102 An example relates to an individual capitalist: 'The ratio between the price which he has to pay for . . . means of production [labour and capital goods], and the produce which they enable him to raise, is the *rate* of *profit*. If he must give for labour and tools four-fifths of what they will produce, the remaining fifth will constitute his profit, and will give him a rate of one in four, or twenty five per cent, on his outlay' (291). But from the aggregative perspective also, capitalists as a whole must give four-fifths of what is produced to labour, since capital is said to reduce entirely to labour, and retain one-fifth so that the profit rate is simply calculated as the ratio: profits/wages.

This initial formulation is immediately amplified in two respects. In the first place, Mill correctly pointed out that Ricardo intended by 'a rise of wages' an increase in the 'value' of wages, that is, in the labour absorbed in the production of wage goods, and not simply an increase in commodity wages; for the latter would not necessarily entail a fall in the profit rate should it occur at the same time as an increase in labour productivity in the wage-goods sector. This formulation identifies an 'increase in wages' with an increase in the proportionate share of labour in output: 'A rise of wages, with Mr. Ricardo, meant an increase in the cost of production of wages; an increase in the numbers of hours' labour which goes to produce the wages of a day's labour; an increase in the *proportion* of the fruits of labour which the labourer receives for his own share; an increase in the ratio between the wages of his labour and the produce of it' (294). Secondly, Mill – in an allusion probably to James Mill's *Elements* – recommends that the Ricardian relationship between profits and wages be defined in per capita terms for the sake of clarity:

Some of Mr. Ricardo's followers, or more properly, of those who have adopted in most particulars the views of political economy which his genius was the first to open up, have given explanations of Mr. Ricardo's doctrine to nearly the same effect as the above, but in rather different terms. They have said that profits depend not on *absolute*, but on *proportional* wages: which they expounded to mean the proportion which the labourers *en masse* receive of the total produce of the country. It seems, however, to be rather an unusual and inconvenient use of language to speak of anything as depending upon the wages of labour, and then to explain that by wages of labour you do not mean the wages of an individual labourer, but of all the labourers in the country collectively. Mankind will never agree to call anything a rise of wages, except a rise of the wages of individual labourers, and it is therefore preferable to employ language tending to fix attention upon the wages of the individual. The wages, however, on which profits are said to depend, are undoubtedly *proportional* wages, namely the proportional wages of one labourer, and (not the whole produce of the country, but) the amount of what one labourer can produce . . . Proportional wages, thus understood, may be concisely termed the cost of production of wages; or, more concisely still, the cost of wages, meaning their cost in the 'original purchase money', labour (294–5).

Mill's formulation of the Ricardian theorem on distribution clearly amounts to the proposition that the rate of profit varies inversely with the fraction of a man's daily labour which is devoted to the production of wage goods.[103]

103 e.g.: If *per capita* wages = 4 corn units; and *per capita* output 5 corn units, then the cost of *per capita* wages – the labour embodied in the wages – equals $4(1/5) = $ four-fifths. In

However, while Mill believed his conception of Ricardo's theory to constitute 'the most perfect form in which the law of profits seems to have been yet exhibited' and 'to be the basis of the true theory of profits' (293, 295), he himself was not prepared to accept it without modification: 'The conclusion . . . cannot be resisted, that Mr. Ricardo's theory is defective: that the rate of profit does *not* exclusively depend upon the value of wages, in his sense, namely the quantity of labour of which the wages of a labourer are the produce; that it does *not* exclusively depend upon proportional wages, that is, upon the proportion which the labourers collectively receive of the whole produce, or the ratio which the wages of an individual labourer bear to the produce of his individual labour' (297). He objected to Ricardo's (supposed) reduction of capital entirely to labour, on the grounds that the repayment of *profits* entered into the outlay of a capitalist upon produced means of production, and not merely wages: 'It is not correct, therefore, to state that all which the capitalist retains after replacing wages forms his profit. It is true the whole return to capital is either wages or profits; but profits do not compose merely the surplus after replacing the outlay; they also enter into the outlay itself. Capital is expended partly in paying or reimbursing wages, and partly in paying the profits of other capitalists, whose concurrence was necessary in order to bring together the means of production' (295).[104] In short, the profit rate is not given simply by the ratio of current surplus after replacing outlay to an outlay amounting solely to the wages of direct and indirect labour.

We shall see presently that Mill was to abandon this objection in the *Principles*, but even at this stage he insisted upon the ease with which the 'defect' in Ricardo's reasoning (as he understood it) might be corrected:

Those political economists, therefore, who have always dissented from Mr. Ricardo's

brief, 80% of the labour of one man is devoted to the production of his wages. (The rate of profit would be calculated as 20/80.)

Mill's predilection for average rather than aggregate terms was based on a desire for clarity. Obviously the preceding percentage figure could be translated to refer to 80% of the total labour force which is devoted to the production of wage goods. Mill did not apparently notice that much more is involved. For in the event of a changing labour force – and thus a changing aggregate value to be shared – there is no necessary identity between the cost of producing wages and labour's proportionate share; the cost may rise, yet the proportion decline. This problem is ruled out by restricting the calculation to per capita terms because the 'value' of per capita output is constant by definition.

104 Cf. Mill's (1945) notes on Senior's *Outline*: what should be intended when profit is included in costs is 'not the profit of the actual producer, but the profit of the producers of his machinery, & other articles of capital which formed part of the price paid, & thence part of his own outlay, on which his profit is to be estimated. I am confirmed in this opinion by remembering that it is chiefly that previous profit which affects price: for it is chiefly in that way (though not wholly so) that profit enters more largely into the conditions of production of one commodity than into those of another' (136).

doctrine, or who, having at first admitted, ended by discarding it, were so far in the right; but they committed a serious error in this, that, with the usual one-sidedness of disputants, they knew no medium between admitting absolutely and dismissing entirely; and saw no other course than utterly to reject what it would have been sufficient to modify.

It is remarkable how very slight a modification will suffice to render Mr. Ricardo's doctrine completely true. It is even doubtful whether he himself, if called upon to adapt his expressions to this particular case, would not have so explained his doctrine as to render it entirely unobjectionable (297).

Mill's 'slight' modification was to relate the rate of profit to the '*cost* of wages' – which includes the profit of the capital-goods producer – as distinct from the '*value* of wages' – which refers specifically to the labour embodied in the wages bill. Unfortunately, in adopting this solution the identity between the cost of wages (the strategic wage variable) and the proportionate share of labour in output was no longer tenable:

Mr. Ricardo's principle, that profits cannot rise unless wages fall, is strictly true, if by low wages we mean not merely wages which are the produce of a smaller quantity of labour, but wages which are produced at less cost, reckoning labour and previous profits together. But the interpretation which some economists have put upon Mr. Ricardo's doctrine, when they explain it to mean that profits depend upon the proportion which the labourers collectively receive of the aggregate produce, will not hold at all.

The only expression of the law of profits, which seems to be correct is that they depend upon the cost of production of wages. This must be received as the ultimate principle (299).[105]

---

105 Mill makes the following assumptions: Producer A's total output = 180 corn units; wages per man = 1 corn unit; 60 men are employed; replacement of fixed capital worn out per period and materials absorb 60 corn units of which 40 represents wages of capital-goods workers and 20 the profits of capital-goods producer B; the remaining 60 corn units then represent A's profits. A's profit rate = 60/120 = 50%; B's profit rate = 20/40 = 50%. (As usual, the rate is calculated on current expenses, not a stock of capital.)

A technical change permitting the abandonment (by A) of fixed capital entirely, at the cost of 40 additional current labourers, is presumed to occur, A's output remaining 180. A's profit rate rises to (60 + 20)/(120 – 20) = 80%: 'Here, therefore, is an undeniable rise of [the rate of] profits. Have wages, in the sense above attached to them, fallen or not? It would seem not' (296). For each worker receives 'the same proportional wages; and the labourers collectively still receive the same proportion . . . of the whole produce' (297). (The *labour* cost of wages is of course unchanged.)

Mill's 'modification' turns on the fact that the *total* cost of wages has actually fallen – including in costs the profits previously paid by A to B – by one-sixth (180 quarters now costing 100 rather than 120). The rate of profit rises as seen above; to prevent the rise,

In all of this can be discerned the probable influence of Robert Torrens, who had insisted that the profit rate depends not only on the proportionate share of wages which governed the ratio of profits to wages, but also on the non-wage capital required. A technological improvement in the wage goods sector permitting the relinquishing of fixed capital or materials – Marx's 'constant' capital – would imply a rise in the profit rate (1829, xv–xviii; cf. Tucker, 1960, 96).

Considering Mill's own insistence that the identity between the cost of production of wages and the proportionate share of labour could no longer be maintained, it seems scarcely justifiable to minimize the significance of the alteration proposed.[106] But doubtless Mill did so because the important corollaries derived by Ricardo from his theory of profits still held good, particularly that 'the rate of profits can never rise but in conjunction with one or other of two changes, 1st, a diminished remuneration of the labourer; or 2ndly, an improvement in production, or an extension of commerce, by which any of the articles habitually consumed by the labourer may be obtained at smaller cost' (299).[107]

* * *

For the next stage in Mill's statement of the Ricardian propositions relating to distribution we turn to the *Principles* itself. Here in the chapter 'Of Profits' (Book II, xv) he discussed the preliminary issue of the source of profits, in a passage which is quite striking as a formulation of what was to become standard Marxian doctrine, namely that profits arise from the fact that labour works for a longer time than is required to reproduce its own subsistence:

wages per man would have to increase to 1 1/5 corn units, or to 120 units for the given work of 100. But the identity between the 'cost of wages' (in the wider sense) and labour's proportionate share is broken.

106 Oddly enough, in the Notes to Senior's *Outline* Mill (1945) retains the proportionality theorem. Thus he appends to Senior's remark that ' . . . confusion has been occasioned by the verbal ambiguity which we have pointed out, and would not have arisen if Mr. Ricardo had used any other adjectives than *high* and *low* to express a larger or smaller proportion': 'It seems to me that the confusion stated to have been occasioned is over estimated. I do not believe that anybody was thinking of low wages in France in reference to the comforts of the labourer, but in reference to the cost of his labour to the employer – which is the only sense in which wages either affect profits or enable a producer to undersell – & the question was, which of the two effects would take place. The cost of wages to the capitalist is a phrase synonymous with proportional wages, or wages in Ricardo's sense' (137).

107 In this context Mill also referred to diminishing returns which 'has a tendency to lower profits' and the 'antagonistic tendency of profits to rise' due to technical progress in the production of wages good, both agricultural and manufactured (300). A reduction in the cost of producing luxury goods lowers their prices, and benefits their consumers but has no effect upon the profit rate.

The cause of profit is, that labour produces more than is required for its support. The reason why agricultural capital yields a profit, is because human beings can grow more food, than is necessary to feed them while it is being grown, including the time occupied in constructing the tools, and making all other needful preparations: from which it is a consequence, that if a capitalist undertakes to feed the labourers on condition of receiving the produce, he has some of it remaining for himself after replacing his advances. To vary the form of the theorem: the reason why capital yields a profit, is because food, clothing, materials, and tools, last longer than the time which was required to produce them; so that if a capitalist supplies a party of labourers with these things, on condition of receiving all they produce, they will, in addition to reproducing their own necessaries and instruments, have a portion of their time remaining, to work for the capitalist (II, 411; the passage was first introduced into the 4th edition of 1857).

This perspective is prologue to a basic methodological proposition – that (to use Marx's terminology) 'appearance' must not be confused with 'essence':

We thus see that profit arises, not from the incident of exchange, but from the productive power of labour; and the general profit of the country is always what the productive power of labour makes it, whether any exchange takes place or not. If there were no division of employments, there would be no buying or selling, but there would still be profit. If the labourers of the country collectively produce twenty per cent more than their wages, profits will be twenty per cent, whatever prices may or may not be. The accidents of price may for a time make one set of producers get more than the twenty per cent, and another less, the one commodity being rated above the natural value in relation to other commodities, and the other below, until prices have again adjusted themselves; but there will always be just twenty per cent divided among them all.

The sharp comments by L. von Bortkiewicz in his famous paper 'Value and Price in the Marxian System' (1907) are pertinent:

Marx comments on this passage that 'Mill here confuses the duration of the working-time with the duration of its products'. In doing so, Marx suppresses the second half of the quotation, which removes any doubt of the fact that Mill deduces profit from surplus value, just as Marx does. Mill goes on to show that profit is conditioned, not by exchange, but by the productive power of labour. Were there no division of labour, says Mill, there would be neither selling nor buying, but profit would continue to exist. One would think that at least with this assertion, Mill would draw a word of approval from his severe critic. But no! Marx pretends to be shocked: 'Here', says he, 'exchange, sale and purchase, the general conditions of capitalist production are thus sheer accessories, and yet there still subsists profit without the sale or purchase of labour power!' It is, however, clear that Mill did not mean the buying and selling of labour power, but merely the buying and selling of *products*. One will not go wrong if one connects the ill will which Marx displays towards Mill, with the circumstance that Mill had, basically, anticipated Marx's theory of surplus value (1952, 52–3n.; see also 35–6n. regarding Marx's negative reaction to Mill's essay).

Doubtless Marx's displeasure can be appreciated in the light of Mill's emphasis on factor productivity. Indeed, in an earlier formulation Mill specifically refers to capital productivity rather than labour productivity (unlike the foregoing passage): 'Looking at things in the aggregate, and permanently, the remuneration of the producer is derived from the productive power of his own capital (*CW*, II, 88).[108] At the same time it is evident that merely to refer profits to excess labour time is inadequate, since we must be given to understand why it is that such excess exists, an exercise that inevitably leads to the productivity dimension as a necessary (though insufficient) condition.[109]

Mill's formulation of the source of profits is very close in substance to the *initial* statement in the essay of 1830 according to which the rate of profit varies inversely with the fraction of a man's labour devoted to the production of his wages. In fact the subsequent objection in that essay to the Ricardian theorem which relates the rate of profit to labour's proportionate share, is withdrawn in the *Principles*, Mill taking the position that since 'our supposed capitalist is not meant to represent a single employment, but to be a type of the productive industry of the whole country', the advances may after all be reduced in the last resort entirely to wages: '[I]n the whole process of production, beginning with the materials and tools, and ending with the finished product, all the advances have consisted of nothing but wages; except that certain of the capitalists concerned have, for the sake of general convenience, had their share of profit paid them before the operation was completed. Whatever, of the ultimate product, is not profit,

108 Here too Mill insists on the need to see through surface manifestations: 'The sale of the produce for money, and the subsequent expenditure of the money in buying other commodities, are a mere exchange of equivalent values for mutual accommodation . . . We cannot too strictly represent to ourselves the operation of exchange, whether conducted by barter or through the medium of money, as the mere mechanism by which each person transforms the remuneration of his labour or of his capital into the particular shape in which it is most convenient to him to possess it; but in no wise the source of the remuneration itself'.

109 In a letter to Cliffe Leslie (May 4 1863) Mill represents profits as 'a mere consequence of increased efficiency in the instruments of production, occasioned by private ownership of them' (XV, 857). The capitalist takes from the workers 'what he first gave them'. That at least is the case assuming a wage that is not unduly depressed, more specifically no less than what could be obtained under 'cooperation'.

is repayment of wages' (II, 412).[110] The rate of profit, Mill concluded, therefore 'depends' upon 'cost of labour' which now, once again, he identified with labour's proportionate share:

The *rate* of profit, the percentage on the capital, depends only on the . . . labourer's proportional share, and not on the amount to be shared. If the produce of labour were doubled, and the labourers obtained the same proportional share as before, that is, if their remuneration was also doubled, the capitalists, it is true, would gain twice as much; but as they would also have had to advance twice as much, the rate of their profit would be only the same as before.

We thus arrive at the conclusion of Ricardo and others, that the rate of profits depends on wages; rising as wages fall, and falling as wages rise. In adopting, however, this doctrine I must insist upon making a most necessary alteration in the wording. Instead of saying that profits depend on wages, let us say (what Ricardo really meant) that they depend on the *cost of labour* (413).

Mill's insistence upon the 'cost of labour' rather than 'wages' was obviously designed to avoid any possible misunderstanding: the profit rate depended on labour embodied in per capita wages and not upon the actual commodity wage.[111]

110 Again: '[T]he advances of the capitalist, or in other words, the expenses of production, consist solely in wages of labour: . . . whatever portion of the outlay is not wages, is previous profit, and whatever is not previous profit, is wages' (428).
111 The 'cost of labour' clearly now refers to labour embodied in wages, since his earlier objection – which led him to insist upon including profits in the cost – had been withdrawn.
　　One problem remains as far as the analysis of *Principles* (Book II) is concerned. In a further elaboration, Mill related the cost of labour to *three* variables, namely 'the efficiency of labour; the wages of labour (meaning thereby the real reward of the labourer); and the greater or less cost at which the articles composing that real reward can be produced or procured' (414). In strict logic only the latter two are relevant; Mill's formulation was criticized by J. E. Cairnes in notes proposing alterations for the sixth edition of the *Principles* on the grounds that to include 'the efficiency of labour' involves double counting since it is covered by the third item (letter to Mill, 29 Nov. 1864, III, 1048–9).
　　Mill did not apparently respond to the criticism. Had he done so, he might have referred to Ricardo's observation in his 'Notes on Malthus' that in dealing with the British profit rate, the value of output must be measured by a 'quantity of our domestic labour' to allow for international differences in the quality of the labour force: 'A day's labour of a Hindoo or a South American . . . cannot be compared with that of an Englishman – was it fair then in Mr. Malthus to suppose that when I was talking of the quantity of labour regulating price and profits I considered it as of no importance whether it was the labour for a given time of a Hindoo, an Irishman, or an Englishman. I apply my doctrine to the same country only, and fix on a standard which is common in that country. I should not estimate profits in England by the labour of a Hindoo . . . unless I had the means of reducing them to a common standard' (1951, II, 272–3). That this is what Mill had in mind is suggested by the various examples given of the determinants of labour efficiency, including bodily vigour, education and the support provided by fixed capital (*CW*, II, 415).

In Mill's final formulation, the so-called 'dependency' of the profit rate upon the proportionate shares in fact constitutes an identity since by the profit rate is intended simply the ratio of surplus to 'capital' the latter referring entirely to wage costs of both direct and indirect labour. It is pertinent, therefore, to allude here to a charge, frequently directed against Ricardo, that the statement of the dependency of the rate of profit upon proportionate wages represents nothing but a 'truism' precisely because capital is reduced entirely to wage advances. This charge is not, in fact, justified in Ricardo's case since capital is certainly not thus reduced to wages; the profit rate is said to be dependent upon, but not identical with, the proportionate shares (Hollander, 1979, 267f). The charge has more validity when directed against Mill's final formulation. This is ironic since it was Mill who in 1830 had taken Ricardo to task for what he believed to have been the practice of reducing capital to wages alone, and had touched upon some of the potential complications arising from the use of a broader conception of capital in the profit rate calculation – illustrating his case by consideration of the consequences of shortening the time period of investment (above, n105).[112] It is doubtful whether Ricardo would have welcomed Mill's final defence of his position, which follows the line of Longfield (1834, 172f.) and Torrens (1844, xxii–xxiii) by representing quantity of capital as accumulated labour (cf. Robbins 1958, 53f.; Hollander, 1977a, 235–6).

* * *

We now arrive at a fundamentally important qualification which puts into proper perspective Mill's earlier downplaying of exchange in analyzing the source of profits. That weighting was designed to warn against focussing unduly on individual sectors when the *general* profit rate is at stake. But it remained true of a capitalist economy that exchange played an indispensable role. Thus the initial analysis of profits was correct but provisional; a complete treatment required a preliminary discussion of value.

In his chapter devoted to value at the outset of Book III Mill emphasized the characteristic interdependence of distribution and pricing, writing of 'the present system of industrial life, in which employments are minutely subdivided, and all concerned in production depend for their remuneration

---

112 Accordingly Marx's strictures in *Theories of Surplus Value* against the identification of the rate of surplus value (the profit–wage ratio) with the rate of profit (the profit–'capital' ratio) would seem to be justified in so far as concerns Mill's final formulation but not the version presented in the essay which amounts precisely to this same point (Marx, 1971, III, 190f).

In *Capital* (1906), Marx criticized Adam Smith for arriving 'by a fundamentally perverted analysis . . . at the absurd conclusion, that even though each individual capital is divided into a constant and a variable part, the capital of society resolves itself only into variable capital, *i.e.*, is laid out exclusively in payment of wages' (1965, 590).

on the price of a particular commodity . . . ' (*CW*, III, 455). It is precisely this vision to which he alluded when describing exchange as constituting 'merely a part of the machinery' for effecting the distribution of the produce. That this view of exchange cannot be read as indicative of a divorce of exchange from distribution and production in any meaningful *analytical* sense, as some commentators believe, is further confirmed by the qualification that follows immediately: 'In a state of society, however, in which the industrial system is entirely founded on purchase and sale, each individual, for the most part, living not on things in the production of which he himself bears a part, but on things obtained by a double exchange, a sale followed by a purchase – the question of Value is fundamental. Almost every speculation respecting the economical interests of a society thus constituted, implies some theory of Value: the smallest error on that subject infects with corresponding error all our other conclusions; and anything vague or misty in our conception of it, creates confusion and uncertainty in everything else' (456).

All of this occurs at the outset of Book III. Evidently the earlier discussion of distribution of Book II was intended to be provisional only. This is in fact intimated in the earlier book as well. For at the close of the discussion of profits we read: 'The evidence of these propositions can be stated generally, though, it is hoped, conclusively, in this stage of our subject. It will come out in greater fulness and force when, having taken into consideration the theory of Value and Price, we shall be enabled to exhibit the law of profits in the concrete – in the complex entanglement of circumstances in which it actually works. This can only be done in the ensuing Book' (II, 415). Similarly, in the final chapter of Book II ('Of Rent') the theory of differential rent is described as 'one of the cardinal doctrines of political economy' without which 'no consistent explanation could be given of many of the more complicated industrial phenomena'; but '[t]he evidence of its truth will be manifested with a great increase of clearness, when we come to trace the laws of the phenomena of Value and Price' (419). At the very close of the second Book a strong caution is given by way of summary: 'We have now completed the exposition of the laws which regulate the distribution of the produce of the land, labour, and capital, as far as it is possible to discuss those laws independently of the instrumentality by which in a civilized society the distribution is effected; the machinery of Exchange and Price. The more complete elucidation and final confirmation of the laws which we have laid down, and the deduction of their most important consequences, must be preceded by an explanation of the nature and working of that machinery – a subject so extensive and complicated as to require a separate Book' (429).[113] Pertinent too is the fact that in the

---

113 At the outset of the third book Mill noted: 'It is true that in the preceding Books we

review of De Quincey's *Logic of Political Economy* (which is warmly commended as reflecting accurately Ricardian distribution theory) Mill expressly writes of value, following the author, as 'the master key to the principal difficulties of the science' (IV, 395).

* * *

We must now clarify the role played by the standard measure of value before turning to the full analysis of distribution as developed in the chapter 'Distribution as affected by Exchange'.

Much of Schumpeter's famous case for the early demise of Ricardianism was based upon the appearance and supposed influence of Samuel Bailey's *Critical Dissertation on Value* (1825): 'Bailey', he wrote, 'attacked the Ricardo–[James] Mill–McCulloch analysis on a broad front and with complete success. His *Dissertation*, which said, as far as fundamentals are concerned, practically all that can be said, must rank among the masterpieces of criticism in our field, and it should suffice to secure to its author a place in or near the front rank in the history of scientific economics' (1954, 486). And it was because, according to Schumpeter, J. S. Mill rejected on Bailey's grounds the conception of a measure of absolute value that he must be excluded from Ricardo's school. The value that really mattered to him, runs the contention, was *relative price*, and since value was a ratio all values could not vary simultaneously. Similarly, there was no such thing as 'the total value of all the services of wealth (or of all wealth) taken as a whole' in contrast to the position adopted by Ricardo and Marx (589). Indeed, 'the energy with which Mill insisted on the relative character of [exchange value] completely annihilated Ricardo's Real Value and reduced other Ricardianisms to insipid innocuousness' (603). The same view of Mill's position has been adopted by Professor Blaug: 'Mill does not derive the theorem about profits and wages, as Ricardo has done, from the concept of an invariable measure of value. Even the standard Ricardian thesis that 'general wages, whether high or low, do not affect values' is entirely divorced from the notion that value is to be measured by an invariant standard . . . There is no mention in Mill's discussion . . . of the important role assigned to the invariable measure of value in Ricardo's system' (1958, 172–3; also Blaug 1978a, 206; and O'Brien, 1975, 94: Mill's 'inverse

have not escaped the necessity of anticipating some small portion of the theory of Value, especially as to the value of labour and of land (III, 455).

It is pertinent that J. E. Cairnes observed to Mill that 'I have never yet succeeded in making the law of profits intelligible to a student till I had first made him familiar with the doctrine of value; and I accordingly now always send my students to your chapter on value before bringing them to grapple with the former problem' (29 Nov. 1864, III, 1050).

relationship of wages and profits is not derived from the invariable measure as we shall see it was for Ricardo').

It is true, as we have seen, that Mill insisted in the *Principles* upon the relativity of exchange value. It is also true that in his first major paper, the defence of McCulloch against Malthus (1825), he rejected (as did Bailey) the notion of a general alteration in exchange values as logically incomprehensible – value was a relative term: '[I]f it is not this, it is nothing: if any one talks about absolute value, or any other kind of value than exchangeable value, we know not what he means. One commodity may rise or fall in value with respect to another; all commodities cannot rise or fall in value, with respect to themselves (*CW*, IV, 36).[114] But Mill did allow for a general change in *prices* which, he pointed out, was 'merely tantamount to an alteration in the value of money' (III, 459; cf. 479), and in the Ricardian manner, he decided to use as *numéraire* a commodity money 'with the proviso that money itself does not vary in its general purchasing power, but that prices of all things, other than that which we happen to be considering, remain unaltered' (458).[115] The constancy of purchasing power is the consequence of a presumed constant cost of producing the monetary metal.

In the light of this procedure we may perhaps better appreciate Mill's formal discussion in the chapter 'Of a Measure of Value' (Book III, xv). Here he rejected (on Bailey's grounds) the notion of a measure of exchange value as a conceptual impossibility but accepted that of a measure of cost

114 But it must not be forgotten that Mill here had Malthus' conception of aggregate demand in mind by his criticism, which does not necessarily imply a rejection of the Ricardian conception.

In his very first excursion into print – two letters in December 1822 to *The Traveller* (owned by Robert Torrens) – Mill came to the defence of his father's formulation of the labour theory of *exchange* value in the *Elements* against criticisms by Torrens in terms not wholly flattering to Ricardo: 'In the first place, if I rightly understand Mr. Mill's chapter on Exchangeable Value, he cannot be said with propriety to have any theory of value – at least, in that sense in which the word theory is applied to Mr. Ricardo's doctrines on this subject. Mr. Ricardo renders the word *value*, as synonymous with *productive cost* – thus introducing a new, and as it appears to me, a needless ambiguity of language. Mr. Mill, on the other hand, never uses the word value in any other than its vulgar [i.e. usual] acceptation' (Mill, 1936 [1822], 15).

Again it is not clear whether Mill here implied a denial of the whole concept of a *measure* of 'value' in the Ricardian sense of the term (namely cost of production) or whether he was simply focussing attention upon exchange value – the 'vulgar' usage – for purposes of his disputation with Torrens on that particular issue. Ricardo (1951) was unimpressed by Mill's juvenile performance (IX, 385f).

115 Mill appreciated the principle (developed in rigorous terms by Hicks, 1939, 33) that if their relative prices are unchanged, commodities can legitimately be treated as a single unit. Mill sought out one commodity as representative of the group wherewith to measure changes emanating in the cost conditions of any single commodity of the system.

of production. The conditions for the measure are stated as follows: '[Economists] have imagined a commodity invariably produced by the same quantity of labour; to which supposition it is necessary to add, that the fixed capital employed in the production must bear always the same proportion to the wages of the immediate labour, and must be always of the same durability: in short, the same capital must be advanced for the same length of time, so that the element of value which consists of profits, as well as that which consists of wages, may be unchangeable' (579).[116] Now such a measure of cost 'though perfectly conceivable, can no more exist in fact, than a measure of exchangeable value', because of the likelihood of changes in the production cost of any commodity chosen. Nevertheless, gold and silver 'are the least variable' and, if used, the results obtained must simply be 'corrected by the best allowance we can make for the intermediate changes in the cost of the production itself'. This is far from an out-of-hand rejection of the measure of absolute value, and in fact represents precisely the position adopted by Ricardo (cf. Hollander, 1979a, 419).

The full analysis of the effects of wage-rate changes is provided in the key chapter 'Distribution, as Affected by Exchange' (Book III, xxvi). Now much is made by commentators of Mill's treatment of production, distribution and exchange in three consecutive books, as indicative of a failure (characteristic of classicism) to envisage any significant analytical relation between value theory and distribution. As already intimated this is a misunderstanding. The early discussion of distribution was clearly only provisional; in the chapter at hand the order is reversed and the problem of distribution is analysed in the light of the theory of exchange value.

When the distribution of national income occurs via the mechanism of exchange and money, Mill argued, the 'law of wages' remains unchanged in so far as the determination of commodity wages is concerned, for this simply depended upon 'the ratio of population and capital' (III, 695). However, as he had already explained at length in Book II, the implications for the employer were more complex since from his point of view it was not *commodity* wages that were relevant, but the 'cost of labour'. Under certain circumstances, Mill now explained, the cost of labour would be reflected accurately by the *money* wages paid – namely when money represented 'an invariable standard': 'Wages in the second sense, we may be permitted to call, for the present, money wages; assuming, as it is allowable to do, that money remains for the time an invariable standard,

---

116 All that is missing is the condition that the metal must be produced by a process representing the mean proportions of those in the economy as a whole. But Mill may have been assuming identical proportions in all commodities, and for this reason neglected the matter.

no alteration taking place in the conditions under which the circulating medium itself is produced or obtained. If money itself undergoes no variation in cost, the money price of labour is an exact measure of the Cost of Labour, and may be made use of as a convenient symbol to express it' (696). Assuming money to be such an invariable measure, the rate of money wages will depend upon both the commodity wage and the prices of wage goods and accordingly productivity in the wage-goods sector.[117] It is upon the 'cost of labour' that the rate of profit depended, as he had concluded earlier in Book II of the *Principles*, but the cost of labour was now identified with 'money' wages.[118]

In the context of a money economy Mill thus defined an inverse relation between the profit rate and the money wage rate. He did not bother formally to equate the 'cost of labour', in contrast to his practice in Book II of the *Principles*, with the proportionate share of the labourer in per capita output. But there is no reason to believe that he no longer maintained this relation, for he in fact referred back in the present context to the earlier analysis of Book II (698). Mill had adopted the 'proportions-measuring' money in terms of which a rise of wages implies an increased share of the labourer in the 'value' of his output and a reduced profit share, and accordingly rate of return on capital.

As in Book II and the earlier essay Mill once again raised the potential complication of an exogenous (economy-wide) reduction in the time period of investment (see example given above, n105), asking whether the relationship between profits and wages, whereby 'the rate of profit and the cost of labour vary inversely as one another, and are joint effects of the same agencies or causes', must now be modified in so far as a reduction in the *time* for which capital is invested in all commodities appears to raise the general rate of return (700).[119] The problem is immediately set aside in the following terms;

117 Though this is not Mill's standard practice, he here identifies wage goods almost entirely with agricultural produce – 'of these articles, food and other agricultural produce are so much the principal, as to leave little influence to anything else' (696-7). Production costs are said to be dependent upon 'the productiveness of the least fertile land, or of the least productively employed portion of capital'; and reference is made to the two countervailing forces at work – diminishing returns with population growth and technological change.

118 Cf. the formulation 698: 'There are two modes in which the Cost of Labour, which is correctly represented (money being supposed invariable) by the money wages of the labourer, may be increased. The labourer may obtain greater comforts; wages in kind – real wages – may rise. Or the progress of population may force down cultivation to inferior soils, and more costly processes; thus raising the cost of production, the value, and the price, of the chief articles of the labourer's consumption. On either of these suppositions, the rate of profit will fall.'

119 More generally: [D]oes not this proposition [the inverse relation between "the rate

in that case, since values and prices would not be affected, profits would probably be raised; but if we look more closely into the case we shall find, that it is because the cost of labour would be lowered. In this as in any other case of increase in the general productiveness of labour, if the labourer obtained only the same real wages, profits would be raised: but the same real wages would imply a smaller Cost of Labour; the cost of production of all things having been, by the supposition, diminished. If, on the other hand, the real wages of labour rose proportionally, and the Cost of Labour to the employer remained the same, the advances of the capitalist would bear the same ratio to his returns as before, and the rate of profit would be unaltered (700–1).

This statement, however, amounts only to the assertion that since the technical change in question applies to all goods, it also applies to wage goods and for this reason the rate of profit is affected inversely.[120] It does not get to grips with the nature of the 'cost of labour' which is the issue in question – whether it is legitimate to conceive thereof in terms solely of the labour costs of producing wages, as Mill does. Indeed, Mill closes the chapter on a confused note by asserting that 'there is nothing in the case in question to affect the integrity of the theory which affirms an exact correspondence, in an inverse direction, between the rate of profit and the Cost of Labour' (701), and referring his reader for amplification to the essay 'On Profits and Interest'; unfortunately, in that essay Mill had introduced the 'slight modification' – now apparently said *not* to be required – relating the general profit rate to the 'cost of wages' including therein profits as well as labour costs.[121]

* * *

Although his exposition was simpler in some respects than that of Ricardo, Mill nevertheless did not reduce the argument to that simplest of distribution models referred to in the literature today as the 'corn profit model' according to which the profit rate is determined within the agricultural sector without reference to value, both numerator and denominator in the expression for

of profit'' and ''the cost of labour''] require to be slightly modified, by making allowance for that portion (though comparatively small) of the expenses of the capitalist, which does not consist in wages paid by himself or reimbursed to previous capitalists, but in the profits of those previous capitalists?'

120 For a similar emphasis upon technical change which affects wages goods as distinct from those which do not, see also III, 742, 724–5, 751.

121 Mill's formulation in the *Principles* is unsatisfactory in another respect. For if fixed capital can be reduced to labour it is not at all clear why there should result the kind of changes in the price structure which, Mill recognized, occur in the case of differential factor proportions upon a change in general wages. This matter is taken up below (p. 359).

the profit rate assumed to consist of the identical physical product, 'corn' (Sraffa, 1951, I, xxxf).[122] It is of particular historical interest that a version of this doctrine was taught by J. E. Cairnes, who in 1864 advised his friend to devote the chapter 'On Profits' in Book II of the *Principles* entirely to the simplest agricultural case reserving 'the full exposition' of the Ricardo–Mill doctrine until after the discussion of value:

I   Take first the simplest conceivable case: – an act of production in which the whole process is performed by labour, and in which the return from that labour is in commodities *the same in kind* as that of which the outlay is composed. For example, 100 quarters of corn are applied to the support of workmen who, while consuming them, produce 120 quarters. Here it is plain the rate of profit, which is obviously 20 per cent, depends upon *two conditions and upon two conditions only* – 1. the real wages necessary to command the labour of the men who produce the 120 quarters; and 2. the productiveness of their industry in raising corn. Diminish the productiveness of their industry, their real wages remaining the same, and you will diminish the rate of profit; and *vice versa*.

II   Take now a slightly more complex case: – another set of workmen, who receive 100 qrs of corn, are employed in producing not corn, but silk: while consuming those 100 qrs they produce, say, 200 lbs of manufactured silk. What will determine the rate of profit in this case? The outlay and the return not being homogeneous, they cannot be directly compared: we must look at them through their values. The rate of profit will plainly depend on the ratio which the *value* of the 200 lbs of silk will bear to the *value* of the 100 qrs of corn which formed the means of effecting their production. What will determine the value of the silk? The cost of its production; but this by hypothesis is equal to the cost of 120 qrs of corn; for it required the same outlay to produce both – viz. 100 qrs of corn. Hence it follows that the rate of profit in the silk manufacture will be the same as in agriculture. And this will be the case whatever may be the productiveness of industry in the former branch of production. For if the silk weavers in the supposed case were only to produce 100 lbs of silk instead of 200, or were to produce 400, this wd not affect the question; since in all cases alike the cost of production being the same, the value of the return, large or small, wd be the same; and, therefore, also the rate of profit.

---

122 There is a nice statement by Mill in 'Newman's Political Economy' (1851) that applies to Ricardo's *Essay on Corn* (1815) where the non-Sraffian perspective is implied: 'Not a few of his criticisms [of the best previous authors] are evidently grounded on imperfect acquaintance with their works. For example, speaking of what is called the Ricardo theory of rent . . . he says "it assumes that wheat is the only agricultural product, and that the value of land is to be measured by capacity of producing it." This is a complete misapprehension. Ricardo's numerical illustrations are expressed in quarters of wheat; but any one, who will take the trouble, can adapt the theory of all other products of land; his successors have partially done so' (V, 447).

It thus appears that the law of profit which we found to operate in the simplest case operates also in that which we may describe as of the first degree of complexity: the rate still depends on the real remuneration of the labourer as compared with the productiveness of his industry in producing his own remuneration (*CW*, III, 1048-9).[123]

Mill did not adopt the suggestion. He agreed that Cairnes's 'mode of putting the doctrine is very good as one among others', and that there was 'no difference of opinion between us' (22 March 1865, XVI, 1018-19; also III, 1095). But he preferred his own formulation which, as we have seen, involved much more complex cases than that represented by the corn model.

There is a further indication of Mill's Ricardian heritage of fundamental importance for the present context. For Ricardo, profits (interest) constituted a surplus in the formal sense only, that it was obtained as the difference between sales proceeds and cost outlays, and not in that sense which implies a divorce of the rate of accumulation from the return on capital; Ricardo thought in terms of 'a natural equilibrium between profits and wages'(1951, I, 226). The standard notion that 'profits depend upon wages' must be understood with this in mind (cf. Hollander, 1979, 265-7). Precisely the same perspective characterizes Mill's work, as is best clarified in the early essay: 'Interest and the wages of superintendance, can scarcely be said to depend upon one another. They are to one another in the same relation as wages and profits are. They are like two buckets in a well: when one rises, the other descends, but neither of the two motions is the cause of the other; both are simultaneous effects of the same cause, the turning of the windlass' (*CW*, IV, 301-2).

XIII   THE FUNDAMENTAL THEOREM GENERALIZED

The general conclusion regarding the inverse wage-profit relation implies that an increase in the cost of labour (or in 'money' wages) cannot be passed on by capitalists: 'The expense of these increased money wages falls wholly on the capitalist. There are no conceivable means by which he can shake it off' (*CW*, III, 699). Thus in common with Ricardo and McCulloch, Mill took the position - as Marx was later to do - that an increase in the real

---

123 Cairnes referred to his proposed formulation as one 'differing from yours and Ricardo's only in form' (III, 1094). In his *Leading Principles* (1874), Cairnes states matters entirely in the Ricardo-Mill fashion: 'The doctrine laid down . . . in the best treatises of Political Economy' includes the proposition 'that general wages (understanding by this general *real* wages, the real remuneration received by the workmen) and general prices have no necessary connection. High wages, we are told, do not make high prices . . . ' This, he maintained, was 'indisputably sound and quite fundamental: yet 'there are few statements in economic science that are more apt to strike an outsider as paradoxical' (200).

costs of wages must be accompanied by a decline in the rate of return on capital, and formulated his argument in the chapter 'Of Distribution as Affected by Exchange' largely upon the assumption that 'money' represents an invariable measure, that is one which accurately reflects the variation in the real costs of wages (or in proportionate wages). In taking this position, Mill divorced himself, in the Ricardian manner, from Adam Smith's conceptions. This comes clearly to the fore during the course of his criticisms of Smith's analysis of the tendency of profits to a minimum which runs in terms of 'competition of capital' (733); and in the context of wage taxation: 'On whom, in this case, will the tax fall? According to Adam Smith, on the community generally, in their character of consumers; since the rise of wages, he thought, would raise general prices. We have seen, however, that general prices depend on other causes, and are never raised by any circumstance which affects all kinds of productive employment in the same manner and degree. A rise of wages occasioned by a tax, must, like any other increase of the cost of labour, be defrayed from profits' (830).

I shall now show that for Mill, as for Ricardo, the substantive prediction that an increase of 'real' wages (that is of labour embodied in wages) is necessarily accompanied by a fall in the rate of return, holds good quite generally – irrespective, that is to say, of the satisfaction by the medium of exchange of the necessary properties required to guarantee its theoretical suitability as invariable standard. The issue pervades the *Principles*.

The 'popular and widely-spread opinion' which Mill set out to refute in his central chapter on the 'Ultimate Analysis of Cost of Production' was 'that high wages make high prices'. In the first place, he insisted, the proposition implied that there could be no 'real rise of wages' for if wages could not rise without a proportional rise of the price of everything, they could not, for any substantial purpose rise at all. This surely is a sufficient *reductio ad absurdam . . .*' (479).[124] Secondly, he made it clear that even were prices to rise following an increase of wages, producers would not benefit therefrom since all their expenses rise: 'It must be remembered too that general high prices, even supposing them to exist, can be of no use to a producer or dealer, considered as such; for if they increase his money returns, they increase in the same degree all his expenses.' Ricardo's identical rendition will be recalled: 'If the price of commodities were permanently

124 Cf. 699: 'The doctrine, indeed, that a rise of wages causes an equivalent rise of prices, is, as we formerly observed, self-contradictory: for if it did so, it would not be a rise of wages; the labourer would get no more of any commodity than he had before, let his money wages rise ever so much; a rise of real wages would be an impossibility. This being equally contrary to reason and to fact, it is evident that a rise of money wages does not raise prices; that high wages are not a cause of high prices. A rise of general wages falls on profits. There is no possible alternative.'

raised by high wages, the proposition would not be less true, which asserts that high wages invariably affect the employers of labour, by depriving them of a portion of their real profits . . . [The employer] would be in no better situation if his money profits had been really diminished in amount, and everything had remained at its former price' (1951, I, 126–7).

These, however, are in the nature of formal arguments. More important is Mill's reliance on the gold-standard mechanism involving the quantity theory to assure the result that wage increases are non-inflationary. Ricardo's formulation may first be noted: 'All commodities cannot rise at the same time without an addition to the quantity of money. This addition could not be obtained at home . . . nor could it be imported from abroad. To purchase any additional quantity of gold from abroad, commodities at home must be cheap, not dear. The importation of gold, and a rise in the price of all home-made commodities with which gold is purchased or paid for, are effects absolutely incompatible' (1951, I, 105).

Mill's own statement to the same effect may be illustrated by reference to the analysis of the inverse profit–wage relation as it appears in the review (1869) of W. T. Thornton's *On Labour*.[125] The reasons given by Mill for the inability of employers to pass on wage increases include the necessity for an increase in the money supply, which would not, under the circumstances supposed, be forthcoming:

There cannot be a general rise of prices unless there is more money expended. But the rise of wages does not cause more money to be expended. It takes from the incomes of the masters and adds to those of the workmen; the former have less to spend, the latter have more; but the general sum of the money incomes of the community remains what it was, and it is upon that sum that money prices depend. There cannot be more money expended on everything, when there is not more money to be expended altogether. In the second place, even if there did happen a rise of all prices, the only effect would be that money, having become of less value in the particular country, while it remained of its former value everywhere else, would be exported until prices were brought down to nearly or quite their former level. But thirdly: even on the impossible supposition that the rise of prices could be kept up, yet, being general, it would not compensate the employer; for though his money returns would be greater, his outgoings (except the fixed payments to those to whom he is in debt) would be increased in the same proportion. Finally, if when wages rose all prices rose in the same ratio, the labourers would be no better off with high wages than with low; their wages would not command more of any

125 While Mill was critical of the Bank Charter Act of 1844, he at no time agreed with Fullarton and Tooke that notes do not affect prices but simply adjust to price movements, or that the money supply is a wholly 'passive' response (Fetter, 1965, 190, 226–7). See for details Chapter 7.

article of consumption; a real rise of wages, therefore, would be an impossibility (*CW*, V, 661).

Precisely the same position was reiterated by Mill in correspondence after the appearance of the review:

I differ from you when you say that a general rise of wages would be of no use to the working classes because it would produce a general rise of prices. A general rise of prices, of anything like a permanent character, can only take place through a general increase of the money incomes of the purchasing community. Now a general rise of wages would not increase the aggregate money incomes, nor consequently the aggregate purchasing power of the community; it would only transfer part of that purchasing power from the employers to the labourers. Consequently a general rise of wages would not raise prices but would be taken out of the profits of the employers: always supposing that those profits were sufficient to bear the reduction (to George Adcroft, 21 June 1870, XVII, 1734–5).

The evidence presented thus suggests a firm conviction on Mill's part of the justness of strict Ricardianism – that is to say a body of doctrine including, first, the relation between the rate of return on capital and the wage rate, identified with the principle of proportionate wages and profits; secondly, the formal derivation of the relation in terms of a money of invariable value, despite an emphasis upon the relativity of exchange value; and thirdly, the view that the inverse relation holds good independently of the measure in that wage-rate increases can be shown to be non-inflationary.

It is this latter characteristic of Mill's work, in conjunction with a too hasty reading of the chapter formally devoted to the measure, which may be responsible for the view that Mill, unlike Ricardo, discussed the inverse relation without making use of an invariable measure. This view is misleading in two regards. First, the measure was in fact used as a conceptual device as we have seen; and second, Ricardo too maintained that the inverse relation holds in the real world independently of the measure. There is a complete identity of opinion between Mill and Ricardo on these fundamental issues. To refer to Mill's maintenance of an 'emasculated' version of Ricardo's system is totally unjustified.

XIV   THE FUNDAMENTAL THEOREM AND ALLOCATION

There is more to Mill's Ricardian approach to distribution. I refer to the case in support of the inverse profit–wage relation based upon the principles of allocation economics alluded to in our discussion of trade. In contrast to an increase in wages affecting one sector where price will rise to assure equality of profit rates across the board, there exists no mechanism whereby prices would be forced upwards in the event of a general increase in wages, since all firms through the system are affected equally by the change. From this perspective the fundamental theorem on distribution is founded squarely upon the theory of allocation:

[A] general rise of wages would not raise prices but would be taken out of the profits of the employers; always supposing that those profits were sufficient to bear the reduction.

The case is different with a rise of wages confined to a single, or a small number of employments. That rise if taken out of profits, would place a particular class of employers at a disadvantage compared with other employers: & as soon as they ceased to hope that the loss would be only temporary, they would withdraw part of their capital, or at all events, all new capital would avoid those trades & go into others. Consequently the supply of these particular articles would fall short, & their prices would rise so as to indemnify the employers for the rise of wages. But this would not happen in case of a rise of all wages, for as all capitalists would be affected nearly alike they could not as a body relieve themselves by turning their capital into another employment (21 June 1870, *CW*, XVII, 1735).

Expenses which affect all commodities equally, have no influence on prices. If the maker of broadcloth or cutlery, and nobody else, had to pay higher wages, the price of his commodity would rise, just as it would if he had to employ more labour; because otherwise he would gain less profit than other producers, and nobody would engage in the employment. But if everybody has to pay higher wages, or everybody to employ more labour, the loss must be submitted to; as it affects everybody alike, no one can hope to get rid of it by a change of employment, each therefore resigns himself to a diminution of profits, and prices remain as they were. In like manner, general low wages, or a general increase in the productiveness of labour, does not make prices low, but profits high. If wages fall, (meaning here by wages the cost of labour), why, on that account, should the producer lower his price? He will be forced, it may be said, by the competition of other capitalists who will crowd into his employment. But other capitalists are also paying lower wages, and by entering into competition with him they would gain nothing but what they are gaining already (III, 692; cf. 699).

International trade provides a nice illustration of this theme – the Ricardian 'dogma' (as Edgeworth called it, 1925, II, 24) that low or high general wages do not affect competitiveness in foreign markets. A standard error that Mill set out to refute in Book III, xxv, relates to the reputed disadvantage under which a 'high-wage' economy operated: 'We continually hear of the disadvantage under which the British producer labours, both in foreign markets and even in his own, through the lower wages paid by his foreign rivals. These lower wages, we are told, enable, or are always on the point of enabling them to sell at lower prices, and to dislodge the English manufacturer from all markets in which he is not artificially protected' (III, 689). Even when allowance is made for labour efficiency and the argument expressed in terms of the 'cost of labour' it is rejected on the Ricardian grounds that only a wage in the export sector kept below the general level would generate a real advantage by lessening the *comparative* cost of those articles relative to others, acting in the manner of a technological innovation: 'no . . . advantage is conferred by low wages when common to all branches of industry' (691–2). To this is attached a qualification which reinforces the underlying allocative logic to the argument: 'It is quite true that if the cost of labour is lower in America than in England, America could sell her cottons to Cuba at a lower price than England, and still gain as high a profit as the English manufacturer. But it is not with the profit of the English manufacturer that the American cotton spinner will make his comparison; it is with the profits of other American capitalists. These enjoy, in common with himself, the benefit of a low cost of labour, and have accordingly a high rate of profit. This high profit the cotton spinner must also have: he will not content himself with the English profit.'[126] A statement of this position will also be found in the context of Mill's discussion of the nature of costs when focussing upon the distinction between costs from an aggregative and industry perspective. It is precisely in this context, involving as it does recognition of the principles of allocation theory, that Mill insisted on the inverse wage–profit relation: 'There is no mode in which capitalists can compensate themselves for a high cost of labour, through any action, on values or prices. It cannot be prevented from taking its place on low profits' (479). It is significant too that during the course of his correspondence with Cairnes regarding the nature of costs (referred to above, pp. 285–6), Mill immediately saw the implications of the argument for the inverse wage–profit relationship (letter, 4 Oct. 1872, XVII, 1909; also in O'Brien, 1943, 285).

The Mill–Ricardo analysis of the inverse profit–wage relation involves

---

126 Torrens (1834) was highly critical of this Ricardian position (as formulated by McCulloch) on the grounds that it presumed full mobility of capital goods between sectors. The assumption is related to 'the fallacy, unfortunately too prevalent amongst economical writers, of confounding distinctions by hasty generalizations' – to 'proneness for general reasoning' – which leads to the 'error of conceiving that because capital, consisting of money, may pass from employment to employment, in order to obtain the customary rate of profit, capital, consisting of buildings and machinery, may be equally locomotive' (67). In the 'Marshallian' short run, the return on fixed capital might fall below the going 'customary' rate. His argument, which involves one of the clearest distinctions between the Marshallian long- and short-run, was not taken up by Mill.

a direct application of the principles of allocation – a general wage change leaves resource allocation and therefore prices unaffected, so that profits are squeezed, whereas a partial wage change generates supply and price adjustments which allow capitalists in the affected sector to be compensated.[127] Now it is unlikely that Mill regarded his demonstration of the theorem on distribution in these terms to be in conflict with the demonstration in terms of the standard measure. It is essential to spell out the status of each of the demonstrations.

The inverse theorem based on the measure constitutes a purely 'logical' or 'formal' argument, and has, strictly speaking, no 'economic' content, for nothing is said of the mechanism whereby the inverse profit variation is achieved. (The same is true of the argument in terms of the quantity theory.) It is perhaps not misleading to say that the theorem is in the nature of a macro-economic statement regarding proportionate shares in a net output of constant value, leaving open the 'causal' explanation of the transition between equilibrium states in a competitive world. (Mill's formulation of the 'cause' of profits (II, 411) will be recalled (see above, p. 341); the 'capitalist' is said to be representative of capitalists as a class.) This explanation is provided by the analysis of the inverse relation in terms of the allocation mechanism: since a general wage variation generates no change in differential cost conditions and thus creates no disturbance to the relative returns on capital, there are no adjustments induced by the disturbance. Producers are either passive recipients of a bonus increase in their profits or passive sufferers of a decrease.

Our discussion thus far has implicitly presupposed a uniformity of capital–labour ratios across the board. But what has been said can, in principle, be generalized for the complex case where differentials in factor proportions exist between industries. Indeed, only in this case can the operation of the allocative mechanism in the event of a general wage change be observed, the ultimate effects of an increase of wages upon relative prices flowing from reactions by capitalists to a disturbed structure of profit rates created by the differential impact upon costs of the initial variation. Specifically, immediately following the wage increase profits are reduced universally at the ruling prices; the allocative effects follow subsequently in consequence of the differential decline in the returns on capital, some industries expanding and others contracting depending upon their factor intensities, with inverse movements in their prices. Equality of profit rates throughout the system at the new lower level emerges along with the new pattern of outputs and new set of long run or 'necessary' cost prices. Can this formulation be attributed to Mill?

127 Here we recall the two modes of reaction to a change in data discussed above (p. 289f). That involving the direct adjustment of price to the new cost level implies a disturbance impinging on one industry.

We have seen in our discussion of cost price that Mill made careful allowance for the effect of a general change in wages upon relative prices (above, p. 282, regarding III, 484, 485). It is true that his formulation implies the immediate emergence of the new lower (equilibrium) rate of profit, upon the hypothesized rise in wages, which is then applied in the calculation of the structure of relative prices, but it is difficult to believe that he could have believed that this sufficed to explain the actual determination of the general profit rate and the price structure at their new levels; we need only recall the central role accorded relative rates of profit in governing the pattern of commodity supplies, and prices, following disturbances to costs (above, p. 282). The process of transition between equilibrium states must, therefore, involve allocative effects set in motion by the disturbed structure of profit rates at original outputs and prices. This interpretation of the inverse profit–wage relation is the only one consistent with Mill's central discussions of demand–supply and profit-rate equalization as indeed it is for those of Ricardo (cf. Hollander, 1979, 299–301).

At the same time historical accuracy requires us to recognize that the above formulation is never explicitly spelled out by Mill. Like Ricardo he frequently sidestepped the issue by supposing (implicitly) uniformity of factor ratios (e.g. XVI, 1127; XVII, 1735). And for both, although the demonstration of the basic distribution theorem in terms of the standard measure was no more than a formal 'predictive' device, the ambiguous methodological status thereof lent itself readily enough to misleading formulations which suggest that capitalists react to the lower general profit rate emerging along with the higher rate of wages, although in reality reactions are solely to the disturbed *structure* of returns on capital.

There is, moreover, the fundamental ambiguity in Mill's analysis already alluded to, namely the proposition in the *Principles* (in contrast to the essay of 1830) that fixed capital can be entirely reduced to labour (above, p. 342). Ricardo for his part had gone to very great lengths in his first chapter to explain why it is that a wage-rate increase has a lesser impact upon the costs of commodities with relatively high fixed-capital coefficients turning upon the fact that the cost of 'machinery' comprises profits as well as wages (1951, I, 61–2; cf. Hollander, 1979, 200–1, 290); without this assumption it is not at all apparent that the supply adjustments and consequent changes in cost prices would be generated by the supposed disturbance. Yet Mill seems to be unaware of the extensive implications flowing from his proposal for his own 'Ricardian' analysis of the impact of the wage change upon the structure of prices.

The same problem emerges when we consider the reverse linkage, and inquire whether an alteration in the configuration of final demand can play back upon distribution. Here we must have in mind Mill's fourth proposition on capital (see below, pp. 371–3). When read literally this proposition appears to rule out any such impact: 'The demand for commodities determines in what particular branch of production the labour and capital shall be employed; it determines the *direction* of the labour; but not the more or less of the labour itself, or of the maintenance or payment of the labour' (II, 78). Indeed, the proposition has been interpreted as a deliberate statement to the effect that 'the configuration of demand was irrelevant to the distribution of the product between profit and wages' – a consequence flowing from the implicit assumption 'like so much of Ricardian reasoning . . . that the proportions between capital and labour were equal in all industries' (Dobb, 1940, 45; cf. also Johnson, 1949, 535; Hollander, 1964. In 1875, W. S. Jevons (1977, VI, 38) criticized Mill for implying by the first proposition on capital 'industry is limited by capital' – of which the fourth is a corollary – a neglect of differential factor ratios notwithstanding Adam Smith's emphasis thereupon.) Yet the existence of differential factor proportions was conspicuous in the Ricardo–McCulloch–Mill formulations of value theory. It is, therefore, possible that Mill assumed uniform factor proportions in formulating his fourth proposition merely because neither the essential principle involved nor the policy conclusions based upon it required a more complex model incorporating the impact on the overall demand for labour of a change in the pattern of final demand. The main theoretical statement was, we shall see, designed to distinguish 'direct' expenditures on labour by capitalists and others from 'indirect' expenditures by final purchasers; so that little purpose would have been served by differentiating between alternative forms of 'indirect' outlay.

In support of this position it should be noted that while Mill said nothing about the effect of a change in tastes upon distribution, he did, following Ricardo, discuss the consequences for wages of exogenous alterations in technology which entailed the 'conversion' of circulating into fixed capital (below, p. 377f). It is difficult to see why an alteration in the distribution of the capital stock brought about by a change in the pattern of tastes cannot be similarly investigated. This is all the more so since Mill, in the same context, alluded to factor proportions which do differ from industry to industry according to circumstances: 'With the proceeds of his finished goods, a manufacturer will partly pay his work-people, partly replenish his stock of the materials of his manufacture, and partly provide new buildings and machinery, or repair the old; but how much will be devoted to one purpose, and how much to another, depends on the nature of the manufacture, and the requirements of the particular moment' (*CW*, II, 99).

Again, while Mill failed to trace through the consequences for distribution of a changed pattern of demand for final goods he did deal with the imposition of a standard *ad valorem* tax in conditions of differential factor ratios, allowing explicitly for an altered pattern of activity:

When the cost of production is increased artificially by a tax, the effect is the same as when it is increased by natural causes. If only one or a few commodities are affected, their value and price rise, so as to compensate the producer or dealer for the peculiar burthen; but if there were a tax on all commodities, exactly proportioned to their value, no such compensation would be obtained: there would neither be a general rise of values, which is an absurdity, nor of prices, which depend on causes entirely different. There would, however, as Mr. M'Culloch has pointed out, be a disturbance of values, some falling, others rising, owing to a circumstance, the effect of which on values and prices we formerly discussed; the different durability of the capital employed in different occupations. The gross produce of industry consists of two parts; one portion serving to replace the capital consumed, while the other portion is profit. Now equal capitals in two branches of production must have equal expectations of profit; but if a greater portion of one than of the other is fixed capital, or if that fixed capital is more durable, there will be a less consumption of capital in the year, and less will be required to replace it, so that the profit, if absolutely the same, will form a greater proportion of the annual returns. To derive from a capital of 1000*l.* a profit of 100*l.*, the one producer may have to sell produce to the value of 1100*l.*, the other only to the value of 500*l.* If on these two branches of industry a tax be imposed of five per cent *ad valorem*, the last will be charged only with 25*l.*, the first with 55*l.*; leaving to the one 75*l.* profit, to the other only 45*l.* To equalize, therefore, their expectation of profit, the one commodity must rise in price, or the other must fall, or both: commodities made chiefly by immediate labour must rise in value, as compared with those which are chiefly made by machinery (III, 838–9; the reference to the disturbance in relative values is presumably to *ibid.*, 485).

It would seem to be self-evident that any change in the pattern of activity thus brought about must play back on the wage rate since the average 'durability of capital' is altered – just as it might be in the case of a simple change in the configuration of demand.

Unfortunately Mill himself did not explicitly draw this conclusion, but brought the matter to an abrupt close by an annoying remark that 'it is unnecessary to prosecute this branch of the inquiry any further'. Moreover, there is the implication flowing from the proposition to reduce fixed capital to labour. Here the example set by McCulloch is instructive. That 'demand for commodities is not demand for labour' was a proposition that did not originate with Mill; it has a Ricardian origin, and it appears in McCulloch's *Principles* (as well as the *Elements* of James Mill). Now McCulloch allowed for effects of alternative consumption patterns upon employment flowing from the existence of differential factor ratios; in his discussion of the proposition that there exists a 'difference in their influence over wages between a demand for labour and for its products' we find an explicit qualification that 'the influence of an increased demand for commodities, or for the produce of labour, is by no means identical with an increased demand for labour, and would depend partly on whether the commodity was wholly or in part the produce of labour or of machinery' (1864, 256). But the issue is immediately dismissed as being 'of little importance; for capital being itself the result of antecedent labour, whatever is expended upon it really goes to replace labour, and in the end is identical in its effects with a direct expenditure upon the latter'. Any effect, therefore, must presumably in McCulloch's view have been short-run only. It cannot be ruled out that J. S. Mill was led to this non-Ricardian position by his adoption of the assumption that capital is wholly 'the result of antecedent labour'.[128]

The absence of an explicit discussion of the matter by Mill combined with the foregoing consideration together suggest that we cannot dismiss the possibility that he ruled out, as a matter of principle, at least the long-run effects upon distribution of a change in the configuration of demand. Against this must be weighed the other considerations outlined above. Here we have a case where caution is the better part of valour in our interpretation.

128 James Mill (1844) for his part argued that 'an increase of demand for labour can arise from two causes only; either from an increase of capital, the fund destined for the employment of labour; or a difference in the proportion between the produce of fixed capital and that of immediate labour' (259–60). Attention was then devoted to changes in the demand for labour resulting from taxation which directed purchasing power to government; in the event that the government's expenditure involved commodities with a peculiarly high labour content compared with commodities otherwise purchased 'there would so far be an increase of demand for labour' (265). But such an effect would, he added, be 'an extraneous circumstance' since anything was possible depending upon the actual patterns of expenditure involved – there was, in short, no general rule which could be formulated.

It is perhaps also significant that Cairnes (1874), in his subsequent treatment of wages and employment, added a qualifying note to his general conclusion that the effects of changes in the pattern of final expenditure would be limited to a mere transference of wealth 'between different groups of workers': 'In theoretical strictness, this position needs qualification. It would only be strictly true, if the Wages fund bore always the same proportion to the capital employed in production, which is not the fact. Supposing, e.g., expenditure were largely directed from clothing to food, and that in consequence capital were transferred from manufactures to agriculture, inasmuch as a given amount of capital employed in agriculture will in general contain a larger element of Wages-fund than the same amount of capital employed in manufactures (owing to the large use of fixed capital in the latter case), it follows that a substitution of a demand for food for a demand for clothes would in this case issue in an increase in the aggregate Wages-fund; as a substitution in the contrary sense would have the opposite effect' (192). Yet Cairnes made little of the qualification, playing it down as inconsequential in contemporary Britain. And this is surprising since he had much to say about the variability of factor proportions from industry to industry.

SIX

Capital, employment and growth

I INTRODUCTION

This chapter sets out by considering Mill's four fundamental propositions regarding capital, with particular attention given to the determinants of aggregate employment. An interpretation of the 1869 reputation of the wage-fund theory consistent with the micro-economic perspective outlined in the previous chapter is suggested; and the relation between wage-fund theorizing and the fundamental theorem on distribution traced out. We then take up the formal contrast between 'statics' and 'dynamics' as preliminary to an analysis of the secular trend of the wage and profit rates both from a theoretical and empirical perspective. Sections on the interconnections between the profit-rate trend and cycle; and the profit-rate trend and the law of markets bring the chapter to a close.

II ON CAPITAL AND CAPITAL MAINTENANCE

Depending on the context so Mill's perspective on the nature of 'capital' varies; there is no single definition. Several of the better-known formulations focus upon the material items constituting a stock of capital goods. Thus capital is 'a stock previously accumulated, of the products of foremen labour', the function of which in production is 'to afford the shelter, protection, tools and materials which the work requires, and to feed and otherwise maintain the labourers during the process. These are the services which present labour requires from past, and from the produce of past, labour' (*CW*, II, 55). In the same context Mill also defined capital as 'wealth appropriated to reproductive employment'; and yet more generally, as 'whatever of the produce of the country is devoted to production' (57).

The first of Mill's four propositions respecting capital – all of which are part and parcel of Ricardian doctrine – asserts that 'industry is limited by capital':

The following are common expressions, implying its truth. The act of directing industry to a particular employment is described by the phrase 'applying capital' to the employment. To employ industry on the land is to apply capital to the land. To employ labour in a manufacture is to invest capital in the manufacture. This implies that industry cannot be employed to any greater extent than there is capital to invest . . . There can be no more industry than is supplied with materials to work up and food to eat. Self-evident as the thing is, it is often forgotten that the people of a country are maintained and have their wants supplied, not by the produce of present labour, but of past. They consume what has been produced not what is about to be produced. Now, of what has been produced, a part only is allotted to the support of productive labour; and there will not and cannot be more of that labour than the portion so allotted (which is the capital of the country) can feed and provide with the materials and instruments of production (63–4).

It is clear that Mill did not intend his proposition to relate solely to a dependency of productive employment upon 'circulating capital' (wage goods and materials). Very specific reference is made to fixed capital.[1] But the 'dependency of productive employment on capital has a dual sense, a duality emphasized in the observation contained within the ellipses in the foregoing passage which reads as follows:

The expression 'applying capital' is of course metaphorical: what is really applied is labour; capital being an indispensable condition. Again, we often speak of 'the productive powers of capital.' This expression is not literally correct. The only productive powers are those of labour and natural agents; or if any portion of capital can by a stretch of language be said to have a productive power of its own, it is only tools and machinery, which, like wind or water, may be said to co-operate with labour. The food of labourers and the materials of production have no productive power; but labour cannot exert its productive power unless provided with them.

An equally clear formulation of the duality appears in the early paper 'On Profits and Interest' (composed in 1830):

1 Capital goods are only played down quantitatively in the case of 'domestic' industry (64n).

Capital, strictly speaking, has no productive power. The only productive power is that of labour; assisted, no doubt, by tools and acting upon materials. That portion of capital which consists of tools and machinery, may be said, perhaps, without any great impropriety, to have a productive power, because they contribute, along with labour, to the accomplishment of production. But that portion of capital which consists of wages, has no productive power of its own. Wages have no productive power; they are the price of a productive power. Wages do not contribute, along with labour, to the production of commodities, no more than the price of tools contribute along with the tools themselves . . .

The proper view of capital, is that anything whatever, which a person possesses, constitutes his capital, provided he is able, and intends, to employ it, not in consumption for the purpose of enjoyment, but in possessing himself of the means of production, with the intention of employing those means productively. Now the means of production are labour, implements, and materials. The only productive power which anywhere exists, is the productive power of labour, implements, and materials.

We need not, on this account, altogether proscribe the expression 'productive power of capital'; but we should carefully note, that it can only mean the quantity of real productive power which the capitalist, by means of his capital, can command. This may change, though the productive power of labour remains the same. Wages, for example, may rise; and then, although all the circumstances of production remain exactly as they were before, the same capital will yield a less return, because it will set in motion a less quantity of productive labour. (*CW*, IV, 290–91).

The references here to physical capital goods (tools and machinery) is appropriate from a 'production function' perspective. Despite an evident hesitancy Mill does allow the part of 'capital' acts as a factor of production, and has productive power, on a par with labour and land. Mill (like Marx) is unwilling to treat wage goods in quite the same manner – although assuming a constant commodity wage, wage-goods capital might at least formally be treated on a par with technological capital in defining a relationship with labour – but in the *Principles*, though not in the essay, he (unlike Marx) also sets apart materials.[2] The contrast between the categories is well defined in a statement regarding fixed capital (machinery, structures and land improvements) that 'in any large increase of capital a considerable portion will generally be thus employed, and will only co-operate with labourers, not maintain them' (*CW*, II, 66).[3]

2 But not consistently so; cf. Mill's 'Marxian' perspective: 'It should be observed further, that the portion of capital consumed in the form of seed or material, though, unlike fixed capital, it requires to be at once replaced from the gross produce, stands yet in the same relation to the employment of labour, as fixed capital does' (99).

3 Marshall (1920), 789, refers critically to (German and Austrian) economists who insist on confining capital to 'auxiliary or instrumental capital' on the grounds that 'to keep

As we shall see, Mill also frequently uses a conception of capital which refers to a fluid fund as distinct from the instruments of production, wage goods and materials in which the fund is embodied. Depending on the context either conception may be appropriate. It is only regrettable that Mill transferred between the two rather too easily; indeed there are suggestions of Fundism in the foregoing statement of the first proposition, which on the whole has a materialist flavour. (For these terms see Hicks, 1974; also Hicks, 1963, 342f.)

In this section we address the issue of capital maintenance, capital envisaged as real goods. The notion of the perpetual 'consumption' of capital according to degree of durability is the conspicuous feature of Mill's account. It governs the distinction between the 'circulating' and the 'fixed' categories and the problem of 'conversion' of the former into the latter. Thus, '[o]f the capital engaged in the production of any commodity there is a part which, after being once used, exists no longer as capital' (91) – a reference to materials and wage goods consumed in a single use, the term 'circulating' conveying the fact that this category is 'constantly renewed by the sale of the finished product, and when renewed is perpetually parted with in buying materials and paying wages; so that it does its works, not by being kept, but by changing hands'. This formulation (which has Smithian pedigree) suggests capital as a circulating money fund, but it is easy enough to transfer as Mill does from the fund to the physical inputs in which it is embodied.

Fixed capital comprises instruments of production 'of a more or less permanent character'; which produce their effect not by being parted with, but by being kept; and the efficacy of which is not exhausted in a single use' (92). Again a Smithian flavour is apparent. The emphasis is upon the two-fold circumstance that 'their function as productive instruments is prolonged through many repetitions of the productive operation', and that 'the return to [them] is spread over a period of corresponding duration'.[4]

The notion of a circulating money fund may seem inappropriate here but it does remain in focus. For Mill had, shortly before, insisted that 'the greater part, in value, of the wealth now existing in England, has been

clear the contrast between production and consumption, nothing which enters directly into consumption should be regarded as a means to production.' Marshall himself takes the eclectic classical view: 'there appears no good reason why a thing should not be regarded in a two-fold capacity.' Clark (1899), 123, is more critical.

4 Mill does not formally emphasize the significance of the annual period. But see his note on Senior's statement that 'the materials and stock-in-trade of an iron founder would be circulating capital according to Smith, and fixed capital according to Ricardo': 'I think Ricardo & my father habitually gave the name fixed capital to all those portions of capital which are not consumed & reproduced within the year – & that they included in each class the same objects as Adam Smith, though their definition may have been different' (Mill, 1945, 135).

produced by human hands within the last twelve months' (73). The point
is elaborated at some length:

A very small proportion indeed of that large aggregate was in existence ten years
ago; – of the present productive capital of the country scarcely any part, except
farmhouses and manufactories, and a few ships and machines; and even these would
not in most cases have survived so long, if fresh labour had not been employed
within that period in putting them into repair. The land subsists, and the land is
almost the only thing that subsists. Everything which is produced perishes and most
things very quickly. Most kinds of capital are not fitted by their nature to be long
preserved . . . If we except bridges and aqueducts (to which may in some countries
be added tanks and embankments), there are few instances of any edifice applied
to industrial purposes which has been of great duration; such buildings do not hold
out against wear and tear, nor is it good economy to construct them of the solidity
necessary for permanency. Capital is kept up from age to age not by preservation,
but by perpetual reproduction: every part of it is used and destroyed, generally
very soon after it is produced, but those who consume it are employed meanwhile
in producing more (73).[5]

With few exceptions, therefore, fixed or durable capital has the same
character as circulating capital. The notion of a revolving fund comes back
into its own, circulating capital constituting a special category of a broader
genus. The empirical allusions have important theoretical implications, for
the principle of profit-rate uniformity is all the more potent the higher is
the annual replacement requirements of the gross capital stock.

Here it is appropriate to refer to the third theorem – that capital 'although
saved, and, the result of saving, is nevertheless consumed' (70), for the
same perspective of capital consumption is much reinforced: 'the word saving
does not imply that what is saved is not consumed, nor even that its
consumption is deferred; but only that, if consumed immediately, it is not
consumed by the person who saves it . . . Part is exchanged for tools or
machinery, which are worn out by use; part for seed or materials, which
are destroyed as such by being sown or wrought up . . . The remainder
is paid to productive labourers, who consume it for their daily wants'.
Similarly, in a restatement of the third theorem in the form 'what is saved
is consumed', Mill points out that

saving (for productive investment), and spending, coincide very closely in the first
stage of their operations. The effects of both begin with consumption; with the
destruction of a certain portion of wealth; only the things consumed, and the persons

---

5 This closing detail is ambiguous, but refers apparently to the fact that productive labour
   replaces its own consumption, and the 'consumption' of the real capital goods used.

consuming, are different. There is, in the one case, a wearing out of tools, a destruction of material, and a quantity of food and clothing supplied to labourers, which they destroy by use: in the other case, there is a consumption, that is to say destruction of wines, equipages, and furniture. Thus far, the consequence to the national wealth has been much the same; an equivalent quantity of it has been destroyed in both cases. But in the spending, this first stage is also the final stage; . . . while, on the contrary, the saving person, during the whole time that the destruction was going on, has had labourers at work repairing it; who are ultimately found to have replaced, with an increase, the equivalent of what has been consumed (71).[6]

And again in an elaboration of the theorem Mill writes that 'everything which is produced is consumed: both what is saved and what is said to be spent, and the former quite as rapidly as the latter' (73), again referring to the notion that 'capital is kept up, not by preservation, but by perpetual reproduction'.[7]

The terms of the discussion thus far have been limited to the maintenance of 'productive' activity in unchanged dimension; more specifically, by 'saving' has been intended the employment, by the provision of appropriate equipment and wage goods, of productive labour – labour engaged with an eye to profit in which venture these capital goods (including wage goods) are used up or 'consumed' more-or-less rapidly. What is involved in the maintenance of activity is well put by the observation that 'as this operation [of supporting productive labour] admits of being repeated indefinitely without any fresh act of saving, a saving once made becomes a fund to maintain a corresponding number of labourers in perpetuity, reproducing annually their own maintenance with a profit' (71). (We shall return to this passage in a subsequent discussion of abstinence; below, p. 436.)

The notion of a perpetual consumption and reproduction of capital coincides with Sir John Hicks's representation of the classical view of capital as 'a sum of values which may conveniently be described as a Fund', a

6 The limitation of the theorem to saving 'for productive investment' excludes hoarding.

7 The 'third theorem' is sometimes understood to refer to the act of adding to capital (e.g. Hayek, 1941a, 272f). This latter issue seems, however, to be the better seen within the confines of the fourth proposition – 'demand for commodities is not demand for labour' (see Section III).

The perpetual consumption and reproduction of capital is graphically described in an analogy between the growth of capital and of population: 'Every individual who is born, dies, but in each year the number born exceeds the number who die: the population, therefore, always increases, though not one person of those composing it was alive until a very recent date' (74). This analogy applies *a fortiori* to the case of constant capital and population.

fund only temporarily embodied in specific assets (1974, 309).[8] This he contrasts with a later 'Materialist' tradition revived more recently with 'the rethinking of capital theory and of growth theory'. But we must be cautious and avoid sharp contrasts. It is not easy to distinguish Mill's fundism from that of Clark (1899).[9] Mill may have hoped, as Hicks suggests, to reconcile the empirical significance of plant and machinery with a fundist outlook on the grounds that 'most kinds of capital are not fitted by their nature to be long preserved' but the notion itself is not much different from the early neo-classicists who were to view interest as the return on 'capital' understood as a 'free' or 'floating' fund, to use Marshall's terms, independent of its specific forms (1920, 411–12, 418–19, 592–3). Conversely, Mill we shall find was sometimes emphatic in adopting a production function perspective, and this with consequences as in the discussion of the source of profits and employment capacity. Neither did he neglect to trace out the implications of long-lived investments.

In most instances capital goods undergo continual depreciation necessitating replacement. But 'permanent' investments entailed 'capital expended once for all, in the commencement of an undertaking, to prepare the way for subsequent operations' (II, 92). Mill here had in mind land improvements in subsoil draining, soil mixtures and the like, the cutting of canals, the construction of roads and docks, the opening of mines and

8 Hicks hypothesizes that the classical schema had its origins in a trading system extended to agriculture where stocks and work-in-progress are the main physical assets, and where the farmer 'turns over his capital' in a process involving the purchase of labour services and sale of the product when completed. This perspective he explains in terms of the accounting practice on the part of merchants; for the merchant's capital is a fund embodied in his physical stock-in-trade, but distinguished from it – the capital appearing on the liabilities side and the real goods on the assets side of the balance sheet. It is the fund that 'circulates' or 'turns over', conceptions which make little sense if applied to the goods themselves (310–11). But when the physical assets took the form of plant and machinery 'a candidate appeared for factor status that did not fit into the classical triad' – that triad involving labour (a flow) working *on* land (a stock) *through* capital (a 'fund') rather than on capital or with capital in the production-function tradition.

9 Clark distinguished his own marginal-productivity approach from that of the Austrians in so far as he recognized 'the difference between permanent capital, or an abiding fund of productive wealth, and particular capital-goods, or instruments of production which perish in the using' (viii; also 116f).

    On the Clarkian notion see Thorstein Veblen (1919), 195f, Veblen argues that the notion of capital 'as a physically "abiding entity" constituted of a succession of productive goods that make up the industrial equipment', breaks down as soon as the problem of 'transfer of capital' between industries is addressed. For Clark (1899) recognized that this does not usually involve a mechanical transfer of physical items, a recognition that contradicts 'the main position, that "capital" is made up of "capital goods". The continuum in which "the abiding entity" of capital resides is a continuity of ownership, not a physical fact . . . "[C]apital" is a pecuniary fact, not a mechanical one.' The criticism is unfair in so far as Clark himself insisted that the fund of capital is 'capital in the abstract . . .' (120n).

so forth. The distinction between the categories is very much emphasized. Implements and buildings 'require, at intervals, partial renewal by means of repairs, and are at last entirely worn out', taking on the character of materials which are renewed in their entirety albeit in a single use. But mines, docks, canals, drains, dykes (while also requiring regular or periodic repair) never require 'entire renewal'. Similarly, land improvements 'produce an increase of return, which, after defraying all expenditure necessary for keeping them up, still leaves a surplus. This surplus forms the return to the capital sunk in the first instance, and that return does not, as in the case of machinery, terminate in the wearing out of the machine, but continues for ever' (93).[10]

In the Ricardian tradition (*Works*, I, 262) the characteristic 'permanence' of roads, canals, docks, and land improvements places them in the same category as the natural properties of the soil, albeit that they are originally man made. Thus the capital 'consumed' in the improvements was 'consumed productively, and has left a permanent result in the improved productiveness of an appropriated natural agent, the land. We may call the increased produce the joint result of the land and of a capital fixed in the land. But as the capital, having in reality been consumed, cannot be withdrawn, its productiveness is thenceforth indissolubly blended with that arising from the original qualities of the soil; and the remuneration for the use of it thenceforth depends, not upon the laws which govern the returns to labour and capital, but upon those which govern the recompense for natural agents' (II, 93). Similarly

with regard to capital sunk in improvements, and not requiring periodical renewal, but spent once for all in giving the land a permanent increase of productiveness, it appears to me that the return made to such capital loses altogether the character of profits, and is governed by the principle of rent. It is true that a landlord will not expend capital in improving his estate, unless he expects from the improvement

10 The real cost of the land 'improvements' – and of course this applies to other permanent investments – is defined as the capital (the wage goods and tools or perhaps the money counterpart thereof) absorbed in their construction, and Mill warns against any misunderstanding on this score: 'The land, thus increased in productiveness, bears a value in the market, proportional to the increase: and hence it is usual to consider the capital which was invested, or sunk, in making the improvement, as still existing in the increased value of the land. There must be no mistake, however. The capital, like all other capital, has been consumed. It was consumed in maintaining the labourers who executed the improvement, and in the wear and tear of the tools by which they were assisted.'

We must reiterate that the third fundamental theorem perceives the 'consumption' of capital as the using up of capital goods in production whether in the consumption ('luxury')-goods sector or the capital-goods sector. But in the present context Mill refers to capital consumption specifically in the production of permanent categories of capital goods.

an increase in income surpassing the interest of his outlay . . . Prospectively, this increase may be regarded as profit; but when the expense has been incurred, and the improvement made, the rent of the improved land is governed by the same rules as that of the unimproved (423–4).

III AGGREGATE EMPLOYMENT CAPACITY: 'DEMAND FOR COMMODITIES IS NOT DEMAND FOR LABOUR'

Thus far we have understood 'saving' to imply the support and maintenance of labour in the production of material wealth, rather than the creation of new capital goods. But the term is also used in this sense. It is the subject matter of the second theorem on capital: 'If all persons were to expend in personal indulgences all that they produce, and all the income they receive from what is produced by others, capital could not increase . . . All that any one employs in supporting and carrying on any other labour than his own, must have been originally brought together by saving; somebody must have produced it and foreborne to consume it. We may say . . . without material inaccuracy, that all capital and especially all addition to capital, are the result of saving' (*CW*, II, 68–9).[11]

In the absence of an initially expanded surplus (in consequence of technical progress) – and one must add assuming no idle resources – an increase in orders for capital goods does imply reduced orders for 'luxuries'. And it is the product thus foregone that constitutes the ultimate cost of creating the new capital; conversely, an actual increase in 'luxury' spending necessarily implies – on the same assumptions – a positive reduction in capital-goods output: 'In proportion as any class is improvident or luxurious, the industry of the country takes the direction of producing luxuries for their use; while not only the employment for productive labour is diminished, but the subsistence and instruments which are the means of such employment do actually exist in smaller quantity' (72). This is the sense we are to understand the celebrated declaration that 'saving . . . enriches, and spending impoverishes, the community, along with the individual'.

In what follows we consider what precisely is involved by a net savings (investment) program. Here we must have in mind Mill's fourth proposition on capital that 'the demand for commodities is not demand for labour' – a direct inference of the first.[12]

---

11 In this context we see very clearly that Mill's main preoccupation is with the capitalist-exchange system. For he himself observes by way of introduction that isolated individuals, or isolated communities of co-operative producers, who consume for their own wants during the production process cannot easily be represented as 'saving'. And although it is true that consumption occurs as part of the production process rather than as 'final' consumption, Mill limits his theorem to the maintenance by the capitalist of 'other labour than his own'.

12 Hayek (1941), 433; also Taussig (1896), 219–20, and Marshall (1920), 828.

The theorem is restated in various forms: 'what supports and employs productive labour, is the capital expended in setting it to work, and not the demand of purchasers for the produce of the labour when completed'; the demand for commodities 'determines in what particular branch of production the labour and capital shall be employed' (*CW* II, 78); 'to purchase produce is not to employ labour; . . . the demand for labour is constituted by the wages which precede the production, and not by the demand which may exist for the commodities resulting from the production . . .' (79); '[i]t is not the money paid by the purchaser, which remunerates the labour; it is the capital of the producer; the demand only determines in what manner that capital shall be employed, and what kind of labour it shall remunerate' (88). The general proposition obviously cannot relate to the demand for particular kinds of labour or labour in particular industries; for Mill insisted repeatedly that a demand for a particular kind of commodity does give rise to a demand for labour to make that commodity (above, p. 261–3). The fourth proposition relates rather to aggregate employment or wages: 'The general principal, now stated is that demand for commodities determines merely the direction of labour, and the kind of wealth produced, but not the quantity or efficiency of labour, or the aggregate of wealth' (87).[13]

Mill had difficulty in expressing his precise intentions.[14] But one particular formulation reveals the essence of the matter and confirms the preoccupation with aggregate employment and earnings, namely the criticism of those economists who argue 'as if a person who buys commodities, the produce of labour, was an employer of labour, and created a demand for it as really, and in the same sense, as if he bought the labour

---

13 An interesting application of the proposition underscores the argument: Mill alluded to a case sometimes made out whereby a technical change in one sector by reducing costs of production and thus the price of the commodity, releases purchasing power for expenditure elsewhere, and thus assures reabsorption of output in the affected industries. This notion Mill rejected on grounds of the fourth proposition: 'It is true, the consumers have now additional means of buying other things; but this will not create other things, unless there is capital to produce them, and the improvement has not set at liberty any capital, if even it has not absorbed some from other employments. The supposed increase of production and of employment for labour in other departments therefore will not take place' (96–7).

   The theorem is also used to refute the idea that wage taxation cannot harm workers 'since the money raised by it, being expended in the country, comes back to the labourers again through the demand for labour' (III, 829).

14 Cf. letter 21 Feb. 1849, XIV, II: 'I am going on revising the book: not altering much, but in one of the purely political economy parts which occurs near the beginning, viz. the discussion as to whether buying goods made by labour gives the same employment to labour as hiring the labourers themselves, I have added two or three pages of new explanation & illustration which I think make the case much clearer.' The hypercritical critic of the classicists, Cannan (1917), failed totally to follow Mill (300).

itself, directly, by the payment of wages . . .'; on the contrary, 'if by the demand for labour, be meant the demand by which wages are raised, or the number of labourers in employment increased, demand for commodities does not constitute demand for labour. I conceive that a person who buys commodities and consumes them himself, does no good to the labouring classes; and that it is only by what he abstains from consuming, and expends in direct payment to labourers in exchange for labour, that he benefits the labouring classes, or adds anything to the amount of their employment' (80).[15]

The fourth proposition carried with it the implication that net investment financed by reduced expenditures upon 'luxuries' generates an increase in the aggregate demand for labour. One of Mill's illustrations (80f.) actually involves an increase in expenditure on housebuilding labour for personal rather than commercial purposes, but the same principle applies; indeed, the general effects upon employment of increased expenditure on service labour (whether or not for commercial ends) financed by reduced consumption of luxury goods will be precisely the same as those of increased investment similarly financed, for the fourth proposition is extended effectively when Mill wrote that the demand for commodities does not determine 'the more or less of the labour itself, or of the maintenance or payment of labour' which depend rather on capital 'or other funds directly devoted to the sustenance and remuneration of labour' (78). The point is that a change in expenditure patterns from luxury consumption to either investment or to services (for personal use or otherwise) entails the redirection of currently utilized resources towards the production of goods which will sooner or later be placed at the disposal of workers in the form of wages or technological capital.

Mill was careful to insist that the mere circumstance that service labour rather than commodities is involved was not the issue at stake. For the employment of service labour entails a net increase in labour demand because the employer is assumed to postpone the date of his own consumption of the results of labour. In the event that he delays his payment of wages, workers would have to draw on some other source for current maintenance by extracting the funds from elsewhere; and in such a case a transfer of expenditure from luxury consumption to services 'does not open a new employment of labour, but merely changes the course of an existing employment' (81n).[16]

---

15 On Ricardo's formulation of the proposition see Hollander (1979), 373–5. For Mill's defence of Ricardo against Senior's strictures, see his notes on the latter's *Outline* (1945), 138. For emphasis upon the savings decision as a prerequisite to the expansion of employment see also Schumpeter (1954), 644; and Blaug (1978), 192.

16 Marshall's (1920) criticism that Mill was unaware of this implication of his main proposition is thus quite unfair (828). See also Pigou (1949), 175, for a similar complaint against Mill.

Part of the difficulty entailed by Mill's formulations on capital derives from the vague treatment of the prerequisites for expanded employment in the productive sector. There is no question that capital goods include not only wage goods but also fixed capital and materials. Yet the fourth proposition focusses largely on wage-goods capital alone. The reason for this seems to lie in an attempt to convey the contrast between the consequences for labour demand of expenditure on final goods for personal consumption on the one hand, and of investment expenditure on the other, in terms of the simplest case conceivable. That this is so is suggested strongly by the predilection for illustrations drawn from the increased employment of unproductive or service labour where, by the character of such employment, 'technological' capital is not required at all or not required to the same extent as in the productive sector (cf. Hollander, 1964)[17]. For all that, one is obliged by the very definition of capital to take seriously the requirement for technological capital in the productive sector.

In the polar extreme case of a given labour supply, an expansion of investment will simply mean that 'the whole of what was previously expended in luxuries, by capitalists and landlords is distributed among the existing labourers, in the form of additional wages' (68);[18] in this case, the least interesting from our present perspective, there will not even be required any transfer of resources from the 'luxury' sector. Evidently only in the event of expanded employment does the requirement for technological capital arise.[19]

In the case of an available labour supply at constant wages, the positive effect of net investment upon employment can only manifest itself after the newly-created (technological) capital goods have been installed. The programme is thus accomplished in two stages: first, a transfer of resources from 'luxury' industries to capital-goods industries, which involves no net effect upon employment (and thus none on aggregate wages) and second, the operation of the capital goods by new entrants into the labour market. It is only at this second stage that the aggregate wage bill rises, implying an expansion in the demand, on the part of the newly employed, for wage goods. Mill was not clear on this kind of complexity which, as before suggested, he was able to bypass by resort to increased demand for service labour.

17 For illustrations strictly involving investment see e.g. *CW*, II, 57, 67–8.
18 This clearly gainsays the view that 'the classicists did not, at any time, apply the fourth proposition to the explanation of wages' (Thompson, 1975, 190).
19 Assuming, for simplicity, an available labour supply at a constant real wage rate, the increase in employment (and accordingly in aggregate wages) generated by any given net investment will be smaller, the higher the fixed capital required per man. See the discussion below (n53).

The expanded production of wage goods constitutes an aspect of the original decision to increase investment. Whether the resources required for this purpose are drawn from the luxury-goods or the capital-goods sector is not certain.[20] But in either event the increased outflow of wage goods is at the expense of other categories of commodities and should not be confused with the net expansion of output resulting from the expanded capacity of the system. Mill seems to have confused the two. He attempts to prove that 'increase of capital gives increased employment to labour, without assignable bounds', since when capitalists abstain and 'turn their income into capital' they do not 'annihilate their power of consumption; they do but transfer it from themselves to the labourers to whom they give employment', so that (in the case of available labour supply) 'the production of necessaries for the new population, takes the place of the production of luxuries for a portion of the old' (67–8). This may be true, but it does not in itself explain the source of purchasing power for the *net* flow of output from the expanded national capacity once in place.

Mill took pains to ward off possible criticisms of his general position by emphasizing its restriction to the long run. For the immediate result of a sudden and unexpected reduction in luxury consumption might merely be a piling up of stocks of goods which by their nature cannot be used in the 'support of labour' (72). From the point of view of accumulation the 'detriment' to society occurs when the luxuries were produced; and 'they continue to be produced as long as there are consumers for them, and are produced in increased quantity to meet an increased demand; the choice made by a consumer to expend five thousand a year in luxuries, keeps a corresponding number of labourers employed from year to year in producing things that can be of no use to production; their services being lost so far as regards the increase of national wealth, and the tools, materials, and food which they annually consume being so much subtracted from the general stock of the community applicable to productive purposes'. As he points out in a related context – of broad relevance to the issue of resource allocation in general – the appropriate assumption is that the supposed fall in demand for luxuries 'be gradual and foreseen', and be 'attended with no waste of capital, the manufacture being discontinued by merely not replacing the machinery as it wears out, and not reinvesting the money as it comes in from the sale of the produce' (79). 'The capital is thus ready for a new employment' (in our case in the capital-goods sector).[21]

20 If capital goods are long lived, so that little maintenance is required, resources would be immediately liberated for wage-goods production. In different circumstances a greater burden on capitalists in the form of reduced luxury purchases would be entailed.

21 In an earlier article I argued that the fourth proposition depended upon the implicit assumption that fixed capital is homogeneous and can be used in any industry (cf. Hollander, 1964, 181n). This is evidently mistaken for the theorem was intended for the long run.

Similarly, it is essential to avoid 'confounding the effects arising from the mere suddenness of a change with the effects of the change itself' (82–3). For us here Mill's intentions might equally well be understood if instead of envisaging a change in the pattern of outlay from consumption to investment we were to compare alternative equilibrium patterns; the higher the rate of investment per period the higher will be the rate of expansion of capacity, and thus the higher the rate of employment.

There remains a further qualification of central importance. The fourth proposition presumed full use of existing capacity. Mill conceded that in the event of available capacity a net expansion in demand for a commodity may stimulate an increase in employment and output in that sector which is not at the expense of activity elsewhere:

The grounds of a proposition, when well understood, usually give a tolerable indication of the limitations of it . . . When labour is supported, but not fully occupied, a new demand for something it can produce, may stimulate the labour thus supported to increased exertions, of which the result may be an increase of wealth, to the advantage of the labourers themselves and of others. Work which can be done in the spare hours of persons subsisted from some other source, can . . . be undertaken without withdrawing capital from other occupations, beyond the amount (often very small) required to cover the expense of tools and materials, and even this will often be provided by savings made expressly for the purpose . . . Employment of this kind may, by the springing up of a demand for the commodity, be called into existence without depriving labour of an equivalent amount in any other quarter (87).

This same qualification had already been made in the context of the first proposition – confirming that the first and the fourth were two sides of the same coin – when Mill also alluded to the case of domestic industry (64). But Mill's various allowances elsewhere for the possibility of unused capacity in advanced capitalist economies (discussed in the next section) suggest a much wider potential applicability of the so-called exception.

IV   ON MACHINERY

The classical problem of 'machinery' (the subject matter of the first part of Ricardo's famous Chapter 31 of that title)[22] is discussed by Mill – more specifically, as an application of the first proposition or its corollary, the

22   On this chapter see Wicksell (1935), I, 137–8, Hayek (1946), 428–9, Hicks (1969), 151–4, 168–71, Beach (1971), 916–22, Hicks (1971), 922–5, Hicks (1973), 97–9, Heertje (1973), 14–20, Hicks (1977), 184–90, Maital and Haswell (1977), Hollander (1979), 339f, Eltis (1984), 223–9. For the historical context see Berg (1980), 65–74.

fourth. It should be clarified at the outset that in the context of capital conversion Mill, like Ricardo, had in mind exogenous technological improvements. There is in this context no suggestion of the adoption of capital-intensive techniques in response to a preceding change in relative factor prices. (See the strictures in O'Driscoll, 1977, 110–11.) Mill applies the general idea not only to 'machinery' but to 'all improvements by which capital is sunk; that is, rendered permanently incapable of being applied to the maintenance and remuneration of labour' (*CW* II, 94).[23]

The problem as formally laid out, arises from the differential rates of depreciation of the two categories of capital:

There is a great difference between the effects of circulating and those of fixed capital, on the amount of the gross produce of the country. Circulating capital being destroyed as such, or at any rate finally lost to the owner, by a single use; and the product resulting from that one use being the only source from which the owner can replace the capital, or obtain any remuneration for its productive employment; the product must of course be sufficient for those purposes, or in other words, the result of a single use must be a reproduction equal to the whole amount of the circulating capital used, and a profit besides. This, however, is by no means necessary in the case of fixed capital. Since machinery, for example, is not wholly consumed by one use, it is not necessary that it should be wholly replaced from the product of that use. The machine answers the purpose of its owner if it brings in, during each interval of time, enough to cover the expense of repairs, and the deterioration in value which the machine has sustained during the same time, with a surplus sufficient to yield the ordinary profit on the entire value of the machine (*CW* II, 93).[24]

The conclusion that 'all increase of fixed capital, when taking place at the expense of circulating, must be, at least temporarily, prejudicial to the interest of the labourers' is technically valid and expresses the main message. Yet it is rather inappropriate since it applies to 'all categories of innovation

23 At one point Mill identifies materials with fixed capital (above, n2). Accordingly, 'what is expected in materials is as much withdrawn from the maintenance and remuneration of labourers, as what is fixed in machinery; and if capital now expended in wages were diverted to the providing of materials, the effect on the labourers would be as prejudicial as if it were converted into fixed capital' (99). But this is said not to be of practical concern, for 'the tendency of improvements in production is always to economize, never to increase, the expenditure of seed or material for a given produce; and the interest of the labourers has no detriment to apprehend from this source'.

24 The main illustration involves a landowner-farmer who initially employs a 2000 quarter corn input in the support of labour which yields a corn output of 2400 quarters, namely a 20% return of 400 that he 'annually consumes'. In the transition period one half the corn input is used to support labour in a land improvement, so that at the beginning of the post-transition year the farmer has a capital invested in land equivalent to 1000

involving conversion, whereas the initial statement focusses specifically on the 'output-reducing' variety. Mill indeed proceeds to play down his concern with the 'temporary' period immediately following the replacement or improvement, to consider the magnitude of the surplus yielded by the new process. In his illustration (which assumes away depreciation of the 'improvement') a gross produce of say, 1500 quarters (raising the rate of profit from 20 per cent to 25 per cent) will motivate its adoption yet be 'very injurious' to the labourers; only an undiminished gross produce would permit 'the loss soon [to] be made . . . up . . . because so enormous an accession of gain will probably induce the improver to save a part, add it to his capital and become a larger employer of labour' (94).

The issue of output-reducing technology again emerges in the course of a discussion of the 'causes' of large-scale activity, which includes adoption of processes requiring 'expensive machinery': 'Expensive machinery supposes a large capital; and is not resorted to except with the intention of producing, and the hope of selling, as much of the article as comes up to the full powers of the machine . . . wherever costly machinery is used, the large system of production is inevitable' (133). This generalization is immediately qualified. The empirical 'test' for efficiency at various scales appropriate where organizational change is at issue – 'the power of underselling' – may fail precisely because some capital-intensive processes are of the output-reducing variety. The pertinent passage contains a nice restatement of the phenomenon:

The power of underselling does not depend on the absolute increase of produce, but on its bearing an increased proportion to the expenses; which, as was shown in the former chapter, it may do, consistently with even a diminution of the gross annual produce. By the adoption of machinery, a circulating capital, which was perpetually consumed and reproduced, has been converted into a fixed capital, requiring only a small annual expense to keep it up: and a much smaller produce will suffice for merely covering that expense, and replacing the remaining circulating capital of the producer. The machinery therefore might answer perfectly well to the manufacturer, and enable him to undersell his competitors, though the effect on the production of the country might be not an increase but a diminution (134).

It is not apparent why the original innovating entrepreneur should wish to undersell his competitors rather than take advantage of his reduced unit costs at the going price. We must not forget, however, that classical

quarters and a corn capital of only 1000. (It is unclear what happens to profits during the transition but presumably they are reduced.) He proceeds to employ the 1000 quarters of wage capital in the support of labour on the improved land, on the assumption that that amount is required for 'the effectual cultivation of his land' (an assumption that implies fixed factor proportions).

competition is not of the 'price-taking' variety. It is conceivable that the innovating firm – even assuming 'output-reducing technology – *expands* its capital stock (with an appreciable positive impact on total output) and that competitors react by following suit; for 'the extension of business which almost certainly follows in any department in which an improvement has been made, affords a strong inducement to those engaged in it to add to their capital' (98). Allowance might also be made for new entrants.

The fact of industry-wide expansion (or at the least constancy) – reflecting the negative sloped demand curve (or in the limit a curve of zero elasticity) – even in the event of 'output-reducing' innovation is much emphasized: 'the article will be sold cheaper, and therefore, of that single article there will probably be not a smaller, but a greater quantity sold'. It emerges very clearly that, as with Ricardo, a reduction in output does *not* occur in the industry where the output-reducing innovation is introduced (cf. Hollander, 1979, 368–9, 371). It is economy-wide contraction accompanied by a reduced aggregate wages fund that is at issue, employment possibly rising in the innovating sector: 'though that particular branch of industry may extend itself, it will be by replenishing its diminished circulating capital from that of the community generally; and if the labourers employed in that department escape loss of employment' – a result hinging obviously on the degree of demand elasticity for the product in question – 'it is because the loss will spread itself over the labouring people at large' (*CW* II, 134).[25]

No attempt was made by Ricardo to justify the supposition relating to output-reducing technology; because a lower yield might suffice to cover depreciation and running costs and yet allow an adequate return, it is presumed that such technologies actually exist. Mill, however, was most sceptical regarding the empirical significance of the case. No examples at all are given from industry. As for agriculture 'the supposition, in the terms of which it has been stated, is purely ideal; or at most applicable only to such a case as that of the conversion of arable land into pasture, which though formerly a frequent practice, is regarded by modern agriculturalists as the reverse of an improvement' (95). A note introduced into the addition of 1865 refers to two recent illustrations of such practices, envisaged as 'improvements',[26] but concludes, in line with earlier editions, that 'the

---

25 In the event of an innovatory industry characterized by zero demand elasticity, there will of course be loss of employment in that industry itself.

26 'The clearing away of the small farmers in the North of Scotland, within the present century, was, however, a case of it; and Ireland, since the potato famine and the repeal of the corn laws, is another. The remarkable decrease which has lately attracted notice in the gross produce of Irish agriculture, is, to all appearance, partly attributable to the diversion of land from maintaining human labourers to feeding cattle; and it could not have taken place without the removal of a large part of the Irish population by emigration or death. We have thus two recent instances, in which what was regarded

effect . . . of all the improvements due to modern science is to increase, or at all events, not to diminish, the gross produce'.

In treating examples of contemporary agricultural innovation two general categories are delineated (180f). The first, which includes but extends beyond land improvement strictly defined, entails land – (and labour) – saving technology, a given area yielding a higher output 'without an equivalent increase of [operating] labour'. The second permits no increase from given land but does allow reduced operating labour. Several of these innovations presumably entail 'conversion' of 'circulating' into 'fixed' capital but in no case is output reduction a characteristic. Other cases – such as improvements in iron smelting which permit the production of cheaper agricultural implements – do not involve 'conversion' in agriculture (but might fall into that category in the capital-goods sector).

* * *

The technologies under discussion thus far are labour displacing for a given value of capital. But here Mill extends the allowance for detrimental welfare effects from the perspective of labour by reference to land improvements which permit an expansion of output from a given area but require retention of the original work force. His purpose is to demonstrate that problems arise even where no labour displacement is involved at the level of the firm.

In both the original and the extended instances one thousand quarters of corn are initially diverted from maintenance of operating labour to its support in undertaking a land improvement (or construction of 'machinery'). But whereas in the original case a smaller crew suffices to operate the original (but improved) land area (the yield therefrom possibly reduced), in the alternative case a smaller work crew can produce nothing. The entire crew must be redirected to act as operating labour. Here, of course, the aggregate yield for the firm must expand if the innovation is to be viable.[27] It is yield

as an agricultural improvement, has diminished the power of the country to support its population. The effect, however, of all the improvements due to modern science is to increase, or at all events, not to diminish the gross product.' (This note replaces a brief statement in the editions from 1849 through 1862 that 'the effect of the agricultural improvements of the present day (of those, at least, which operate on the soil itself), is to increase, not to diminish the gross produce'.) The formulation is ambiguous. It is not made absolutely clear, whether population reduction was the 'cause' of the conversion program, its consequence, or both. (Eltis, 1984, 328, reads Mill as maintaining that the conversion involved induced substitution against labour following the potato famine and the necessity to pay wages sufficient to purchase more costly foodstuffs.) See further, below (pp. 954–5).

27 For example: Assume an initial outlay of 2000 quarters of corn, and zero maintenance costs:

in value terms that is relevant, but the issue of an adequate market for the expanded output is set aside for the while by the assumption that 'the greater produce, which by means of the improvement can be raised from the soil with the same labour, is all wanted, and will find purchasers'. The problem that arises again reflects capital scarcity, but whereas in the standard case the ratio of 'fixed' to 'circulating' capital rises within a *given* total, now the increase in the ratio requires a larger total capital and this implies – assuming capital in the economy as a whole to be unchanged (and one should add full use of capacity) – transfers from other sectors:

The improver will in that case require the same number of labourers as before, at the same wages. But where will he find the means of paying them? He has no longer his original capital of two thousand quarters disposable for the purpose. One thousand of them are lost and gone – consumed in making the improvement. If he is to employ as many labourers as before, and pay them as highly, he must borrow, or obtain from some other source, a thousand quarters to supply the deficit. But these thousand quarters already maintained, or were destined to maintain, an equivalent quantity of labour. They are not a fresh creation; their destination is only changed from one productive employment to another; and though the agriculturist has made up the deficiency in his own circulating capital, the breach in the circulating capital of the community remains unrepaired (95).

The contrast can be summarized thus: Simple conversion involves labour displacement by firms of given size (in terms of capital value); complex conversion necessitates a higher capital by the firm but no labour displacement. In both categories the net outcome is industry expansion, at least in the general case where quantity demanded increases upon a fall in price, but at the expense of activity elsewhere. By implication if demand elasticity is very low there will be employment or wage contractions in the innovating sector itself.

The extended formulation illustrates particularly clearly the need to consider the economy-wide implications of the adoption of technical change in any one sector. In fact, precisely at this juncture Mill interjects a warning

	original	intermediate	final
operating costs	2000	1000	2000
construction costs	X	1000	0
value of 'machinery' or 'improvement'	X	X	1000
maintenance of 'machinery' or 'improvement'	X	X	0
surplus	400		750
rate of return on capital	$\frac{400}{2000} = 20\%$		$\frac{750}{3000} = 25\%$

against drawing unjustified conclusions from observations of industry expansion and cost-price reduction which characterize technologically progressive industries:

> The argument relied on by most of those who contend that machinery can never be injurious to the labouring class, is that by cheapening production it creates such an increased demand for the commodity, as enables, ere long, a greater number of persons than ever to find employment in producing it. This argument does not seem to me to have the weight commonly ascribed to it. The fact, though too broadly stated, is, no doubt, often true. The copyists who were thrown out of employment by the invention of printing, were doubtless soon outnumbered by the compositors and pressmen who took their place; and the inventions of Hargreaves and Arkwright, . . . employs a far larger circulating capital . . . than at any former time. But if this capital was drawn from other employments; if the funds which took the place of the capital sunk in costly machinery, were supplied not by any additional saving consequent on the improvements, but by drafts on the general capital of the community; what better were the labouring classes for the mere transfer? In what manner was the loss they sustained by the conversion of circulating into fixed capital made up to them by a mere shifting of part of the remainder of the circulating capital from its old employments to a new one? (96)

Attention must, therefore, always be focussed upon the ultimate source of any observed increase in funds for the maintenance of labour in a particular industry. Unless they emanate from net accumulation, expansion is necessarily at the expense of employment elsewhere. This amounts to a direct application of the first fundamental proposition on capital. The argument is also restated in terms of the corollary of the first proposition, namely the fourth proposition that 'demand for commodities is not demand for labour':

> All attempts to make out that the labouring classes as a collective body *cannot* suffer temporarily by the introduction of machinery, or by the sinking of capital in permanent improvements, are, I conceive, necessarily fallacious. That they would suffer in the particular department of industry to which the change applies is generally admitted, and obvious to common sense; but it is often said, that though employment is withdrawn from labour in one department, an exactly equivalent employment is opened for it in others, because what the consumers save in the increased cheapness of one particular article enables them to augment their consumption of others, thereby increasing the demand for other kinds of labour. This is plausible, but, as was shown in the last chapter, involves a fallacy; demand for commodities being a totally different thing from demand for labour. It is true, the consumers have now additional means of buying other things; but this will not create the other things, unless there is capital to produce them, and the improvement

has not set at liberty any capital, if even it has not absorbed some from other employments. The supposed increase of production and of employment for labour in other departments therefore will not take place; and the increased demand for commodities by some consumers, will be balanced by a cessation of demand on the part of others, namely, the labourers who were superseded by the improvement, and who will now be maintained, if at all, by sharing, either in the way of competition or of charity, in what was previously consumed by other people (96–7).

The foregoing passages are important for the micro-economic adjustments presumed to be at play following conversion. The industry price falls to reflect the higher productivity implied by conversion, a conclusion which, we have seen, applies not only to cases where fixed capital is required which cannot be operated by a smaller work crew than originally and so cannot generate a fall in 'gross produce' at the level of the firm, but also to innovation of the 'output-reducing' variety.

* * *

The formal statements of the conversion problem all entail a wage-fund model involving literal pre-accumulations of agricultural produce (or wage goods). This is also true of Ricardo's analysis. But Ricardo made clear that the 'real-world' counterpart of the reduced wage fund takes the form of contractions in agricultural output in response to a lower demand for food on the part of those labourers adversely affected by the adoption of new technology (Hollander, 1979, 341, 372–3). It is to be noted too that Ricardo was very conscious of the methodology: 'The case which I have supposed' – a case involving an innovating farmer – 'is the most simple that I could select; but it would make no difference in the result, if we supposed that the machinery was applied to the trade of any manufacturer . . .' (*Works*, I, 390–1), because of the negative effects on the global wages fund. His analysis in brief was not intended to be limited to agricultural (wage-goods) production. But presumably in the event of new technology entailing conversion which is introduced in agriculture the same principles apply: if agricultural output expands it is at the expense of employment or earnings elsewhere (assuming global capital unchanged) and therefore will be accompanied by a reduced demand for food by displaced or worse-paid manufacturing labour. (In the case of zero demand elasticity, there will be negative employment or earnings effects within agriculture itself.)

As for Mill we shall encounter (below, p. 394) statements of economic organization which avoid literal preaccumulations of food: 'Food and clothing for his operatives, it is not the custom of the present age that [the capitalist] should directly provide . . . each capitalist has money, which he pays to his workpeople, and so enables them to supply themselves' (*CW*,

II, 55-6). There is therefore every reason to believe that, like Ricardo's, his simplified model was designed as a convenient mental exercise to convey the essential forces at play rather than as a representation of the workings of an actual capitalist-exchange system. If this is indeed so then what is made explicit by Ricardo may be presumed to be acceptable to Mill, namely that as employment opportunities fall in consequence of the adoption of new technology – assuming it not to impinge on agriculture – the demand for 'food' declines and the agricultural sector contracts in consequence. This then is what the reduction in the aggregate wages fund amounts to. If we presume agricultural innovations – and Mill was indeed much interested in land-improvements – we again are faced, as in Ricardo's discussion, by the complex implication that output expansion of the farm sector will be accompanied by demand contractions reflecting the loss of employment elsewhere – always assuming constant capital in the economy as a whole.

It should further be remarked that global technological unemployment as such was not Mill's concern. Rather it was 'injury' to labour in the broadest sense, even allowing for reabsorption at reduced wages. For the problem was to assure the employment of 'as many labourers as before, and pay them as highly' (above, p. 381); similarly in the context of complex conversion 'if the condition of the labouring classes enables them to bear a temporary reduction of wages, and the superseded labourers become absorbed in other employments, their labour is still productive, and the breach in the gross produce of the community is repaired, though not the detriment to the labourers' (134). To this extent the potency of Wicksell's famous strictures is somewhat reduced.[28]

* * *

Capital absorbing innovation necessarily has a negative effect on the global 'wages fund' and, therefore, on either employment or per capita wages or both. That is the alpha and omega of the Ricardian analysis of machinery. The whole problem will, however, be 'temporary' in the event of net savings sufficient to compensate for the short fall in employment opportunities wherever created. The extent and speed of adoption of such technologies relative to the rate of capital accumulation – the generation of new savings – becomes the key issue. And here, Mill, like Ricardo, was optimistic. For

28 Wicksell (1935, I, 137-8) based himself on Ricardo's neglect of expansionary effects induced by wage reductions. His modifications were actually applied to contemporary Swedish problems (see Lars Jonung, 1981). The paper presenting these ideas was rejected for publication in 1924 by the *Economic Journal* , then edited by Keynes, on the grounds 'that any treatment of this topic [unemployment] at the present day ought to bring in various modern conceptions for handling the problem and that the time has gone by for a criticism of Ricardo on purely Ricardian lines'.

a variety of reasons, the adverse impact of new technology involved 'a case abstractedly possible [rather] than one which is frequently realized in fact' (134).

As for extent and rapidity of adoption relative to net accumulation:

I do not believe that as things are actually transacted, improvements in production are often, if ever, injurious, even temporarily, to the labouring classes in the aggregate. They would be so if they took place suddenly to a great amount, because much of the capital sunk must necessarily in that case be provided from funds already employed as circulating capital. But improvements are always introduced very gradually, and are seldom or never made by withdrawing circulating capital from actual production, but are made by the employment of the annual increase. There are few if any examples of a great increase of fixed capital, at a time and place where circulating capital was not rapidly increasing likewise. It is not in poor or backward countries that great and costly improvements in production are made. To sink capital in land for a permanent return – to introduce expensive machinery – are acts involving immediate sacrifice for distant objects; and indicate, in the first place, tolerably complete security of property; in the second, considerable activity of industrial enterprise; and in the third, a high standard of what has been called the "effective desire of accumulation:" which three things are the elements of a society rapidly progressive in its amount of capital. Although, therefore, the labouring classes must suffer, not only if the increase of fixed capital takes place at the expense of circulating, but even if it is so large and rapid as to retard that ordinary increase to which the growth of population has habitually adapted itself; yet, in point of fact, this is very unlikely to happen, since there is probably no country whose fixed capital increases in a ratio more than proportional to its circulating (97).[29]

The conclusion of this passage, incidentally, stands in striking contrast with Karl Marx's prognosis – his prediction of a declining growth rate of labour demand due to an increase in the ratio of fixed to circulating (rather, 'constant' to 'variable') capital in the face of on-going population expansion (cf. Hollander, 1984a).

Further reason for optimism lies in the positive stimulus to savings engendered by the effects of new technology – effects on both the 'ability' and 'motive' to save:

29 Mill reasons that entrepreneurs take great care before committing funds to long-lived projects, knowing full well the potential difficulties of extricating sunk capital: 'If the whole of the railways which, during the speculative madness of 1845, obtained the sanction of Parliament, had been constructed in the times fixed for the completion of each, this improbable contingency would, most likely, have been realized; but this very case has afforded a striking example of the difficulties which oppose the diversion into new channels, of any considerable portion of the capital that supplies the old: difficulties generally much more than sufficient to prevent enterprises that involve the sinking of capital, from extending themselves with such rapidity as to impair the sources of the existing employment for labour' (97–8).

To these considerations must be added, that even if improvements did for a time decrease the aggregate produce and the circulating capital of the community, they would not the less tend in the long run to augment both. They increase the return to capital; and of this increase the benefit must necessarily accrue either to the capitalist in greater profits, or to the customer in diminished prices; affording, in either case, an augmented fund from which accumulation may be made, while enlarged profits also hold out an increased inducement to accumulation. In the case we before selected, in which the immediate result of the improvement was to diminish the gross produce from two thousand four hundred quarters to one thousand five hundred, yet the profit of the capitalist being now five hundred quarters instead of four hundred, the extra one hundred quarters, if regularly saved, would in a few years replace the one thousand quarters subtracted from his circulating capital. Now the extension of business which almost certainly follows in any department in which an improvement has been made, affords a strong inducement to those engaged in it to add to their capital; and hence, at the slow pace at which improvements are usually introduced, a great part of the capital which the improvement ultimately absorbs, is drawn from the increased profits and increased savings which it has itself called forth (98).[30]

In all of this Mill was at one with Ricardo and, one may add, with William Ellis whose work he so much admired.[31]

Karl Marx somehow managed in his account in *Capital* totally to distort J. S. Mill's position (1965, 438). He represents Mill, along with various other 'bourgeois political economists', as maintaining 'that all machinery that

30 In our contrived example regarding 'output-expanding' technology, 1000 quarters of corn must be transferred to the industry to allow operation of the new technique. The surplus has increased by 350 quarters and the rate of return by five percentage points so that both the 'ability' and 'motive' to save have risen. It will take several 'years' to make up the short-fall.

31 Cf. Ellis (1826), 119: 'The grand source of all the false reasoning upon machinery is to be found in the supposition that every new application of capital to other purposes than that of paying wages is a deduction from the fund devoted to that purpose. To be convinced of the groundlessness of that assumption it is merely necessary to bear in mind, that every improvement in the arts of production is uniformly attended with an increase of profit, which acts as a stimulus to an increase of capital; or, more correctly, it is attended with an increase of capital by which the rise in the rate of profit is anticipated. The capital, therefore, attracted to a new and more profitable employment, is not drawn from that fund to which the labourers look for support, but from fresh savings. The increase in the capital of a country keeps pace with the improvements in the arts of production. "Discoveries," to use the words of Mr. Ricardo, "rather operate in determining the employment of the capital which is saved and accumulated, than in diverting capital from its actual employment" [1951, I, 395]'. Cf. also *ibid.*, 122 regarding the motive for machinery in (expectation of) 'a profit above the ordinary rate': 'Fresh capital would be accumulated and flow into this new opening for its profitable employment . . . the capital employed in the new implements &c. is entirely drawn from an independent fund.'

displaces workmen, simultaneously and necessarily sets free an amount of capital adequate to employ the same identical workmen'. By contrast, he insisted, was his own model wherein 'the labourers, when driven out of the workshop by the machinery, are thrown upon the labour-market, and there add to the number of workmen at the disposal of the capitalists' (440). Marx proceeds to summarize his position thus:

In Part VII of this book it will be seen that this effect, which . . . is represented to be a compensation to the working-class, is on the contrary a most frightful scourge . . . The labourers that are thrown out of work in any branch of industry, can no doubt seek for employment in some other branch. If they find it, and thus renew the bond between them and the means of subsistence, this takes place only by the intermediary of a new and additional capital that is seeking investment; not at all by the intermediary of the capital that formerly employed them and was afterwards converted into machinery.

There is nothing here that Mill, following Ricardo, had not already maintained; reabsorption hinged for him as for Ricardo on net accumulation. Interestingly enough, Marx excluded Ricardo from his charges against the bourgeois literature, referring to his 'scientific impartiality and love of truth' (438n). We will refrain from speculating on why this allowance was not similarly applied, where appropriate, to other 'bourgeois' writers.

V   THE WAGES-FUND THEORY AND ECONOMIC ORGANIZATION

It is a common presumption that classical economists utilized a conception of economic activity which runs in terms of discontinuous output. The wages-fund theory is said to fall into this category of model. Contrasting with such conceptualizations are those which emphasize the continuity of production according to which approach, while it remains true that current production is the result of inputs applied in the past, it is legitimate to conceive of the income of such inputs as deriving directly from the current flow of production. Matters turn out to be rather more complex.

There is certainly no question that the time-consuming character of economic activity caught the eye of the classicists. As Mill phrased the matter:

Except the labour of the hunter and fisher, there is scarcely any kind of labour to which the returns are immediate. Productive operations require to be continued a certain time, before their fruits are obtained. Unless the labourer, before commencing his work, possesses a store of food, or can obtain access to the stores of some one else, in sufficient quantity to maintain him until the production is completed, he can undertake no labour but such as can be carried on at odd intervals, concurrently with the pursuit of his subsistence . . . A country like England or

France is only able to carry on the agriculture of the present year, because that of past years has provided, in those countries or somewhere else, sufficient food to support their agricultural population until the next harvest. They are only enabled to produce so many other things besides food, because the food which was in store at the close of the last harvest suffices to maintain not only the agricultural labourers, but a large industrious population besides (*CW*, II, 33).

Other passages, including the first proposition on capital (63–4, cited above, p. 364) and qualification thereto, also point to an 'advance' model, with emphasis upon discontinuity. Thus, for example, the observation that industry does not always come up to the limit of capital' with special reference to the 'unproductive consumption of productive labourers':

[T]here may not be as many labourers obtainable, as the capital would maintain and employ . . . There are many persons maintained from existing capital, who produce nothing, or who might produce much more than they do. If the labourers were reduced to lower wages, or induced to work more hours for the same wages, or if their families, who are already maintained from capital, were employed to a greater extent than they now are in adding to the produce, a given capital would afford employment to more industry. The unproductive consumption of productive labourers, the whole of which is now supplied by capital, might cease, or be postponed until the produce came in; and additional productive labourers might be maintained with the amount' (65).

that large portion of the productive capital of a country which is employed in paying the wages and salaries of labourers, evidently is not, all of it, strictly and indispensably necessary for production. As much of it as exceeds the actual necessaries of life and health (an excess which in the case of skilled labourers is usually considerable) is not expended in supporting labour, but in remunerating it, and the labourers could wait for this part of their remuneration until the production is completed; it needs not necessarily pre-exist as capital: and if they unfortunately had to forego it altogether, the same amount of production might take place. In order that the whole remuneration of the labourers should be advanced to them in daily or weekly payments, there must exist in advance, and be appropriated to productive use, a greater stock, or capital, than would suffice to carry on the existing extent of production: greater, by whatever amount of remuneration the labourers receive, beyond what the self-interest of a prudent slave-master would assign to his slaves. In truth it is only after an abundant capital had already been accumulated, that the practice of paying in advance any remuneration of labour beyond a bare subsistence, could possibly have arisen: since whatever is so paid, is not really applied to production, but to the unproductive consumption of productive labourers, indicating a fund for production sufficiently ample to admit of habitually diverting a part of it to a mere convenience (58–9).[32]

32 See also references (*ibid.*) to the advance of taxes and rent.

Yet many of Mill's utterances on substantive matters relating to capital, investment and production indicate a model more consistent with synchronized activity. It is my impression that his formal accounts involving discontinuities were designed to bring to the fore as clearly as possible the time-consuming character of economic activity. But synchronization economics, properly understood, does not gainsay this particular circumstance (although it may disguise its presence) since it remains true that the flow of current input is responsible for the flow of future production (and decisions regarding the current use of input must be made on the basis of expectations regarding the future); similarly, it remains true that the current flow of output is the consequence of input use in the past. These facts, of course, only become conspicuous within the terms of the model when consideration is given to an expansion of capacity from period to period. If this indeed represents Mill's position – the evidence will presently be laid out – the greatest care is required in understanding what he had in mind by the wages-fund doctrine. To this matter we now turn.

It will be convenient to have before us that strong version of the doctrine wherein a specific annual wage bill is 'destined' – no more and no less – to be paid out to labour, upon which assumption is based the celebrated labour-demand curve of unitary elasticity. The most explicit statement is by Mill himself at the time of his retraction of belief in the doctrine in 1869. In his review of Thornton on labour Mill laid out what he conceived to be the received doctrine:

The demand for labour consists of the whole circulating capital of the country, including what is paid in wages for unproductive labour. The supply is the whole labouring population. If the supply is in excess of what the capital can at present employ, wages must fall. If the labourers are all employed, and there is a surplus of capital still unused, wages will rise. This series of deductions is generally received as incontrovertible. They are found, I presume, in every systematic treatise on political economy, my own certainly included.

The theory rests on what may be called the doctrine of the wages fund. There is supposed to be, at any given instant, a sum of wealth, which is unconditionally devoted to the payment of wages of labour. This sum is not regarded as unalterable, for it is augmented by saving, and increases with the progress of wealth; but it is reasoned upon as at any given moment a predetermined amount. More than that amount it is assumed that the wages-receiving class cannot possibly divide among them; that amount, and no less, they cannot but obtain. So that, the sum to be divided being fixed, the wages of each depend solely on the divisor, the number of participants. In this doctrine it is by implication affirmed, that the demand for labour not only increases with the cheapness, but increases in exact proportion to it, the same aggregate sum being paid for labour whatever its price may be (*CW* V, 643–4).

This is an ambiguous statement, since it is not specified whether the 'circulating capital' is in real or money terms.[33] But the latter was probably intended. For while the rationale for this conception of the labour market involves a form of discontinuous production – periodic 'rounds of business operations' – it is with specific reference to 'the capitalist's pecuniary means' that the argument runs:

In the common theory, the order of ideas is this. The capitalist's pecuniary means consist of two parts – his capital, and his profits or income. His capital is what he starts with at the beginning of the year, or when he commences some round of business operations: his income he does not receive until the end of the year, or until the round of operations is completed. His capital, except such part as is fixed in buildings and machinery, or laid out in materials, is what he has got to pay wages with. He cannot pay them out of his income, for he has not yet received it. When he does receive it, he may lay by a portion to add to his capital, and as such it will become part of next year's wages-fund, but has nothing to do with this year's (644).[34]

Let us now gather evidence from Mill's *Principles* to evaluate the accuracy of this retrospective view. The picture which emerges bears little resemblance to that portrayed in 1869.

Statements relating to wage-rate determination in the *Principles* are relatively few and surprisingly ambiguous. The most important appears at the outset of the chapter 'Of Wages' and deals with the general return to labour (including service or unproductive labour):

Wages, then, depend mainly upon the demand and supply of labour; or as it is often expressed, on the proportion between population and capital. By population is here meant the number only of the labouring class, or rather of those who work for hire; and by capital only circulating capital, and not even the whole of that, but the part which is expended in the direct purchase of labour. To this, however, must be added all funds which, without forming a part of capital, are paid in exchange for labour, such as the wages of soldiers, domestic servants, and all other unproductive labourers. There is unfortunately no mode of expressing by one familiar term, the aggregate of what has been called the wages-fund of a country:

---

33 Cf. Taussig (1896), 230f, on the characteristic ambiguity of Mill's statements regarding this matter.
34 An earlier statement bearing a close resemblance is given in 'On Profits and Interest' (1836), *CW*, IV, 301.

and as the wages of productive labour form nearly the whole of that fund, it is usual to overlook the smaller and less important part, and to say that wages depend on population and capital. It will be convenient to employ this expression, remembering, however, to consider it as elliptical, and not as a literal statement of the entire truth. With these limitations of the term, wages not only depend upon the relative amount of capital and population, but cannot under the rule of competition be affected by anything else. Wages (meaning, of course, the general rate) cannot rise, but by an increase of the number of the competitors for hire; nor fall, except by a diminution in the number of labourers to be paid (II, 337–8).

This is the most important formal statement of the principle of wage-rate determination in the entire work. Its crude inadequacies are such that it is hardly unfair to say that, from a theoretical viewpoint, we are scarcely carried beyond the assertion that 'wages are what wages are' (see the dismissive comment by Cannan, 1917, 301). For it begs a host of questions, including the precise determination of the breakdown of aggregate capital between its components by reference to some kind of production function. Yet Mill evidently believed, and perhaps justifiably so, that the formulation sufficed for his purposes, basing upon it a barrage of conclusions regarding labour policy. It is difficult to avoid the impression that the theoretical details relating to the demand for labour simply did not concern him deeply in this context; it was application, based upon a minimal theoretical structure, that was the major preoccupation, the intention being to demonstrate that the condition of the labouring class 'can be bettered in no other way than by altering that proportion [between capital and population] to their advantage; and every scheme for their benefit, which does not proceed on this as its foundation, is, for all permanent purposes, a delusion' (340–3; also 354).[35] It is pertinent that the larger part of the chapter 'Of Wages' itself, apart from two subsequent chapters on 'Popular Remedies for Low Wages', focusses upon the implications of the Malthusian population doctrine – labour supply – rather than the nature of the demand for labour. Thus a change in the cost of wage-goods, 'when of a permanent character, and capable of being calculated on beforehand' (340), works its effects upon wages first by impinging upon labour supply.[36] There can

35 Cf. Taussig (1896), 236, for this same perspective; also Bowley (1973), 213, Stigler (1968), 98–9.

36 This requires emphasis. One case sometimes made out against the wage fund, interpreted in terms of a predetermined sum of money to be devoted to wage payments, is that 'since taxes on wage goods apparently raised wages *in the short run*, by an amount sufficient to pass on the tax, the size of the fund was indeterminate' (O'Brien, 1975, 113). This is not a legitimate statement, for a change of this kind operates only in the long run and upon labour supply. See also *CW*, XIV, 52 regarding a fall in food price which may lead to reduced wages but does so 'only gradually' by 'giving a stimulus to population, unless there is already a surplus of unemployed labourers supported by charity'.

be little question that Mill's primary concern was with issues of this order.[37]

A second formal statement of the doctrine in the same context dealing with the equilibrating function of wage movements is equally vague:

Goods can only be lowered in price by competition, to the point which calls forth buyers sufficient to take them off; and wages can only be lowered by competition until room is made to admit all the labourers to a share in the distribution of the wages-fund. If they fell below this point, a portion of the capital would remain unemployed for want of labourers; a counter-competition would commence on the side of capitalists, and wages would rise (356).

This passage might be read as assuming a rigidly pre-determined wages bill; but it is also not inconsistent with a totally different version of the wages-fund theory wherein the wages bill is not a pre-determined sum but the equilibrium outcome of a market-clearing process (below, pp. 400–1). Once again, there is too little theoretical detail to be sure of Mill's intentions regarding strict analysis. The formulation served the purpose of an elementary exposition of the notion of an equilibrium wage rate designed to counter popular remedies for low wages (such as minimum-wage legislation) in which context precisely the extract appears.

Yet for all that there comes to light, upon closer examination of the qualifications allowed by Mill to the main statement, some profoundly interesting theoretical insights. A brief discussion of approaches to wage determination 'in apparent contradictions to this doctrine' yields a fairer appreciation of what is at stake.

I have in mind Mill's qualifications in the present context to the 'law of markets'. Mill recognized the possibility of slack periods in particular trades when available capital is kept idle – a circumstance which could still be formally absorbed into the doctrine, for 'capital which the owner does not employ in purchasing labour, but keeps idle in his hands, is the same thing to the labourers, for the time being, as if it does not exist' (338). More significant, the allowance is extended to the aggregate labour market, albeit somewhat grudgingly:

When there is what is called a stagnation . . . then work people are dismissed, and those who are retained must submit to a reduction of wages: though in these

37 For example, one of Mill's objectives was to counter the view 'that the government without providing additional funds, could create additional employment' (*CW*, II, 64) by means of tariffs or other protective measures. Cannan (1917), 93, regarded Mill's proposition 'industry is limited by capital' as no more than a 'useful catchword with which to attack the protectionist fallacy of giving employment or "creating an industry" '. See also Marshall (1920), 826.

cases there is neither more nor less capital than before . . . If we suppose, what in strictness is not absolutely impossible, that one of these fits of briskness or of stagnation should affect all occupations at the same time, wages altogether might undergo a rise or a fall. These, however, are but temporary fluctuations: the capital now lying idle will next year be in active employment, that which is this year unable to keep up with the demand will in its turn be locked up in crowded warehouses; and wages in these several departments will ebb and flow accordingly: but nothing can permanently alter general wages, except an increase or a diminution of capital itself (always meaning by the term, the funds of all sorts, devoted to the payment of labour) compared with the quantity of labour offering itself to be hired (338–9).

Further allowances for excess capacity will be found in the formal discussion of capital in the first Book. Thus 'a fund may be seeking for productive employment, and find none, adapted to the inclinations of its possessor: it then is capital still, but unemployed capital. Or the stock may consist of unsold goods, but susceptible of direct application to productive uses, and not at the moment, marketable: these, until sold, are in the condition of unemployed capital' (57). And again: 'capital may be temporarily unemployed, as in the case of unsold goods, or funds that have not yet found an investment: during that interval it does not set in motion any industry' (65).

Idle capital in these contexts apparently refers not only to unsold stocks of goods but also to money funds available for investment in wage payments or other disbursements. What is involved is a well-considered qualification to the proposition that 'industry is limited by capital' making it apparent that Mill intended to supplement the basic doctrine regarding aggregate employment by some function describing the state of aggregate demand for final goods – or what is equivalent, the net excess demand for money (see below, p. 520).

The significance of the qualification extends beyond its linkage of monetary and employment theory, important though this is. Most relevant for us here is that the qualification points away from any notion of an aggregate sum of wealth unconditionally 'destined' for the payment of wages; this obviously cannot be so if variations in the aggregate demand for commodities can lead to variations in capacity usage. In the light of all this it appears all the more likely that the wages-fund doctrine was simply not designed in its pristine form for the short run; it was, as Mill himself put it in one of our foregoing citations, a theory relating to 'permanent' wages – assuming full equilibrium as far as concerns aggregate demand for commodities – and particularly relevant for an appreciation of the general problems of population or the inability to generate increased employment by protective measures.

Mill did not then subscribe to the notion of an inelastic aggregate wage bill in the short run. Quite apart from the explicit denial of a lower bound to the wage bill flowing from recognition of unused capital there is much else pointing to the same conclusion. What we have to say next is pertinent both to the question of a strict upper bound to the wage bill, and that of whether literal advances out of preaccumulated stocks are intended.

The function of capital, as we have seen, is 'to afford the shelter, protection, tools and materials which the work requires, and to feed and otherwise maintain the labourers during the process'. But Mill was very careful indeed to specify that stocks of various kinds, whose function it is to fulfil the latter tasks of 'maintaining' labour, need not actually take the form of wage goods;

> Food and clothing . . . it is not the custom of the present age that [the capitalist] should directly provide; and few capitalists, except the producers of food or clothing, have any portion worth mentioning of their capital in that shape. Instead of this, each capitalist has money, which he pays to his workpeople, and so enables them to supply themselves: he also has finished goods in his warehouses, by the sale of which he obtains more money, to employ in the same manner, as well as to replenish his stock of materials, to keep his buildings and machinery in repair, and to replace them when worn out. His money and finished goods, however, are not wholly capital, for he does not wholly devote them to these purposes . . . what then is his capital? Precisely that part of his possessions, whatever it be, which is to constitute his fund for carrying on fresh production. It is of no consequence that a part, or even the whole of it, is in the form in which it cannot directly supply the wants of labourers (55–6).[38]

Similarly in the celebrated paper 'On the Influence of Consumption on Production' (composed in 1830):

> The capital, whether of an individual or of a nation, consists, we apprehend, of all matters possessing exchangeable value, which the individual or the nation has

---

38 Mill's fixed-circulating capital categories thus proved rather clumsy: 'Since all wealth which is destined to be employed for reproduction comes within the designation of capital, there are parts of capital which do not agree with the definition of either species of it; for instance, the stock of finished goods which a manufacturer or dealer at any time possesses unsold in his warehouses. But this, though capital as to its destination, is not yet capital in actual exercise: it is not engaged in production, but has first to be sold or exchanged, that is, converted into an equivalent value of some other commodities; and therefore is not yet either fixed or circulating capital; but will become either one or the other, or be eventually divided between them. With the proceeds of his finished goods, a manufacturer will partly pay his work-people, partly replenish his stock of the materials of his manufacture, and partly provide new buildings and machinery, or repair the old; but how much will be devoted to one purpose; and how much to another, depends on the nature of the manufacture, and the requirements of the particular moment' (II, 99).

in his or its possession for the purpose of reproduction, and not for the owner's unproductive enjoyment. All unsold goods, therefore, constitute a part of the national capital, and of the capital of the producer or dealer to whom they belong. It is true that tools, materials, and the articles on which the labourer is supported, are the only articles which are directly subservient to production: and if I have a capital consisting of money, or of goods in a warehouse, I can only employ them as means of production in so far as they are capable of being exchanged for the articles which conduce directly to that end (IV, 266–7).

The reason for this position lies in the supposed flexibility of the system which permits, by exchange or by production, the easy and rapid transformation of commodities into a form suitable for workers' consumption.

An example – of particular significance in the light of the fourth proposition on capital – illustrates the perspective. A decision by a capitalist to increase investment implies a fall off in his demand for luxuries ('plate and jewels') hitherto financed from the sale of his product (iron goods) and a corresponding increase in expenditure upon productive labour; this entails appropriate increases in money wages and accordingly in the demand by labour for food (57).[39] What, however, of expanded production of food to meet the new demand?

Increased food supplies might be obtained immediately by importation, presumably in exchange for the luxuries hitherto consumed by capitalists, or for goods produced by means of the resources made available by the reduction in luxury consumption. But if increased importation is not possible, then 'labourers will remain for a season on their short allowance: but the consequence of this change in the demand for commodities, occasioned by the change in the expenditure of capitalists from unproductive to productive, is that next year more food will be produced, and less plate and jewellery. So that . . . without having had anything to do with the food of the labourers directly, the conversion by individuals of a portion of their property, no matter of what sort' – in this case stocks of iron goods – 'from an unproductive destination to a productive, has had the effect of causing more food to be appropriated to the consumption of productive labourers.' True enough, in the absence of increased food imports it may take a 'season' for food supplies to be expanded, but there is little question that Mill intended to minimize the significance of any such delay. The ease of achieving expansions of the food supply explains the conclusion which follows, that what distinguishes capital 'does not lie in the kind of commodities, but in the mind of the capitalist – in his will to employ them for one purpose rather than another; and all property, however ill-adapted in itself for the use of

39 As explained before the net increase in labour demand where labour supply is elastic only becomes effective once the 'real' capital goods have been constructed. In the case of inelastic labour supply, however, the increase becomes effective immediately.

labourers, is a part of capital, so soon as it, or the value to be received from it, is set apart for productive reinvestment. The sum of all the values so destined by their respective possessors, composes the capital of the country. Whether all those values are in a shape directly applicable to productive uses, makes no difference. Their shape, whatever it may be, is a temporary accident: but once destined for production, they do not fail to find a way of transforming themselves into things capable of being applied to it' (cf. also 67–8, 82–3).[40] Or again, as in 1830 'the food, machinery &c., which will ultimately be purchased with the goods in my warehouse, may at this moment not be in the country, may not be in existence. If, after having sold the goods, I hire labourers with the money, and set them to work, I am surely employing capital, though, the corn, which in the form of bread those labourers may buy with the money may be now at Dantzic or perhaps not yet above the ground' (IV, 267).

The less significant is the distinction between wage goods and luxury goods, of course, the greater the flexibility of the productive system and the ease of expanding the former at the expense of the latter. In point of fact the notion of a sharp distinction between the two categories broke down at an early stage in Mill's exposition. The example he devised involving an increased investment programme deliberately assumed for the sake of the argument that workers initially were 'like the Irish peasantry, only half employed and half fed' (II, 56). This was to assure that their increased money wages would in fact be devoted to goods distinct from those consumed by capitalists; but it was not the normal case. The significance for us of the recognition that workers normally earn a 'surplus' over subsistence requirements can best be seen by reverting to the case of increased investment with given labour supply. In this event 'the whole of what was previously expended in luxuries, by capitalists and landlords, is distributed among the existing labourers, in the form of additional wages. We will assume them already sufficiently supplied with necessaries. What follows? That the labourers become consumers of luxuries; and the capital previously employed

---

40 See also Mill's defence of his definition in correspondence (26 Oct. 1854): 'I hold . . . that whether any given portion of wealth is capital or not, is solely a question of the intentions of its owner: just as it is wholly a question of the intentions of the owner whether a given bushel of wheat is seed or food' (*CW*, XIV, 242). Similarly the early notes on Senior's *Outline* (1945): 'I would call capital, what anyone possesses which he designs for productive use either in itself or in what it will purchase. If all the food in this country were to be destroyed to-night, & every capitalist determined still to keep up the extent of his productive operations, I should say our capital was undiminished though much of it must change its shape before it could be used as capital. If half the plate-glass was destroyed & the remainder sold for as much as the whole did before, I should say the capital of the country was *not* diminished: the whole loss would fall on the revenue of the purchasers, & next year's productive operations would be impaired. I think a peculiar use of the word capital confuses exceedingly the latter part of this Treatise' (137).

in the production of luxuries, is still able to employ itself in the same manner: the difference being, that the luxuries are shared among the community, generally, instead of being confined to a few' (68). There can be no question of a rigid upper bound to the wages fund.[41]

* * *

From all this there emerges a rather clear picture of Mill's vision of economic process in an advanced capitalist-exchange system. It is far removed from a primitive agricultural economy for which the wages-fund theory rigidly interpreted might be appropriate, namely one wherein workers consume a distinct class of commodities, produced in annual jets, and opportunities for carry-over from period to period are limited.[42] According to the picture that has emerged workers are paid in money, not in kind, and enter the market to purchase commodities at retail like any other consumers; there is no distinction in this regard between consumption by labour, by capitalists or by landlords. The 'wages-fund' is thus expressed in money but has a real counterpart in the flow of goods currently made available in the commercial sector.[43]

41 Cf. the discussion by Taussig (1896), 233, pointing to the same conclusion: 'In the chapter specifically devoted to wages, the passages quoted show no stress on the rigidity of the fund, and indeed hardly give an indication one way or the other as to Mill's opinion . . . We have just seen how often, in other passages than those which were expressly concerned with wages, he discussed the relations between capitalists and labourers as if the essential thing were the advance of money funds or proceeds by the individual employers. On this basis, he could hardly have entertained the notion of any rigid source of wages; for he had set forth that these funds would shrink or swell with the capitalist's change of intention, and had implied that they varied with his control over immediate money funds. In the main there is thus little direct indication in the body of the *Political Economy* of any iron-clad doctrine, and certain proof that such a doctrine, if entertained at all, was far from prominent in Mill's own thinking.'

42 A view of the doctrine recently described, for example, by Ekelund (1976), 66–85; the entire paper turns upon this erroneous view. See also the 'relativist' interpretation of the doctrine suggested by Pigou (1949), 178–9, and Scott Gordon (1973), 16–17.

43 In an important passage in his *Treatise on Money*, J. M. Keynes (1930), while critical of Mill's formulations of the wages-fund doctrine added that 'it has proved injurious to clearness of thought to demolish this doctrine without putting anything into its place. For if "circulating capital" is identified not merely with liquid capital or "goods in stock", but with liquid stocks *plus* the flow of *available* income accruing during the period of process, then the Doctrine of the Wages Fund embodies an important truth, without which the nature of the productive process through time and its relation to capital and to saving cannot be understood. It is the flow of income, available for consumption by the factors of production, which constitutes the true *Wages Fund*; and it is the distribution of this Fund between relatively productive and relatively unproductive consumption which determines the volume of employment and of output' (II, 129). As explained in our text, something very like this conception seems to be what Mill himself appreciated well enough.

For current production to proceed smoothly at any particular rate it is obviously necessary for the 'pipelines' to be of suitable dimension – the stock of technological capital including machinery and structures and its distribution between sectors to correspond with the pattern of consumer demand (including demand by labourers) must be appropriate. The attention paid to 'capital' as the result of *past* labour accords with this conception even in so far as concerns the real goods purchased by labour; for while the production process only ends upon final sale – so that labourers and others, strictly speaking, buy goods currently produced – the bulk of activity on any batch of goods must have been undertaken in the past.[44]

The great emphasis on the 'unproductive consumption of productive labourers' is pertinent to the flexibility of the system in allowing rapid increases in real wages upon an alteration in the investment decisions by capitalists or their decisions to increase outlays on service labour. In some cases no change at all is required in the kinds of goods produced. But even where this is not the case, so that some time must pass before the flow of wage goods can be increased, Mill did not emphasize the problem which would be no greater than that posed by any alteration in the pattern of final demand. There is no emphasis whatsoever upon rigidity of the real wage bill.

All this is further confirmed by Mill's early account of the fourth proposition in his defence of Ricardo against Senior:

Ricardo seems to me quite right . . . If I ceased to buy (suppose) silks, and with the money they cost, hired servants, I should add to the fund for the purchase of labour. The silk manufacturer's capital remains as before; & I come into the market too, as a competitor with him. The commodities on which the labourers will chose to spend their increased wages, are not yet increased, it is true; but (I contend) that increase always *follows* instead of preceding & causing the rise of wages. Rise of wages means, the labourer's having more given him – more commodities, more

---

44 Cf. Jacob Viner's rejection of the critical comments on the wages-fund theory by Edwin Cannan in the latter's *A Review of Economic Theory* (1929): 'The wages-fund theory is criticized on the familiar, but I believe thoroughly erroneous, ground that wages are paid out of current production, and that no stock of provision has to be accumulated . . . The great bulk of the process of producing what labor consumes as real wages in any one year must have been undergone long before . . . Assuming continuous production, no stock of provision wholly ready for consumption need to be accumulated in advance of payment of wages. But current wages are paid out of a flow of finished products to which current labor has contributed only the finishing touches. What current labor produces is in the main future and not current real wages' (Viner, 1958, 402). Cf. also the comment by Samuelson (1978), 1425–6, on the 'vulgar view of Henry George that production of outputs by inputs is instantaneously and automatically synchronized' and the literature, *pro* and *con.*, there cited. (For George's criticisms of Mill see the chapter 'Wages not drawn from capital, but produced by labour', 1879, 33f).

exchangeable value, for his labour: his tastes & wants & the demand they generate, *afterwards*, determine what commodities shall be produced to meet that demand. Thus the series of facts is not, 1st, increase of necessaries & labourer's luxuries 2d, consequent rise of wages. It is, 1st, increase of commodities of any kind destined by their owners for productive employment, 2d, rise of wages, 3d, increased demand for necessaries & labourer's luxuries; 4th, increased production of them (1945, 138).

What, though, of the conception of interest as the reward for 'abstinence', the so-called 'waiting' theory of interest (below, p. 427)? It is sometimes presumed that this approach necessarily presumed discontinuous production, and a corresponding pre-accumulation of stocks and advance to labour (e.g. Blaug, 1978, 189). This view seems to involve a misunderstanding, as will be clear from Mill's treatment of the fourth proposition on capital. When capitalist employers make an investment decision and abstain from current consumption, they do not directly provide wage goods to tide over additional labourers until the next season; rather, they abstain from using their own claim to purchase output currently forthcoming at retail outlets and (ultimately) place this purchasing power at the disposal of labourers: 'I have consumed [wine] that much less, and made over my consuming power to them . . . I have . . . postponed my consumption, and have turned over part of my share of the present produce of the community to the labourers. If after an interval I am indemnified, it is not from the existing produce but from a subsequent addition made to it. I have therefore left more of the existing produce to be consumed by others; and have put into the possession of labourers the power to consume it' (II, 83–4).[45] In this sense it is legitimate to talk of 'advances' in Mill's system, and understand the fourth proposition that a 'demand [for labour] delayed until the work is completed, and furnishing no advances, but only reimbursing advances made by others, contributes nothing to the demand for labour; and that what is so expended, is, in all its effects, so far as regards the employment of the labouring class, a mere nullity; it does not and cannot create any employment except at the expense of other employment which existed before' (85).

It is, however, easy enough to confuse the latter conception of 'advances' with one entailing literal pre-accumulations of real wage goods by the immediate employer especially if one is seeking to formulate a simplified version of the doctrine; or to confuse the fact that capitalists receive their income only upon sale of their product, as the difference between proceeds and outlay, while their workers are paid 'in advance' of the sale, with a conception of economic organization involving discrete production periods and literal pre-accumulations and advances. This is the trap into which Mill

45 Cf. also an early account (1824) of what is entailed by saving, emphasizing the postponement of consumption now in order to increase consumption later out of the higher (absolute) yield generated by expanded capacity (IV, 18).

himself seems to have fallen not only in the essay on profits and interest and the 1869 review but, on occasion, in the *Principles* as well.

The foregoing picture of economic process in its sophisticated version corresponds to that outlined by David Ricardo (see Hollander, 1979, 326f). The conceptual framework of an annual cycle governing the demand for labour, which was characteristic of James Mill, had been unacceptable to him. And all the standard implications for policy (including policy regarding unions) drawn from the wages-fund theory in its narrow version – that involving a stock of wage goods (largely food) on hand for disbursement during the year to labour, no more and no less – had been rejected by him. J. S. Mill was merely continuing a Ricardian tradition.

VI   THE WAGES FUND AS EQUILIBRIUM SOLUTION

In his recantation of the wages-fund doctrine in 1869 Mill criticized what he claimed to be the technological inability, deriving from the discontinuity of the production process, to alter the magnitude of the wages bill – apparently even the 'pecuniary' wages bill – during the course of the 'year'. This criticism appears unjustified if directed against the position actually developed in the *Principles* where there is little to suggest any such rigidity of the wages bill – in either money or real terms. Yet there is a sense in which the wages bill can be said to be 'predetermined' – the sense implied by any determinate solution to a problem of competitive pricing. For the demand and supply curves must be stable for such solution to be meaningful, their stability reflecting investment plans by capitalists and plans regarding work and leisure by labourers. It would seem that Mill erred in 1869 by confusing the two in his description of his original position. (In any event, what is important in so far as concerns the ability of unions to raise wages without loss of employment – the matter raised by Thornton in his work upon which Mill was commenting – is not the technical possibility of expanding the flow of real wages within a very brief time span, but whether the planned rate of consumption by capitalists can be permanently depressed by union pressure without negative effects on labour demand or whether the consequence will be a reduced rate of capital accumulation.)

What then of the precise nature of the demand for labour in Mill's *Principles*? In most secondary versions of the doctrine the wages bill together with the given work force yield the wage rate. The discussions of the manner whereby the classics supposedly envisaged the 'pre-allocation' of the total capital stock between technological and wage-goods capital imply that the size of the wages fund is a datum of the analysis once aggregate capital is known: 'Like all "classical" leaders he [J. S. Mill] took the relationship between technological and wage capital as a datum, so that in the final result saving would increase both of them in proportion' (Schumpeter 1954, 642).

In this approach wage capital is treated on a par with technological capital, a practice which makes little sense unless it is assumed that the real wage is a constant so that a given stock of wage goods invariably 'supports' a certain quantity of labour. But in the event of variability of wages – and the theory was evidently intended to deal with such cases – the 'dependency' of labour upon wages and its 'dependency' upon technological capital (structures, machinery, materials) are quite distinct in nature. The distinction is expressed with admirable clarity in one of Mill's earlier essays (above, pp. 364–5).

There is to be found in the classical literature a version of the wages-fund theory which formally avoids the treatment of wages on the same footing as technological capital. According to this approach the wage bill is not a datum of the model but a dependent variable. The analysis to which I refer is to be found explicitly laid out in the work of Robert Torrens (1834) and that of J. E. Cairnes (1874).

To assure the employment of any given work force, it is presumed that a specific quantum of non-labour inputs must be provided. An increase in the capital stock therefore will take the form of wage payments entirely if the labour force is unchanged. The wages bill appears as the 'residual' portion of the capital stock, and is the equilibrium outcome in a model the data of which are aggregate capital and the factor proportions (which together generate the labour demand 'curve') and the working population which, as usual, constitutes labour supply.[46]

Robert Torrens' version is formulated in the following terms:

For when the farmer, in order to extend his cultivation, makes an addition to this capital, he will require a greater number of hands, and will seek to tempt them into his employ by the offer of higher wages. But as the increase of capital is supposed to be general, all other capitalists will require additional hands as well as our farmer, and will be offering higher wages also. All the capitalists will be unwilling to let their additional capital lie idle for want of hands, and, with the two-fold object of retaining their own labourers, and of obtaining those of their neighbours, will go on advancing wages, until the whole of their additional capital is absorbed. Assuming that all the labourers are already employed, and that no addition is made to their numbers, it is morally certain, that the whole of every new accumulation of capital will assume the form of increased wages, until the reward of the labourer has reached its maximum. New accumulations of capital are made for the sake

---

46 Consideration of the section on wages in James Mill (1844), 40f, will show that it is by no means certain that the wages-fund theory involving a predetermined fund was invariably in mind. There is much to suggest that James Mill too was groping towards a Torrens–Cairnes type version of the theory, where the wage rate is not obtained simply by dividing a given wages bill by the working population.

of obtaining advantage therefrom. But it is impossible that new accumulations of capital should be advantageously employed, unless labourers can be procured. The new capital, accumulated for the purpose of gaining an advantage by the employment of labourers, comes into the market and bids for hands; the old capital, in order to retain its hands, is compelled to bid against the new, and this process goes on until the whole existing capital is invested in wages, seed, materials, and machinery. But as a given number of hands can use only a given quantity of seed, materials, and machinery, these ingredients or component parts of capital cannot be increased, while the quantity of labour remains the same; and therefore it is only in the form of increased wages that the new accumulations of capital can appear. When the number of labourers remains the same, nothing can prevent new accumulations of capital from appearing in the form of increased wages, except such an intimate understanding and concert among capitalists, as would induce each individual of the class, instead of seeking for additional hands, to allow all his new accumulations of capital to remain idle and unproductive . . . If such new accumulations are made, it is in order that they may be employed; and if they are employed the quantity of labour, and the state of applying mechanical power remaining the same, there is no form in which they can appear, except in that of increased wages (1834, 16–18).[47]

The observation supporting fixed technical proportions – 'a given number of hands can use only a given quantity of seed, materials, and machinery' – must be understood as relevant to the productive sector only since service labour does not require the support of other inputs: 'The effectual demand for labour . . . consists merely in the offer of an adequate quantity of subsistence. Productive labour, indeed, cannot be put into operation unless the labourers, in addition to their subsistence, are furnished with tools and material. But an increased supply of the necessaries of life is of itself sufficient to enable us to engage an increased number of menial servants and unproductive retainers' (1821, 355).[48]

Cairnes's formulation is almost identical:

Assuming a certain field for investment, and the prospect of profit in this such as to attract a certain aggregate capital, and assuming the national industries to be of a certain kind, the proportion of this aggregate capital which shall be invested in wages is not a matter within the discretion of the capitalists, always supposing

47 Torrens's model assumed full capacity usage. Elsewhere he allowed for variations in the intensity of operation of fixed capital (1834, 63f). A somewhat less explicit statement by Torrens of complementarity appears in his *Colonization of South Australia* (1835), 23–4. An excellent account of these contributions – and of the entire classical wages-fund literature – is given by Stuart Wood (1890), 426–61. For a discussion of Wood's account see Stigler (1965), 287f.

48 See below note 64, for some of the implications of Torrens's position for the efficacy of union activity.

they desire to obtain the largest practical return upon their outlay. To accomplish this, the instruments of production, labor, fixed capital and raw material must be brought together in certain proportions – a condition which requires . . . the supply of labor being given – a distribution of the aggregate capital in certain proportions among those instruments. Supposing, now, capitalists to succeed in forcing down the rate of wages below the point at which, having regard to the number of the laboring population, the amount, which the fulfilment of this condition would assign to the payment of wages, was absorbed – either the capital thus withdrawn from the Wages-fund must remain uninvested and therefore unproductive, or if invested, and not invested in wages, it would take the form of fixed capital or raw material. But by hypothesis the fixed capital and raw material were already in due proportion to the labor force, and they would consequently now be in excess of it. A competition among capitalists for labor would consequently ensue; and what could this end in but a restoration to the Wages-fund of the amount withdrawn from it? (1874, 186–7)

Similarly, Cairnes limited the discussion to productive labour, the characteristic of which is the requirement of strictly complementary inputs:

It will be remembered that in the enunciation which I quoted from Mr. Mill of the wages question, the Wages-fund is stated to consist of two distinct parts – one, the largest and by much the most important, constituting a portion of the general capital of the country; while the other is derived from that part of the nation's wealth which goes to support unproductive labor, of which Mr. Mill gives as an example the wages of soldiers and domestic servants. In proceeding to deal with the wages question, it will be convenient to omit for a time all consideration of the latter part: this will be more easily dealt with when we have ascertained the causes which govern the main phenomenon.

Restricting our view then for the present to that portion of the general Wages-fund which goes to support productive labor, we have, in the first place, to observe that the hiring of labor for productive purposes is an incident of the investment of capital. A capitalist engages and pays a workman from precisely the same motives which lead him to purchase raw material, a factory, or a machine (168).

Although Cairnes promised to take up the problem of service labour, he did not return to examine the issue.

According to this version of the theory, the amount of wage payments is not given independently of the size of the working population. An exogenous increase in the work force will actually lead to a smaller equilibrium wages bill since a larger fraction of total capital is required to take the form of fixed capital. The percentage decline in the wage rate thus exceeds the percentage increase in population; the demand curve for labour is relatively inelastic, in contrast to the unitary elasticity of the 'standard' case.

We may formalize the Torrens–Cairnes version of the wages-fund theory by means of simple equations. In their treatment the quantity demanded of labour in aggregate P, is dependent upon the aggregate capital stock C, the technically determined ratio λ between labour and non-labour inputs or 'fixed capital', and the wage rate w, and may be written as

$$P = \frac{C}{1/\lambda + w} \tag{1}$$

This is derived directly from the constraint that total capital is made up of fixed capital M and wages capital wP, that is $C = M + wP$. Replacing M by P/λ we obtain expression (1).

If the supply of labour S is given, equilibrium will be assured by the full-employment condition

$$P = S \tag{2}$$

and the wage rate may be obtained by inserting expression (1) into (2) since w is the only unknown to be determined. Thus

$$\frac{C}{1/\lambda + w} = S$$

and

$$w = C/S - 1/\lambda \tag{3}$$

Expression (3) may be compared with the 'standard' wages-fund theory wherein

$$w = C'/S$$

C' referring to wage capital only, rather than aggregate capital.

The characteristic inelasticity of demand

$$\frac{dPw}{dwP} = \frac{1}{1/\lambda w + 1} < 1$$

is recognized only by Cairnes (173–4) who described it as 'an unexpected consequence, not, so far as I know, before adverted to . . .' (173–4).[49]

---

49 But we find Cairnes playing down the negative effect on aggregate wages of increases in the working population: 'This occurs, I say, where labor is of a kind to be employed

Cairnes' analysis is usually regarded as peculiar to him alone and atypical of classical doctrine. This was Marshall's opinion: 'After a while Cairnes, in his *Leading Principles*, endeavoured to resuscitate the Wages-fund theory by expounding it in a form which he thought would evade the attacks that had been made on it. But, though in the greater part of his exposition, he succeeded in avoiding the old pitfalls, he did so only by explaining away so much which is characteristic of the doctrine, that there is very little left in it to justify its title' (1920, 825; cf. Hutchison, 1953, 260). Similarly, Schumpeter comments that Cairnes 'interpreted the wages fund in a way that left little to defend' (1954, 670n).

It is doubtless Mill's formal account of the doctrine in 1869 which did more than anything to establish the content of the 'classical' wages-fund theory. Cairnes himself expressed surprise at Mill's statement because in his view Mill was mistaken in ascribing to the *Principles* the notion of a predetermined wages bill:

The law of the supply of labor is no longer called in question; but several able writers have within a few years, in dissertations directed against what is known as the 'Wages-fund' doctrine, challenged the view hitherto received as to the law of its demand. Foremost among these have been Mr. Thornton . . .; nor is it possible to deny the ability and skill with which the assault has been conducted, when we find that he can boast, as among the first-fruits of his argument, no less a result than the conversion of Mr. Mill . . .

. . . I must own myself unconvinced by Mr. Thornton's reasonings, strengthened and reinforced though these have been by the powerful comments of Mr. Mill. Not indeed that I am prepared to defend all that has been written on what, for convenience, I may call the orthodox side of this question, but I believe the view maintained by those who have written on that side, and pre-eminently the view maintained by Mr. Mill himself – taking it as set forth in his original work, not as explained in his retraction – to be substantially sound, though needing, as it seems to me, at once fuller development and more accurate determination than it has yet received (1874, 157–8).[50]

in conjunction with fixed capital and raw material; and, it may be added, that the effect would only assume sensible dimensions where those agencies constituted a substantial proportion of the whole capital invested. Indeed it would be a mistake to regard this particular condition – the supply of labor considered as a cause affecting, not the rate of wages, but the aggregate Wages-fund – as under any circumstances more than a subordinate and modifying influence in the case. The point is one of theoretic rather than of practical importance'.

50 Yet in a letter of 23 May 1869 Cairnes had praised the review article: 'All that you have said on the subject of the wages-fund seems to me excellent. The conception, as now delineated, is, so far as I can see, invulnerable; while it retains all that is required to serve as a basis for a theory of wages' (quoted in George O'Brien, 1943, 283). Needless to say, Mill was delighted (23 June 1869): 'You may imagine how gratifying it is to me that you give so complete an adhesion to the view I take of the wages fund' (*ibid.*, 284; also *CW*, XVII, 1616); Some time after May 1869 Cairnes had second thoughts.

Now it seems to me that Cairnes was justified in taking this position. That this is so can best be seen by consideration of two case studies drawn from the *Principles*.

I have in mind, first, Mill's discussion of the consequences for wages of increased investment. Mill admitted that an excess supply of commodities due to a temporary excess demand for money was to be recognized. This proposition, however, was strictly separated from the argument put forward by Malthus and others that it was possible for investment to be carried on at an excessive rate. Even if population is presumed to be constant there would be no problem; the attempt by capitalists to expand the real capital stock will be thwarted by the lack of a sufficient labour supply, and competition for the given labour force will cause the wage rate to rise until the additional investment outlays simply take the form of increased labour income. The income distribution will turn in favour of labour, but at no point will there occur a deficiency of purchasing power (*CW*, II, 68, cited above, p. 375). Quite clearly the distribution of capital between its constituent elements is not a technological *datum* given independently of labour supply – the key characteristic of the 'standard' version of the doctrine; the wages bill is the outcome of an equilibration process. It is essential to reiterate that we are not dealing here with an exceptional statement. Rather it reflects a position central to Ricardian theory (Hollander, 1979, 326f.), and must be regarded as representative of Mill's considered opinion.

Mill also considered the reverse case – that of a war loan raised by the government, where the funds are assumed to be derived from reductions in investment outlays, in essence a case of disinvestment:

We will suppose the most unfavourable case possible: that the whole amount borrowed and destroyed by the government, was abstracted by the lender from a productive employment in which it had actually been invested. The capital, therefore, of the country, is this year diminished by so much. But unless the amount abstracted is something enormous, there is no reason . . . why next year the national capital should not be as great as ever. The loan cannot have been taken from that portion of the capital of the country which consists of tools, machinery, and buildings. It must have been wholly drawn from the portion employed in paying labourers: and the labourers will suffer accordingly. But if none of them are starved; if their wages can bear such an amount of reduction, or if charity interposes between them and absolute destitution, there is no reason that their labour should produce less in the next year than in the year before. If they produce as much as usual, having been paid less by so many millions sterling, these millions are gained by their employers. The breach made in the capital of the country is thus instantly repaired, but repaired by the privations and often the real misery of the labouring class (II, 76).

It is apparent, in the present instance, that employment is dependent upon fixed capital; it is only because the community's fixed capital stock remains unchanged that it is possible to maintain the same level of employment as in previous periods. The wages fund is reduced by the 'disinvestment' but there occurs no reduction in the fixed capital stock and employment.[51]

This is the sequence of events which must follow from the analysis of Cairnes and Torrens. Full employment requires the support of a specific fixed-capital complex and any given capital outlays must – in equilibrium – allow for such necessary expenditures, so that only the remainder is available for wage payments. Hence, a reduction in capital outlays for any reason will simply reduce the 'residual' available for wages.

References to our case studies from the *Principles*, which imply the use of a 'Torrens–Cairnes' model, will show that Mill's concern too was with the productive sector only. By contrast, Mill in the *Fortnightly Review* opens his attack with a formal statement of 'received doctrine' which includes a service sector – 'The demand for labour consists of the whole circulating capital of the country, including what is paid for unproductive labour' (above, p. 389).[52] This distinction between the theory implied by the formal accounts (where a service sector is recognized) and that implied in Mill's consideration of specific issues relating to the effects of investment which deal only with a 'productive' sector is not accidental. The Torrens–Mill–Cairnes version of the wages-fund theory – where the wages fund appears as the outcome of a competitive supply-demand equation – is applicable to the productive sector only because the use of technological capital was regarded by all three economists as irrelevant to the service sector.

* * *

A word now regarding the (theoretical) scope of the Torrens–Cairnes–Mill approach. In order to be assured of the competitive solution a labour-demand curve of negative slope is required. To obtain such a demand curve when

51 The same line is followed in a discussion of the speed of recovery from war damage. A given working population requires the support of a certain fixed capital complex. If the necessary prerequisites are undamaged then employment will be unaffected; even though aggregate wages may be reduced, the economy is capable of the same industrial performance (75).

52 That Mill's concern, in stating that the average wage rate is determined by the ratio of 'capital' to 'population', extended beyond the issue of employment in the industrial sector was in fact stated clearly in a revealing footnote attached to the 1848 edition of the *Principles*: 'Although, in this place, where the subject under discussion is the causes and remedies of low wages, the question of population is treated chiefly as a labourer's question, the principle contended for includes not only the labouring classes, but all persons, except the few who being able to give to their offspring the means of independent support during the whole of life, do not leave them to swell the competition for employment . . .' (II, 372n.; cf. 419n). Professor J. M. Robson has pointed out to me that this is the one passage of the manuscript which *may* be in Harriet Taylor's hand.

constant proportions between labour and fixed capital is the rule, it is necessary to allow for *hypothetical* variations in the aggregate quantity of technological capital. For an excess demand for labour at the initial wage, which must be supposed generates the upward pressure following an increase in investment, implies that employers are able to provide additional labour with the capital equipment. In point of fact the wage rate is forced upwards towards a new equilibrium position by the scarcity of labour services so that additional technological capital is not actually constructed, but the logic of the process requires that firms be able to make additions to their fixed capital stocks. Thus the analysis turns out to be applicable only to the Marshallian long run. Similarly, an hypothetical rise in the wage rate implies a reduction in the quantity of labour demanded. This is because, for each demand 'curve', aggregate capital C is assumed unchanged so that the higher wage payment must be 'financed' from that part of the stock hitherto reserved for fixed capital M; given M/P ($= 1/\lambda$) employment capacity falls. Again it is clear that the analysis logically applies to the 'long run' since a variation in the quantity of fixed capital, albeit within a given aggregate, must be accommodated. In the short run the demand for labour will be completely inelastic; and in fact, once the fixed capital is in place, there is no mechanism to assure that the 'residual' portion of the capital stock will be devoted to wage payments. These severe problems were not recognized by either Mill, Cairnes or Torrens.

We shall, however, take the model on its own terms. When the work force is assumed constant, as above, the entire net increase in capital is supposedly devoted to wage payments. But the general argument can be applied to the case of an expanding labour force. For Mill also stated: 'I do not mean to deny that capital, or part of it, may be so employed as not to support labourers, being fixed in machinery, buildings, improvements of land, and the like. In any large increase of capital, a considerable portion will be thus employed' (66). This latter qualification can only refer to the case where population is rising; it is again evident that the division of total capital, and accordingly the size of the wages bill, depends upon the work force itself, and is not given independently. Assuming the simplest case of a fixed real wage rate, the model yields the result that an increase in aggregate capital entails a proportionate expansion in the net demand for labour (employment capacity) even allowing for the need to provide the expanded work force with 'machinery' provided the constant-proportions axiom is maintained.[53]

* * *

53 Labour supply will be given by $w = \bar{w}$. Knowledge of aggregate capital C and the machinery–labour ratio yields the quantity of labour demanded (or employment capacity) by equation (1) above.

Let us now formulate a conclusion regarding the account of the wages-fund doctrine given by Mill in 1869. It is significant that in the *Principles* itself Mill gave no indication whatsoever that the wages-fund theory entailed unitary elasticity of demand for labour; the only statement, to my knowledge, of this property is the *ex post* attribution of 1869. When this fact is considered in conjunction with the case studies in the *Principles* which imply that the relationship between technological- and wage-capital is not a technical datum; with Mill's appreciation of the sharp distinction between the 'dependency' of labour upon technological capital and its 'dependency' upon wages; with the fact that all the statements of the wages-fund theory in the *Principles* was consistent with the alternative version of the doctrine formulated explicitly by Torrens and Cairnes which recognizes these features of the labour market; and finally with the positive allowances in that work for variability of the wage bill during the course of a single 'production period', one is led to conclude that Mill erred by his formulation of 'received doctrine' in 1869. Cairnes was correct in his reaction. But it is also difficult to avoid the conclusion that, from the outset in the 1840s, Mill could not have been fully aware of all the properties of his own model. For while he did not spell out the unitary elastic version in 1848 and thereafter, it is also true that he did not spell out the alternative version as clearly as Torrens had done and Cairnes was to do.

VII THE WAGES-FUND THEORY:
THE RECANTATION INTERPRETED

Mill's precise intentions by his retraction of support from the wages-fund doctrine constitutes one of the most difficult problems in the history of economics. A word first about Thornton's position to which Mill had responded in his 1869 article.

Much of the relevant material in Thornton's book *On Labour* (1869) was repeated from his article entitled 'What Determines the Price of Labour or Rate of Wages?' which had appeared in the *Fortnightly Review* for May 1867. Here he had presented a general criticism of simple supply–demand equilibrium analysis turning upon complications created by the presence of completely inelastic schedules, severe discontinuities, and the like, to

There is one further theoretical matter that requires note. Formally, the model is restricted by the assumption of fixed factor proportions. To accommodate the possibility of variable proportions – at least by way of variation in the commodity mix assuming differential ratios from industry to industry – we must introduce a functional relationship tying the *average* machinery–labour ratio to the wage rate, a requirement which in turn implies knowledge of the coefficients of production of all commodities in the economy and the system of demand curves (below, p. 421).

which we have alluded earlier.[54] The argument was applied to the labour market, as well as to commodity markets, on the grounds principally that both the labour demand and supply schedules were, at least over wide ranges, completely inelastic. Since employers' combinations were easily formed and workers were badly organized it followed that in practice the wage rate was determined by the employers' *dictat*: '[employers] both can and do force them [the labourers] to take as little more than the bare means of subsistence as it pleases them to offer' (1867, 564). Such an arbitrary process was not necessarily to the workers' detriment, since in cases of excess labour supply the wage might be set at a higher level than would have resulted were employers to remain passive and workers to compete for scarce jobs; in the absence of the employers' intervention the wage rate would be driven down to the lowest conceivable level, the excess supply remaining apparently uncorrected (561, 563). But in the case of excess demand such combinations function to prevent competition between the employers and upward pressure on the wage rate. Employers realize that it is 'better for them to go without part of the labour they desire, rather than, for the sake of obtaining that portion, to incur the obligation of paying a greatly increased price for all the labour they employ' (560–1). Again the divergence between demand and supply would not apparently have been corrected by changing wage rates. Some reference is also made to coincidental schedules. In such instances the wage rate will in practice also be determined by the employers (561).

The practical outcome of Thornton's treatment is that workers' unions have a significant role to play: 'counter combinations may be potent enough to unsettle and resettle' the wage rate determined otherwise by one-sided 'artificial' and 'capricious' rulings (565). The 'new view' led Thornton to reject the notion of a predetermined wages fund. But very little attention was in fact paid to the issue in the first edition of *On Labour*; no more than a lengthy footnote was devoted to the wages-fund theory.[55] The wages bill, Thornton argued, cannot be regarded as a fixed cost to each individual employer; his demand is for a certain specific quantity of labour, and there is no reason for him to spend the maximum sum which he may have available for wage payments:

---

54 As a result of criticism by Thornton on similar grounds many years earlier, Mill had altered, in his third edition (1852), the analysis of the determination of equilibrium in international exchange to meet the objection 'that several different rates of international value may all equally fulfil the conditions of this law' (*CW*, III, 608).

55 The particular formulation under attack was that by H. Fawcett, *The Economic Position of the British Labourer* (1865), 120, quoted in *On Labour*, 84n.: 'The circulating capital of a country is its wages fund. Hence, if we desire to calculate the average money wages received by each labourer, we have simply to divide the amount of capital by the number of the labouring population'.

But has any individual any such fund? Is there any specific portion of any single individual's capital which the owner must necessarily expend upon labour? Of course, there is a certain amount which every effectual employer can afford to spend upon labour, as also there is in every instance a certain limit to that amount which cannot possibly be exceeded. But must the amount, so limited, which is thus applicable to the purchase of labour, be necessarily so applied? Does any farmer or manufacturer or contractor ever say to himself, I can afford to pay so much for labour: therefore, for the labour I hire, whatever the quantity be, I will pay so much? Does he not rather say, '*So much labour I require*, so much is the utmost I can afford to pay for it, but I will see for how much less than the utmost I can afford to pay, I can get all the labour I require'? (1869, 84–5n.; cf. 1867, 564).[56]

Since wages are not a fixed cost to the individual employer, Thornton concluded rather too hastily, there could be no fixed aggregate wages fund.

These arguments were in general accepted by Mill, who pleaded guilty to having maintained the notion of a predetermined wages fund. But it was Mill, not Thornton, who spelled out that logically according to received doctrine the demand curve for labour is of unitary elasticity. Secondly, the rationale for a labour demand curve of zero elasticity was given more attention by Mill than by Thornton in the first edition of his book.

In the review Mill emphasized the special case of completely inelastic and apparently coincidental schedules of supply and demand 'where neither sellers nor buyers are under the action of any motives derived from supply and demand to give way to one another'; in this case, 'which the law of equality between demand and supply does not provide for, because several prices all agree in satisfying that law . . . the question between one of those prices and another will be determined by causes which operate strongly against the labourer, and in favour of the employer . . . nothing but a close combination among the employed can give them even a chance of successfully contending against the employers' (*CW*, V, 643).[57] A more positive outcome from labour's perspective is also contemplated:

---

56 In this passage Thornton presumed that there exists a determinate *maximum* to the wages bill. But some of his examples do imply some response of the employer to wage-rate variations.

57 In this regard the labour market differed from commodity markets. As we have seen above (pp. 275–6) in competitive commodity markets, assuming zero demand elasticity, there would be a rough balance between dealers (sellers) and buyers. But in the case of labour the buyer takes the initiative and the initial presumption that the advantage lay with the price-setting party was substantiated: 'Whatever advantage can be derived

It has made it necessary for us to contemplate, not as an impossibility, but as a possibility, that employers, by taking advantage of the inability of labourers to hold out, may keep wages lower than there is a natural necessity for; and *e converso*, that if work-people can by combination be enabled to hold out so long as to cause an inconvenience to the employers greater than that of a rise of wages, a rise may be obtained which, but for the combination, not only would not have happened so soon, but possibly might not have happened at all. The power of Trade's Unions may therefore be so exercised as to obtain for the labouring classes collectively, both a larger share and a larger positive amount of the produce of labour; increasing, therefore, one of the two factors on which the remuneration of the individual labourer depends (646).

Let us now trace out the argument made for zero elasticity of labour demand. The precise order of the argument must be carefully followed.

The case begins by the account of the wages-fund doctrine in terms of a unitary elastic demand curve for labour as a whole – the theory was applied to aggregate wages (643–4; see above, p. 389). At the next stage, however, Mill referred to the individual employer of labour in making what seems to be his *main* argument against the notion of unitary elasticity – indeed against any response at all to wage variation:

Does the employer require more labour, or do fresh employers of labour make their appearance, merely because it can be bought cheaper? Assuredly, no. Consumers desire more of an article, or fresh consumers are called forth, when the price has fallen: but the employer does not buy labour for the pleasure of consuming it; he buys it that he may profit from its productive powers, and he buys as much labour and no more as suffices to produce the quantity of his goods which he thinks he can sell to advantage. A fall of wages does not necessarily make him expect a larger sale for his commodity, nor, therefore, does it necessarily increase his demand for labour (644).

There is little ambiguity about this statement: Mill's case relates to the *derived demand* for labour. Since a fall in wage costs does not 'necessarily' lead to expectations of greater final sales for the product it does not 'necessarily' lead to an increase in demand for the factor.

At first sight this is an extraordinary statement. Thornton believed that commodity markets are typically characterized by totally inelastic demand – at least over significant ranges – but it is clear from the passage itself that Mill

from the initiative is, therefore, on the side of the employer. And in that context of endurance buyer and seller, by which alone, in the expected case, the price so fixed can be modified, it is almost needless to say that nothing but a close combination among the employed can give them even a chance of successfully contending against the employers.'

did not. Moreover, he had just written a few pages earlier that 'it is the next thing to impossible that more of the commodity should not be asked for at every reduction of price' (637; see above, p. 276, for the general context). The response of quantity demanded to price played an important role in Mill's economics as we have seen at length.

Yet on reflection his position is not difficult to understand. In the *Principles* itself Mill had alluded to the properties of 'derived' and 'joint' demand with regard to input use (above, p. 263). these characteristics were the source of technical rigidities of production and consumption which tended to delay the fall of price to new cost levels:

Whether there will be a greater permanent supply of the commodity after its production has been cheapened, depends on quite another question, namely, on whether a greater quantity is wanted at the reduced value. Most commonly a greater quantity is wanted, but not necessarily. 'A man', says Mr. De Quincey, 'buys an article of instant applicability to his own purpose the more readily and the more largely as it happens to be cheaper. Silk handkerchiefs having fallen to half-price, he will buy, perhaps, in threefold quantity; but he does not buy steam-engines because the price is lowered. His demand for steam-engines . . . is almost always predetermined by the circumstances of his situation. So far as he considers the cost at all, it is much more the cost of working this engine than the cost upon its purchase. But there are many articles for which the market is absolutely and merely limited by a pre-existing *system*, to which those articles are attached as subordinate parts or members. How could we force the dials or faces of timepieces by artificial cheapness to sell more plentifully than the inner works or movements of such timepieces? Could the sale of wine-vaults be increased without increasing the sale of wine? Or the tools of shipwrights find an enlarged market whilst shipbuilding was stationary? (*CW*, III, 474).

But such rigidities would ultimately be overcome: 'Nobody doubts, however, that the price and value of all these things would be eventually lowered by any diminution of their cost of production; and lowered through the apprehension entertained of new competitors, and an increased supply.' It is thus Mill's position that a reduction in the price of an input will generate an expansion of demand for the input if sufficient time is allowed to overcome the various 'technical' rigidities in question, including that of assuring an expansion in final demand.

It seems probable, therefore, that when in the passage cited above from the recantation Mill stated that 'a fall of wages does not necessarily make [the employer] expect a larger sale for his commodity, nor, therefore, does it necessarily increase his demand for labour' he had in mind so short a period that an expansion of sales may not be taken seriously by employers who accordingly refrain from immediately increasing their demand for labour. This position does not, however, rule out expanded demand for input when a longer period is allowed. And in Part II of the review Mill recognized such reaction.[58]

The logic of the argument, strictly speaking, is of the partial-equilibrium variety. It implies therefore a concern with a variation in the wages paid to a particular category of labour in a single industry, other wage rates and product prices held constant. Yet the wages-fund doctrine under attack involves the aggregate demand for labour, and suggests a concern with a variation in general wages. Mill was on dangerous grounds when he

---

58 Stigler (1968), 99, is quite unjustified in his very critical remarks on Mill's position: 'Mill earned an hour in purgatory with this passage, because lower wages make for lower costs and larger sales of the product, and hence for more employment – this is a conclusion which is essentially exceptionless.'

According to a recent interpretation, Mill's position in Part II reveals that he had 'clearly abandoned the special (vertical demand curve) cases' (West and Hafer, 1978, 612; also *ibid.*, 604–5: 'Mill himself explicitly gave up the contention that the demand curve is typically of zero elasticity in the very same review of Thornton that is traditionally taken as the main locus of Mill's recantation'). This view of the matter is not convincing. Mill's recantation turns upon the unprofitability of offering additional employment upon wage reductions in the event of poor or non-existent opportunities for expanded final sales in the short run; but when full allowance is made for the working out of all responses to a change in wages, the responsiveness of labour demand to wage variations had to be recognized. Nothing was 'given up'; all depended on the time period assumed. The authors further maintain (610–11) that Mill was merely entertaining a theoretical possibility of zero elasticity, temporarily assuming – merely for the sake of argument – that Thornton's special circumstances involving small numbers applied in the labour market. This too does not seem convincing since Mill's analysis of zero elasticity in the short run was quite general; the key passage discussed above is not restricted to the small-numbers case.

There are other aspects of the interpretation that do not ring true. West and Hafer (1981) (following Taussig, (1896), 248) maintain that in his review article of 1869 Mill adopted Thornton's habit of analysing the wages fund in monetary instead of real terms: '[here] is another case where Mill was not so much recanting as changing the assumptions of the classical doctrine' (608). The basic classical model, attributed also to Ricardo, is said (following Ekelund, 1976) to assume rigidly differentiated consumption patterns between capitalists and workers and, more strongly yet, to assume a given stock of real wage goods which will all be used up in a discrete time period (605, 608). On this view, an increase in money wages 'will merely bid up the prices of a fixed, predetermined amount of "corn"; for the real wage bill is technologically determinate in a very strict sense. The authors assert that Mill's new insistence upon the indeterminacy of the (real?) wage bill follows from his decision to alter the assumption and henceforth analyse the wages fund in money terms – presumably (the authors are not clear on this matter) because this implies also the abandonment of the rigidly distinguishable consumption patterns. But we have shown at length – and these remarks apply to Ricardo too – that Mill had quite habitually worked in monetary terms in discussing the offer of employment, and had never insisted on a rigid distinction between 'wage and goods' and 'luxuries'. Mill's new insistence upon an indeterminate wage bill – money and real – seems to have nothing to do with any alteration in perspective such as the authors describe.

(implicitly) transferred from the former to the latter context. In short he based his case for a zero elasticity of aggregate labour demand upon an argument not strictly applicable to that area of discourse. That he was troubled by the issue is probable; for he himself raised the possibility that while the demand for labour by an employer in a particular industry may be totally inelastic upon wage reductions – a rendition suggesting a partial wage change – yet the capital released may be invested elsewhere in the system so that the whole of the wages-fund will be paying wages as before' (V, 644).

No direct answer to the foregoing question is given, but precisely at this juncture Mill denied the existence of a 'pre-determined' aggregate wages bill: 'Exists there any fixed amount which, and neither more nor less than which, is destined to be expended in wages?' The wages bill 'cannot exceed the aggregate means of the employing classes' (after allowance for their personal maintenance), but 'short of this limit, it is not, in any sense of the word, a fixed amount'. For the capitalist is under no obligation to spend a specific sum upon labour; each employer (and therefore, Mill implies, *all* employers) can be obliged to spend more than expected on wages or may enjoy a windfall gain even during the brief period before old plans can be revised and new plans put into operation:

This distinction, however, between the relation of the capitalist to his capital, and his relation to his income is wholly imaginary. He starts at the commencement with the whole of his accumulated means, all of which is potentially capital: and out of this he advances his personal and family expenses . . . If we choose to call the whole of what he possesses applicable to the payment of wages, the wages fund, that fund is co-extensive with the whole proceeds of his business, after keeping up his machinery, buildings and materials, and feeding his family; and it is expended jointly upon himself and his labourers. The less he expends on the one, the more may be expended on the other, and *vice versa*. The price of labour, instead of being determined by the division of the proceeds between the employer and the labourers, determine[s] it. If he gets his labour cheaper, he can afford to spend more upon himself. If he has to pay more for labour, the additional payment comes out of his own income . . . There is no law of nature making it inherently impossible for wages to rise to the point of absorbing not only the funds which he had intended to devote to carrying on his business, but the whole of what he allows for his private expenses, beyond the necessaries of life. The real limit to the rise is the practical consideration, how much would ruin him or drive him to abandon the business: not the inexorable limits of the wages-fund.

In short, there is abstractly available for the payment of wages, before an absolute limit is reached, not only the employer's capital, but the whole of what can possibly be retrenched from his personal expenditure; and the law of wages, on the side of demand, amounts only to the obvious proposition, that the employers cannot pay away in wages what they have not got (644–5).[59]

We are now in a position to draw the threads of the argument together. On close inspection it will be seen that there are two distinct aspects to Mill's case. First, the argument that in the very short run firms may not respond to a fall (rise) in the wage rate, because of low expectations of increased (decreased) final sales; this, in fact, was the only rationale offered for zero demand elasticity for an input. In the event of zero elasticity a variation in the wage rate entails a variation in the total industry wage bill. The second part of Mill's case – which, as the above passage clarifies, also applies to the short-run period – urges that such alterations in the wage bill are indeed conceivable, since the received notion of a technical inability to vary its size was groundless. The latter aspect of the argument does not, however, in itself provide a rationale for zero demand elasticity for labour.[60] Finally, it must be reiterated that though the main argument implies partial-equilibrium reasoning, it was Mill's intention to make a case for zero elasticity of the *aggregate* demand for labour, in the short period, upon variation in *general* wages.[61]

* * *

In Part II of his review, Mill considered further the implications of upward wage pressure by a union in a particular industry: 'it still requires to be asked, whether Unionists are justified in seeking a rise of wages for themselves, which in all probability produces a fall of wages,

59 As for supply 'the law as laid down by economists remains intact. The more numerous the competitors for employment, the lower *ceteris paribus*, will wages be'.

60 In an earlier article (1968) I erred in arguing otherwise. By interpreting the second part of the case as providing a *rationale* for zero elasticity of demand it appeared to me that Mill based his case on a *technological inability* to vary the rate of application of labour, given 'machinery, buildings and materials', turning upon fixed factor proportions. See West and Hafer (1981), 607, for valid criticism of this view, namely that the relevant text–'a fall of wages does not necessarily make him expect a large sale for his commodity' – does not lend itself convincingly to this interpretation.

Professor Schwartz (1972), 275, argues that the 'rejection of the yearly period of production lies . . . at the origin of the recantation' (also 95–6). In my view, as explained in the text, Mill had never in practice been restricted by the yearly period of production despite the retrospective attribution of 1869; while the case made out for the possibility of rapidly increasing the real wage bill was supplementary to the primary intention of the review.

61 But cf. Marshall's complaint that in stating the limit to the rise of wages to be 'the practical consideration how much would ruin [the employer], or drive him to abandon the business, not the inexorable limits of the Wages-fund', Mill failed to clarify whether the 'immediate or ultimate effects . . . short periods or long' were intended, though 'in either case it appears untenable' (1920, 825).

or loss of employment, to other labourers, their fellow-countrymen . . . For (as Mr. Thornton recognizes) there is no keeping up wages without limiting the number of competitors for employment' (662). Nothing that is said here, however, precludes reduced employment in the unionized industry itself – and as we have suggested the argument of Part I in favour of zero demand elasticity appears to be a special case applicable only to the very short-run period – but since Mill's concern was with the justice of union activity the emphasis is upon the potential loss of employment elsewhere.

The analysis is based upon the interrelationship between the commodity and labour markets. The primary concern is a case where demand elasticity for the product of the unionized industry is less than unity. Here total expenditure on the product will rise in consequence of the higher price charged by firms to cover increased wage costs – implying a fall in the demand for labour in the unionized industry – and thus, since aggregate incomes are no larger, outlays on other products will be reduced with adverse effects on employment therein (661).

This is a remarkable argument; for no use is made of the proposition that 'industry is limited by capital' maintained in all editions of the *Principles*, and in fact applied to *counter* an argument of the same kind regarding the impact of technical change (above, pp. 382–3 regarding *CW*, II, 96–7). If this is taken seriously much more than the recantation of the wages-fund doctrine would be at stake. It is difficult to believe that Mill fully realized this implication. It does emerge, however, that Mill recognizes a long-run impact of wage increase upon labour demand. The potential accorded a union in Part I is considerably diluted in Part II.

In the following year Mill went yet further. In reviewing Leslie on the *land question* he allowed for the possibility that a reduction in labour supply – his analysis here refers to the market as a whole – might result in no wage increase at all, the (long-run) demand being infinitely elastic. This view has most profound implications for population policy (below, p. 954).

VIII   THE WAGES-FUND THEORY AND THE FUNDAMENTAL THEOREM ON DISTRIBUTION

We have been preoccupied with two general aspects of Mill's approach to aggregative distribution – the wages-fund doctrine in this chapter and the fundamental theorem on distribution in the preceding chapter. We must now set the perspectives face to face.

It has been said of Mill's position in 1869 that it is directly based on the fundamental (Ricardian) theorem of distribution.[62] There is much to this, though we must be cautious in the precise formulation. The Ricardian theorem maintains that a rise in the rate of wages entails a fall in the return on capital; but it leaves quite open whether or not a rise in the wage rate is possible in the first place. Once Mill had taken for granted the possibility (during the short period) of a real wage-rate increase – even one that involves a rise in aggregate real wages – then the fundamental theorem indeed comes fully into play. It must not be forgotten that the fundamental theorem retained its central position throughout the 1860s and early 1870s; indeed it was used in support of the position that real-wage increases are attainable at the expense of profits even before the appearance of Thornton's book or his article of 1867.[63]

62 Miller (1940), 111–12: 'If one looks to what Mill believed about wages, as expressed in this review, it will be seen that he still clings to the fundamental proposition of the theory, i.e., that labor is paid from a fund in the hands of the capitalist; only now, wages are not rigidly predetermined, they are a residuum along with profits for each individual entrepreneur; wages can be raised at the expense of profits. This is a reversal to the strict Ricardian position on the relation between wages and profits.'

Schwartz (1972), 101, writes of Mill's 'adoption of the Ricardian doctrine that "profits depend upon wages" in place of the wages fund doctrine', but implies that the Ricardian position was adopted for the first time in 1869. This is quite inaccurate – Mill had subscribed to it from the outset of his career.

63 The particular formulation under attack by Thornton was that by Fawcett, and Mill himself in 1865 had already alluded to his objections (on Ricardian grounds) to that formulation: 'The chapter which on the whole I least like is the one on wages, though it will probably be more praised than any of the rest: but I think I could shew that an increase of wages at the expense of profits would not be an impracticility on the true principles of political economy. It might doubtless send capital to other countries but we must recollect that the movement for higher wages and shorter working hours is now common to all the industrious nations' (to Fawcett, 1 Jan. 1866; *CW*, XVI, 1130–31).

Thornton's book merely provided an opportunity to express publicly the reservations against formulations of the wages-fund doctrine such as Fawcett's which Mill had already had in mind for several years. Cf. Mill's remark to Cairnes when eliciting criticisms of the 1869 review: 'I am on the point of sending to the Fortnightly the first part of a review of Thornton's book; the purely economical part. I shall be very desirous of knowing whether you agree with my judgment of the book from the purely scientific point of view. I feel pretty sure you will concur in what I have written on the so-called wages-fund, a subject on which I expressed myself in my Political Economy as inaccurately as other people, and which I have only within the last two or three years seen in its proper light' (9 April, 1869, *ibid.*, XVII, 1587).

The first concerted attack on the wages-fund doctrine by Francis D. Longe – *A Refutation of the Wages Fund Theory of Modern Political Economy as Enunciated by Mr. Mill and Mr. Fawcett* (1866) – elicited no response at all from Mill in print or correspondence, although a copy was sent to him by the author. (On Longe, see Breit, 1967, esp. 513–17, and Taussig, 1896, 241f.) But Mill's reservations evidently preceded 1866 (cf. Schwartz, 1972, 92, on this matter).

We can go further. It is only if we have in mind the 'primitive' version of the wages-fund doctrine – the wages bill as a predetermined sum at the beginning of the production period implying an aggregate labour demand of unitary elasticity – that a major inconsistency of approach arises, since (once an increase in the wage rate is allowed) the fundamental theorem implies an increase in the wage bill at the expense of profits with no fall-off in employment in the short run, that is before economic activity can be reordered in favour of relative 'capital-intensive' industries. On the other hand, the more sophisticated version of the wages-fund doctrine, which envisages the wages bill as the outcome of a market-clearing process, presumes variation in fixed capital to be technically possible – it is of long-run relevance alone; in the short period the aggregate demand for labour is inelastic (for which reason there is no assurance that the 'residual' portion of capital will be devoted to wage payments). From this perspective Mill's rejection in 1869 of the primitive version of the wages-fund theory on grounds of the ability of workers to force a real increase of wages at the expense of profits, with no loss of employment, implies a reinforcement of the sophisticated version.[64] Instead of constituting a challenge to the very foundation of orthodox doctrine it effectively amounted to its defence, although this may not have been the formal intention. The episode of 1869 loses much of its apparent theoretical significance as indicative of a collapse of confidence in the central features of classical doctrine.

What remains troublesome is the loud breast-beating; Mill insisted on himself taking responsibility for inaccurate statements in his *Principles* (on a par with Fawcett's). That there are statements that can be so read is undoubtedly true – but the formulations in the *Principles* as a whole do not point in that direction. To appreciate this aspect of the record allowance may have to be made for a degree of genuine confusion on Mill's part, leading him to identify the decision to 'abstain' from present consumption by the capitalist-employer – also bearing in mind the fact that his income is received as the difference between sales revenue and outlay – with

---

64 Ekelund (1976), 73, argues that Mill was in error by 'implying an elasticity in the real allocation over a given period of production', because 'the real wage cannot be altered *vis-à-vis any* exogenous redistribution of money funds between capitalists and labourers. Union bargains, in other words, or any other factors which affected the redistribution of money funds could have no appreciable effect upon real wages over the short run.' This kind of criticism breaks down completely before the evidence of Mill's allowance for the consumption by labourers of 'luxuries'.

It is relevant that Torrens, who explicitly laid out the sophisticated version of the theory (above, pp. 401–2) gave no indication of a rigid barrier to real-wage increase by way of union pressure; indeed, in *Wages and Combination*, he emphasized the possibilities of securing an immediate rise of wages at the expense of profits by universal combination (cf. Taussig, 1896, 211–12, Robbins, 1958, 49–50).

a conception of the economy entailing the literal preaccumulation of wage goods to tide over the period of production.[65]

* * *

We return to the consistency of the main themes in Mill's aggregate distribution theory. According to the Torrens–Cairnes–Mill version of the wages-fund theory, an increase in the wage rate will, in the long run at least, bring about a reduction in the demand for labour. This is true despite the assumption of uniform factor proportions between industries implied by the presumed constancy of the overall labour-machinery ratio in the model. Yet in the same case the prediction generated by the inverse profit-wage relationship interpreted in allocative terms is very different. For assuming uniform factor proportions the reduction in profits at going outputs and prices following the wage increase generates no allocative consequences and, accordingly, the overall quantity of labour demanded remains unchanged.

The source of this distinct difference in outcome is not difficult to discern. The contrast arises from the fact that in the allocative model the wage increase is 'financed' by a reduction in profits rather than by a transfer of funds hitherto invested in fixed capital. There occurs, as it were, a *forced* rise in aggregate capital (taking the form of the increased wage payments) which is not allowed for in the wages-fund model.[66]

In terms of the economic quality of the respective models the allocative analysis seems to have the advantage. For it is directly based upon the behaviour patterns of individual employers in the face of a wage change. The wages-fund model has no behavioural implications whatsoever. Above all, there is no explanation of the decision on the part of individual employers to reduce their M quotas; indeed if factor ratios are everywhere identical, as is the supposition, then no motive exists to vary output or employment in any single industry and therefore in the economy as a whole.[67] The conflicting outcomes of the two approaches to distribution can, however, be in part reconciled. For the fall in profits upon an increase in the wage

---

65 There is also Mill's 'personal regard' for his friend Thornton, for whom 'he was disposed to make every possible concession' (Taussig, 1896, 248).

66 In the case of an unchanged labour force, an increase in net investment will be devoted entirely to wage payments according to the wages-fund theory, a result assured by competitive pressure in the labour market. But this involves a shift in the negatively-sloped labour demand curve, and cannot be identified with the vertical 'curve' implied by the allocation analysis.

67 A second and related, defect of the wages-fund model is the fact that, since the wage increase is accommodated by reductions in M, the impression is given that the profit rate on total capital remains unchanged; but this matter is left unstated by the expositors of the model.

rate allowed by the fundamental theorem is not the end of the story. Since the return on capital has fallen it is to be expected that the rate of accumulation will be adversely affected, in which event the 'forced' increase in capital will ultimately be, at least partially, corrected.

We turn next to the case of differential factor ratios between industries. We must first set aside a perennial charge of indeterminacy directed against the classical literature, going back at least as far as Wicksell. The Ricardo–Mill tradition does not preclude allowance for differential factor ratios; indeed, precisely in consequence of such differentials a change in the wage rate will affect the configuration of relative prices by altering the cost structure and the pattern of output. But such allowances, it is said, fly in the face of that version of the wages-fund theory treating the wage bill as a predetermined quantity which, together with the 'working population', determines the wage rate. For if the wage bill is a *datum*, knowledge of which is required to determine the wage rate, then the wage rate cannot be permitted in its turn to affect the wage bill.[68] This charge can be dismissed not only because of the limited relevance of the 'primitive' version of the wages-fund doctrine, but also because – even taking that version for granted – no allowance is made by critics for the fact that the allocative consequences of a change in the wage rate are *long-run* while the supposed constancy of the aggregate wage bill applies only for a single 'production period'. In what follows we shall have in mind only the sophisticated version of the wages-fund doctrine.

Formally, of course, this version also rules out the case of differential factor ratios.[69] But it is possible to accommodate the more complex case by tying the average machinery–labour ratio to the wage rate assuming knowledge of the pattern of demand curves in the system and the factor ratios in each industry. Making allowance for a variation in the (average) machinery–labour ratio λ the demand for labour will fall off in consequence of a wage increase more sharply than if λ is held constant.

The analysis in terms of allocation theory also generates a negatively sloped aggregate demand curve for labour by way of a change in activity to the detriment of labour-intensive industries. But the wages-fund theory assumes

68 See for example, Sidgwick (1883), 315–16, Stephen (1900), III, 215f, Stigler (1941), 283–5 and the comments on Knut Wicksell's critique.

69 D. P. O'Brien (1975), 116, has criticized my interpretation of the Cairnes–Torrens–Mill model on grounds, *inter alia*, that these writers recognized differential factor ratios from industry to industry. This fact does not, however, demonstrate the illegitimacy of attributing to them that version of the wages-fund theory which implicitly assumes uniform factor ratios. It is quite reasonable for certain purposes to set such complications aside–and this the economists seem to have done. But since it is a fact that differential ratios were recognized in other contexts it is a pertinent question whether the two sets of assumptions can be reconciled. More generally, O'Brien's criticisms cannot cause the model to disappear. The three economists clearly formulated it whether consistently with other of their statements or not.

given capital, while according to the theory of allocation, as already explained, no such constraint is imposed, since the wage increase will be financed from reduced profits. The elasticities of demand for labour derived by way of the two mechanisms will not therefore generally be the same.

Mill's comprehension of this kind of complication may be appreciated by reverting back to an earlier discussion of the fourth fundamental proposition on capital (above, pp. 359–60). There we argued that while the proposition, taken literally, ruled out any effect on distribution of a change in the configuration of final demand – a result depending upon an implicit assumption of uniform factor ratios – the evidence suggested that Mill took this position as a matter of convenience rather than principle. If this is indeed so, it is possible for us to extend his wages-fund model by allowing for differential factor ratios and yet remain faithful to Mill's view of economic process.

IX 'STATICS' AND 'DYNAMICS'

The trend paths of the distributive variables are treated in Book IV: 'The Influence of the Progress of Society on Production and Distribution', a Book prefaced with the famous distinction between 'Statics' and 'Dynamics':

The three preceding Parts include as detailed a view as our limits permit, of what, by a happy generalization of a mathematical phrase, has been called the Statics of the subject. We have surveyed the field of economical facts, and have examined how they stand related to one another as causes and effects; what circumstances determine the amount of production, of employment for labour, of capital and population; what laws regulate rent, profits, and wages; under what conditions and in what proportions commodities are interchanged between individuals and between countries. We have thus obtained a collective view of the economical phenomena of society, considered as existing simultaneously. We have ascertained, to a certain extent, the principles of their interdependence; and when the state of some of the elements is known, we should be able to infer, in a general way, the contemporaneous state of most of the others. All this, however, has only put us in possession of the economical laws of a stationary and unchanging society. We have still to consider the economical condition of mankind as liable to change, and indeed (in the more advanced portions of the race, and in all regions to which their influence reaches) as at all times undergoing progressive changes. We have to consider what these changes are, what are their laws, and what their ultimate tendencies; thereby adding a theory of motion to our theory of equilibrium – the Dynamics of political economy to the Statics (*CW*, III, 705).

Aspects of a theory of growth do appear in earlier parts. It is to Book I that one turns, for example, for Mill's detailed observations on the labour and capital supply functions and the determinants of productivity in agriculture and manufacturing. But all this amounted to a setting of the stage. The formal growth model with its analysis of the trend path of wages and profits only comes in Book IV.

This conceptual division is of the first importance. It implies that the discussion of 'long-run' or 'natural' price formation turning on costs of production – which, as we know, simply amounts to satisfaction of the wage- and profit-rate uniformity conditions – applied to artificially imposed states of stationarity respecting capital, population and knowledge. (Indeed, J. B. Clark was later to point out that the classical notion of 'natural' value, wages and interest 'was unconsciously employed as an equivalent of the term static' (1899, vi; cf. 69–70).)

Clark himself was preoccupied with an assumed 'paralysis of the forces of progress' – 'what would be the rate of wages, if labour and capital were to remain fixed in quantity, if improvements in the mode of production were to stop, if the consolidating of capital were to cease, and if the wants of consumers were never to alter' (vi–vii). Lord Robbins, in his historical account (1930), is undecided whether Mill in his description of a static state intended this Clarkian concept or the traditional stationary state – the one involving an artificially imposed paralysis of the forces of progress, and the other the end point of the process of accumulation and population growth (a distinction also clarified in Knight, 1921, 142–3n):

Now it is probable that, in writing this passage, Mill had nothing more in mind than the existing notion of the stationary state, and the difference between the phenomena of such a state and the phenomena of society which is still advancing. That at any rate is what seems to emerge from a study of the actual content of the subsequent chapters of the section [Book IV].

But it is clear that, taken apart from its general context, it is capable of another interpretation. Taken as it stands, it seems to say, 'We have studied what happens when the factors of production are constant. Now we must proceed to ask what causes their numbers to change.' That at least is suggested by the passage, 'we have still to consider the condition of mankind as liable to change . . . we have to consider what these changes are and what are their laws' . . . [It] is easy to see how reading this passage, and being impressed by the verbal distinction, one might come to divide the statics and dynamics of the subject on this plan. The statics should deal with what happens when the factors are given. The dynamics, with the laws of change in the quantity of the factors' (1930, 202–3).[70]

The weight of circumstantial evidence, however, does seem to point to

---

70 But throughout his paper, Lord Robbins presumes that stationarity implied for the classics – unlike Clark – 'the resultants of the equilibrating process' (204; cf. 206–7).

a Clarkian intention. For it was standard already in the classical literature at least implicitly. Thus one recalls Adam Smith's notion of natural price as a sum of wage, interest and rental costs per unit, each of the factor returns at its 'natural' level, in the very limited sense of a presupposed average level – a known general return to each factor – which presumes constant capital and population but not necessarily at 'stationary state' levels. On this basis he traced out the effects of a change in the pattern of demand with the average factor returns assumed unchanged. And so did Ricardo with the absence of an allowance for rent. (On Smith and Ricardo, cf. Hollander 1973a, 122; 1979, 287–8.)

What is true of an artificial 'paralysis' of the forces of progress in regard to wage- and profit-rate uniformity holds *a fortiori* for the stationary state proper – the very long run. But severe interpretive problems arise when we attempt to define the precise status of the points of any path of wages (or profits) during the course of progress towards stationarity. What must be clarified is whether the points on a wage path constitute a *locus* of static states within each of which wage- and profit-rate uniformity ('long-run' equilibrium) is achieved; or whether, in this context of 'progress', we abandon the artificial 'static' construct and consider it pertinent only to the stationary state proper. In the first case, we can say nothing of the transfer between the points (Hicks, 1983a, 39). In the second, the system is continually 'out of equilibrium' on the path to stationarity, the non-uniformity of factor returns itself constituting an essential part of the driving mechanism of the growth process.

This matter, as far as concerns Mill, must be approached with caution. The location of Mill's formal distinction between statics and dynamics might be said to imply that the progressive state was indeed envisaged by him as a locus of static states; why else lay out the distinction precisely at the point of transfer to the problem of progress? But this is not necessarily a justified presumption. There are broad ranges of problems that may best be treated by assuming constancy of aggregate capital, population size and knowledge. It is by no means an unreasonable procedure to provide a framework for study of such problems, before proceeding to the distributive and pricing implications of variable capital, population and knowledge – and without any presumption that the secular path of wages (and profits) is to be treated as a locus of static states.

Clark, it is true, supposes this was not the 'Ricardian' procedure:

Had the Ricardians recognized the fact that they were trying to study a static world and then studied it consistently, they would have made even their own system more realistic. Boldly suppressing one set of actual forces, in order to study more easily another set, must result in reaching conclusions that are partial but are not necessarily unreal. If these early students had later done what they never tried to do, and had completed their system by separately examining the dynamic forces, they would have attained a complete and realistic science (1899, 70).

But it is not clear why Clark makes no reference to Mill's formulation, since it is certainly consistent with precisely this intention.[71]

More positively, Mill in his passage distinctly required 'a theory of motion' to supplement his 'theory of equilibrium'. This suggests that a stringing together of a series of equilibria would not have sufficed for his purposes. Having in mind the evidence in Chapter 5 of Mill's 'Marshallian' perspective on the role of demand – and his explicit references to the 'growing demand [for food] of so rapidly increasing a population as that of Great Britain' (*CW*, III, 745), it is likely that for him, as for Ricardo, the wage path (the falling wage path) reflects extensions of cultivation under pressure of increased demand for food; while the lesser pressure of demand for workers' 'conveniences' and 'luxuries' rules out profit-rate uniformity until the stationary state is achieved.[72] In brief, market and natural prices coincide in stationary state equilibrium alone once the (artificial) assumption of constant capital and population is relaxed. This conclusion – spelled out by Clark (29) – was never drawn explicitly by either Ricardo or Mill.

The argument may be extended. Under conditions of steady-state wages most features characterizing stationarity still apply. This is a point Marshall made in his *Principles*:

The Stationary state has just been taken to be one in which population is stationary. But nearly all its distinctive features may be exhibited in a place where population

71 In correspondence with Clark, Marshall objected to this procedure (15 Dec. 1902): 'I cannot conceive of any Static state, which resembles the real world closely enough to form a subject of profitable study, and in which the notion of change is set aside even for an instant' (1925, 415). But in this letter Marshall defined a Static state as 'a position of rest due to the equivalence of opposing forces which tend to produce motion . . . In my view there may be no change in fact; but only because the forces tending to make change are (or for the purposes of a particular argument or illustration are supposed to be) equal and opposite.' This is not the static concept that either Clark or Mill had in mind. Furthermore, Marshall himself in his own *Principles* (1920) does exactly what Mill does: devotes a chapter to 'The Influence of progress on Value' (Book VI: Ch. xii) and argues that a vast variety of problems can be dealt with on the assumption of given data of various kinds – 'impounded in *ceteris paribus* – including capital and population. Indeed, 'in the relatively short-period problem no great violence is needed for the assumption that the forces not specially under consideration may be taken for the time to be inactive', 369, 379n). See also Robbins (1930), 201: 'Marshall is quite clear that the stationary state is a fiction, an analytical instrument simply. But the basic conception is there, and that, of course, in a form whose effectiveness is vastly enhanced by all Marshall's own analytical improvements.'

72 For Mill as for Ricardo the demand for 'food' was of zero price- and income-elasticity unlike the demand for other items.

and wealth are both growing provided they are growing at about the same rate, and there is no scarcity of land: and provided also the methods of production and the conditions of trade change but little; and above all, where the character of man himself is a constant quantity. For in such a state by far the most important conditions of production and consumption, of exchange and distribution will remain of the same quality, and in the same general relations to one another, though they are all increasing in volume (1920, 368; cf. 1925, 315).

We shall show in this chapter that, following Malthus, Mill went much further than Marshall. For he provided a model of steady-state wages even in the presence of land scarcity (and assuming given technology) – the growth rate of population declining in line with that of labour demand. Here uniformity of the return on capital can be maintained continuously as all sectors expand proportionately. In brief, market prices equal natural prices, outputs expanding simply because of cross-the-board increase of capital independently of any allocative motivation. Thus he transposed to the 'dynamic context' a characteristic formally applicable to stationarity – whether 'artificially' imposed or an end point – alone. The range of conclusions derived on the basis of 'static' analysis is in this way much extended.

Sir John Hicks has argued recently that although it is only in Book IV that Mill's notions on the stationary state come to the fore explicitly, they yet influenced the whole structure of the work in that the investigation of 'equilibrium positions', for which static theory is appropriate, takes precedence over the 'path to equilibrium', a weighting reflected in his concern for a desirable pattern of distribution as an end in itself (1983b, 69). Mill's concern with equilibrium positions and desirable distribution is for us not in question. But it is going too far to conclude with Hicks that Mill 'disposed of the old growth economics' (70). The stationary state was, of course, everything equal, that much closer with population control than without – a fact that comes out very clearly in the Malthus–Mill prudential wage path (below, p. 445) – and there is no gainsaying the new look at distribution. Yet Mill was thoroughly aware that both technological progress and increases in the propensity to save – the 'effective desire for accumulation' – were in practice proceeding apace, so that stationarity was not in sight; indeed, in the Britain of his day steady-state wages (and a constant return on capital) were possible even without population control because of the extremely high rate of capital accumulation reflecting new technology. We must be very careful how we understand the celebrated observation that the return on capital in contemporary Britain was 'within a hand's breadth of the minimum' (*CW*, III, 738). In any event Mill allows for technical change in his stationary state; even there 'growth' of sorts is by no means ruled out.

X CAPITAL-SUPPLY CONDITIONS AND
THE MINIMUM RATE OF RETURN

We will be concerned in the following sections with the isolation of the building blocks that are required in the reconstruction of Mill's models of growth and their application to the circumstances ruling in contemporary Britain. Our starting point in this exercise is capital supply.

The chapter treating the 'Law of the Increase of Capital' conceives the source of savings to be the 'surplus' which defines the maximum 'ability to save', while the 'disposition' or motive to save governs the actual rate of savings: 'Since all capital is the product of saving, that is, of abstinence from present consumption for the sake of a future good, the increase of capital must depend upon two things – the amount of the fund from which saving can be made, and the strength of the dispositions which prompt it' (*CW*, II, 160):

The fund from which saving can be made, is the surplus of the produce of labour, after supplying the necessaries of life to all concerned in the production: (including those employed in replacing the materials, and keeping the fixed capital in repair). More than this surplus cannot be saved under any circumstance. As much as this, though it never is saved, always might be. This surplus is the fund from which the enjoyments, as distinguished from the necessaries, of the producers are provided; it is the fund from which all are subsisted, who are not themselves engaged in production; and from which all additions are made to capital. It is the real net produce of the country. The phrase, net produce, is often taken in a more limited sense, to denote only the profits of the capitalist and the rent of the landlord . . . But this is too narrow an acceptation of the term. The capital of the employer forms the revenue of the labourers, and if this exceeds the necessaries of life, it gives them a surplus which they may either expend in enjoyments, or save. For every purpose for which there can be occasion to speak of the net produce of industry, this surplus ought to be included in it (160–1).[73]

The surplus is carefully defined to include wage income exceeding subsistence. This is an important theme, frequently alluded to. For example, in the discussion of circulating capital: 'such portion of it as the workmen consume no longer exists as capital at all: even if they save any part, it may now be more properly regarded as a fresh capital, the result of a second act of accumulation' (91); similarly, part of the capitalist's funds are paid

---

73 Mill defines the net surplus more specifically as the excess over 'the physical necessaries of the producers' (161). But this is somewhat misleading for the 'subsistence' wage for Mill is without question culturally determined. (See also the discussion above of national accounting, pp. 659–60).

to productive labourers 'who consume it for their daily wants; or if they in their turn save any part, this also is not, generally speaking hoarded, but (through savings banks, benefit clubs, or some other channel) re-employed as capital, and consumed ['productively']' (70).[74]

The 'abstinence' involved in saving need not, however, literally entail 'privation'. Ricardo had made the point, neglected by James Mill, that additions to capital might be financed out of a surplus increased by technical change (Hollander, 1979, 321–2). As J. S. Mill put it:

If it were said . . . that the only way to accelerate the increase of capital is by increase of saving, the idea would probably be suggested of greater abstinence, and increased privation. But it is obvious that whatever increases the productive power of labour, creates an additional fund to make savings from, and enables capital to be enlarged not only without additional privation, but concurrently with an increase of personal consumption. Nevertheless, there is here an increase of saving, in the scientific sense. Though there is more consumed, there is also more spared. There is a greater excess of production over consumption . . . To consume less than is produced is saving; and that is the process by which capital is increased; not necessarily by consuming less absolutely (67–70).[75]

Actual savings out of the maximum sum potentially available depends in part on prospective yield – 'the greater the profit that can be made from capital, the stronger is the motive to its accumulation',[76] but also on a variety of personal, sociological and institutional considerations, including the state of national security and life expectation, intellectual development and consciousness of the future, and the strength of other, as distinct from self-regarding, interests (161–2). In all this Mill relied heavily upon John Rae's 'effective desire of accumulation'.[77]

74 Elsewhere Mill specified the inclusion of the surplus wages of 'unproductive' as well as 'productive' labour (cf. IV, 288).

75 Cf. further on the implications for saving of 'improvements': 'They increase the return to capital; and of this increase the benefit must necessarily accrue either to the capitalist in greater profits, or to the customer in diminished prices; affording, in either case, an augmented fund from which accumulations may be made, while enlarged profits also hold out an increased inducement to accumulation' (98). Conceivably savings might rise even if 'profits' are unchanged.

76 Mill should have written 'interest' since profit includes a return to the risk involved in investment of saved funds and to superintendence.

77 On Rae's contribution and his influence see Spengler (1959), James in Rae (1965), I. Cf. the account in Robbins (1930): 'With Smith and Ricardo . . . if we exclude variations due to variations of risk, the supply price of capital is treated as more or less invariable.' But with Mill 'it is quite clear that he had learned from John Rae the notion of a natural rate of profit which varies according to what Fisher would call the prevalent conditions of time preference, and what Rae called the effective desire of accumulation' (200).

It is not clarified beyond a shadow of doubt whether (assuming all other conditions unchanged) an increase in the surplus during the course of development, the yield falling, will allow not only increased absolute savings but possibly generate an increased rate of savings. This had been Adam Smith's view: 'a great stock, though with small profits, generally increases faster than a small stock with great profits. Money, says the proverb, makes money' (1937, 93). But whatever the theoretical possibilities, Mill did insist – possibly assuming that a falling return coincided with a falling surplus available for accumulation – upon a regular positive relation: '[it is] an almost infallible consequence of any reduction of profits' – a reference to the rate of return – 'to retard the rate of accumulation' (*CW*, III, 843); equally strongly: 'it seems impossible that, if capital is accumulating at all, its accumulation should not be in some degree retarded by the abstraction [by taxation] of a portion of its profit' (828); 'when profits fall' (again an allusion to the profit *rate*) 'increase of capital slackens' (II, 344); 'the progress of accumulation would no doubt be considerably checked if the returns on capital were to be reduced still lower than at present' (172).[78]

---

78 Consistently with this view 'labour' and 'abstinence' are placed on a par: 'Since the human race has no means of enjoyable existence, or of existence at all, but what it derives from its own labour and abstinence, there would be no ground for complaint against society if every one who was willing to undergo a fair share of this labour and abstinence could attain a fair share of the fruits' (III, 714).

Critical reference is made to Continental socialists who, unlike their British counterparts, 'deny the legitimacy of deriving an income in any form from property apart from labour' (708–9). See also the praise in the *Principles* for socialists who did not call for 'robbing the capitalists of what they or their predecessors had acquired by labour and preserved by economy . . . ' (775).

The evidence given before the Select Committee on the Savings of the Middle and Working Classes (June 1850) is also instructive. Here Mill made a case for the creation by labourers of 'associations' as an experiment. He observed that while an 'astonishingly' small proportion of the price of a commodity represented a return to labour, the divergence was not basically created by interest on capital. For the rate of return was low and even labourers organized in associations would require that rate as an inducement to postpone present consumption: 'Now one thing is very important to remember in itself, and it is important that the working classes should be aware of it; and that is, that this does not arise from the extravagant remuneration of capital. Capital, when the security is good, can be borrowed in any quantity at little more than three per cent, and I imagine there is no co-operative association of working-people who would find it in their interest to allow less than that remuneration, as an inducement to any of their members who, instead of consuming their share of the proceeds, might choose to save it, and add it to the capital of the association' (V, 411). More generally, Mill implied the justness of interest payments made by the association to capitalist lenders: 'it is very natural that the working classes should wish to try whether they could not contrive to get this portion of the produce of their labour for themselves, so that the whole of the proceeds of an enterprize in which they are engaged might be theirs, after deducting the real remuneration of the capital they may require from others, which, we know does not

A complexity arises, however, in some of Mill's later statements. For in a letter of 1867 he writes as if the savings rate will be adversely affected only by a major reduction in the return on capital:

On the mere economic question, 'is there any way in which an annual tax may be collected from capital without leaving to the latter an opportunity to collect it back from labor?' I should answer that capital will not have this opportunity. I do not believe that the burthens laid on the capitalist ever fall on labor, except in one way, viz: if the burthens are so heavy as to check the accumulation of capital, and prevent it from keeping up with the increase of population; in which case, without doubt, wages would fall, unless the increased numbers migrated to a less crowded field for the employment of labor, such as your Western Territories (28 Dec. 1867, XVI, 1338–9).

Similarly, Mill wrote regarding the impact of union activity in terms suggesting a presumption against any reduction in savings or the rate of accumulation upon wage increases, except in extreme cases:

Of course neither Trade Unions, nor anything else can permanently raise wages so high as not to leave a rate of profit sufficient to encourage accumulation. But if they limit their attempt within reasonable bounds, I do not see why they should not in many cases succeed, both in raising wages, and in (what is equivalent) diminishing the hours of labour (22 Dec. 1867, ibid., 1335).

Most interesting in this regard is the 1869 review of Thornton. We have shown earlier that the limits to wage increase were defined with short-run labour demand in mind. But Mill also briefly alluded to the long run:

What is true is, that wages might be so high as to leave no profit to the capitalist, or not enough to compensate him for the anxieties and risks of trade; and in that case labourers would be killing the goose to get at the egg. And, again, wages might be so low as to diminish the numbers or impair the working powers of the labourers, and in that case the capitalist also would generally be a loser . . .
Between the limits just indicated – the highest wages consistent with keeping up the capital of the country, and increasing it *pari passu* with the increase of people, and the lowest that will enable the labourers to keep up their numbers with an increase sufficient to provide labourers for the increase of employment – there is

in general, when the security is good, much exceed three per cent' (411–12). (The 'portion' of the produce which labour is justified to obtain is that which formally is designated profit or the wages of superintendence which, Mill agrees, is frequently determined monopolistically.)

an intermediate region within which wages will range higher or lower according to what Adam Smith calls 'the higgling of the market' (V, 657).

There is a strong impression here of an exogenous rate of accumulation and labour demand though within limits (and similarly an exogenous rate of population growth again within limits), which plays havoc with the standard growth model set forth in the *Principles* involving regular functional relationships both of savings and population growth.

If indeed there did occur a late change in opinion regarding the nature of the savings–interest nexus, the motivation behind the revision requires attention.[79] But the 'new' position, if such it is, is not carried through consistently, since the standard version appears in all the later editions of the *Principles*.[80]

It is pertinent also that the regular functional relation is generalized to a Socialist system. This emerges in Mill's observation of 1869 that the required interest rate – the context involves Fourier's proposals – would be dictated by the chosen rate of growth of the national economy: 'the remuneration for capital is to be such as is found sufficient to induce savings from individual consumption, in order to increase the common stock to such a point as is desired' ('Chapters on Socialism', c. 1869, *CW*, V, 747).

A further complexity relates to the empirical observation in 1850 that 'capital, when the security is good, can be borrowed in any quantity at little more than three per cent' (411). In 1869 Mill reiterated the observation adding that it had been at that level 'for many years' (734). In the *Principles* too he refers to the 'astonishingly' high rate of accumulation under way:

---

79 A contemporary critic of the review, James Sterling, complained in 1870 that Mill's position 'rests on the monstrous assumption, that it is possible to take from profits and add to wages, without weakening the effective desire of accumulation' (cited Schwartz, 1972, 98).

80 This should be qualified. In the preface to the seventh edition Mill states that while there had recently taken place 'some instructive discussion on the theory of Demand and Supply, and on the influence of Strikes and Trade Unions on wages, by which additional light has been thrown on these subjects', yet the results were 'not yet ripe for incorporation in a general treatise on Political Economy' (*CW*, II, Ch. xciv). None the less, the review *did* have an impact. In a discussion of the laws against combination of workmen Mill had written in all of the first six editions that if (in the unlikely event of a union encompassing the entire work force) labourers 'aimed at obtaining actually higher wages than the rate fixed by demand and supply – the rate which distributes the whole circulating capital of the country among the entire working population – this could only be accomplished by keeping a part of their number permanently out of employment' (III, 930). But in the seventh edition this statement is altered, Mill now conceding that workers would 'have a limited power of obtaining, by combination, an increase of general wages at the expense of profits'. Yet the 'limited power' is emphasized: 'the limits of this power are narrow; and were they to attempt to strain it beyond those limits, this could only be accomplished by keeping a part of their number permanently out of employment'.

In the more prosperous countries of Europe, there are to be found abundance of prodigals; in some of them (and in none more than England) the ordinary degree of economy and providence among those who live by manual labour cannot be considered high: still, in a very numerous portion of the community, the professional, manufacturing and trading classes, being those who, generally speaking, invite more of the means with more of the motives for saving than any other class, the spirit of accumulation is so strong, that the signs of rapidly increasing wealth meet every eye: and the great amount of capital seeking investment excites astonishment, whenever peculiar circumstances turning much of it into one channel, such as railway construction or foreign speculative adventure, bring the largeness of the total amount into evidence (II, 170).

In this same context we also find the observation that in England as in Holland 'for a long time past', and elsewhere in Europe in more recent times, 'the desire of accumulation does not require to make it effective, the copious returns which it requires in Asia, but is sufficiently called into action by a rate of profit so low, that instead of slackening, accumulation seems now to proceed more rapidly than ever' (172). Can all this be said to reverse the positive relation between rate of accumulation and the return on capital, and to define the Smith-Jones-Marx position?[81]

Considering all that has been said above pointing to an upward sloping schedule relating the growth rate of savings to the interest rate these observations cannot possibly imply an infinitely elastic curve, not to speak of the Smithian relationship. Mill might in these contexts be exaggerating somewhat and attempting to express a very high elasticity (though not infinite); but it is also highly probable that he was allowing for shifts in the function with secular increases in the 'effective desire to accumulate'. This tendency is much emphasized in the discussion of 'social progress' in Book IV of the *Principles*:

[I]t is . . . one of the consequences of civilization that mankind become less the slaves of the moment, and more habituated to carry their desires and purposes forward into a distant future. This increase of providence is a natural result of the increased assurance with which futurity can be looked forward to; and is, besides, favoured by most of the influences which an industrial life exercises over the passions and inclinations of human nature. In proportion as life has fewer vicissitudes as

---

81 Mill (see pp. 434–5) maintained that interest should be taxed at a higher rate than profits on the grounds that the former represents a 'permanent' income. And in general, the placing of rent and interest within the same category for relatively severe treatment compared with wages and profits neglects any negative effects for the supply of savings. But not too much weight should be placed on this fact considering that his specific concern was with the 'normative' problems of fair taxation rather than the 'positive' consequences of his recommendations.

habits become more fixed, and great prizes are less and less to be hoped for by any other means than long perseverance, mankind become more willing to sacrifice present indulgence for future objects . . . The present kind of social progress . . . decidedly tends, though not perhaps to increase the desire of accumulation, yet to weaken the obstacles to it, and to diminish the amount of profit which people absolutely require as an inducement to save and accumulate (III, 738).

The extracts cited earlier are thus to be read as allowing for changes in *ceteris paribus* conditions such that at a return of approximately 3 % the rate of increase of capital was rising over time.

* * *

Let us now take a close look at the return to capital in stationarity: 'When a country has carried production as far as in the existing state of knowledge it can be carried with an amount of return corresponding to the average strength of the effective desire of accumulation in that country, it has reached what is called the stationary state; the state in which no further addition will be made to capital, unless there takes place either some improvement in the arts of production, or an increase in the strength of the desire to accumulate' (II, 169). As a factual matter, Mill (V, 734) took for granted that the minimum at which savings would cease was approximately 1 per cent. But the question arises whether a positive interest rate was seen as a *necessary* feature of the full stationary state.

This issue should be seen in appropriate perspective. The 'problem of interest', made famous by J. A. Schumpeter, involves the presumption that in the very long run – the stationary state – there will be no net return to produced capital goods, their value falling to the land and labour required to produce them – the ultimate factors of production. Interest, for Schumpeter, is associated with change (1934, 29–30). To this kind of argument it has been objected that interest will arise even in full stationarity 'if abstinence is to be regarded, not only as a refusal to consume in the present in order that the sources of future income may be *enlarged*, but also as a refusal to enhance one's consumption now in order that the income of the future may not be depleted' (Robbins, 1930, 213). For 'if produced means of production are not productive of a net product, why devote resources to maintaining them when these resources might be devoted to providing present enjoyment? . . . It is, in short *an* interest rate, which, other things being given, keeps the stationary state stationary – the rate at which it does not pay to turn income into capital or capital into income.

If interest were to disappear the stationary state would cease to be stationary.'[82]

Schumpeter, however, was not unique in taking his position. Clark, for one, had much earlier insisted that the maintenance of given capital required no special incentive:

[The] making of a new instrument, to take the place of an old one, imposes on the owner no such sacrifice as that involved in making the original one; for the reason that the instrument virtually, though not literally, makes its own successor. The loom in the factory that is worn out and is about to be replaced has, during its career, earned its share of dividends for the stockholders of the mill and, besides this, has earned for them a sum that will buy a new loom. It is not necessary, therefore, to take the cost of the new loom out of the stockholders' incomes. That would impose on them the necessity for a genuine act of abstinence, and that only would do so. If the loom had not done what well-selected machines always do, – if it had not created a fund to replace itself, – then it might have been necessary to assess the stockholders for the cost of new machinery. This would have made them abstainers; for it would have caused them to trench upon their incomes and forego some consumers' goods (1899, 133).[83]

Where did J. S. Mill stand on this complexing matter? Various motives to saving, unrelated to any inducement offered by the rate of interest, emerge in the taxation context. In his evidence before the Select Committee on Income and Property Taxation of May 1852 and June 1861 (*CW*, V, 463f., 549f.) and in the *Principles*, Mill recommended that, in principle, taxes should be imposed only on that part of income which is spent rather than saved. His rule of thumb proposed a higher rate of taxation in the case of 'permanent' incomes including rent and interest (that is, income from property) than in the case of 'temporary' or 'life' incomes including wages, salaries and profits, on the grounds that the 'necessity to save' (for retirement and support of children) is lower for the first group.[84] Even within a single class, differences are noted which it would be desirable to recognize in taxation. For example: 'The great mass of professional men are aiming

---

82 For further discussion of this issue see Harrod (1948), 35f, Hayek (1936), (1941a), 229f, Keynes (1973), VII, 220–1, Pigou (1943, 1947).

83 Joan Robinson (1942), 54–5, complained that Marshall's treatment was obscured by a presumption that 'waiting' implies saving rather than simply owning (and not eating into) capital.

84 Some categories of income from land and 'realized property' are for life; but in most of these cases too 'the obligation or necessity for saving, though it exists, is of a considerably less binding character. For example, in the case, which is one of the strongest, that of provisions for younger children, I do not think that the necessity of saving in that case is on the average nearly so great as in the case of industrial incomes' (*CW*, V, 558–9).

probably at little except a moderate provision for their children and for their old age while successful traders are mostly aiming at making fortunes and passing into a superior class altogether' (V, 560; cf. III, 814f). But in the *Principles* Mill insisted that any savings made by individuals at certain periods of their earnings cycle, even at zero interest, was of small relevance for secular or growth considerations:

There would be adequate motives for a certain amount of saving, even if capital yielded no profit [interest]. There would be an inducement to lay by in good times a provision for bad; to reserve something for sickness and infirmity, or as a means of leisure and independence in the latter part of life, or a help to children in the outset of it. Savings, however, which have only these ends in view, have not much tendency to increase the amount of capital permanently in existence. These motives only prompt persons to save at one period of life what they purpose to consume at another, or what will be consumed by their children before they can completely provide for themselves. The savings by which an addition is made to the national capital, usually emanate from the desire of persons to improve what is termed their condition in life, or to make provision for children or others, independent of their exertions. Now, to the strength of these inclinations it makes a very material difference how much of the desired object can be effected by a given amount and duration of self-denial; which again depends on the rate of profit. And there is in every country some rate of profit, below which persons in general will not find sufficient motive to save for the mere purpose of growing richer, or of leaving others better off than themselves (III, 737).

Yet more specifically, 'any accumulation . . . by which the general capital is increased, requires as its necessary condition a certain rate of profit; a rate which an average person will deem to be an equivalent for abstinence, with the addition of a sufficient insurance against risk'. When this rate which varies geographically and temporally rules, 'no further increase of capital can for the present take place'.

All this points to the necessity for a positive rate of interest to assure *net* accumulation,[85] not necessarily to prevent decumulation. There is,

---

85 But in his critique of the 'over-production' theorists Mill allows, though as an exceptional case, some accumulation even at zero interest: 'Driven to his last retreat, an opponent may perhaps allege, that there are persons who produce and accumulate from mere habit; not because they have any object in growing richer, or desire to add in any respect to their consumption, but from *vis inertiae*. They continue producing because the machine is ready mounted, and save and re-invest their savings because they have nothing on which they care to expend them. I grant that this is possible, and in some few instances probably happens . . . ' (573).

One economist known to Mill (and respected by him) assumed positive savings even at zero interest. William Ellis, in his analysis of the propensity to save not only emphasized

after all, the celebrated proposition encountered earlier (above, p. 368) that as the operation of maintaining labour 'admits of being repeated indefinitely without any fresh act of saving, a saving once made becomes a fund to maintain a corresponding number of labourers in perpetuity, reproducing annually their own maintenance with a profit'.[86] It is true enough that 'improvidence' might result in decumulation, and that capital maintenance is not guaranteed 'independently of human decision' (Hayek, 1941b, 48–9), but this need not mean that a special reward is required for that purpose. Add to this an observation that 'where capital is stationary because there is no accumulation' a tax on profits 'might not prevent the old capital from being kept up through habit, or from unwillingness to submit to impoverishment . . .' (*CW*, III, 828). All would be clear sailing, but for the definition of interest given in the chapter 'Of Profits' – 'all a person is enabled to get by merely abstaining from the immediate consumption of his capital, and allowing it to be used for productive purposes by others' (II, 400), or 'an equivalent to the owner of the capital for forbearing to consume it . . .' (402). These are not casual formulations in which 'capital' is loosely used in place of 'income'. Mill seems very deliberately to be choosing his terms to indicate a positive return to assure against capital depletion. Equally suggestive is the formal justification given in the discussion 'Of Property' for an interest payment out of the product of current labour – namely the notion of interest as 'equivalent . . . for the antecedent labour, and for the abstinence by which the produce of that labour, instead of being expended in indulgencies, has been reserved for this use', where Mill explicitly recognizes that while existing capital is not necessarily 'created' by the current owner, 'the abstinence at least must have been continued by each successive owner, down to the present' (215–16). In the formal discussion of the stationary state, allowance seems to be made for a class of capitalists (III, 755).[87] On balance the evidence thus suggests Mill's adherence to the Robbins rather than the Clark–Schumpeter view: a positive interest rate was a necessary feature of the stationary state.[88]

inducements existing 'totally independent of any additions which might be made to the savings themselves', but allowed that while they were 'no doubt considerably strengthened by the knowledge that the sum saved will be doubled . . . even where no profit is superadded . . . savings would still be made in conformity with the inducements we have mentioned' (1826, 106–7; cf. 112). Ellis much emphasized, by contrast, the role of the return on capital as a stimulus to *investment* (cf. 108, 128–9).

86 Cf. 1852: income invested productively 'reproduces itself, and becomes the means of supporting a number of persons in perpetual succession' (*CW*, V, 494).

87 But note Mill's reference to the 'annihiliation' of profits in his paraphrase of Wakefield (735–6).

88 This conclusion is somewhat reinforced by the position in the essays on religion that 'the power of sacrificing a present desire to a distant object' is not a 'natural' virtue but has to be learned (*CW*, X, 395), which is in sharp contrast to Adam Smith's declaration that 'the principle which prompts to save . . . comes with us from the womb . . . ' (1937, 324).

We now turn from the *propensity to save* to the *inducement to invest*. The rate of profit allows not only for 'the effective desire to accumulate savings' but also for the productive application of those savings, the minimum profit rate at which net investment is reduced to zero exceeding the minimum rate of interest by a risk premium:

There is at every time and place some particular rate of profit, which is the lowest that will induce the people of that country and time to accumulate savings, and to employ those savings productively. This minimum rate of profit varies according to circumstances. It depends on two elements. One is, the strength of the effective desire of accumulation; the comparative estimate made by the people of that place and era, of future interests when weighed against present. This element chiefly affects the inclination to save. The other element, which affects not so much the willingness to save as the disposition to employ savings productively, is the degree of security of capital engaged in industrial operations . . . [In] employing any funds which a person may possess as capital on his own account, or in lending it to others to be so employed, there is always some additional risk, over and above that incurred by keeping it idle in his own custody. This extra risk is great in proportion as the general state of society is insecure: it may be equivalent to twenty, thirty, or fifty per cent, or to no more than one or two; something, however, it must always be: and for this, the expectation of profit must be sufficient to compensate (736-7).[89]

---

89 The posthumous 'Chapters on Socialism' provides a summary of the empirical orders of magnitude involved: 'Even of [the capitalist's] own share a small part only belongs to him as the owner of capital. The portion of the produce which falls to capital merely as capital is measured by the interest of money, since that is all that the owner of capital obtains when he contributes nothing to production except the capital itself. Now the interest of capital in the public funds, which are considered to be the best security, is at the present prices (which have not varied much for many years) about three and one-third per cent. Even in this investment there is some little risk – risk of repudiation, risk of being obliged to sell out a low price in some commercial crisis.

Estimating these risks at 1/3 per cent., the remaining 3 per cent. may be considered as the remuneration of capital, apart from insurance against loss' (V, 734).
Industrial profits exceed 3% by an appropriate risk premium and by a return to superintendence (in the case of the self-employed): 'Of what he obtains beyond 3 per cent. a great part is insurance against the manifold losses he is exposed to, and cannot safely be applied to his own use, but requires to be kept in reserve to cover those losses when they occur. The remainder is properly the remuneration of his skill and industry – the wages of his labour of superintendence' (735).

The 3% figure has something of a pedigree. For example, Torrens had referred to it as 'the lowest rate of profit which will induce men to embark in business' (1821, 219), and 'the lowest rate of return, for the sake of which the capitalist will risk his property in production' (216). For Mill, we have seen, 3% was the going rate of interest narrowly defined.

In 1826 Mill referred to an interest rate on best security of approximately 4% over

The minimum profit rate in the stationary state thus exceeds the return required by savers: 'Any accumulation . . . by which the general capital is increased, requires as its necessary condition a certain rate of profit; a rate which an average person will deem to be an equivalent for abstinence, with the addition of a sufficient insurance against risk'. Having in mind that some insurance against risk is required even on investment by the capitalist 'on his own account' we are faced by the implication that, at low ranges of the interest rate, savings will occur which will not yet be absorbed into industrial projects, a conclusion not noted by Mill.

* * *

The import of the sociological conditions that lie behind the savings schedule is enhanced by Mill's recognition that there may occur divergences between the paths taken by the interest and the profit rates. In correspondence of 1852 with John Lalor, Mill denied any fall in the return on capital understood as the rate of profit, at least since 1840, but conceded a fall in the rate of interest (XIV, 91). This decline was unrelated to deteriorating agricultural productivity (indeed Mill did not believe such deterioration had occurred). Already in his early essay 'On Profits and Interest' Mill had noted that 'although the rate of profit is one of the elements which combine to determine the rate of interest, the latter is also acted upon by causes peculiar to itself, and may either rise or fall, both temporarily and permanently, while the general rate of profits remains unchanged' (IV, 305). Here Mill had in mind features of the social structure pertinent to the relationship between the lending and borrowing classes of the community – a perspective very reminiscent of that of A.R.J. Turgot (1770):

The amount of borrowers being given, (and by the amount of borrowers is here meant the aggregrate sum which people are willing to borrow at some given rate,) the rate of interest will depend upon the quantity of capital owned by people who are unwilling or unable to engage in trade. The circumstances which determine this, are, on the one hand, the degree in which a taste for business, or an aversion to it, happens to be prevalent among the classes possessed of property; and on the other hand, the amount of the annual accumulation from the earnings of labour. Those who accumulate from their wages, fees, or salaries, have of course, (speaking generally) no means of investing their savings except by lending them to others:

the period 1783–93 as the rate 'warranted by the average rate of profit upon capital'. Setting aside an increase during the war period due to 'unproductive' expenditures and immense pressure on loanable funds and looking at the trend, Mill indicates an interest rate of about 3 ½ % as the 'justified' rate in the post-war period up to 1826 and relates the secular fall to a higher money wage and consequently lower profits due in turn to population expansion under conditions of agricultural protection (CW, IV, 103).

their occupations prevent them from personally superintending any employment (303–4).[90]

Mill next focusses on the impact of government borrowing on the market, and subsequently on the impact of banking, specifically the possibility of 'forced saving' (307).

The same issue arises again conspicuously in the 1860s. Despite all that Mill had himself said on the matter he had still taken for granted in his *Principles* that movements in the interest rate reflected movements in the profit rate, observing that the profit rate in the US was evidently higher than in London since the interest rate stood at 6 per cent and 3¼ per cent respectively in these centres (II, 414). In a letter to Cairnes late in 1864 Mill admitted that 'supposing profits to be lower than in Europe instead of higher, it is yet quite intelligible that interest might be higher', because of the fact that 'there is . . . in America, scarcely any unoccupied class, living on interest: almost everybody is in active business, needing all his own capital and more too' (Dec. 1, 1864, XV, 968). Mill feared he may have been 'misleading' his readers and indeed the observation regarding the evidence for a differential profit rate was withdrawn in the edition of 1865. Following discussion with Cairnes formal allowance is made for permanent changes in circumstances which alter the ratio of the interest-receiving to the profit-receiving class, including the impact of gold discoveries and joint-stock organization (III, 651–2; 1062f).[91]

---

90 In this passage the demand for loans is represented as interest elastic. A few paragraphs earlier Mill had declaimed that the 'class of borrowers may be considered as unlimited. There is no quantity of capital that could be offered to be lent, which the productive classes would not be willing to borrow, at any rate of interest which would afford them the slightest excess of profit above a bare equivalent for the additional risk, incurred by that transaction, of the evils attendant on insolvency' (302). But this strong statement is qualified in various ways, leading to the final conclusion that there is, after all, 'a practical limit to the demands of borrowers at any given instant; and when these demands are all satisfied, any additional capital offered on loan can find an investment only by a reduction of the rate of interest' (303).

91 Cf. also Mill's evidence of 1867: *'Quelles sont, dans un pays, les causes régulatrices du taux de l'intérêt?* Le taux moyen de l'intérêt dans un pays dépend de deux causes générales, savoir: 1° le taux moyen du profit industriel; 2° la proportion qui existe dans ce pays entre la classe industrielle, qui fait valoir elle-même ses capitaux, et ce qu'on peut nommer la classe prétante, comme aussi entre les capitaux dont ces deux classes disposent. Aux États-Unis par exemple, où la classe des rentiers vivant de l'intérêt de capitaux prêtés est extrêmement restreinte, le taux de l'intérêt est naturellement beaucoup plus élevé, relativement au profit moyen du commerce et de l'industrie, que dans la plupart des pays de l'Europe. Je ne parle ici que du taux moyen et normal de l'intérêt. Ses variations passagères dépendent de toutes les vicissitudes de la production, de la consommation et de la spéculation' (V, 602).

XI LABOUR-SUPPLY CONDITIONS

Our principal concern is with the secular conditions of labour supply.[92] But a word first regarding the short run. In the course of the critique of the 'overproduction' theorists, where it is assumed that rapid capital accumulation transfers the surplus to the labouring class, there is a suggestion that at least beyond a point, a rise in the commodity wage might be accompanied by a decline in the supply forthcoming from a given population:

Now, will that class . . . not know what to do with it [the surplus]? Are we to suppose that they . . . have their wants perfectly satisfied, and go on labouring from mere habit? Until this is the case; until the labouring classes have . . . reached the point of satiety – there will be no want of demand for the produce of capital, however rapidly it may accumulate: since, if there is nothing else for it to do, it can always find employment in producing the necessaries or luxuries of the labouring class. And when they too have had no further desire for necessaries or luxuries, they would take the benefit of any further increase of wages by diminishing their work; so that the over-production which then for the first time would be possible in idea, could not even then take place in fact, for want of labourers (*CW*, III, 573-4).[93]

Conversely the opening of foreign trade has a reverse impact: 'making [a people] acquainted with new objects, or tempting them by the easier acquisition of things which they had not previously thought attainable, sometimes works a sort of industrial revolution in a country whose resources were previously underdeveloped for want of energy and ambition in the people: inducing those who were satisfied with scanty comforts and little work to work harder for the gratification of their new tastes' (593-4). As in consumption theory the interpersonal dimension enters to complicate the wage-effort relationship. Thus 'a state of complete equality of fortunes would not be favourable to active exertion for the increase of wealth. Speaking of the mass, it is as true of wealth as of most other distinctions – of talent, knowledge, virtue – that those who already have, or think they have, as much of it as their neighbours, will seldom exert themselves to acquire more' (890). By implication an across-the-board increase would have a different impact on labour supply than a more localized increase.

92 Mill nowhere much emphasized the possibility of drawing in labour from the unproductive into the productive sector at a constant wage – the basis for W. A. Lewis' celebrated attribution to the classicists of growth with 'unlimited supplies of labour' (Lewis, 1954).

93 An upward-sloping segment of the supply function cannot be positively excluded, but Mill is not at all clear about this matter either in this passage or elsewhere.

Mill approached the population variable by insisting upon the continued usefulness of the Malthusian conception of a maximum physiological capacity of a human population to double itself in approximately two decades, drawing for evidence upon 'the most favourable circumstances known to exist, which are those of a fertile region colonized from an industrious and civilized community' (II, 155; even the critic Carey admitted to a doubling in three decades). Under usual conditions this geometrical ratio is not, however, encountered and it is with the nature of the 'checks' that Mill is preoccupied in his chapter on the 'Law of the Increase of Labour' – 'the unlimited extent of its natural powers of increase and the causes owing to which so small a portion of that unlimited power is for the most part actually exercised' (159).[94]

Carey's denial of a divergence between the population and the food maximum growth rates is firmly rejected at the outset of the argument: Malthus had not taught, Mill observed, that the increase in population was limited by the power of increase of food, but rather 'by the limited quantity of the land on which [food] can be grown' which in the last resort constrained food supplies. Where (as in North America) land was not a limiting factor, the food supply was indeed forthcoming at 'the higher rate natural to it' permitting population also to expand at its maximum: 'When Mr. Carey can show, not that turnips and cabbages, but that the soil itself, or the nutritive elements contained in it, tend naturally to multiply, and that too at a rate exceeding the most rapid possible increase of mankind, he will have said something to the purpose' (156n.; see also below, p.453, regarding 353–4).

The check to the maximum population growth rate was not, however, necessarily imposed by the death rate ['only' by the death rate was added in 1852] (war and disease, and insufficiency of food), with the birth rate achieving the maximum level physiologically possible. Human reproduction was also to a greater or lesser degree 'restrained by the fear of want rather than by want itself', so that even where starvation was not an issue, 'many' ['most' in ms. and eds. 1848, 1849] were yet influenced by 'the

94 Mill rejected both a contemporary suggestion that fecundity was 'in inverse proportion to the quantity of nutriment' – 'whatever the limit to the increase of population among the richer classes in Great Britain may be, it certainly is not the smaller number of births to a marriage' (although he did concede that a 'positive excess of nutriment' may be detrimental), and Carey's view that reduced fecundity would accompany the advance of civilization in consequence of a lower physiological need for sexual intercourse: any such trend towards reduced fecundity 'if ever realized', would have this effect 'rather by rendering physical self-restraint easier, than by dispensing with its necessity'. On the contrary since rapid population growth was consistent with 'a very sparing employment of the multiplying power', physical self-restraint was all the more significant (155–6n).

apprehension of losing what have come to be regarded as the decencies of their situation in life'; acting from motives of 'prudence' or the 'social affections' people married at such age and had that number of offspring 'consistent with maintaining themselves in the condition of life which they were born to, or were accustomed to consider as theirs' (157).[95]

The constraints on population growth by way of deaths and of prudential control on births designed to maintain living standards, are said to be the sole forces 'hitherto . . . found strong enough, in the generality of mankind, to counteract the tendency to increase' – that is, the tendency to increase at the maximum rate possible. Limitation of family size to the end of actually raising living standards is conceded amongst members of the middle class; but 'such a desire is rarely found, or rarely has that effect, in the labouring classes. If they can bring up a family as they were themselves brought up, even the prudent among them are usually satisfied.' And although Mill adds that 'too often they do not even think of that, but rely on fortune, or on the resources to be found in legal or voluntary charity', which implies an absence of 'prudence,' he concludes by contrasting contemporary Asian conditions or those of medieval Europe, where population was held down by starvation, with modern societies wherein 'few, even among the poorest of the people, are limited by actual necessaries, and to a bare sufficiency of those: and the increase is kept within bounds, not by excess of deaths, but by limitation of births' (157) – again a reference to restrictions on the growth rate of population below its maximum.

'Prudence' refers, so it might seem, to 'conscientious self-restraint' with an eye on the maintenance of living standards, examples of which behaviour are provided by the Norwegian and Swiss cases where population growth was 'very slow' in consequence of a low birth rate and living standards were comfortable (158).[96] But reference is also made to societies where the exercise of foresight is imposed by the State, particularly control of the minimum age of, and the material conditions to be satisfied prior to, marriage, and where standards were satisfactory and illegitimate births relatively few (cf. also 346f. on the role of the State). Eighteenth-century Britain provided an instance of yet a different pattern of constraint, namely, prudential behaviour imposed by 'accidental habit', a reference to the customary need for private cottages by married couples and the unwillingness

95 Cf. 'There is a condition, to which the labouring people are habituated; they perceive that by having too numerous families, they must sink below that condition or fail to transmit it to their children; and this they do not choose to submit to' (157; also 287, 340, 341–2, 396).

96 Mill discreetly suggests also the absence of Malthusian *vice* and thus venereal diseases in those societies: The low birth rate 'tends directly to prolong life, by keeping the people in comfortable circumstances; and the same prudence is doubtless exercised in avoiding causes of disease, as in keeping clear of the principal cause of poverty'.

of landowners to provide them (circumstances which were transformed with the outbreak of the Revolutionary Wars and the encouragement given on 'patriotic grounds' to a large population (58; cf. 350).

All that is required for the wage to remain unchanged is that population and capital 'increase with equal rapidity' (III, 723; see below, pp. 446–7); and Mill's notion of a culturally-determined standard of living implies that an acceleration in the rate of accumulation relative to that of population will push the wage above that level generating an appropriate adjustment to the population growth rate designed to bring the wage back into line (conversely in the event of a deceleration in accumulation), while an unchanged labour demand requires an unchanged population to satisfy wage constancy. This is expressed in the opening sentence of the following passage – at least with regard to a wage increase – although unfortunately Mill then proceeds to expound the argument in terms implying variation in absolute population size, as distant from the growth rate of population, to assure a tendency of wages to the subsistence level:

Whatever the causes by which population is anywhere limited, to a comparatively slow rate of increase, *an acceleration of the rate very speedily follows any diminution of the motives to restraint*. It is but rarely that improvements in the condition of the labouring classes do anything more than give a temporary margin, speedily filled up by an increase of their numbers . . . Unless either by their general improvement in intellectual and moral culture, or at least by raising their habitual standard of comfortable living, they can be taught to make a better use of favourable circumstances, nothing permanent can be done for them; the most promising schemes end only in having a more numerous, but not a happier people. By their habitual standard, I mean that (*when any such there is*) down to which they will multiply but not lower (II, 159; my emphasis).[97]

The last sentence is most revealing. It is important to note how in the same passage, Mill switched between two conceptions of the notion of a 'habitual standard', one appropriate in a growth context and the other appropriate in a stationary state – a complexity which we frequently encounter. (It must, however, be remembered that the theme is not the explanation of zero population growth but a growth rate below the physiological maximum – 'a comparatively slow rate of increase'.)

---

97 The first italicized phrase was introduced in 1852, and read thus in the manuscript and the first two editions (1848 and 1849): 'there is always an immense residuary power behind, ready to start into activity as soon as the pressure which restrained it is taken off'. The alteration implies a greater emphasis on growth rates. The second italicized phrase was added in 1852. Mill may have had in mind the downward flexibility of the floor in some societies (see below, p. 457).

We shall see presently that the practice of 'prudential' behaviour is represented as limited to the class of skilled labourers. Reproduction in the case of the unskilled was at the maximum rate possible. This need not imply that we must treat as otiose Mill's statements regarding 'the generality of mankind' or the 'many' even 'most' workers who attempt to maintain their normal (culturally governed) standards. For such statements are appropriate for a 'static' world disturbed by some event that temporarily raises the wage above the initial culturally-determined minimum, to which labourers 'people up' in response. The industrial sector of contemporary Britain, by contrast, was one characterized by very rapid growth of capital, with the wage rate 'initially' *above*, not *at*, 'subsistence' (culturally defined). Mill's fear was that, upon any deceleration of accumulation, wages would be forced down to the (cultural) minimum unless the unskilled learned to treat the going rate as the minimum acceptable rate. To this matter we turn next.

XII THE WAGE PATH

In his basic exposition of growth theory Mill followed a Ricardian rather than a Malthusian procedure (above, Chapter 1, Section V) in that any downward wage trend results from deviations between the factor growth rates to labour's disadvantage. For the 'test of equality' between the factor growth rates is that 'each labourer obtains the same commodities as before, and the same quantity of those commodities' (*CW*, III, 723, cited also below p.447).

A formal contrast is made between circumstances where land scarcity is not yet manifest and where an increase in population at its 'maximum' potential rate (namely a doubling in a single generation) can proceed with no downward pressure on the real wage rate, and situations where this possibility is ruled out. The former included specifically North America and the Australian colonies which enjoyed advanced European technology and savings habits and were not subject to land scarcity, so that the rate of accumulation proceeded at least as rapidly as population at its maximum and the real wage was maintained above subsistence at a steady, perhaps increasing level:

In countries like North America and the Australian colonies, where the knowledge and arts of civilized life, and a high effective desire of accumulation, co-exist with a boundless extent of unoccupied land, the growth of capital easily keeps pace with the utmost possible increase of population, and is chiefly retarded by the impracticability of obtaining labourers enough. All, therefore, who can possibly be born, can find employment, without overstocking the market: every labouring family enjoys in abundance the necessaries, many of the comforts, and some of the luxuries of life . . . (II, 343-4).

In the absence of this special confluence of circumstances and assuming given technology it would be impossible for population to expand at the maximum rate without downward pressure on wages and this because of impediments to the rate of accumulation. These impediments might act directly: 'those circumstances . . . in which population can with impunity, increase at its utmost rate, are rare, and transitory. Very few are the countries presenting the needful union of conditions. Either the industrial arts are backward and stationary, and capital therefore increases slowly; or the effective desire of accumulation being low, the increase soon reaches its limit.' But even where these direct constraints on accumulation are absent, the effects of land scarcity – a falling marginal product – will bear upon the rate of accumulation: 'The increase of capital is checked, because there is not fresh land to be resorted to, of as good quality as that already occupied' (344).

The assumption of unrestricted population growth in this 'Ricardian' analysis, be it noted, is not required for the wage decline to occur. That is a limiting case only; and it is enough that there exists downward pressure on g_K for deviations between g_K and g_L to be generated, even if g_L is related positively to w.

We shall show in what follows that Mill also developed a constant-wage growth model reflecting in his case not only displacements of a given curve but the incorporation of 'prudence' within the labour supply function itself. On this view workers as a class may deliberately constrain the population growth rate to prevent the wage falling below some designated 'conventional' level (w_c). In effect, the labour supply function becomes horizontal at w_c, and this wage traces out the 'dynamic equilibrium' path. For population growth to fall in order to *prevent* any wage decline implies ascribing to labour prescience of the steady decline in g_K. Since the model was designed to portray the effect of an instillation of responsible habits this strict assumption makes considerable sense at least as a limiting case.

The possibility of ongoing factor expansion, with the wage constrained to a level exceeding subsistence strictly defined, emerges in various contexts. A version of the argument will be found in the chapter 'On Wages' (Book II, Ch. xi) where Mill discusses the effect of secular changes in the real price of food on the money wage, with an eye especially to the implications of diminishing returns. He assumes first that wages are initially at 'subsistence' so that higher food prices generate an increased death rate, and secondly that wages are initially at some 'culturally' determined level the upward trend of food prices acting on the birth rate:

dearness or cheapness of food, when of a permanent character, and capable of being calculated on beforehand, may affect wages. In the first place, if the labourers have, as is often the case, no more than enough to keep them in working condition, and enable them barely to support the ordinary number of children, it follows that if food grows permanently dearer without a rise of wages, a greater number of children will prematurely die; and thus wages will ultimately be higher, but only because the number of people will be smaller, than if food had remained cheap. But secondly, even though wages were high enough to admit of food's becoming more costly without depriving the labourers and their families of necessaries; though they could bear, physically speaking, to be worse off, perhaps they would not consent to do so. They might have habits of comfort which were to them as necessaries, and sooner than forego which, they would put an additional restraint on their power of multiplication; so that wages would rise, not by increase of deaths but by diminution of births (II, 340).

This passage is not free from ambiguity. What, for one, are we to make of 'the ordinary number of children' in the first case? While it seems that Mill had in mind the definition of the stationary state wherein any rise in the price of necessaries generates an absolute fall in population size,[98] the further question arises why food prices are supposed to be rising secularly since if wages are initially at 'subsistence' in the foregoing sense, there would – as Malthus pointed out so clearly (above, pp. 47–8) – be no stimulus for the expansion to marginal land implied by the argument. But it is the second part of the passage that is most important for us. Here there is little question of initial ongoing population growth, and a refusal by 'labour' – as a class – to allow its real wage to fall below the culturally determined level, a refusal manifested in deliberate restriction of the birth rate.

Formulations of this kind suggest that Mill recognized a culturally determined minimum to the functional relation $g_L - w$.[99] The introduction to the passage further elaborates on his precise intention: the relation between food prices and the 'money' wage 'in so far as true, in no way affects the dependence of wages on the proportion between capital and labour: since the price of food, when it affects wages at all, effects them through that law'. Now in logic it is clear that for the real wage to remain unchanged under conditions of land scarcity the growth rate of population (labour supply) must decline in line with that of capital (labour demand). Although

98 On the minimum at which population is constant, see also 396: 'Where the wife of a labouring man does not by general custom contribute to his earnings, the man's wages must be at least sufficient to support himself, a wife, and a number of children adequate to keep up the population . . . And even if the wife earns something, their joint wages must be sufficient to support not only themselves, but (at least for some years) their children also.'

99 The presumption in our passage that the rise in the price of food is 'capable of being calculated on beforehand' is important. The constant wage path depends on it. If, however, allowance is made for a lag in the response of g_L the actual wage path will sag below the horizontal, and will not be one of 'dynamic equilibrium'.

Mill does not spell this out at the present juncture, he does so with eminent clarity in the central chapter on the distributive trends: 'Influence of the Progress of Industry and Population, on Rents, Profits, and Wages' (Book IV, Ch. iii). Here, given land scarcity and unchanged technology, population and capital are presumed to proceed at the same rate, which presumption alone assures a constant commodity wage: 'We shall suppose them . . . to increase with equal rapidity; the test of equality being, that each labourer obtains the same commodities as before, and the same quantity of those commodities' (III, 723). (The assumption of equal growth rates is made since a more rapid expansion of either factor can be analysed in terms of one or other of the simpler hypothetical disturbances dealt with earlier in the chapter – population increasing with capital stationary which implies a falling wage; or capital increasing with population stationary when wages rise.) Mill then proceeded to trace out the consequence of a growth path entailing a given commodity wage:

Population having increased, without any falling off in the labourer's condition there is of course a demand for more food. The arts of production being supposed stationary, this food must be produced at an increased cost. To compensate for this greater cost of the additional food, the price of agricultural produce must rise . . . Rent will rise both in quantity of produce and in cost [labour embodied]; while wages, being supposed to be the same in quantity, will be greater in cost. The labourer obtaining the same amount of necessaries, money wages have risen; and as the rise is common to all branches of production, the capitalist cannot indemnify himself by changing his employment, and the loss must be borne by profits.

It is thus Mill's view that capital accumulation and population growth can proceed at the same rate, with the return on capital declining – so that, considering the accumulation-interest relationship always insisted upon, the common factor growth rate must also be decelerating – but the wage rate constant at a level above 'subsistence', that wage at which population is constant.

Mill's formulations in the chapter 'Of Taxes on Commodities' (Book V, Ch. iv) prove to be highly pertinent to the growth theme, particularly the deceleration of the capital growth rate during the course of expansion – a central feature of a comparison made between two economies identical in all respects bar the imposition of a tithe in one alone:

Though the untithed island is always verging towards the point at which the price of food would overtake that in the tithed island, its progress towards that point naturally slackens as it draws nearer to attaining it; since – the difference between the two islands in the rapidity of accumulation depending upon the difference in the rates of profit [higher in untithed] – in proportion as these approximate, the movement which draws them closer together, abates of its force. The one may not actually overtake the other, until both islands reach the minimum of profits: up to that point, the tithed island may continue more or less ahead of the untithed island in the price of corn: considerably ahead if it is far from the minimum, and is therefore accumulating rapidly; very little ahead if it is near the minimum, and accumulating slowly (845).

Nothing is here said of a declining real wage during the course of ongoing expansion, and this despite the slowing down in the rate of accumulation, so that on Mill's own terms one must suppose population growth to be declining in line with capital.[100]

That a falling real wage is not a *sine qua non* of the growth process, despite a slowing down in capital accumulation, similarly comes to the fore in the earlier chapter (Book V, Ch. iii) 'Of Direct Taxes':

At its first imposition, the tax falls wholly on profits, but the amount of increase of capital, which the tax prevents, would, if it had been allowed to continue, have tended to reduce profits to the same level; and at every period of ten or twenty years there will be found less difference between profits as they are, and profits as they would in that case have been; until at last there is no difference, and the tax is thrown either upon the labourer or upon the landlord. The real effect of a tax on profits is to make the country possess at any given period, a smaller capital and a smaller aggregate production, and to make the stationary state be attained earlier, and with a smaller sum of national wealth . . .
Even in countries which do not accumulate so fast as to be always within a short interval of the stationary state, it seems impossible that, if capital is accumulating at all, its accumulation should not be in some degree retarded by the abstraction of a portion of its profit; and . . . it is inevitable that a part of the burthen will be thrown off the capitalist, upon the labourer or the landlord. One or other of these is always the loser by a diminished rate of accumulation. If population continues to increase as before, the labourer suffers: if not, cultivation is checked in its advance, and the landlords lose the accession of rents which would have accrued to them (827–8).

Here it is apparent that a slackening in the population growth rate – presumably to keep in line with that of capital – ensures against any fall of wages; only to the extent that population growth proceeds unchecked will wages fall.

The famous chapter 'Of the Stationary State' (Book IV, Ch. vi) contains

---

100 The absence of a falling real wage is clear in a statement, shortly before, that 'the effect of accumulation, when attended by its usual accompaniment, an increase of population, is to increase the value and price of food, to raise rent, and to lower profits' (843).

a passage which conveniently encapsulates the entire argument to this point. Most significant is the sharp awareness of the need for prudence in a growing system as well as in a state of stationarity to assure constancy of the wage. It is noteworthy too that Mill suggests that the appropriate restraint might be more easily generated in a slowly growing economy or even a zero-growth economy:

Even in a progressive state of capital, in old countries, a conscientious or prudential restraint on population is indispensable, to prevent the increase of numbers from outstripping the increase of capital, and the condition of the classes who are at the bottom of society from being deteriorated. Where there is not, in the people, or in some very large proportion of them, a resolute resistance to this deterioration – a determination to preserve an established standard of comfort – the condition of the poorest class sinks, even in a progressive state, to the lowest point which they will consent to endure. The same determination would be equally effectual to keep up their condition in the stationary state, and would be quite as likely to exist. Indeed, even now, the countries in which the greatest prudence is manifested in the regulating of population, are often those in which capital increases least rapidly. Where there is an indefinite prospect of employment for increased numbers, there is apt to appear less necessity for prudential restraint. If it were evident that a new hand could not obtain employment but by displacing, or succeeding to, one already employed, the combined influences of prudence and public opinion might in some measure be relied on for restricting the coming generation within the numbers necessary for replacing the present (753).[101]

It remains to note that Mill attributed the constant-wage growth model to Ricardo (II, 340). He was, however, critical. Ricardo, he believed, had only allowed for the two cases of constant real wages alluded to above (pp. 445–6): 'Mr. Ricardo considers these two cases to comprehend all cases. He assumes, that there is everywhere a minimum rate of wages: either the lowest with which it is physically possible to keep up the population, or the lowest with which the people will choose to do so. To this minimum

---

101 In the constant-wage dynamic-equilibrium model labourers exert a deliberate effort to keep long-run wages unchanged by appropriately reducing the rate of population growth – by way of variation in the marriage and birth rates – in the face of a falling rate of accumulation, itself the result of a steadily declining return on capital. A possible rationalization for such a hypothesis is given by Caravale (1982), 149n, whereby 'la crescita della popolazione è funzione decrescente del livello assoluto della popolazione stessa; una sorta di spontaneo e collettivo controllo delle nascite motivato ad esempio da ragioni ecologiche'. For Mill, however, the model was designed for exhortatory purposes. Workers were to be inculcated with the principle that where there were manifestations of slackening accumulation, it was in their hands to prevent (permanent) wage reductions and/or general excess labour supplies by appropriate reductions in family size.

he assumes that the general rate of wages always tends; that they can never be lower, beyond the length of time required for a diminished rate of increase to make itself felt, and can never long continue higher' (340–1). Here we have the source *par excellence* for a Ricardian constant wage, although broad enough to incorporate the 'dynamic equilibrium' version. But the extraordinary historiographical fact emerges of an apparent failure to recognize the downward *secular* path which is the essence of Ricardo's growth model:

This assumption [the 'tendency' of the commodity wage to a given level] contains sufficient truth to render it admissable for the purposes of abstract science; and the conclusion which Mr. Ricardo draws from it, namely, that [money] wages in the long run rise and fall with the permanent price of food, is, likely almost all his conclusions, true hypothetically, that is granting the suppositions from which he sets out. But in application to practice, it is necessary to consider that the minimum of which he speaks, especially when it is not a physical, but what may be termed a moral minimum, is itself liable to vary. If wages were previously so high that they could bear reduction, to which the obstacle was a high standard of comfort habitual among the labourers, a rise in the price of food, or any other disadvantageous change in his circumstances, may operate in two ways: it may correct itself by a rise of [money] wages brought about through a gradual effect on the prudential check to population; or it may permanently lower the standard of living of the class, in case their previous habits in respect of population prove stronger than their previous habits in respect of comfort. In that case the injury done to them will be permanent, and their deteriorated condition will become a new minimum, tending to perpetuate itself as the more ample minimum did before (341).

We shall see (pp. 730–1) that a declining-wage model was actually adopted in Mill's early papers on the corn trade. It is regrettable that he failed to record the full Ricardian argument in the *Principles* itself, and he must be held partly responsible for thereby misleading commentators ever after.

* * *

The stationary state as far as concerns population growth was not for Mill an immediate prospect: 'There is room in the world, no doubt, and even in old countries, for a great increase of population, supposing the arts of life to go on improving, and capital to increase' (III, 756) – with commodity wages presumably unchanged since Mill focusses only on the disadvantages of poorer amenity (including space). But he does speculate on the nature of a stationary state. Here we recall our earlier conclusion that, for Mill, a positive interest rate would rule in a state of stationarity. Starting out from stationarity, an increase in the 'effective desire of accumulation' – a reduction in the minimum supply price of capital r^* – would create the potential for new growth, as will technological 'improvement': the stationary state is 'the state in which no further addition will be made to capital, unless there takes place either some improvement in the arts of production, or an increase in the strength of the desire to accumulate' (above, p. 433; see also II, 98). Considering the recognition of a psychological supply price for capital (as for population) 'there is now the possibility of an infinite diversity of stationary equilibria with natural wages and profits at different levels' (Robbins, 1930, 200). And the specific message of the chapter 'Of the Stationary State' is that any new potential for expansion should ideally be taken out in higher real wages, pre-eminently in the form of increased leisure: '[A] stationary condition of capital and population implies no stationary state of human improvement . . . Even the industrial arts might be as earnestly and as successfully cultivated, with this sole difference, that instead of serving no purpose but the increase of wealth, industrial improvements would produce their legitimate effect, that of abridging labour' (*CW*, III, 756).

The possibility of technological progress despite zero capital formation raises, incidentally, the shadow of 'conversion' with its potentially adverse effects on labour but Mill is silent on that issue. (But see below, pp. 468–70.) And while 'it would be an odd series of inventions which had no influence on the supply of savings' (Robbins, 206n.), Mill (as noted) is hopefully presuming that any initial increase in the return on capital will be absorbed by an upward shift of the supply price of labour, preferably a shorter working day.

It remains to note that in the absence of innovation a constant population could not be maintained permanently at constant wage and profit rates in consequence of exhaustibility of natural resources: 'The only products of industry, which, if population did not increase, would be liable to a real increase of cost of production, are those which, depending on a material which is not renewed, are either wholly or partially exhaustible; such as coal, and most if not all metals' (*CW*, III, 712).[102] The full-fledged stationary state – constant technology, population, and capital with unchanged wage and profit rates – was thus an approximation.

XIII   APPLICATION TO THE CONTEMPORARY LABOUR MARKET

Sharp differences are manifest in Mill's various accounts of contemporary British conditions between population mechanism in the case of the skilled

---

102 In 1866 Mill proposed paying off the national debt before Britain's coal supplies were exhausted (*CW*, I, 277). Jevons (1906) had made the same proposal the previous year (448). Cf. Mill's commendation of Jevons's *Coal Question* in his House of Commons speech, 17 April 1866.

and unskilled workers, and between the secular pattern of demand for various categories of unskilled labour. It transpires that 'prudential' behaviour was practised only by the skilled class – that 'the restraining principle lies in the very great proportion of the population composed of the middle classes and the skilled artizans, who in this country almost equal in number the common labourers, and on whom the prudential motives do, in a considerable degree, operate' (*CW*, II, 346) – and that the maintenance of high and steady wages in the manufacturing centres was possible despite uncontrolled population growth on the part of the unskilled only because of the high and steady rate of capital accumulation in those centres.

A word first, however, regarding agriculture. Here because of the total absence of the 'prudential' motive, population was constrained by variation in the death rate alone:

In England . . . I much doubt if the generality of agricultural labourers practise any prudential restraint whatever. They generally marry as early and, have as many children to a marriage, as they would or could do if they were settlers in the United States. During the generation which preceded the enactment of the present Poor Law, they received the most direct encouragement of this sort of improvidence . . . Under such prompting, the rural labourers acquired habits of recklessness, which are so congenial to the uncultivated mind that in whatever manner produced, they in general long survive their immediate causes . . . It does . . . seem that if the rate of increase of population depended solely on the agricultural labourers, it would, as far as dependent on births, and unless repressed by deaths, be as rapid in the southern counties of England as in America (346).

Indeed, 'the checks to population' in the case of the common agricultural labourer – prudential checks acting on the birth rate – 'may almost be considered as non-existent' and the living conditions in the Southern counties were consequently 'painful to contemplate. The labourers of these counties, with large families, and eight or perhaps nine [7 or 8 till 1857] shillings for their weekly wages when in full employment, have for some time been ['lately became' till 1857] one of the stock objects of popular compassion: it is time that they had the benefit also of some application of common sense' (351). The implications for policy were clear:

I ask, then, is it true or not, that if their numbers were fewer they would obtain higher wages? This is the question, and no other; and it is idle to divert attention from it, by attacking any incidental position of Malthus or some other writer, and pretending that to refute that, is to disprove the principle of population . . . And though its pressure [of population on the means of subsistence] diminishes, the more the ideas and habits of the poorest class of labourers can be improved, to which it is to be hoped that there is always some tendency in a progressive country, yet since that tendency has hitherto been, and still is extremely faint, and (to descend to particulars) has not yet extended to giving the Wiltshire labourers higher wages than eight shillings per week, the only thing which it is necessary to consider is, whether that is a sufficient and suitable provision, for a labourer? for if not, population does, as an existing fact, bear too great a proportion to the wages-fund [ms. 48, 49: means of subsistence]; and whether it pressed still harder or not quite so hard at some former period is practially of no moment, except that, if the ratio is an improving one, there is better hope that by proper aids and encouragements it may be made to improve more and faster (353–4).

We return to the manufacturing sector. Mill's formal analytical categories included the cases of North America and the colonies where land scarcity was not yet manifest and which enjoyed a growth rate of capital at least equal to the maximum physiological growth rate of population such that the real wage was maintained at a high (and even rising) level. Yet contemporary Britain, albeit an 'old' country, enjoyed some of the characteristic features of the North American scene. For in consequence of an extraordinary rate of accumulation in the cotton-manufacturing centres – a consequence of new technology – it had been possible for wages to be there maintained over several decades despite unchecked (urban) population growth (and even for agricultural wages in areas near the cities to be pulled up):

A similar advantage [to that experienced in North America], though in a less degree, is occasionally enjoyed by some special class of labourers in old countries, from an extraordinarily rapid growth, not of capital generally, but of capital employed in a particular occupation. So gigantic has been the progress of the cotton industry since the inventions of Watt and Arkwright, that the capital engaged in it has probably quadrupled in the time which population requires for doubling. While, therefore, it has attracted from other employments nearly all the hands which geographical circumstances and the habits or inclinations of the people rendered available; and while the demand it created for infant labour has enlisted the immediate pecuniary interest of the operatives in favour of promoting, instead of restraining, the increase of population; nevertheless wages in the great seats of manufacture are generally ['still' prior to 1865 ed.] so high, that the collective earnings of a family amount, on an average of years, to a very satisfactory sum; and there is, as yet, no sign of permanent [added, 1865 ed.] decrease, while the effect has also been felt in raising the general standard of agricultural wages in the counties adjoining (344).[103]

103 Cf. the strong statement in the first two editions that 'from successive improvements in machinery, and a progressive cheapening of the manufactured article, the expansion of factory employment has for half a century outstripped even the rapid growth of the factory population' (394n). Why this was subsequently deleted is not clear; but Mill may have feared the formulation clashed with wages-fund theorizing.

The high rate of accumulation and constancy of wages, it may be added, proceeded 'consistently with not forcing down profits to a lower level' (III, 742; below, p. 460).

As already remarked, 'prudential' behaviour was, in contemporary circumstances, practised only by the skilled class of labourers. Evidence of behaviour in the British case consistent with an attempt to maintain living standards is found in 'the diminished number of marriages in the manufacturing districts in years when trade is bad' (II, 159); conversely 'according to all experience, a great increase invariably takes place in the number of marriages, in seasons of cheap food and full employment' (342). But Mill also alluded to evidence of a changing minimum standard implying attempts on the part of labour, by restraint on population growth, to *raise* average real earnings.[104]

French data since the Revolution are referred to often and cited as evidence that the actual experience by labour of an improvement in living standards may itself stimulate an altered conception, a higher conception, of the 'minimum' wage in the event that the improvement is sufficiently marked to affect significantly the rising generation:

To produce permanent advantage, the temporary cause operating upon them must be sufficient to make a great change in their condition – a change such as will be felt for many years, notwithstanding any stimulus which it may give during one generation to the increase of people. When, indeed, the improvement is of this signal character, and a generation grows up which has always been used to an improved scale of comfort, the habits of this new generation in respect to population become formed upon a higher minimum, and the improvement in their condition becomes permanent (342).

The same phenomenon, though less conspicuous, is said (following Malthus) to have characterized the English case in the years 1715–65, a period experiencing a succession of remarkably good harvests: 'So considerable an improvement in the condition of the labouring class, though arising from an accident of seasons, yet continuing for more than a generation, had time to work a change in the habitual requirements of the labouring class; and this period is always noted as the date of "a marked improvement of the quality of the food consumed, and a decided elevation in the standard of their comforts and conveniencies" ' (343n).

However, such cases were rare, and Mill in a famous statement played down 'the repeal of the corn laws, considered merely as a labourers' question, or . . . any of the schemes . . . for making the labourers a very little better

---

104 These allusions can perhaps best be interpreted as inward shifts of the g_L-w relationship.

off. Things which only affect them a very little, make no permanent impression upon their habits and requirements, and they very soon slide back into their former state' (341–2). And even major improvements need not have the desired effect (cf. 188).

Drawing again upon illustrations from the British and French cases, Mill alluded also to the consequences of changes in the minimum standard unrelated to the actual experience of higher wages:

> Every advance they make in education, civilization, and social improvement, tends to raise this standard; and there can be no doubt that it is gradually, though slowly, rising in the more advanced countries of Western Europe. Subsistence and employment in England have never increased more rapidly than in the last forty years, but every census since 1821 showed a smaller proportional increase of population than that of the period preceding; and the produce of French agriculture and industry is increasing at a progressive ratio, while the population exhibits in every quinquennial census, a smaller proportion of births to population (159; on the French data see also 152, 287f., 342).

The reference to 40 years of particularly rapid growth of labour demand was inserted in 1862 (correcting appropriately the figure in the two preceding editions, 1852 and 1857); but the manuscript and the first two editions (1848 and 1849) read '16 years' placing the rapid increase in 'subsistence and employment' only from the early 1830s, and suggesting a realization in about 1850 that the particularly striking upward trend in capital accumulation commenced earlier (dating at least to the early 1820s) than Mill had originally imagined. That the census data revealed a declining population growth rate since 1821 appears unchanged throughout all editions.

The recognition of a slackening in the growth rate of population relative to that of capital, and this over an extended period of at least four decades, is striking and comes as a surprise considering a statement in the same context that attempts to raise living standards are 'rarely found' in the labouring classes, and that 'hitherto' restraints on population expansion had reflected at best the insistence upon unchanged standards but no more (157; cf. above, p. 442). What reasons can be offered for Mill's apparent hesitation to allow formally for the evidence of secular real wage increase.[105] To the extent that the improvements are unrelated to prior

---

105 Cf. Samuelson (1978), 1417: 'Even Mill is not realistic enough in his modelling of innovation and the lagging supply of population in advanced economies. What observers like Kuznets have observed this past century is that the growth of technology has been enough *to keep the real wage growing at something like an exponential rate*, with the growth in population and saving not being fast enough to wipe out the rising trend in real wages.'

wage increases in standards there is, strictly speaking, no problem; cultural or sociological changes of this nature are on a par with technological changes and it is legitimate to set them aside for some analytical purposes. Setting aside technological change also allows a focus upon land scarcity; setting aside cultural change allows a focus on labour reproducibility. In any event the contemporary British case indicated only a 'gradual' change in the minimum standard. Changes induced by the magnitude of the wage are more problematic. But even here the problem is only apparent. Mill does not, after all, propose a regular functional relation whereby a rising real wage acts to raise the minimum standard; the wage increase must be a discrete one and significantly so, and such cases were few and far between.

Most important, however, is Mill's observation that prudential behaviour directed towards the achievement of higher wages characterized only skilled labourers in the British case. For the high earnings of the unskilled town workers had not encouraged a noticeable alteration in their conception of the minimum standard *so that future slackening in the growth rate of capital would necessarily lead to a fall in wages to the level of the common farm labourer.*[106] This was the ever-present shadow, and it is scarcely surprising that Mill minimized labour scarcity – albeit manifested in the class of skilled workers – under these conditions.

The shadow was also cast over the agricultural sector where to the extent wages had been kept up above subsistence, it was due to the attraction of labour into the towns, and without which attraction wages were likely to fall into the Irish level:

If the growth of the towns, and of the capital there employed, by which the factory operatives are maintained at their present average rate of wages notwithstanding their rapid increase, did not also absorb a great part of the annual addition to the rural population, there seems no reason in the present habits of the people why they should not fall into as miserable a condition as the Irish ['previous to 1846'; added 1862]; and if the market for our manufactures should, I do not say fall off, but even cease to expand at the rapid rate of the last fifty years, there is no certainty that this fate [a decline of wages to the Irish level] may not be reserved for us (351).

We must also keep in mind here that while 'the permanent remuneration of the labourers essentially depends on what we have called their habitual

---

106 In addition to the falling profit rate during the growth process and its effect on accumulation, various other disturbances are also relevant. See, for example, letter dated 28 Dec. 1867, *CW*, XVI, 1338–9: 'I do not believe that the [tax] burthens laid on the capitalist ever fall on labor, except in one way, viz: if the burthens are so heavy as to check the accumulation of capital, and prevent it from keeping up with the increase of population; in which case, without doubt, wages would fall. . . . '

standards' it was 'much more difficult to raise than to lower, the scale of living which the labourer will consider as more indispensable than marrying and having a family' (III, 727; II, 342). There was thus always the danger of a deterioration in the 'habitual standard' itself, a danger leading Mill to question 'all propositions ascribing a self-repairing quality to the calamities which befall the labouring classes':

There is considerable evidence that the circumstances of the agricultural labourers in England have more than once in our history sustained great permanent deterioration, from causes which operated by diminishing the demand for labour, and which, if population had exercised its power of self-adjustment in obedience to the previous standard of comfort, could only have had a temporary effect: but unhappily the poverty in which the class was plunged during a long series of years, brought that previous standard into disuse; and the next generation, growing up without having possessed those pristine comforts, multiplied in turn without any attempt to retrieve them (341).[107]

XIV 'THE TENDENCY OF PROFITS TO A MINIMUM'

On either version – the 'standard' or the 'prudential' – the profit rate declines until the minimum is achieved. But this is a *tendency* only. The chapter spelling out the path of profits (Book IV, Ch. iv) provides a splendid illustration of the sense of the notion 'tendency' to refer not to a necessarily observable trend but rather to one force amongst other and possibly conflicting forces on a particular variable:

We now arrive at the fundamental proposition which this chapter is intended to inculcate. When a country has long possessed a large production, and a large net income to make savings from, and when, therefore, the means have long existed of making a great annual addition to capital; (the country not having, like America, a large reserve of fertile land still unused;) it is one of the characteristics of such a country, that the rate of profit is habitually within, as it were, a handsbreadth

107 According to one formulation not even technical improvement at its most rapid conceivable rate could, in 'old' countries, prevent a fall in real wages except a population growth restrained below the physiological maximum. Thus though 'improvement may during a certain space of time keep up with, and even surpass, the actual increase of population' (190), as indeed was the case in Britain where technical advance (even prior to corn law repeal) had 'so materially lightened, for the time being, the pressure of population upon production', yet 'it assuredly never comes up to the rate of increase of which population is capable; and nothing could have prevented a general deterioration in the condition of the human race, were it not that population has in fact been restrained.' Here he may have been somewhat carried away for it appears that there had in fact been no restraint on the supply of unskilled labour in the towns, and yet wages had not declined; see above.

of the minimum, and the country therefore on the very verge of the stationary state. By this I do not mean that this state is likely, in any of the great countries of Europe, to be soon actually reached, or that capital does not still yield a profit considerably greater than what is barely sufficient to induce the people of those countries to save and accumulate. My meaning is, that it would require but a short time to reduce profits to the minimum, if capital continued to increase at its present rate, and no circumstances having a tendency to raise the rate of profit occurred in the meantime. The expansion of capital would soon reach its ultimate boundary, if the boundary itself did not continually open and leave more space (*CW*, III, 738–9).

It must always have been seen, more or less distinctly, by political economists, that the increase of wealth is not boundless: that at the end of what they term the progressive state lies the stationary state, that all progress in wealth is but a postponement of this, and that each step in advance is an approach to it. We have now been led to recognise that this ultimate goal is at all times near enough to be fully in view; that we are always on the verge of it, and that if we have not reached it long ago, it is because the goal itself flies before us. The richest and most prosperous countries would very soon attain the stationary state if no further improvements were made in the productive arts, and if there were a suspension of the overflow of capital from those countries into the uncultivated or ill-cultivated regions of the earth (752).

In the contemporary British case: 'the mere continuance of the present annual increase of capital, if no circumstance occurred to counteract its effect, would suffice in a small number of years to reduce the rate of net profit to one per cent'. Similarly: 'in such a country as England, if the present annual amount of savings were to continue, without any of the countervailing circumstances which now keep in check the natural influence of those savings in reducing profit, the rate of profit would speedily attain the minimum, and all further accumulation of capital would . . . cease' (741).

Mill then specifies more fully the conditions for a satisfactory test of the theory – 'to fulfil the conditions of the hypothesis':

To fulfil the conditions of the hypothesis, we must suppose an entire cessation of the exportation of capital for foreign investment. No more capital sent abroad for railways or loans; no more emigrants taking capital with them, to the colonies, or to other countries; no fresh advances made, or credits given, by bankers or merchants to their foreign correspondents. We must also assume that there are no fresh loans for unproductive expenditure, by the government, or on mortgage, or otherwise; and none of the waste of capital which now takes place by the failure of undertakings which people are tempted to engage in by the hope of a better income than can be obtained in safe paths at the present habitually low rate of profit. We must suppose the entire savings of the community to be annually invested in really productive employment within the country itself; and no new channels opened by industrial inventions, or by a more extensive substitution of the best known processes for inferior ones (739).

A short while later a further condition is added: 'In old countries like England, if, in addition to supposing all improvement in domestic agriculture suspended, we suppose that there is no increased production in foreign countries for the English market, the fall in profits would be very rapid' (740).

What assumption is being made regarding population growth and the wage rate? A slower growth rate of population than capital would speedily reduce the rate of profit assuming given technology – and also, Mill adds, assuming a given distribution of productive and unproductive labour (yet a further condition added to the *ceteris paribus* basket): '[a]n augmentation of capital, much more rapid than that of population, must soon reach its extreme limit, unless accompanied by increased efficiency of labour (through inventions and discoveries, or improved mental and physical education), or unless some of the idle people, or of the unproductive labourers, became productive'. But even '[i]f population did increase with the increase of capital, and in proportion to it the fall of profits would still be inevitable' in consequence of the rising cost of agricultural produce. Any such decline, Mill proceeds to clarify, would be 'retarded' – though not prevented – in the event of a falling commodity wage – a possibility that is not much emphasized:

If both these avenues to an increased supply of food were closed [domestic technological change, easier food imports], and population continued to increase as it is said to do, at the rate of a thousand a day, all waste land which admits of cultivation in the existing state of knowledge would soon be cultivated, and the cost of production and price of food would be so increased, that, if the labourers received the increased money wages necessary to compensate for their increased expenses, profits would very soon reach the minimum. The fall of profits would be retarded if money wages did not rise, or rose in a less degree; but the margin which can be gained by a deterioration of the labourers' condition is a very narrow one: in general they *cannot* bear much reduction; when they can, they have also a higher standard of necessary requirements, and *will* not (740–1; 'cannot' italicized through ed. 1865).

That wages are above subsistence in the strict sense of the term is obvious – how else could population be growing at 'the rate of a thousand a day'?

The account is complicated in one respect. In the passages cited above a constant rate of accumulation is assumed ('if capital continued to increase at its present rate', 'if the present annual amount of savings were to continue'). Nothing is formally said of the deceleration in accumulation in the face of falling profits (and thus the necessary deceleration in population growth that must occur for wage constancy to be assured). There is, however, a reason for this apparent lacuna.

As we shall see Mill had much to say in the present context regarding 'commercial revulsions' (741f.) envisaged as one of the 'countervailing circumstances' which in actuality check the downward trend in profits – that a portion of the annual savings are, so to speak, lost. Notwithstanding, major net accumulation characterized contemporary English conditions – the fact that capital 'does increase greatly and rapidly' and, similarly, population. Indeed, the most striking feature of Mill's account of British conditions over several decades (at least in the manufacturing sector) is constancy of the profit rate – the 'tendency of profits to a minimum' was not manifest despite rapid accumulation and rapid population growth – and constancy of wages (perhaps even an upward trend) at a level above 'subsistence':

[Rapid accumulation] is shown by the increasing productiveness of almost all taxes, by the continual growth of all the signs of national wealth, and by the rapid increase of population, while the condition of the labourers is certainly not declining, but on the whole improving [1867; till 1862: certainly is not on the whole declining]. These things prove that each commercial revulsion, however disastrous, is very far from destroying all the capital which has been added to the accumulations of the country since the last revulsion preceding it, and that invariably, room is either found or made for the profitable employment of a perpetually increasing capital, consistently with not forcing down profits to a lower level (742).

The *actual* circumstances thus stood in sharp contrast with the *theoretical* in this respect, that both capital and population were proceeding at a very rapid (roughly constant) pace with the profit rate unchanged over time and the wage rate at worst unchanged and perhaps rising. All this the result of a variety of 'counteracting circumstances' including new technology.[108] Now in the absence of such circumstances even the *same* rate of factor growth as experienced in actuality – *and no greater* – must 'soon' force profits to the minimum. Although in his chapter on the profit rate trend Mill did not focus on the deceleration in accumulation that can be expected as profits fall (and in population growth if the wage was to be maintained) there is

---

108 We must presume the innovations, albeit in cotton, to be (indirectly) land augmenting; cf. Samuelson (1978), 1416: 'Mill went on to emphasize that technological innovation, continued in the long-run steady state, would imply *rising output forever*; we can show on Mill's behalf that, if the technical change is *land-augmenting* at a steady exponential rate, then labor and capital will grow forever at the same balanced exponential rate, just enough to match the growth of land measured in "efficiency units" and with the long-run wage rate and profit rate each just high enough above their respective bare minima to elicit the implied growth rates of the factors'.

no reason to believe that he turned his back on his own repeatedly reiterated principle that accumulation is a regular function of the return on capital. (In fact the savings function is again restated, 747.)

We conclude this section with a brief statement by Mill encapsulating the argument, namely that land scarcity alone prevented population and capital growth from proceeding at going rates of return on a steady-state growth path: 'The progress of accumulation would no doubt be considerably checked, if the returns to capital were to be reduced still lower than at present. But why should any possible increase of capital have that effect . . . The limitation to production, not consisting in any necessary limit to the increase of the other two elements, labour and capital, must turn on the properties of the only element which is inherently, and in itself, limited in quantity' (II, 172). Needless to say, by technical progress (or easier importation) even land scarcity might be overcome to allow for ongoing factor growth without pressure on the returns – the contemporary situation precisely.

XV SECULAR TREND AND CYCLE

The regularity of cyclical fluctuations was a matter already emphasized in an early monetary paper composed in 1830 (see below, pp. 498–508). There it is asserted that 'except during short periods of transition, there is almost always either great briskness of business or great stagnation'. In the *Principles* Mill reiterates of speculative periods that 'all times are so, more or less' (*CW* III, 512). But now more attention is paid to the 'quiescent' period itself and its place in the cycle. In the chapter 'Of the Regulation of a Convertible Paper Currency' (662–3) Mill asserts that during the quiescent state 'nothing [tends] to engender in any considerable portion of the mercantile public a desire to extend their operations. The producers produce and the dealers purchase only their usual stocks, having no expectation of a more than usually rapid vent for them'. In this account he contrasts quiescence with the 'unusual extension' of the speculative state where there exists some stimulus which 'exciting more than usual hopes of profit, gives increased briskness to business'; and by the term 'unusual extension' is intended the exogenous disturbances such as short crops, import restrictions, new foreign markets so much emphasized by Tooke and Fullarton. But there is a further issue which Mill now emphasizes with implications for the relationship between the process of growth and the cycle – the fact that inquiescence itself allows for expansion: 'Each person transacts his ordinary amount of business, and no more; or increases it only in correspondence with the increase of his capital or connexion, or with the gradual growth of the demand for his commodity, occasioned by the public prosperity.'

That a quiescent period entails expansion rather than stationarity in fact constitutes the necessary condition for the generation of regular cyclical fluctuations. Here we allude to the cyclical consequences flowing from the downward 'tendency' of the profit rate.[109] And as we shall see the relationship is a mutual one, for while the profit rate trend engenders speculation and the cycle, various capital losses associated with the cycle play back on the profit rate itself.

That cyclical fluctuations are a direct outcome of the falling profit rate had already been alluded to in the essays of 1826 and 1844 (below, pp. 497, 512).[110] The notion appears in the *Principles* with particular emphasis on the *periodicity* of cycles. Thus in the chapter on the 'Rate of Interest' (Book III, Ch. xxiii):

Except at such periods [of 'speculation' and 'revulsion'], the amount of capital disposable on loan is subject to little other variation than that which arises from the gradual process of accumulation: which process, however, in the great commercial countries is sufficiently rapid to account for the almost periodic recurrence of these fits of speculation; since when a few years have elapsed without a crisis, and no new and tempting channels for investment have been opened in the meantime, there is always found to have occurred in those few years so large an increase of capital seeking investment, as to have lowered considerably the rate of interest, whether indicated by the prices of securities or the rate of discount on bills; and this diminution tempts the possessor to incur hazards in hopes of a more considerable return (651).

The references to additions to capital 'seeking investment' or variation in 'the amount of capital on loan' suggest accumulations available for

109 Cf. Link (1959), 178: 'The cycle exists [for Mill] primarily because of the secular trend.' Moreover, while 'the idea of an endogenous cycle, with distinct phases, one merging into another in semiautomatic fashion, was not entirely new . . . the concept seems more developed by Mill than in the case of any earlier writer' (149).

James Wilson, who contributed much to the early analysis of sources of instability, emphasized rather exogenous disturbances. On his discussion of agricultural sources of instability, see Link (1959), 104–14. On his treatment of the 1847 crisis, which applies Ricardo's doctrine of the conversion of circulating into fixed capital to the railway boom, see *ibid.*, 154 and Boot (1983). Although Mill in effect approves of Wilson's analysis (*CW*, II, 97–8, III, 543) it is none the less a secondary consideration in his approach to cycles.

110 See also 'Civilization' (1836), XVIII, 136 for allusions to the Marx-like consequences of a low return on capital for industrial structure: 'In Great Britain especially (which so far surpasses the rest of the world in the extent and rapidity of the accumulation of wealth) the fall of profits, consequent upon the vast increase of population and capital, is rapidly extinguishing the class of small dealers and small producers, from the impossibility of living on their diminished profits, and is throwing business of all kinds more and more into the hands of large capitalists – whether these be rich individuals, or joint-stock companies formed by the aggregation of many small capitals.' (The reference to Britain surpassing 'the world' is qualified to read 'the old world' in 1859).

investment but not yet committed. A modification of the first sentence introduced in the edition of 1865, however, points to an actual expansion of capacity with a resultant fall in the return on capital due to upward pressure on 'proportional' wages in the Ricardian manner rather than a fall in interest merely because of the increased availability of loanable funds. ('In the intervals between commercial crises, there is usually a tendency in the rate of interest to a progressive decline, from the gradual process of accumulation: which process . . .')

In the chapter on the 'Tendency of Profits to a Minimum' itself the periodicity of depressions is again related specifically to the secular decline in profits:

that . . . revulsions are almost periodical, is a consequence of the very tendency of profits which we are considering. By the time a few years have passed over without a crisis so much additional capital has been accumulated, that it is no longer possible to invest it at the accustomed profit: and public securities rise to a high price, the rate of interest on the best mercantile security falls very low, and the complaint is general among persons in business that no money is to be made. Does this not demonstrate how speedily profits would be at a minimum, and the stationary condition of capital would be attained, if these accumulations went on without any counteracting principle? But the diminished scale of all safe gains, inclines people to give a ready ear to any projects which hold out, though at the risk of loss, the hope of a higher rate of profit; and speculations ensue, which, with the subsequent revulsions, destroy or transfer to foreigners, a considerable amount of capital, produce a temporary rise of interest and profit, make room for fresh accumulations, and the same round is recommenced (742).[111]

Now we have seen that the *ceteris paribus* conditions upon which the trend is predicated include constant technology, and the absence of capital 'losses' of various kinds. Let us consider Mill's treatment of these phenomena more closely, first those losses relating to cyclical fluctuations – the 'waste of capital

111 In the *Principles* Mill expressed some doubt whether social progress would 'moderate' those price fluctuations arising 'from miscalculation, and especially from the alternations of undue expansion and excessive contraction of credit, which occupy so conspicuous a place among commercial phenomena . . . Such vicissitudes, beginning with irrational speculation and ending with a commercial crisis, have not hitherto become less frequent or less violent with the growth of capital and extension of industry. Rather they may be said to have become more so: in consequence, as is often said, of increased competition; but, as I prefer to say, of a low rate of profits and interest, which makes capitalists dissatisfied with the ordinary course of safe mercantile gain' (III, 718).

The weight of emphasis regarding the factual question of instability altered somewhat during the 1860s. Cf. 'Chapters on Socialism' (c. 1869), V, 728; and evidence on 'Currency and Banking' (1867), V, 601, where Mill denied that the cyclical problem was worsening in terms either of depth or frequency.

in periods of over-trading and rash speculation, and in the commercial revulsions by which such times are always followed' (741).

The precise form of these wastages include unsustainable capital projects, during speculation periods, and 'unproductive' consumption during the depressions which follow:

It is true that a great part of what is lost at such periods is not destroyed, but merely transferred, like a gambler's losses, to more successful speculators. But even of these mere transfers, a large portion is always to foreigners, by the hasty purchase of unusual quantities of foreign goods at advanced prices. And much also is absolutely wasted. Mines are opened, railways or bridges made, and many other works of uncertain profit commenced, and in these enterprises much capital is sunk which yields either no return, or none adequate to the outlay. Factories are built and machinery erected beyond what the market requires, or can keep in employment. Even if they are kept in employment, the capital is no less sunk; it has been converted from circulating into fixed capital, and has ceased to have any influence on wages or profits. Besides this, there is a great unproductive consumption of capital, during the stagnation which follows a period of general over-trading. Establishments are shut up, or kept working without any profit, hands are discharged, and numbers of persons in all ranks, being deprived of their income, and thrown for support on their savings, find themselves, after the crisis has passed away, in a condition of more or less impoverishment.[112]

Here we have a supplementing of the inventory cycle of the essays by a cycle in fixed capital formation; as Link observes, 'Mill appears to have believed . . . that in most historical cycles, both factors were involved' (1960, 154).

Now we are faced by a methodological complexity. A 'tendency' which itself encourages the counteracting force renders entirely questionable the designation of the latter as a 'disturbing cause' bearing as it does a connotation of independence. This problem can be overcome if the response in question is to an historically low rate of return as distinct from an actual or contemporaneous decline. Although some of Mill's formulations do

---

112 Cf. Link (1959), 166–7: 'Thus there was a period, following the crisis, that involved a number of recognized features of business contraction: reduced employment, low profits (apparently meaning profits on sunk capital, not profits on circulating capital), and reduced production. The low profits on sunk capital and the low level of employment both force consumers to live on accumulated savings, so that savings are reduced. This seemed to be another way of saying that capital is consumed unproductively. As a result circulating capital shrinks and the rate of profit on circulating capital rises.' Mill attributes the notion of capital losses to Chalmers and Sismondi but disagrees with their interpretations of these events (574, 735). For they denied the 'law of markets' in its secular guise, a principle to which Mill always adhered.

indeed suggest that he had this in mind, there are others which point to actual declines in the return to which speculators are supposed to respond.[113]

There arises a related doctrinal problem. Mill, we know, had established that over recent decades the secular return on capital had remained unchanged at a little over 3 per cent. We must try then to appreciate to what his speculators are supposed to respond. Specifically, we must presume exogenous technical change to have sufficed to counteract only partially the force of land scarcity putting downward pressure on the return on capital which pressure induces wastages and a resumption of the roughly constant trend path (cf. Link, 178).[114]

Similar issues are raised by capital exportation which contributes to prevent profits 'from reaching the minimum' (*CW*, III, 745).[115] But here the emphasis is on an historically low level compared to the level ruling in foreign countries so that such outflows might legitimately be treated as 'disturbing causes': 'the perpetual overflow of capital into colonies or foreign countries, to seek higher profits than can be obtained at home [has] . . . been for many years one of the principal causes by which the decline of profits in England has been arrested . . . it does what a fire, or an inundation, or a commercial crisis would have done:

---

113 e.g. 739, 752 (cited above, p. 458), 746 (below, p. 465) in the first category and 742 (above, p. 463) in the second.

114 Some cases of speculation seem perhaps to be treated as 'exogenous', and as moderating the impact of a falling profit-rate trend. Thus 'the railway gambling of 1844 and 1845 probably saved the country from a depression of profits and interest, and a rise of all public and private securities, which would have engendered still wider speculations, and when the effects came afterwards to be complicated by the scarcity of food, would have ended in a still more formidable crisis than was experienced in the years immediately following' (750). But more likely Mill should be understood as saying that even the events of 1844–45 were stimulated by a low return on capital, keeping the short-run interest rate higher than otherwise and thus dampening the stimulus for subsequent bouts of speculation including, pre-eminently, that of 1847. What occurred is a spreading of the speculation over several years rather than its more dangerous concentration. At all events Mill concludes that '[t]he railway operations of the various nations of the world may be looked upon as a sort of competition for the overflowing capital of the countries where profit is low and capital abundant . . . The English railway speculations are a struggle to keep our annual increase of capital at home; those of foreign countries are an effort to obtain it'.

115 The importance of capital exportation was 'a fact now beginning to be recognized' – 'that the passage of the precious metals from country to country is determined much more than was formerly supposed, by the state of the loan market in different countries, and much less by the state of prices' (515). Following Göschen, Mill in 1865 further elaborated upon increased capital mobility: 'the loan market of the whole commercial world is rapidly becoming one. The rate of interest, therefore, in the part of the world out of which capital most freely flows, cannot any longer remain so much inferior to the rate elsewhere, as it has hitherto been' (652n).

it carries off a part of the increase of capital from which the reduction of profits proceeds' (746).[116]

As in our discussion of domestic losses of capital, it is apparently Mill's position that in the absence of such outflows the impact of technical progress would not have sufficed to balance that of land scarcity and allow 'a tolerably equal struggle against the downward tendency of profits' (741). At all events, the *net* outcome of the various conflicting tendencies actually at work, however designated, assures a constant trend path of the return on capital with fluctuations about it of a cyclical order.

Most striking however, are some of Mill's allusions in our present context to technical change. For the most part the argument proceeds as if technical change constitutes a purely exogenous disturbance:

[The] tendency of improvements in production to cause increased accumulation, and thereby ultimately to increase the gross produce, even if temporarily diminishing it, will assume a still more decided character if it should appear that there are assignable limits both to the accumulation of capital, and to the increase of production from the land, which limits once attained, all further increase of produce must stop; but that improvements in production, whatever may be their other effects, tend to throw one or both of these limits farther off. Now, these are truths which will appear in the clearest light in a subsequent stage of our investigation. It will be seen, that the quantity of capital which will, or even which can, be accumulated in any country, and the amount of gross produce which will, or even which can, be raised, bear a proportion to the state of the arts of production there existing; and that every improvement, even if for the time it diminish the circulating capital and the gross produce, ultimately makes room for a large amount of both, than could possibly have existed otherwise. It is this which is the conclusive answer to the objections against machinery; and the proof thence arising of the ultimate benefit to labourers of mechanical inventions even in the existing state of society, will hereafter be seen to be conclusive (II, 98–9; cf. III, 742–5).

Yet allowance is sometimes also made for induced technology:

When the capital accumulated is so great and the rate of annual accumulation so rapid, that the country is only kept from attaining the stationary state by the

116 Yet the inappropriateness of the notion 'disturbing cause' in our present context comes to the fore in the following passage: 'We have now seen that the lowness of profits is a proof that the spirit of accumulation is so active, and that the increase of capital has proceeded at so rapid a rate, as to outstrip the two counter-agencies, improvements in production, and increased supply of cheap necessaries from abroad: and that unless a considerable portion of the annual increase of capital were either periodically destroyed, or exported for foreign investment, the country would speedily attain the point at which further accumulation would cease, or at least spontaneously slacken, so as no longer to overpass the march of invention in the arts which produce the necessaries of life' (747).

emigration of capital, or by continual improvements in production; any circumstance [such as a profits tax] which virtually lowers the rate of profit cannot be without a decided influence on these phenomena. It may operate in different ways. The curtailment of profit, and the consequent increased difficulty in making a fortune or obtaining a subsistence by the employment of capital, may act as stimulus to inventions, and to the use of them when made. If improvements in production are much accelerated, and if these improvements cheapen, directly or indirectly, any of the things habitually consumed by the labourer, profits may rise, and rise sufficiently to make up for all that is taken from them by the tax (827).

True enough, Mill's specific concern is with a tax on profits and he does not here explicitly relate the innovative or inventive effort to the secular fall in the return on capital. But there is no reason why the secular decline itself should not act as a stimulus in the same way as an artificial reduction. In any event, Mill has in mind the case where the system is operating at a 'low' range of profit. The treatment of technical progress purely as a 'disturbing cause' is clearly inappropriate.[117] It is worth noting that Mill himself speaks of the profit-technology relation as a 'tendency', thus according it the same status, with all its implications, as he accords the pressure on profits of scarce land.[118]

XVI SECULAR TREND AND THE LAW OF MARKETS

The deductions for practice drawn from the analysis in Book V, Ch. v; 'Consequences of the Tendency of Profits to a Minimum', are remarkable: 'The theory of the effect of accumulation on profits, laid down in the preceding chapter, materially alters many of the practical conclusions which might otherwise be supposed to follow from the general principles of Political Economy, and which were, indeed, long admitted as true by the highest authorities on the subject' (*CW*, III, 747). Specifically, in the absence of capital loss of various kinds, the rate of accumulation would be so great (on Mill's empirical estimate) as to force down the return on capital since technical progress could not in practice be relied upon to counteract such heavy pressure on scarce land. The first deduction Mill draws from the fact of a highly active 'spirit of accumulation' is that 'a sudden abstraction of capital, unless of inordinate amount', need not be feared, for 'after a

---

117 There is a suggestion here of the notion of Leibenstein's 'X-efficiency' conception (1966). Only under pressure do firms adopt or approach already known optima. See also 739 regarding the adoption of known techniques (above p. 458).

118 '[T]he artificial abstraction of a portion of profits would have a real tendency to accelerate improvements in production, no considerable improvements might actually result, or only of such a kind as not to raise general profits at all, or not to raise them so much as the tax had diminished them'.

few months or years, there would exist in the country just as much capital as if none had been taken away'. At most 'the abstraction, by raising profits and interest, would give a fresh stimulus to the accumulative principle, which would speedily fill up the vacuum' (747–8). But more likely 'the only effect that would ensue, would be that for some time afterwards less capital would be exported, and less thrown away in hazardous speculation'.

The conclusion altered the perspective towards government expenditure. The standard warnings by orthodox writers against measures which might reduce the capital stock, or its rate of accumulation, were no longer pertinent. In discussing a case for increased government spending – within moderation be it noted – Mill writes indeed as if capital is no longer to be treated as a scarce factor:

In the first place, then, this view of things greatly weakens, in a wealthy and industrious country, the force of the economical argument against the expenditure of public money for really valuable, even though industriously unproductive, purposes. If for any great object of justice or philanthropic policy, such as the industrial regeneration of Ireland, or a comprehensive measure of colonization or of public education, it were proposed to raise a large sum by way of loan, politicians need not demur to the abstraction of so much capital, as tending to dry up the permanent sources of the country's wealth, and diminish the fund which supplies the subsistence of the labouring population. The utmost expense which could be requisite for any of these purposes, would not in all probability deprive one labourer of employment, or diminish the next year's production by one ell of cloth or one bushel of grain. In poor countries, the capital of the country requires the legislator's sedulous care; he is bound to be most cautious of encroaching upon it, and should favour to the utmost its accumulation at home, and its introduction from abroad. But in rich, populous, and highly cultivated countries, it is not capital which is the deficient element, but fertile land; and what the legislator should desire and promote, is not a greater aggregate saving, but a greater return to savings, either by improved cultivation, or by access to the produce of more fertile lands in other parts of the globe. In such countries, the government may take any moderate portion of the capital of the country and expend it as revenue, without affecting the national wealth: the whole being either drawn from that portion of the annual savings which would otherwise be sent abroad, or being subtracted from the unproductive expenditure of individuals for the next year or two, since every million spent makes room for another million to be saved before reaching the overflowing point. When the object in view is worth the sacrifice of such an amount of the expenditure that furnishes the daily enjoyments of the people, the only well-grounded economical objection against taking the necessary funds directly from capital, consists of the inconveniences attending the process of raising a revenue by taxation, to pay the interest of a debt (748).

The brief reference in this passage to government funding of emigration in fact refers to one of the most important of Mill's deductions: 'The same considerations enable us to throw aside as unworthy of regard, one of the common arguments against emigration to the colonies as a means of relief for the labouring classes.[119] Again, Mill proceeds to insist that the capital required for 'the most extensive colonization' was likely to be moderate (748–9); adding however that even in the highly unlikely event of an absolute decline in the domestic capital stock, the fresh stimulus afforded savings would rapidly make up the deficiency.[120]

119 Mill may have had his father in mind (for James Mill's hostility towards colonization see 'Colonies', 1824, 262). Ricardo too might have been a candidate although he wrote little on the issue. Ricardo supported Wilmot-Horton's plan for assisted emigration but only given the ruling system of poor relief; otherwise he was apparently opposed (Sraffa, *Works*, XI, xvi).

J. S. Mill's support for colonies was expressed in the press and journals of 1834 with reference to the proposals by Wakefield, Spring Rice and others for a South Australian project. The Ricardian growth model involving relative labour–land scarcity stands at the core: the project was expected 'to afford a sensible relief to the overcrowded labourers and capitalists of the mother country' (*The Examiner*, 29 June, 403); 'whatever relief it can allow to the pressure of population against subsistence in our own country, will be clear gain – pure, unalloyed good' (*ibid.*, 6 July, 420); '[I]n England land is the deficient element; labour and capital, relatively to land, are both in excess . . . low wages and low profits are the consequence' (*ibid.*, 20 July, 453). (For a further reference at the same period to a low return on capital see 'Civilization' 1836, *CW*, XVIII, 136.)

As explained in the *Examiner* for July 6, the expenses of transportation would be financed out of the surplus generated in the colonies: '[I]f a portion of our labourers could be removed from the country, where they are now earning a scanty and precarious subsistence, and placed in a new and fertile country, under the best arrangements which could be desired for giving the greatest possible productiveness to their labour, the surplus of what they would there produce, above what they can produce in their present situation, would form a fund sufficient, in a year or two at farthest, to repay with interest the whole expense of their emigration' (419).

On this plan there would be no drain on domestic savings to finance the costs of transportation – apart from that implied by a loan (against the security of the future funds) required to finance the *initial* flow of emigration to establish the colony. The colonial project itself was, however, expected to involve capital export: the prospect of available labour in the colonies 'induces capitalists to emigrate' (419). Yet no objection is made to the scheme on that score. This early position is in line, therefore, with that of the *Principles* although the full logic is not spelled out.

120 This perspective dictated Mill's response to critics of his proposal in favour of the heavy taxation of legacies and inheritances which (following Ricardo) he conceded are 'taxes on capital'. This was not an objection in a country 'where capital abounds, and the spirit of accumulation is strong'. He responded similarly to proposals to pay off the national debt. See also Mill's evidence in 1852 before the Select Committee on the Income and Property Tax (V, 493–4). Although he favoured savings over consumption expenditure, on the grounds that by savings a 'fund in perpetuity [is created] for maintaining labour', a still better choice would be endowing a school. This evaluation might perhaps bear on the present issue.

The issue is also graphically formulated in a comment on fixed-capital construction: 'Since even the emigration of capital, or its unproductive expenditure, or its absolute waste, do not in such a country, if confined within any moderate limits, at all diminish the aggregate amount of wages fund' – the actual and potential increase in capital being so great – 'still less can the mere conversion of a like sum into fixed capital, which continues to be productive, have that effect. It merely draws off at one orifice what was already flowing out at another; or if not, the greater vacant space left in the reservoir does but cause a greater quantity to flow in . . . the sums so applied [e.g. to railways] are mostly a mere appropriation of the annual overflowing which would otherwise have gone abroad, or been thrown away unprofitably, leaving neither a railway nor any other tangible result' (749–50).[121]

* * *

The question immediately arises whether Mill's favourable attitude towards expenditure of public money 'for really valuable, even though industrially unproductive purposes', has genuine Keynesian overtones. Was he, in brief, advocating 'government spending of redundant capital (idle funds) to offset a secular decline in the profit rate (marginal efficiency of capital)' (Petrella, 1970, 162n). This argument has been very strongly asserted:

Mill is here contrasting the economics of new underdeveloped countries with those applicable to advanced industrialized societies, such, apparently, as Britain in 1848. For the latter, Mill appears to cast doubt on Adam Smith's profoundly influential doctrine on the unconditional benefits of parsimony, which Mill had previously seemed to hail with such enthusiasm as destroying the erroneous view (of Malthus, Sismondi, and the 'Mercantilists') that capital might possibly accumulate too fast. Furthermore, Mill refutes, by implication, the dogma, propagated by Ricardo, which later came to be called the 'Treasury View', of a fixed capital fund and that employment on public works would simply be subtracted from employment in private industry (Hutchison, 1953, 354).

Hutchison concludes that 'Mill's earlier dogmatic castigations of Malthus and eighteenth-century ideas become all the more obviously regrettable in the light of this later chapter in the *Principles*'.

Clearly no issue can be of greater importance for an appreciation of Mill's macro-economics. The general issue is, of course, central to his entire perspective on the law of markets from a secular perspective: 'The point

121 Further applications will be found in the discussions of capital in the first book (II, 75–6).

is fundamental; and difference of opinion on it involves radically different conceptions of Political Economy, especially in its practical aspect. On the one view, we have only to consider how a sufficient production can be combined with the best possible distribution; but on the other there is a third thing to be considered – how a market can be created for produce, or how production can be limited to the capabilities of the market' (*CW*, III, 575). But the point to note is that in our present context itself, precisely the same strong position is reiterated – the downward 'tendency' of the profit rate was at all costs to be divorced from any notion of lack of markets:

Few persons would hesitate to say, that there would be great difficulty in finding remunerative employment every year for so much new capital, and most would conclude that there would be what used to be termed a general glut; that commodities would be produced, and remain unsold, or be sold only at a loss. But the full examination which we have already given to this question ('Of Excess of Supply'), has shown that this is not the mode in which the inconvenience would be experienced. The difficulty would not consist in any want of a market. If the new capital were duly shared among many varieties of employment, it would raise up a demand for its own produce, and there would be no cause why any part of that produce should remain longer on hand than formerly. What would really be, not merely difficult, but impossible, would be to employ this capital without submitting to a rapid reduction in the rate of profit (739–40).

The chapter on the downward trend in profits opens indeed with a severe criticism of the Smithian doctrine regarding 'competition of capitals' allowing only that Smith was on 'the very verge of grasping the complete theory of the subject' in the chapters on colonies (734–5).

Consistently with this view we have also seen that Mill distinguished between established 'general principles of Political Economy' and the practical conclusions drawn therefrom by 'the highest authorities', insisting that his new position related to the latter. We would in fact be faced with a remarkable situation indeed were it the case that while insisting upon adherence to orthodoxy on matters of principle Mill proceeded to reject that very orthodoxy.

However, Mill's general argument is far more consistent than might appear at first sight.[122] The potential problem, as we have seen, was excessive capital accumulation forcing down the return on capital *in the Ricardian fashion* – excessive in the sense that the pressure on land exceeds the counteracting force of new technology. Such a decline in the return was in practice temporary, however, in consequence of capital losses. Those losses include ill thought out 'speculative' additions to the real capital stock

---

122 Kittrell (1966), 349, has phrased the matter nicely: 'if the classical system was not monolithic . . . [at least] it was not schizoid'.

which prove untenable in quiescent periods – the speculation induced to some degree by the fall in the profit rate – and the drawing down of saved funds for consumption purposes in depression, the inevitable sequel to speculative periods. To this extent there is no question of leakages from the income stream by the non-investment of savings: savings are lost in the sense only of being unproductively used up. Mill's allowance for higher government spending thus amounts to a recommendation to tap the flow of savings, thereby preventing their excessive accumulation, the pressure on land and the fall in the return on capital – the Ricardian consequences – but also the various cyclical consequences of that fall which include wastage of capital. In effect, Mill was calling for opera houses in place of a superfluous network of railways and 'unproductive' private consumption. This is not a 'Keynesian' secular perspective.

That Mill recognized excess capacity and excess supplies of labour and commodities in depression we shall demonstrate. But government expenditures were certainly not envisaged as counter-cyclical measures, for cyclical depression was seen to be self-corrective.[123] So from this perspective too, we are not in a Keynesian world. Only indirectly would government spending be effective, for by imposing a floor to the return on capital it checks the 'speculative fever' from which depression ultimately proceeds.

Apart from capital 'losses' of the order discussed thus far, attention is, it is true, also given to genuine leakages from the expenditure stream, in the form of exorbitant payments for imports during speculative periods, and – more important for us – capital exportation. But here it is essential to distinguish outflows due to a relatively low domestic return on capital, from those due to an actual decline thereof. As for the latter, to the extent that government spending absorbs savings that would otherwise be invested in domestic ventures with a negative effect on profits – the reduction reflecting a rise in proportionate wages in the Ricardian manner – outflows abroad are actually discouraged. Government spending can be said to involve the absorption of funds otherwise leaking from the system only in so far as capital export reflects a relatively low domestic level of profits and is 'actually' rather than 'potentially' underway.

This is the full extent of Mill's abandonment of the treasury view. Yet his formal adherence to the full-fledged law of markets (at the secular level) can easily be appreciated. For that doctrine applies, strictly speaking, to a closed economy. There is no reason to believe that Ricardo (or even James Mill) would have denied that once allowance is made for international capital flows a deviation between rates of return might generate savings which

---

123 Although Bank of England monetary intervention was invoked, as we shall see, to help the process.

are lost to the domestic economy.[124] We shall return to this matter presently.

We must also recall Mill's evaluation of contemporary circumstances. There was no question of the leakages abroad generating actual problems for the maintenance of activity. The rate of net domestic investment was positive and markedly so, capital export entailing not an absolute reduction in domestic employment but merely a restraint on its growth rate, so that increased government spending was not conceived as absorbing otherwise unemployed resources. Applying the principles of Mill's first book to include the increased requirements for 'unproductive' workers in the public sector, such expenditures would assure at most a rather faster growth rate in labour demand.[125] But their primary object was not the employment effect at all, but rather the desirable public projects that could be undertaken, in circumstances allowing a simultaneous expansion of the private sector.

Circumstances might be contemplated wherein capital exportation carried with it ominous implications. This would be so in a stationary state (as far as concerns net domestic investment) should savings continue to be made only to flow abroad. Mill himself points to a case where a floor to the return on domestic capital is imposed by returns abroad above the minimum rate corresponding to zero net savings such that 'all further accumulations would go abroad':

In countries which are further advanced in industry and population, and have therefore a lower rate of profit, than others, there is always, long before the actual minimum is reached, a practical minimum, viz. when profits have fallen so much below what they are elsewhere, that, were they to fall lower, all further accumulations would go abroad. In the present state of the industry of the world, when there is occasion, in any rich and improving country, to take the minimum of profits at all into consideration for practical purposes, it is only this practical minimum that needs be considered. As long as there are old countries where capital increases very rapidly, and new countries where profit is still high, profits in the old countries will not sink to the rate which would put a stop to accumulation; the fall is stopped at the point which sends capital abroad. It is only, however, by improvements in production, and even in the production of things consumed by labourers, that the capital of a country like England is prevented from speedily reaching that degree of lowness of profit, which would cause all further savings to be sent to find employment in the colonies, or in foreign countries (746).

124 On Ricardo's position see Sraffa, *Works*, I, 136-7, 247-8, 396-7, *CW*, IV, 16n, V, 32-3, 38-9, 187-8, VII, 171, VIII, 358.

125 They would not even have that consequence if the capital losses attached to the speculative and depression stages of the cycle were correspondingly reduced.

But, to repeat, this was not the issue in contemporary Britain where net domestic investment was proceeding rapidly, the return on capital held roughly constant above the 'practical' minimum.

Mill's 'failure' to investigate the full implications of capital exportation from a 'Keynesian' perspective can thus be largely appreciated as a reflection of a buoyant environment. In fact his thoroughly optimistic outlook extends far beyond the account already given of the maintenance of the return on capital notwithstanding rapid net domestic investment at a constant level above even the 'practical' minimum. It is true that he introduced a moderately concerned tone in discussing the implications of corn law repeal – 'how far this resource can be counted upon, for making head during a very long period against the tendency of profits to decline as capital increases' having in mind the maintenance of a population 'increasing at its present rate' (745). European agriculture was technologically backward; colonial and American agriculture had already adopted the latest advances; extension of agriculture in Europe would be a slow process; and America's own population growth equalled the rate of expansion of its agricultural produce: 'This limited source of supply', Mill concludes with reference to foreign imports, 'unless great improvements take place in agriculture, cannot be expected to keep pace with the growing demand of so rapidly increasing a population as that of Great Britain; and if our population and capital continue to increase with their present rapidity, the only mode in which food can continue to be supplied cheaply to the one is by sending the other abroad to produce it'. But clearly any 'pessimism' on Mill's part regarding the future is highly conditional: first, on a presumption of an unchanged population growth rate, whereas a central policy objective was to assure a reduction thereof by prudential checks and emigration; secondly, on a presumption of sluggish technological advance, whereas nothing in recent experience, as Mill himself evaluated the evidence, justified that expectation; and thirdly, on a neglect of the full implications of capital exportation; and thirdly, on a neglect of the full applications of capital exportation whether to the colonies or elsewhere.

* * *

A further word regarding capital exportation, with special reference to the colonies. One aspect of Mill's support for colonial activity turns on the check to any reduction in the domestic profit rate due to pressure on scarce land by providing a source of cheap wage goods:

[T]he capital so carried off is not lost, but is chiefly employed either in founding colonies, which become large exporters of cheap agricultural produce, or in extending and perhaps improving the agriculture of older communities. It is to the emigration of English capital, that we have chiefly to look for keeping up a supply of cheap food and cheap materials of clothing, proportional to the increase of our population; thus enabling an increasing capital to find employment in the country, without reduction of profit, in producing manufactured articles with which to pay for this supply of raw produce. Thus, the exportation of capital is an agent of great efficacy in extending the field of employment for that which remains: and it may be said truly that, up to a certain point, the more capital we send away, the more we shall possess and be able to retain at home (746).

For this positive outlook towards capital exportation in circumstances of buoyant capital supply Mill paid warm tribute to a paper of 1826 by William Ellis.[126] Mill's formulation of the positive advantages of the outflow – 'up to a certain point, the more capital we send away, the more we shall possess and be able to retain at home' – was a straightforward paraphrase of Ellis who had written that 'the act of exporting capital would be the means of increasing the capital of the country from which it was exported' (1826, 129), a reference to the improved terms of trade whereby imported wage goods from newly developed areas might be acquired in exchange for manufactured exports and which by acting positively on the domestic return on capital would encourage renewed net investment in a country otherwise in a state of stationarity in consequence of the pressure of population on scarce land.

In the latter regard, as well as with respect to the basic forces acting on the profit rate, Ellis was thoroughly Ricardian, at no time relating the downward trend to a failure of markets. Mill's approval is easy to appreciate although he himself was not immediately concerned with stationarity. But what of his tribute, albeit rather more qualified, in the same context, to Gibbon Wakefield – whose position on the 'field of employment' for capital is often associated with a Smithian perspective on the falling profit-rate trend?

The first point to note is that, on Mill's reading, Wakefield adopted a full-fledged Ricardian position; his only complaint was that Wakefield failed to appreciate that his doctrines indeed were simply 'corollaries' from 'the principles of the best school of preceding political economists':

126 Cf. 736: 'The most scientific treatment of the subject which I have met with, is in an essay on the effects of Machinery, published in the *Westminster Review* for January 1826, by Mr. William Ellis; which was doubtless unknown to Mr. Wakefield, but which had preceded him, though by a different path, in several of his leading conclusions. This essay excited little notice, partly from being published anonymously in a periodical, and partly because it was much in advance of the state of political economy at the time. In Mr. Ellis's view of the subject, the questions and difficulties raised by Mr. Wakefield's speculation and by those of Dr. Chalmers, find a solution consistent with the principles of political economy laid down in the present treatise.'

Mr. Wakefield's explanation of the fall of profits is briefly this. Production is limited not solely by the quantity of capital and of labour, but also by the extent of the 'field of employment'. The field of employment for capital is two-fold; the land of the country, and the capacity of foreign markets to take its manufactured commodities. On a limited extent of land, only a limited quantity of capital can find employment at a profit. As the quantity of capital approaches this limit, profit falls; when the limit is attained, profit is annihilated; and can only be restored through an extension of the field of employment, either by the acquisition of fertile land, or by opening new markets in foreign countries, from which food and materials can be purchased with the products of domestic capital. These propositions are, in my opinion, substantially true; and, even to the phraseology in which they are expressed, considered as adapted to popular and practical rather than scientific uses, I have nothing to object. The error which seems to me imputable to Mr. Wakefield is that of supposing his doctrines to be in contradiction to the principles of the best school of preceding political economists, instead of being, as they really are, corollaries which, perhaps, would not always have been admitted by those political economists themselves (735–6; cf. also IV, 394).[127]

Mill may indeed have been guilty of neglecting that part of Wakefield's position that does have an underpinning in Smith's 'competition of capitals'. But Wakefield must take part of the blame for the confusion, since there is reason to believe that he misconceived the nature of the Ricardian doctrine he purported to counter. The extension of the field of employment by way of new markets as a means of raising the return on capital, in so far as it relates to improvements in the terms of trade in favour of British manufactures and the openings thus made (albeit indirectly) for an inflow of cheaper imports of wage goods, is thoroughly Ricardian, as Mill insisted. Ricardo had allowed the case himself.[128] Wakefield, however, may have

127 Cf. also 735: 'Mr. Wakefield, in his Commentary on Adam Smith [1843], and his important writings on colonization, takes a much clearer view of the subject [than Thomas Chalmers, 1832], and arrives, through a substantially correct series of deductions, at practical conclusions which appear to me just and important; but he is not equally happy in incorporating his valuable speculations with the results of previous thought, and reconciling them with other truths.' In a note attached to the text of 1848 but later dropped (742) Mill indicates that he was confused by Wakefield's position on improvements – whether they were included in his 'field of employment'. For references to Wakefield's extensive writings on colonies see the bibliography in Winch (1965), 174.

128 Ricardo, Works, II, 288–91. Malthus had argued in his Principles (1820) that if home demand for foreign food increases more rapidly than foreign demand for home manufactures, the terms of trade will turn unfavourable and the prices of manufactured goods would fall relative to costs. Ricardo, in his commentary, accepted this view and insisted on its consistency with his position: 'Profits in all countries must mainly depend

believed erroneously that the Ricardian opposition to Smith's process – Smith's very different process – whereby the discovery of new markets generates an increase in the return on capital, applied to this terms-of-trade case.[129]

There still remains to consider a complaint that Mill 'missed the main point', since Wakefield's fear of the advent of the stationary state included considerations which Ricardo did not have in mind, specifically the danger of savings running to waste in stationary conditions:

Ricardo and his associates feared the stationary state because they thought that accumulation would cease before the labouring classes had learnt the habits which would lift the equilibrium rate of wages above the bare subsistence level. Wakefield and Torrens feared it because they thought that, in stationary conditions, although there was no longer any profitable outlet for investment, the disposition to accumulate would continue and so lead to a general slackness in the economic system. The Ricardians feared the declining rate of profit because they thought it would stop accumulation, Torrens and Wakefield feared it because they thought that nevertheless accumulation would be attempted but would find no realization in investment (Robbins, 1958, 248).[130]

Lord Robbins, however, proceeds with great caution in his interpretation of Torrens (and Wakefield) conceding that 'it would certainly be difficult to find in their works any very explicit explanation of depression in terms of savings running to waste', and pointing out only that 'the language they employ is not such as to exclude such an explanation; and their general position, much more than Malthus's, seems to point directly to it. For good

upon the quantity of labour given for corn, either when grown on their own land, or embodied in manufactures and with them bought from other countries.' The terms of trade argument and its relevance for the profit rate appears *inter alia* in Torrens (1821), 258–9, (1835), 231–70, and (1844), 79–102.

129 Cf. Tucker (1960), 183–4: 'It is a peculiar feature of Wakefield's writings that he regarded his analysis of the causes of the low rate of profit in England as in direct contradiction of the theories of Ricardo, James Mill and McCulloch . . . It is difficult to resist the conclusion that Wakefield simply did not understand the theories that he criticized so confidently' (but see Corry, 1961, 210, for a less critical view of Wakefield).

For the notion of pressure of labour and capital on land reflected in low productivity, see Wakefield (1833), I, 120–6, 120–30. For his full case see II, note xii: 'The Art of Colonization', 83–111. The outflow of capital and labour would reduce the 'competition for employment' at home by raising factor productivity; and provide new markets for British manufactured goods, and a steady supply of cheap imports. On his reading of Wakefield Mill linked the latter two consequences understanding them as allusions to the terms-of-trade case.

Wakefield makes clear his view that capital was already flowing out. It was desirable to attract these flows to the colonies and assure a complement for labour. The outflow is related to the relatively good opportunities available in North America.

130 See also Winch (1965), 77f, 1966, 341f.

or for bad, their position appears to be in many respects an anticipation of modern stagnation theory.'[131]

This caution is well advised. We can point to conspicuous instances of an author totally unaware of the potential for stagnation introduced by a combination of zero net investment and positive savings.[132] But there are more conclusive considerations than this in Mill's case. As we have seen, the problem of leakages abroad under circumstances of zero domestic investment did not present itself to him as an actual issue, considering the very rapid rate of domestic accumulation under way. Secondly, as we have also intimated earlier, even an allowance for a loss of capital abroad would not be 'un-Ricardian' in circumstances of differential domestic and foreign rates of return. Allowing the abandonment of the standard closed-economy axiom, there is no reason to believe that Ricardo – who indeed normally maintained that profits can decline readily to the minimum at which net investment ceases without disturbance – would have been 'horrified' by any such conclusion (Robbins, 1968, 68).

For similar reasons we cannot accept the view that Mill 'acceded to the policy implications of Wakefield's theory while attempting to retain a thin veneer of consistency with his account of the fall of profit in Ricardian terms, and his defence of Say's Law' – implications which include the new position on the effects of government expenditure (his repudiation of the Treasury View), and his championship of capital export and state support of emigration (Winch, 1965, 139–40).[133] His defence of Say's Law is scarcely

---

131 Torrens summarized his position in his *Essay on the External Corn Trade* referring to 'the important and fundamental principles, that the rate of profit falls, as the cost and the value of raw produce rises; and that a densely-peopled country, where additional supplies of food and material cannot be raised without resorting to soils of less and less fertility, must, unless she obtains supplies of foreign produce, speedily approach that stationary state in which additional accumulations cannot be productively employed' (1829, 322–3). This might be read as a reference to conditions of zero net domestic investment with net savings proceeding notwithstanding, particularly in the light of an earlier reference to the general principle 'that an artificial elevation in the value of food and of the materials of other necessaries lowers the rate of profit [and] forces capital abroad', and to 'the natural order of events' whereby capital flows 'from countries where it is relatively abundant to those in which it is relatively deficient' (284–5).

132 William Ellis, whose paper of 1826 was so much admired by Mill, is the case in point. Essential to his position is the insistence that the interest rate is not the predominant variable playing upon savings decisions–this we know Mill did not accept–so that even at the minimum return on capital at which net domestic investment ceases, net accumulation of money funds proceeds, some of which finds its way abroad (Ellis, 1826, 128–9). Yet nowhere in his paper did Ellis indicate concern that this divergence implied domestic stagnation.

133 For this view see also Winch (1963), 397, (1966), 344–5, Hutchison (1953), 353, Corry (1961), 209–10. For a position consistent with our own see Kittrell (1965), 189: '[T]here was no need for any . . . substantial change in the classical aggregative economic analysis

a 'thin veneer', since the 'repudiation' of the treasury view applied only to the special conditions outlined above – international capital mobility in circumstance of differential returns, the differential accounted for in purely Ricardian terms. It must also be repeated that government spending was not recommended to replace domestic investment in conditions of stationarity, and in the absence of which the system would suffer from general slackness. Government spending domestically would provide for important social projects partly at the expense of misconceived private projects and unproductive private consumption (the cyclical features). To the extent that capital exports were reduced by the absorption of funds, there might obviously result a net injection of spending power.[134] But this did not enter into Mill's account. There was certainly no actual threat to the level of activity emanating from capital leakages that government outlays were designed to compensate, considering the contemporaneous heavy domestic investment and corresponding growth in demand for labour.

* * *

The same positive approach towards government expenditure taken in the discussion of the tendency of profits to a minimum is reiterated in the formal treatment of taxation. We have first a statement fully in line with the position that where the magnitude of the surplus and the strength of the sociological forces defining the 'spirit of accumulation' are high (and the return on capital low) the standard case against government expenditure is weakened. For taxation merely draws on funds that would otherwise have been 'wasted' in outflows abroad or in the course of cyclical movements; accordingly both categories of loss are avoided or at least diminished:

and taxes . . . are in some sense partly paid out of capital, and in a poor country it is impossible to impose any tax which will not impede the increase of the national wealth. But in a country where capital abounds, and the spirit of accumulation is strong, this effect of taxation is scarcely felt. Capital having reached the stage in which, were it not for a perpetual succession of improvements in production, any further increase would soon be stopped – and having so strong a tendency even to outrun those improvements, that profits are only kept above the minimum by emigration of capital or by a periodical sweep called a commercial crisis; to take from capital by taxation what emigration would remove, or a commercial crisis destroy, is only to do what either of these causes would have done, namely, to make

in order to cope with the colonization question . . . [and] the theory did not undergo any such transformation'.

134 *Might*, but need not – for some forms of government spending on colonial ventures would entail capital export.

a clear space for further saving . . . The amount which would be derived, even from a very high legacy duty, in each year, is but a small fraction of the annual increase of capital in such a country; and its abstraction would but make room for saving to an equivalent amount; while the effect of not taking it, is to prevent that amount of saving, or cause the savings, when made, to be sent abroad for investment. A country which, like England, accumulates capital not only for itself, but for half the world, may be said to defray the whole of its public expenses from its overflowings; and its wealth is probably at this moment as great as if it had no taxes at all. What its taxes do is, to subtract from its means, not of production, but of enjoyment; since whatever anyone pays in taxes, he could, if it were not taken for that purpose, employ in indulging his ease, or in gratifying some want or taste which at present remains unsatisfied (*CW*, III, 823–4).

Now a sharp distinction must be made between forms of taxation that merely eat into funds available for accumulation, and taxation which has a disincentive effect by reducing the return on capital. The distinction reflects the earlier proposition (748, above, p. 468) that 'in rich, populous, and highly cultivated countries . . . what the legislator should desire and promote, is not a greater aggregate saving, but a greater return to savings . . .' for a greater return will automatically generate an increased flow of funds available for domestic investment. In these terms we can appreciate why Mill – notwithstanding his positive approach towards government spending within moderate bounds – continued to put much weight on the need to avoid laying 'a tax on industry and economy' (811) by, for example, progressive income taxation.[135] Rapid capital accumulation remained a key objective of policy and doubtless would so remain at least as long as population growth

---

135 Similarly, to the objection that 'any advantage given to saving' by tax arrangement would favour the rich, Mill responded that 'in proportion as they divert their income from the supply of their own wants, to a productive investment . . . it is distributed in wages among the poor' (816). More specifically, a tax impinging directly or indirectly on profits would be 'detrimental to the public wealth, and consequently to the means which society possesses of paying any taxes whatever' (830).

In evidence given before the Select Committee on the Income and Property Tax in 1852 Mill observed that 'ordinarily speaking, a person does a work of public utility who saves money to employ it productively' (V, 494) – although there were yet more desirable forms of expenditure. In 1861 he objected to what he saw as 'the double taxation' of saving by referring to the peculiar burden placed on a use of income 'which it [should] rather be public policy to encourage' (565); 'it is contrary to the canon of equity, and contrary to it in the worst way, because it makes that mode of employing income which it is public policy to encourage, a subject of discouragement' (570). In short, policy (as well as justice) dictated that no advantage should be given – as would be given by a progressive taxation – 'to self-indulgence over industry and economy, even though the effect may be to give some advantage, or rather, not to interfere with the natural advantage of the rich over the poor' (567).

proceeded at a rapid pace. This latter qualification is important. For without it the criticism in the chapter on the 'Stationary State' of political economists for identifying 'all that is economically desirable with the progressive state', – 'with McCulloch, for example, prosperity does not mean a large production and a good distribution of wealth, but a rapid increase of it; his test of prosperity is high profits . . .' (752) – would be troublesome for he himself called for 'a rapid increase of wealth'.

A problem emerges, however, in the present context regarding Mill's evaluation of current circumstances. We know that Mill did not believe the return on capital to be actually at the 'practical' minimum – that rate at which *all* savings flow abroad – considering the massive net domestic investment underway of which he was so conscious. Yet not only is dire warning given of the consequences which 'might' result from profits taxation when the going return is actually at the 'practical minimum', but this condition is said actually to rule:

It is possible that a tax on profits might even diminish the existing capital of a country. If the rate of profit is already at the practical minimum, that is, at the point at which all that portion of the annual increment which would tend to reduce profits is carried off either by exportation or by speculation; then if a tax is imposed which reduces profits still lower, the same causes which previously carried off the increase would probably carry off a portion of the existing capital. A tax on profits is thus in a state of capital and accumulation like that in England, extremely detrimental to the national wealth. And this effect is not confined to the case of a peculiar and therefore intrinsically unjust, tax on profits. The mere fact that profits have to bear their share of a heavy general taxation, tends, in the same manner as a peculiar tax, to drive capital abroad, to stimulate imprudent speculations by diminishing safe gains, to discourage further accumulation, and to accelerate the attainment of the stationary state. This is thought to be the principal cause of the decline of Holland, or rather of her having ceased to make progress (827–8).

It can only be surmised that Mill's fear of reducing the net return on capital under conditions of rapid population expansion,[136] led him to overstate his case and represent England to be already *at* the 'practical minimum'.[137]

136 The potential danger to labour, assuming unchanged population growth, is quite clear in our present context. 'If population continues to increase as before, the labourer suffers . . . ' (828).

137 That Mill went overboard is suggested by the fact that only a few lines before he had referred to a profits tax which (in the absence of technical change) would tend to lower the net return *towards* the 'practical minimum': '[T]he rate of profit would be brought closer to that practical minimum, to which it is constantly approaching: and this diminished return to capital would either give a decided check to further accumulation, or would cause a greater proportion than before of the annual increase to be sent abroad, or wasted in unprofitable speculations' (827).